DATE DUE

Biological Process Design for Wastewater Treatment

PRENTICE-HALL SERIES IN ENVIRONMENTAL SCIENCES

Granville H. Sewell, editor

Biological Process Design for Wastewater Treatment

Larry D. Benefield
Auburn University

Clifford W. Randall
Virginia Polytechnic Institute and State University

Prentice-Hall, Inc., Englewood Cliffs, NJ 07632

Library of Congress Cataloging in Publication Data

Benefield, Larry D
Biological process design for wastewater treatment.

Includes bibliographies and index.
1. Sewage—Purification—Biological treatment.
2. Sewage disposal plants—Design and construction.
I. Randall, Clifford W., joint author. II. Title.
TD755.B36 628′.3 79-13745
ISBN 0-13-076406-X

Interior design and editorial/production supervision: STEVEN BOBKER
Cover design: WANDA LUBELSKA
Manufacturing buyer: GORDON OSBOURNE

10 9 8 7 6 5 4 3 2 1

Printed in the United States of America

Prentice-Hall International, Inc., *London*
Prentice-Hall of Australia Pty. Limited, *Sydney*
Prentice-Hall of Canada, Ltd., *Toronto*
Prentice-Hall of India Private Limited, *New Delhi*
Prentice-Hall of Japan, Inc., *Tokyo*
Prentice-Hall of Southeast Asia Pte. Ltd., *Singapore*
Whitehall Books Limited, *Wellington, New Zealand*

To our wives, parents, and children for their love and encouragement

Contents

Preface xiii

1 Fundamentals of Process Kinetics 1

1-1 Reaction Rates 2
Zero-Order Reactions 3
First-Order Reactions 4
Second-Order Reactions 5
Enzyme Reactions 7
Effects of Temperature on Reaction Rate 11

1-2 Reactor Analysis 14
Completely Mixed Batch Reactor 14
Continuous-Flow Stirred Tank Reactor 15
Continuous-Flow Stirred Tank Reactors in Series 18
Plug-Flow Reactor 19
Plug Flow with Dispersion (Arbitrary Flow) 21

Problems 22

References 23

2 Fundamentals of Microbiology 24

2-1 Nutritional Requirements 25
Microbial Enzymes 26

2-2 Environmental Effects on Microbial Growth 28
Temperature Effects 28
Oxygen Requirements 29
pH Effects 30

2-3 Metabolism of Microorganisms 31
Energy Metabolism of Heterotrophs 33
Energy Budget 40
Energy Metabolism of Autotrophs 40

2-4 Kinetic Relationships from Microbiology 43
Biomass Growth Rate 45
Growth Yield 47
Substrate Utilization in Microbial Culture 48
Effect of Substrate Concentration on Growth Rate 49
Energy and Carbon Source Requirements 49
Maintenance as Endogenous Respiration 53

2-5 Kinetic Relationships Applied in Process Design 55

Problems 58

References 60

3 Wastewater Characteristics and Flows 62

3-1 Organic Materials 62
Biochemical Oxygen Demand 63
Chemical Oxygen Demand 73
Total Organic Carbon 74

3-2 Inorganic Materials 77
pH and Alkalinity 78
Nitrogen 87
Phosphorus 98

3-3 Solid Content 99

3-4 Wastewater Composition and Flow 103

3-5 Wastewater Flow Gauging and Sampling 113
Data Analysis 117

Problems 125

References 126

4 Activated Sludge and Its Process Modifications 129

4-1 Mixing Regime 131

4-2 Kinetic Model Development 131

4-3 Process Modifications 152

Conventional Activated Sludge Treatment 152
Tapered Aeration 154
Step-Aeration 154
High-Rate Activated Sludge 159
Complete Mixing 160
Extended Aeration 161
Contact-Stabilization 163
Sludge Reaeration 170
Pure Oxygen Aeration 174
Oxidation Ditches 181
Other Process Modifications 184

4-4 Process Design Considerations 184

Loading Criteria 184
Excess Sludge Production 186
Sludge Viability 189
Oxygen Requirement 190
Nutrient Requirement 195
Temperature Effects 197
Solids–Liquid Separation 201
Effluent Quality 208

4-5 Evaluation of Biokinetic Constants 210

4-6 Nitrification 218

Nitrification with Suspended-Growth Systems 222

4-7 Biological Denitrification 249

Microbiological Aspects of Denitrification 249
Denitrification in a Suspended-Growth System 251

4-8 Anaerobic Contact Process 256

Fundamental Microbiology 257
Process Kinetics: The Rate-Limiting Step Approach 259
Gas Production 263
Environmental Factors 264
Process Design Considerations 268

Problems 272

References 274

5 Aeration 281

5-1 Fundamentals of Gas Transfer 281

5-2 Factors Affecting Oxygen Transfer 285

Oxygen Saturation 285
Temperature 287
Wastewater Characteristics 287
Turbulence 288
Oxygen Transfer Rates 289

5-3 Determination of K_La and α Values 290

5-4 Design of Aeration Systems 293

Diffused Aeration 294
Submerged Turbine Aeration 308
Surface Aerators 310
Mixing Considerations 317

Problems 318

References 319

6 Treatment Ponds and Aerated Lagoons 322

6-1 Aerobic Ponds 323

Diurnal Variations in Aerobic Ponds 323
Design Relationships 324
Design Considerations 332

6-2 Facultative Ponds 338

Design Relationships 340
Design Considerations 345
Flow Patterns 348
Performance 351

6-3 Anaerobic Ponds 353

6-4 Polishing Ponds 358

6-5 Aerated Lagoons 359

6-6 Aerobic Lagoons 360

Mixing and Aeration 361
Design Relationships 365
Temperature Effects 368
Design Considerations 374

6-7 Facultative Lagoons 382

Design Considerations 383

Problems 385

References 386

7 Attached-Growth Biological Treatment Processes 391

7-1 Trickling Filters 391
Filter Media 393
Types of Filters 393
Popular Design Equations 396
The Trickling Filter as a Treatment Alternative 409

7-2 Rotating Biological Contactors 410
Design Relationships 413

7-3 Activated Biofilters 420
Design Considerations 422

7-4 Anaerobic Filters 423
Design Considerations 425

7-5 Nitrification in Attached-Growth Systems 429
Nitrification in Trickling Filters 429
Nitrification in Submerged Filters 431
Nitrification with Rotating Biological Contactors 436
Nitrification with the Activated Biofilter 441

7-6 Denitrification in Attached-Growth Systems 442
Submerged Rotating Biological Contactors 443
Submerged High-Porosity Media Columns 443
Submerged Low-Porosity Fine-Media Columns 448
Fluidized Bed Denitrification Columns 449

Problems 452

References 453

8 Sludge Digestion 457

8-1 Anaerobic Digestion 460
Process Description 461
Kinetic Relationships 464
Gas Production and Heating Requirements 466
Sludge Characteristics 467
Design Considerations 470
Process Modeling and Control 477

8-2 Aerobic Digestion 479
Kinetic Relationships 482
Temperature Effects 490
Oxygen Requirements 492
Mixing Requirements 493
Design Considerations 495
Laboratory Evaluation 496
Autothermal Thermophilic Aerobic Digestion 507
Energy Considerations for Sludge Digestion 507

Problems 508
References 511

Index 515

Preface

The authors have observed during their years of experience that the majority of the biological wastewater treatment processes actually put into operation are not designed using fundamental biological treatment principles, even though such principles have been available for many years. The result is that many of the installations do not operate properly and considerable time and expense must be spent diagnosing design problems and correcting them before satisfactory operation can be achieved. We believe that much of the problem exists because the design engineers do not have available to them a thorough yet relatively concise compilation of the fundamental and design aspects of biological waste treatment, presented so that the underlying principles are truly learned. Such learning is required if the engineer is to become a process design decision maker rather than a user of formulas. This book is our attempt to fill this perceived gap in available information.

Even though biological processes are essential components of most wastewater treatment systems, no text is available which covers in detail both the theoretical and design aspects of such components. Our objective in this book is to integrate both of these aspects into a single text which can be used by both the student and the practicing engineer. To achieve this objective, the book is developed in the following format: (a) process fundamentals are presented; (b) wherever possible, these fundamentals are used to develop design relationships for a particular process; (c) a design procedure using these relationships is illustrated using example problems which typify the calculations required in each process application; (d) finally, essential reference material is included at the end of each chapter. Throughout the text, it is emphasized that process design criteria should be obtained from laboratory investigations. However, in many instances practicing consultants do not have the time or resources for laboratory studies. In this regard, design information for specific treatment situations is presented which can be particularly valuable when comparing process alternatives for facility planning purposes.

A word of appreciation is due to Elizabeth Stimmel, Pam Murdock, and Donna Griffith who typed the manuscript for publication and to the many graduate students who were particularly helpful in suggesting improvements to the original draft for this book.

Larry D. Benefield
Clifford W. Randall

Biological Process Design for Wastewater Treatment

1

Fundamentals of Process Kinetics

All biological wastewater treatment processes take place in a volume defined by specific boundaries. Such a volume is commonly termed a *reactor*. Changes in the composition and concentration of materials that occur while the wastewater is retained in the reactor are important factors in wastewater treatment. These changes are caused by hydraulic transport of materials into and out of the reactor as well as by reactions that occur within the reactor. To fully define a reactor system and design similar ones it is necessary to know the rate at which the changes occur and the extent of the changes.

The engineer who is designing a biological process is generally interested in the rates at which various components (such as organic material) are removed from the wastewater and the rate at which biomass is produced in the reactor. Such rates of change are important because they directly affect the size of the reactor required for a specific degree of treatment.

1-1
Reaction Rates

Chemical reactions may be classified in one of the following ways:

1/ On the basis of the number of molecules that must react to form the reaction product.

2/ On a kinetic basis by reaction order.

It is the latter classification that is useful in describing the kinetics of most biological processes.

When reactions are classified on a kinetic basis, different reaction orders may occur for variations in organisms, substrates, or environmental conditions.

The relationship among rate of reaction, concentration of reactant, and reaction order, n, is given by the expression

$$\text{rate} = (\text{conc})^n \tag{1-1}$$

or by taking the log of both sides of the equation,

$$\log \text{rate} = n \log (\text{conc}) \tag{1-2}$$

By applying equation 1-2, experimental results may be interpreted to establish a reaction order and rate. For any constant-order reaction, if the log of the instantaneous rate of change of reactant concentration at any time is plotted as a function of the log of the reactant concentration at that instant, a straight line will result and the slope of the line will be the order of the reaction (see Figure 1-1). The zero-order reaction results in a horizontal line, and the rate of reaction is concentration-independent or the same at any reactant concentration. For the first-order reaction the rate of reaction is directly propor-

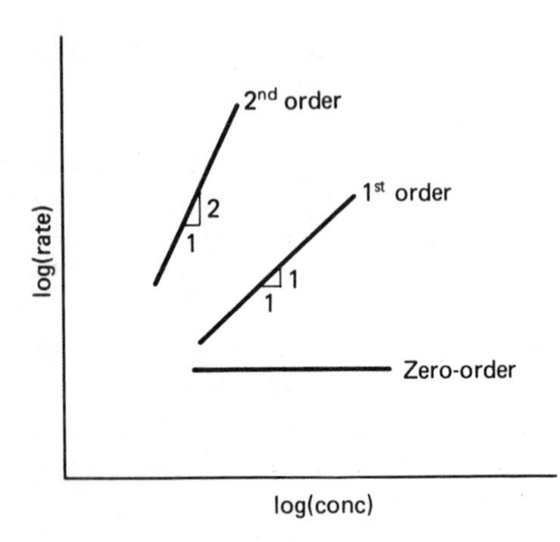

Figure 1-1. Determining Reaction Order by Log Plotting.

tional to the reactant concentration, and with second-order equations the rate is proportional to the concentration squared. Fractional reaction orders are possible, especially in mixed biological cultures, but for the solution of many rate problems, an integer value for the reaction order is determined or assumed. With this condition, a more detailed evaluation of integer-order rate equations can be made as a function of reaction elapsed time.

Zero-Order Reactions

Zero-order reactions are those reactions that proceed at a rate independent of the concentration of any reactant. As an example, consider the conversion of a single reactant to a single product:

$$\underset{\text{reactant}}{A} \longrightarrow \underset{\text{product}}{P}$$

If such a conversion follows zero-order kinetics, the rate of disappearance of A is described by the rate equation

$$-\frac{d[A]}{dt} = K[A]^0 = K$$

where $-\frac{d[A]}{dt}$ = rate of disappearance of A

K = reaction-rate constant

If C represents the concentration of A at any time, t, then the rate equation can be expressed as

$$-\frac{dC}{dt} = K \qquad (1\text{-}3)$$

where $-\frac{dC}{dt}$ = rate of change in concentration of A with time, mass volume^{-1} time^{-1} (the negative sign indicates that the concentration of A decreases with time; if a positive sign were given, this would indicate an increase in concentration with time)

K = reaction-rate constant, mass volume^{-1} time^{-1}

Integrating equation 1-3 gives the formulation

$$C = -Kt + \text{constant of integration} \qquad (1\text{-}4)$$

The constant of integration is evaluated by letting $C = C_0$ at $t = 0$. This implies that

$$C_0 = \text{constant of integration}$$

and shows that the integrated rate law has the form

$$C - C_0 = -Kt \qquad (1\text{-}5)$$

A plot of concentration versus time for a zero-order reaction is illustrated in Figure 1-2. Note that the response is linear when the plot is made on arithmetic paper.

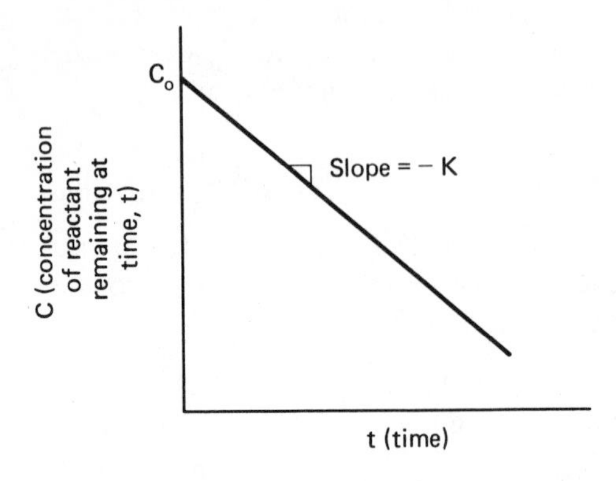

Figure 1-2. Arithmetic Plot of the Course of a Zero-Order Reaction.

First-Order Reactions

First-order reactions are those reactions that proceed at a rate directly proportional to the concentration of one reactant. Since the rate of the reaction depends on the concentration of the reactant and since the concentration of the reactant changes with time, an arithmetic plot of the variation in the concentration of the reactant with time will not give a linear response as it did for a zero-order reaction. Such a graph is presented in Figure 1-3.

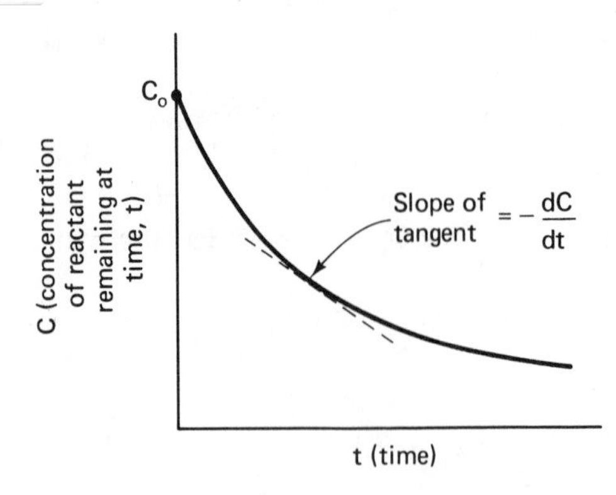

Figure 1-3. Arithmetic Plot of the Course of a First-Order Reaction.

Again consider the conversion of a single reactant to a single product,

$$\underset{\text{reactant}}{A} \longrightarrow \underset{\text{product}}{P}$$

If first-order kinetics are followed, the rate of disappearance of A is described by the rate equation

$$-\frac{dC}{dt} = K(C)^{-1} = KC \tag{1-6}$$

where $-\frac{dC}{dt}$ = rate of change in the concentration of A with time, mass volume^{-1} time^{-1}

C = concentration of A at any time, t, mass volume^{-1}

K = reaction-rate constant, time^{-1}

Integrating equation 1-6 and letting $C = C_0$ at $t = 0$ gives an integrated rate law of the form

$$\ell n\left(\frac{C_0}{C}\right) = Kt \tag{1-7}$$

or, in the more familiar form,

$$\log\left(\frac{C_0}{C}\right) = \frac{Kt}{2.3} \tag{1-8}$$

Equation 1-8 suggests that a plot of log C versus time for a first-order reaction will give a linear trace, as shown in Figure 1-4.

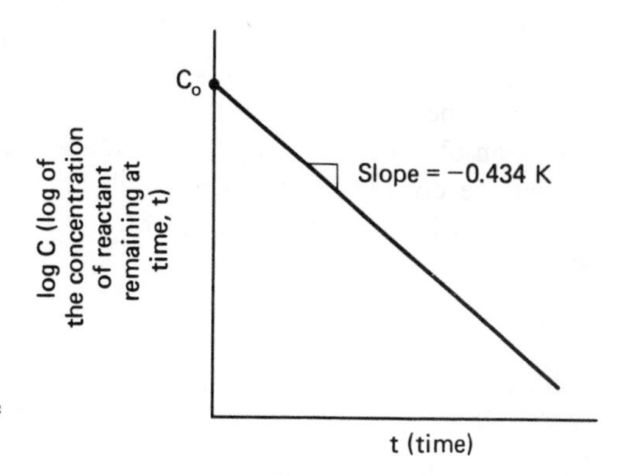

Figure 1-4. Semilog Plot of the Course of a First-Order Reaction.

Second-Order Reactions

Second-order reactions are those reactions that proceed at a rate proportional to the second power of a single reactant. For the reaction describing the conversion of a single reactant to a single product,

$$\underset{\text{reactant}}{2A} \longrightarrow \underset{\text{product}}{P}$$

The rate of disappearance of A, for a second-order reaction, is described by the rate equation

$$-\frac{dC}{dt} = K(C)^2 \tag{1-9}$$

where K = reaction-rate constant, mass^{-1} volume time^{-1}

The integrated rate law for a second-order reaction has the form

$$\frac{1}{C} - \frac{1}{C_0} = Kt \qquad (1\text{-}10)$$

Figure 1-5 indicates that an arithmetic plot of $1/C$ versus time will give a linear trace, the slope of which yields the value of K.

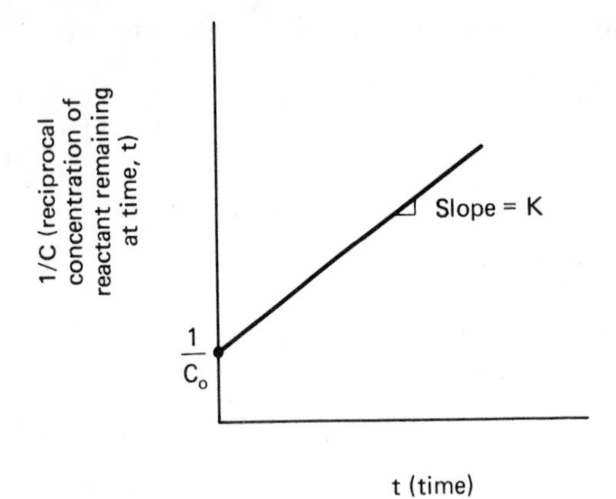

Figure 1-5. Plot of the Course of a Second-Order Reaction.

For a given set of experimental values of C and t, equations 1-5, 1-8, and 1-10 can be used to test for a particular reaction order. This is accomplished by making the appropriate concentration versus time plot and noting any deviation from linearity.

Example 1-1

Glucose was added to a batch culture of microorganisms and removal was measured over time. The following data were obtained:

Glucose concentration measured as COD (mg/ℓ)	*Time (min)*
180	0
155	5
95	12
68	22
42	31
26	40

Determine the reaction order of the removal process by curve fitting.

solution

Make the appropriate concentration versus time plots and note any deviation from linearity.

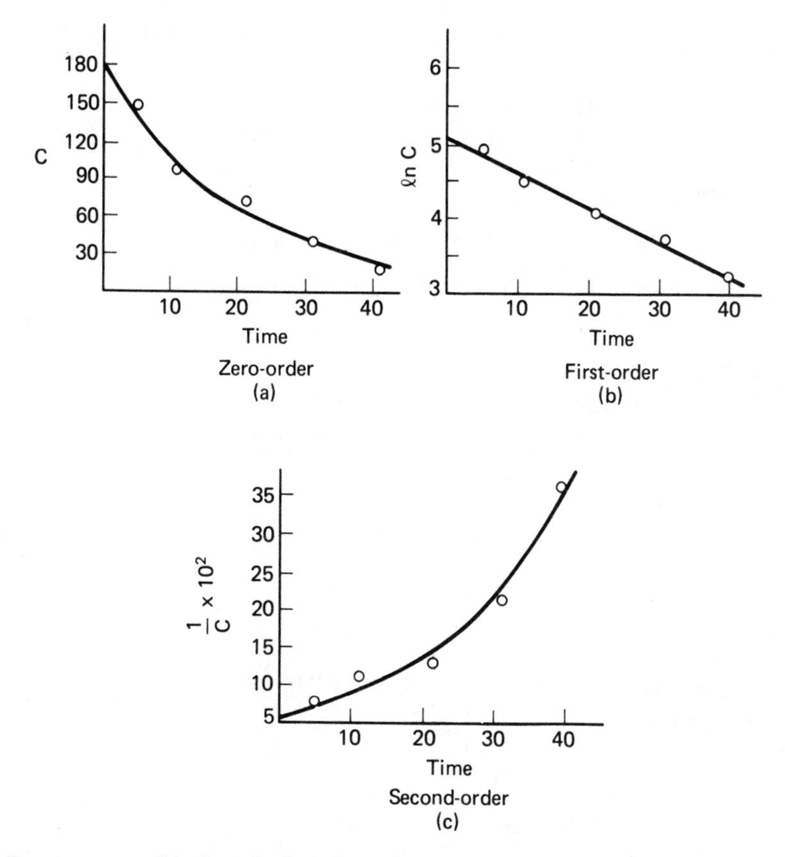

Because curve (b) gives the best fit to the experimental data, glucose removal is assumed to follow first-order kinetics.

Enzyme Reactions

The overall rate of a biological reaction is dependent on the catalytic activity of the enzymes in the prominent reaction. Enzyme kinetics has been defined by Michaelis and Menten for a single reaction involving a single substrate. It has been shown that the same form of equations can be used in many cases to simulate the kinetics observed in the multisubstrate and mixed-culture reactions occurring in wastewater treatment processes.

It is generally assumed that an enzyme-catalyzed reaction involves the reversible combination of the enzyme, E, and substrate, S, to form an enzyme/substrate complex, ES, and then the irreversible decomposition of the complex to form free enzyme and product, P:

$$E + S \underset{K_2}{\overset{K_1}{\rightleftharpoons}} ES \xrightarrow{K_3} E + P$$

In the enzyme reaction, K_1, K_2, and K_3 represent rate constants for the reactions designated.

When the concentration of the ES complex appears constant, a dynamic steady-state condition prevails, where the rate at which the complex is being formed is equal to the rate at which it is breaking down. This steady-state condition for the ES complex can be expressed as

$$[\text{rate of formation}] = [\text{rate of decomposition}] \tag{1-11}$$

From the law of mass action, equation 1-11 may be written as

$$K_1[\text{E}][\text{S}] = K_2[\text{ES}] + K_3[\text{ES}] \tag{1-12}$$

where [E] = free enzyme concentration, mass volume^{-1}
[S] = substrate concentration, mass volume^{-1}
[ES] = enzyme/substrate complex concentration, mass volume^{-1}

Equation 1-12 can be rearranged into the form

$$\frac{[\text{E}][\text{S}]}{[\text{ES}]} = \frac{K_2 + K_3}{K_1} = K_m \tag{1-13}$$

where K_m represents the Michaelis constant and is numerically equal to $(K_2 + K_3)/K_1$.

The maximum reaction rate for formation of product will occur when all the enzyme present is associated with the enzyme/substrate complex, so that

$$R_{\text{max}} = K_3[\text{E}_{\text{total}}] \tag{1-14}$$

where R_{max} = maximum rate at which product is formed, mass volume^{-1} time^{-1}
$[\text{E}_{\text{total}}]$ = total concentration of enzyme in system, mass volume^{-1}

At any other stage of enzyme saturation the reaction rate, r, for the formation of product is

$$r = K_3[\text{ES}] \tag{1-15}$$

The total amount of enzyme in the system is given by the mass-balance expression

$$[\text{E}_{\text{total}}] = [\text{E}] + [\text{ES}] \tag{1-16}$$

Substituting from equations 1-14 and 1-15 into equation 1-16 gives

$$[\text{E}] = \frac{R_{\text{max}}}{K_3} - \frac{r}{K_3} \tag{1-17}$$

A further substitution from equation 1-17 for [E] in equation 1-13 gives

$$\frac{[\text{S}]}{K_3[\text{ES}]}(\text{R}_{\text{max}} - r) = K_m \tag{1-18}$$

A final substitution from equation 1-15 for K_3 in equation 1-18 will yield the expression

$$\frac{[\text{S}]}{r}(\text{R}_{\text{max}} - r) = K_m \tag{1-19}$$

which can be rearranged into the Michaelis–Menten equation,

$$r = \frac{R_{max}[S]}{K_m + [S]} \tag{1-20}$$

K_m is also termed the *saturation constant* and is equal to the substrate concentration when the reaction rate is equal to $R_{max}/2$. Equation 1-20 is illustrated graphically in Figure 1-6.

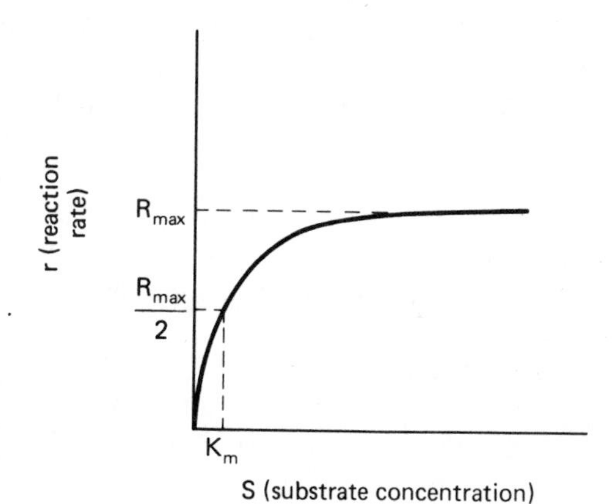

Figure 1-6. Graphical Representation of the Michaelis–Menten Equation.

Example 1-2

Compute the ratio of the substrate concentration required for 80% of R_{max} to the concentration required for 20% of R_{max}.

solution

1/ Determine $[S]_{80}$ and $[S]_{20}$ in terms of K_m using equation 1-20.

$$\frac{r}{R_{max}} = \frac{[S]}{K_m + [S]}$$

or

$$\frac{0.8}{1} = \frac{[S]_{80}}{K_m + [S]_{80}}$$

which reduces to

$$[S]_{80} = 4K_m$$

Similarly, it can be shown that

$$[S]_{20} = 0.25K_m$$

2/ The ratio $[S]_{80}/[S]_{20}$ is computed using the values obtained in step 1.

$$\frac{[S]_{80}}{[S]_{20}} = \frac{4}{0.25} = 16$$

Figure 1-6 illustrates a substrate saturation phenomenon which is only observed in enzymatic reactions. This figure shows that the rate of an enzyme-catalyzed reaction is proportional to the substrate concentration at low substrate concentrations; that is, the reaction is approximately first-order with respect to substrate concentration. However, a zone is soon reached where the increase in reaction rate falls off as the substrate concentration is increased. This is referred to as the *region of mixed order*. On further increase in substrate concentration, the reaction rate becomes constant and independent of substrate concentration. In this region the reaction rate is zero-order with respect to substrate concentration. Such observations imply two special cases of the Michaelis–Menten equation.

When the substrate concentration is much greater than the value of K_m, K_m can be neglected in the sum term found in the denominator of equation 1-20. For this situation equation 1-20 reduces to

$$r = R_{\max} \tag{1-21}$$

This shows that the rate of the reaction is constant and is equal to the maximum reaction rate. Under these conditions the reaction follows zero-order kinetics; that is, the rate of the reaction is independent of substrate concentration.

When the substrate concentration is much smaller than the value of K_m, [S] can be neglected in the sum term found in the denominator of equation 1-20. For this situation equation 1-20 reduces to

$$r = \frac{R_{\max}}{K_m}[S] \tag{1-22}$$

However, since both $R_{\max}$ and K_m are constants, a new term can be defined:

$$K = \frac{R_{\max}}{K_m} \tag{1-23}$$

K represents a reaction-rate constant which has the units of reciprocal time. Substituting from equation 1-23 for $R_{\max}/K_m$ in equation 1-22 gives

$$r = K[S] \tag{1-24}$$

Thus, for this case, the rate of the reaction is proportional to the substrate concentration. Under these conditions the reaction follows first-order kinetics.

Segel (1968) states that zero-order kinetics may be assumed when $[S] \geq 100K_m$ and that first-order kinetics may be assumed when $[S] \leq 0.01K_m$. However, Goldman et al. (1974) suggest that for all practical purposes, first-order kinetics may be assumed when $[S] \leq K_m$. At intermediate ranges of $[S]/K_m$ the reaction is termed *mixed order* and results in fractional exponents.

The Michaelis–Menten equation represents a continuum for defining enzyme-catalyzed reactions. If an experiment was started with a large amount of substrate, with no new substrate added and the reaction kinetics followed with time, the reaction would be initially zero-order, as there would be excess food and the reaction rate would be limited by the ability of the enzymes. As the food became used up the reaction would begin to become substrate-limited, and a fractional-order reaction would emerge. When the food level became quite low, the rate of finding the substrate by the organisms would become controlling and first-order kinetics would result.

The Michaelis–Menten equation is a general expression. Wastewater treatment processes are designed to use a high microorganism concentration, and this results in a substrate-limiting condition that can be defined in many cases by first-order kinetics.

Later discussion will show that under certain conditions microbial substrate utilization rate and specific growth rate can be represented by expressions that have the same form as equation 1-20.

Effect of Temperature on Reaction Rate

The rate of any simple chemical reaction is increased when the temperature is elevated, provided that the higher temperature does not produce alterations in the reactants or catalyst. Biological reactions have the same tendency except that a decrease in the rate of an enzyme-catalyzed reaction may be observed at high temperatures because the enzyme may be denatured at such temperatures. This response is illustrated in Figure 1-7.

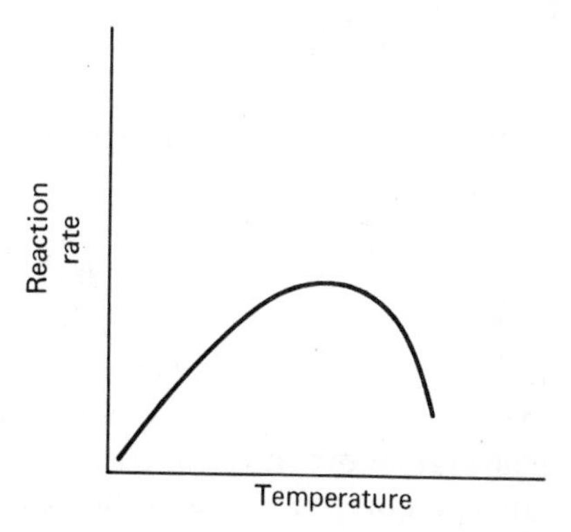

Figure 1-7. Response of Enzyme-Catalyzed Reaction to an Increase in Temperature.

To describe the variation in reaction rate with temperature of a biological reaction, a frequently quoted approximation known as the *van't Hoff rule* states that the reaction rate doubles for a 10°C temperature rise.

Arrhenius proposed that the effect of temperature on the reaction-rate constant in a chemical reaction may be described by equation 1-25:

$$\frac{d(\ell n\, K)}{dt} = \frac{E_a}{R}\frac{1}{T^2} \qquad (1\text{-}25)$$

where K = reaction-rate constant
E_a = activation energy, cal/mole
R = ideal gas constant, 1.98 cal/mole-deg
T = reaction temperature, °K

Equation 1-25 can be integrated to give the expression

$$\ell n\, K = -\frac{E_a}{R}\frac{1}{T} + \ell n\, B \qquad (1\text{-}26)$$

where B represents a constant. Plotting experimental data according to equation 1-26 is useful when it is desired to determine the activation energy associated with a particular reaction. The plot necessary for such a determination is given in Figure 1-8. This figure shows that the slope of the line given when $\ell n\, K$ is plotted versus $1/T$ is $-E_a/R$.

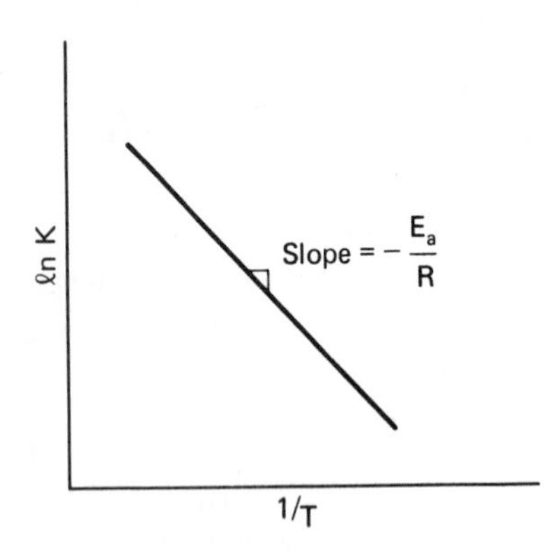

Figure 1-8. Arrhenius Plot for Determining Activation Energy.

When equation 1-25 is integrated between the limits of T_1 and T_2, equation 1-27 is obtained.

$$\ell n\left(\frac{K_2}{K_1}\right) = \frac{E_a}{R}\frac{T_2 - T_1}{T_2 T_1} \qquad (1\text{-}27)$$

If K_1 is known for T_1, then using equation 1-27, it is possible to compute K_2 for the temperature change from T_1 to T_2 if the activation energy for the reaction is known. The equation has been used to estimate the effect of temperature over a limited range for biological reactions. Metcalf and Eddy (1972) report that the activation energy, E_a, for biological wastewater treatment processes generally will fall within the range 2000 to 20,000 cal/mole.

For most situations encountered in biological wastewater treatment processes, the quantity E_a/RT_1T_2 located on the right side of equation 1-27 may be considered constant. Thus, equation 1-27 may be approximated by the expression

$$\ell n\left(\frac{K_2}{K_1}\right) = \text{constant}\ (T_2 - T_1) \qquad (1\text{-}28)$$

Equation 1-28 may then be written in the form

$$\frac{K_2}{K_1} = e^{\text{constant}\,(T_2 - T_1)} \qquad (1\text{-}29)$$

It is common practice to introduce a temperature characteristic term, Θ, which has a value equal to that given by e^{constant}. Thus, equation 1-29 becomes

$$\frac{K_2}{K_1} = \Theta^{T_2 - T_1} \tag{1-30}$$

A later chapter will present data which will show that many biological processes do not respond to temperature variations according to equation 1-30; while some processes may follow equation 1-30 only within a narrow temperature range.

Example 1-3

Given the following experimental data:

Temperature (°*C*)	*Reaction-rate constant, K* (*day*$^{-1}$)
15.0	0.53
20.5	0.99
25.5	1.37
31.5	2.80
39.5	5.40

Determine the value of the temperature characteristic, Θ.

solution

1/ Plot data according to equation 1-26 and determine E_a/R. To simplify the graphical construction, all K values will be multiplied by 10. Such a manipulation will not change the slope of the line.

$K \times 10$	$\ell n\,(K \times 10)$	$1/T$
5.3	1.67	0.00347
9.9	2.29	0.00340
13.7	2.62	0.00335
28.0	3.33	0.00328
54.0	3.99	0.00320

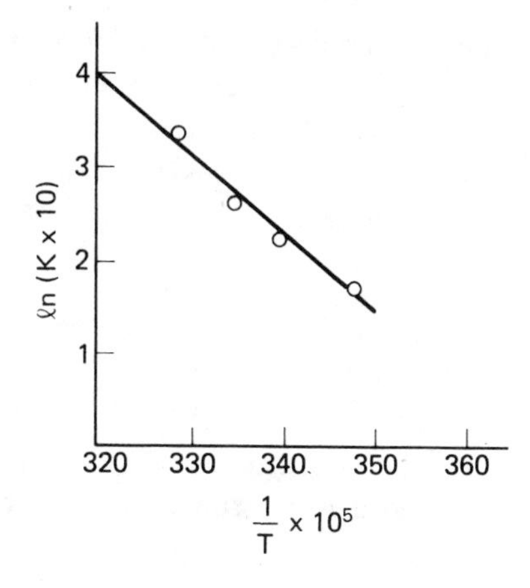

$$\text{slope} = -\frac{E_a}{R} = \frac{3.99 - 1.67}{0.00347 - 0.00320} = -8593; \text{ or } \frac{E_a}{R} = 8593$$

2/ Estimate Θ using the value of 8593 for E_a/R. Assume that Θ is valid over the temperature range 15 to 40°C. Then for any temperature variation within this range, Θ is given by

$$\Theta = e^{(8593/T_1T_2)}$$

where T_1 and T_2 are the respective temperatures in °K.

However, one value of Θ is generally used for a range of temperatures. In this case, if $T_1 = 15 + 273 = 288$°K and $T_2 = 40 + 273 = 313$°K, then

$$\Theta = e^{(8593/(288)(313))} = 1.100$$

1-2
Reactor Analysis

Up to this point the discussion has focused on the development of specific formulations to describe reaction rates. These rate-law expressions can be combined with the hydraulic characteristics of a particular reactor system to develop mathematical expressions that can be used to predict the degree to which a reaction will proceed or to compute the reactor volume required to give a particular extent of reaction.

Reactor systems that will be reviewed are the completely mixed batch reactor, the completely mixed continuous-flow reactor, the plug-flow reactor, and the plug-flow reactor with longitudinal dispersion.

Completely Mixed Batch Reactor

The *completely mixed batch* (CMB) *reactor* is a closed system. The reactants are added to the empty vessel, and the contents are withdrawn after the reaction has proceeded to the desired degree. The composition of such a system varies with time. However, at any one time the composition can be regarded as uniform throughout the reactor.

Since there is no flow into or out of the reactor during the period alloted for reaction time, a material-balance equation for a specific reactant can be expressed as

$$\begin{bmatrix}\text{rate of change}\\ \text{in the mass of}\\ \text{reactant A within}\\ \text{the reactor}\end{bmatrix} = \begin{bmatrix}\text{rate of the}\\ \text{reaction of}\\ \text{A within the}\\ \text{reactor}\end{bmatrix} \tag{1-31}$$

If C represents the concentration of reactant A at any time, t, and V represents the volume of the reactor, and if it is assumed that the rate of the reaction is described by first-order kinetics, equation 1-31 can be expressed mathematically as

$$V\left(\frac{dC}{dt}\right)_{\text{net}} = V\left(\frac{dC}{dt}\right)_{\text{reaction}} = V(KC) \tag{1-32}$$

Canceling the volume terms, equation 1-32 reduces to

$$\frac{dC}{dt} = KC \tag{1-33}$$

If the concentration of the reactant of interest decreases with time, the left-hand term in equation 1-33 is given a negative value, whereas a positive value is assigned if the concentration increases with time.

To determine the reaction time to realize a desired reactant concentration, equation 1-33 can be integrated between the limits C_0 and C_d, where C_0 represents the initial concentration of reactant A and C_d represents the desired concentration of reactant A. Such a mathematical manipulation results in equation 1-34:

$$t = \frac{1}{K} \ell\text{n}\left(\frac{C_0}{C_d}\right) \tag{1-34}$$

Completely mixed batch reactors have very limited use in field-scale biological wastewater treatment processes, although they should be considered for some operations in small plants and for sludge digestion. Use of batch reactors is found in bench-scale experimental studies.

Continuous-Flow Stirred Tank Reactor

A *continuous-flow stirred tank reactor* (CFSTR), also referred to as a *completely mixed continuous-flow reactor*, operates under steady-state conditions so that the properties throughout the system do not vary with time. The reactants flow continuously into the reactor and the products flow continuously out while a uniform concentration is maintained throughout the reactor volume. A flow schematic of such a system is presented in Figure 1-9. In this figure V represents the reactor volume, Q the volumetric flow rate into and out of the reactor, C_0 the initial reactant concentration in the inlet stream (influent), and C_e the concentration of the reactant in the exit stream

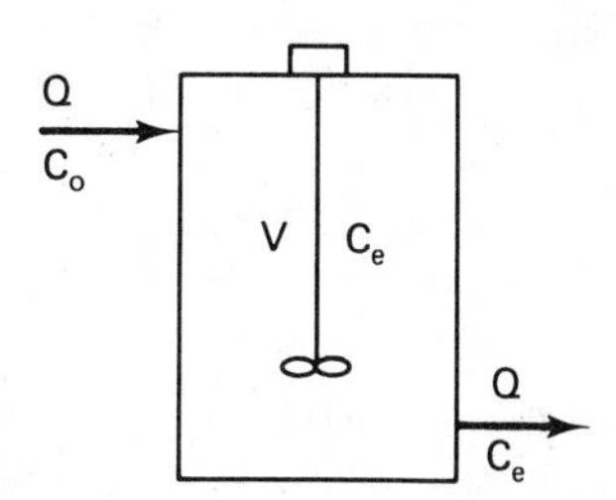

Figure 1-9. Schematic of a Continuous-Flow Stirred Tank Reactor.

(effluent). For a CFSTR the reactant concentration in the exit stream is the same as the reactant concentration at any point within the reactor volume.

The CFSTR represents the extreme case of back mixing or longitudinal dispersion. The initial high driving force (concentration of reactant) at the reactor inlet is instantaneously reduced to the final low driving force present at the reactor exit. A completely mixed batch reactor has a mean driving force somewhere between the high initial value and the low final value, which means that a CFSTR must have a larger volume than a CMB to accomplish the same amount of reaction. However, CFSTRs have a number of advantages, which will be discussed when the various activated sludge process modifications are presented.

Since there is continuous liquid movement into and out of a CFSTR, a material balance must consider not only changes that occur as a result of reactions within the reactor but also must include those changes resulting from the hydraulic characteristics of the system as well. In reactor analysis the rate of change in the mass of reactant with both time and reactor position are required in a material expression. However, because the reactant concentration is constant throughout the reactor volume for a CFSTR, there is no requirement to consider the change in mass of reactant with reactor position. Thus, a material balance for the rate of change in the mass of reactant A within the reactor can be expressed as

$$\begin{bmatrix}\text{net rate of change}\\ \text{in the mass of}\\ \text{reactant A within}\\ \text{the reactor}\end{bmatrix} = \begin{bmatrix}\text{rate of increase}\\ \text{in the mass of}\\ \text{A due to its}\\ \text{presence in the}\\ \text{influent}\end{bmatrix} - \begin{bmatrix}\text{rate of decrease}\\ \text{in the mass of}\\ \text{A due to removal}\\ \text{in the effluent}\end{bmatrix} - \begin{bmatrix}\text{rate of decrease}\\ \text{in the mass of}\\ \text{A due to the}\\ \text{reaction of A}\\ \text{in the reactor}\end{bmatrix} \qquad (1\text{-}35)$$

The last term on the right-hand side of equation 1-35 has been assigned a negative value because it is assumed that the reaction of A within the reactor results in a decrease in the quantity of A. Such is the case for most biological wastewater treatment applications where substrate utilization is considered. However, if the reaction of A within the reactor results in an increase in the quantity of A, a positive value should be assigned to this term. This is generally the case when a material balance for biomass is formulated for a particular biological reactor system. It should also be noted that the first term on the right-hand side of equation 1-35 is assigned a positive value because it is assumed that the influent will tend to increase the quantity of A within the reactor, whereas the second term has been assigned a negative value because it is assumed that the effluent will tend to remove A from the system.

Equation 1-35 can be expressed mathematically as

$$V\left(\frac{dC}{dt}\right)_{\text{net}} = QC_0 - QC_e - V\left(\frac{dC}{dt}\right)_{\text{reaction}} \tag{1-36}$$

If first-order reaction kinetics are assumed for the reaction of A within the reactor, equation 1-36 becomes

$$V\left(\frac{dC}{dt}\right)_{\text{net}} = QC_0 - QC_e - VKC_e \tag{1-37}$$

For steady-state conditions the net rate of change in the mass of reactant A within the reactor is zero. Under such conditions equation 1-37 reduces to

$$0 = QC_0 - QC_e - VKC_e \tag{1-38}$$

Equation 1-38 can then be rearranged into the form

$$\frac{C_e}{C_0} = \frac{1}{1 + K(V/Q)} \tag{1-39}$$

If the nominal hydraulic retention time in the CFSTR is defined as

$$t_{\text{CFSTR}} = \frac{V}{Q} \tag{1-40}$$

equation 1-39 can be expressed as

$$\frac{C_e}{C_0} = \frac{1}{1 + Kt_{\text{CFSTR}}} \tag{1-41}$$

The reaction time required to achieve a desired reactant concentration can then be found by rearranging equation 1-41.

$$t_{\text{CFSTR}} = \frac{1}{K}\left(\frac{C_0}{C_e} - 1\right) \tag{1-42}$$

Example 1-4

The concentration of a particular reactant is reduced from 200 mg/ℓ to 15 mg/ℓ as the liquid flow containing the reactant passes through a CFSTR. If the reaction in the reactor follows first-order kinetics and the liquid flow rate is 1 million gallons per day (1 MGD), what is the volume of the reactor. Assume that the reaction-rate constant has a value of 0.4 day^{-1}.

solution

Equating equation 1-40 and 1-42 gives

$$\frac{V}{Q} = \frac{1}{K}\left(\frac{C_0}{C_e} - 1\right)$$

or

$$\frac{V}{1} = \frac{1}{0.4}\left(\frac{200}{15} - 1\right)$$

$$V = 30.8 \text{ MG}$$

Continuous-Flow Stirred Tank Reactors in Series

Figure 1-10 is a schematic illustrating a system of two CFSTRs of equal volume connected in series. The concentration of reactant in the effluent from the first and second reactor is represented by C_1 and C_2, respectively. If a first-order reaction is assumed, a steady-state balance for reactant A across the first reactor gives

$$\frac{C_1}{C_0} = \frac{1}{1 + Kt_{\text{CFSTR}}} \tag{1-43}$$

where t_{CFSTR} represents the nominal hydraulic retention time in the first reactor. Similarly, a steady-state material balance for reactant A within the second reactor gives

$$\frac{C_2}{C_1} = \frac{1}{1 + Kt_{\text{CFSTR}}} \tag{1-44}$$

where t_{CFSTR} represents the nominal hydraulic retention time in the second reactor.

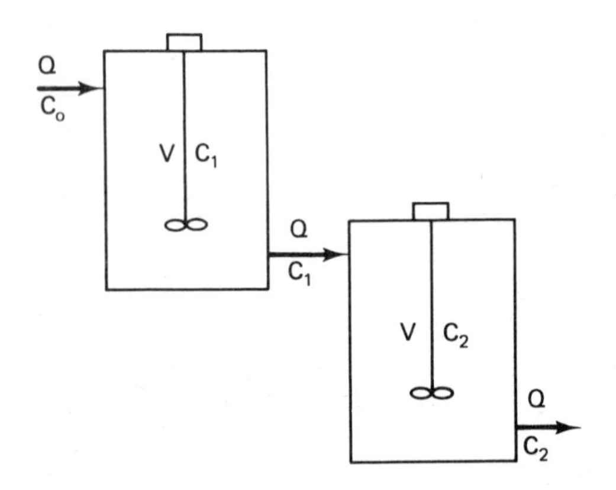

Figure 1-10. Schematic of Two Continuous-Flow Stirred Tank Reactors Connected in Series.

An expression relating the concentration of reactant A in the effluent from the system to the influent reactant concentration can be developed for uniform-sized reactors by multiplying equations 1-43 and 1-44 together.

$$\frac{C_2}{C_0} = \frac{C_1}{C_0}\frac{C_2}{C_1} = \left(\frac{1}{1 + Kt_{\text{CFSTR}}}\right)^2 \tag{1-45}$$

A similar relationship can be developed for n number of equal-sized CFSTRs connected in series. Such a relationship has the form

$$\frac{C_n}{C_0} = \left(\frac{1}{1 + Kt_{\text{CFSTR}}}\right)^n \tag{1-46}$$

where C_n represents the reactant concentration in the effluent from the nth or last reactor in the series. Equation 1-46 can then be rearranged into a form that gives the nominal retention time of the total reactor system when multiplied by n.

$$nt_{\text{CFSTR}} = \frac{n}{K}\left[\left(\frac{C_0}{C_n}\right)^{1/n} - 1\right] \tag{1-47}$$

For a given value of C_n, the greater the number of reactors in series, the smaller will be the total reactor volume required for the system. Such a system approaches a plug-flow reactor.

Plug-Flow Reactor

In the CFSTR an effort is made to keep the contents uniform, but in the *plug-flow* (PF) *reactor* the aim is to avoid mixing. Plug flow assumes that no longitudinal mixing occurs between adjacent fluid elements. Each element of fluid in this type reactor is analogous to a completely mixed batch reactor moving along a time axis; that is, the position variable in a plug-flow reactor corresponds to the time variable in a completely mixed batch reactor. Thus, in a plug-flow reactor the variation in the concentration of reactant A in both space and time is of interest. In other words, not only is it necessary to know how the concentration of A varies with time but also how it varies along the length of the reactor. This situation is illustrated in Figure 1-11.

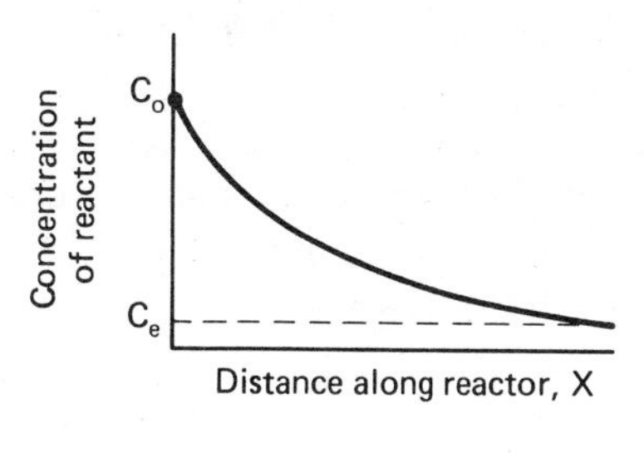

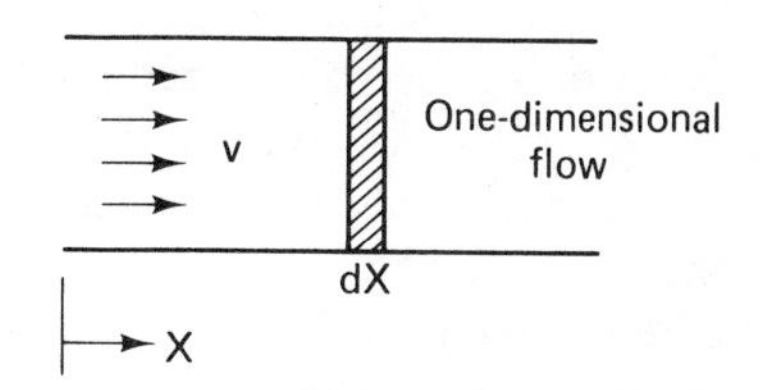

Figure 1-11. Schematic of Ideal Plug-Flow Reactor. (After Weber, 1972.)

In the analysis of a PF reactor the primary concern is with the relationship among three variables: concentration, time, and distance (position of a fluid element in the reactor relative to the inlet). Using time as the reference variable, a steady-state condition for the rate of change in the concentration of reactant A can be represented by

$$\begin{bmatrix}\text{change in the} \\ \text{concentration of A} \\ \text{due to the reaction} \\ \text{of A in the dif-} \\ \text{ferential time, } dt\end{bmatrix} = \begin{bmatrix}\text{change in the concen-} \\ \text{tration of A due to} \\ \text{the change in position} \\ \text{of the fluid element in} \\ \text{the differential time, } dt\end{bmatrix} \tag{1-48}$$

Assuming first-order kinetics, where the concentration of A decreases as a result of the reaction of A, equation 1-48 can be expressed as

$$-\frac{dC}{KC} = \frac{dx}{v} \tag{1-49}$$

where v represents the velocity of flow through the reactor and dx represents a differential change in distance along the length of the reactor. Integrating the left-hand side of equation 1-49 between the limits of C_0 and C_e and integrating the right-hand side of equation 1-49 over the entire length, L, of the reactor gives

$$-\int_{C_0}^{C_e} \frac{dC}{KC} = \int_0^L \frac{dx}{v} \tag{1-50}$$

or, as shown by Weber (1972),

$$\frac{1}{K}\left[\ell n \left(\frac{C_0}{C_e}\right)\right] = \frac{L}{v} = \frac{LR}{vR} = \frac{V}{Q} \tag{1-51}$$

where R represents the cross-sectional area of the reactor.

By recalling that the nominal hydraulic retention time is given by V/Q, equation 1-51 can be modified to give the retention time required to obtain a desired reactant concentration in the effluent:

$$t_{PF} = \frac{1}{K}\left[\ell n \left(\frac{C_0}{C_e}\right)\right] \tag{1-52}$$

As noted earlier for CFSTRs in series, to produce identical effluent concentrations, the volume required for a PF reactor would be much less than the volume required for a single CFSTR. The reason that the volume of the PF reactor would be less is that the average concentration of reactant A in the reactor, which is the reaction driving force, is higher than that for the CFSTR. Hence, the average reaction rate is higher, suggesting that the required reaction time would be less.

Table 1-1 presents nominal hydraulic retention-time equations for reactions of different orders in CFSTR and PF reactors, while Table 1-2 gives the operating characteristics of CMB reactors, CFSTRs, and PF reactors.

TABLE 1-1

NOMINAL HYDRAULIC RETENTION-TIME EQUATIONS FOR REACTIONS OF DIFFERENT ORDER IN CFSTRs AND PF REACTORS

	Nominal hydraulic retention time	
Reaction order	t_{CFSTR}	t_{PF}
0	$\frac{1}{K}(C_0 - C_e)$	$\frac{1}{K}(C_0 - C_e)$
1	$\frac{1}{K}\left(\frac{C_0}{C_e} - 1\right)$	$\frac{1}{K}\left[\ell n\left(\frac{C_0}{C_e}\right)\right]$
2	$\frac{1}{KC_e}\left(\frac{C_0}{C_e} - 1\right)$	$KC_0\left(\frac{C_0}{C_e} - 1\right)$

Source: After Weber (1972).

TABLE 1-2

OPERATING CHARACTERISTICS OF DIFFERENT REACTOR SYSTEMS

Reactor type	*Variation of composition with time*	*Variation of composition with position in reactor*
CMB reactor	Yes	No
CFSTR	No	No
PF reactor	Yes	Yes

Plug Flow with Dispersion (Arbitrary Flow)

Complete mixing and plug flow represent the extremes of mixing. However, seldom will either situation be observed in practice. Instead, some intermediate amount of intermixing will generally occur. To acount for such an effect, a dispersion model has been developed by Wehner and Wilhem (1958) which approaches complete mixing when the degree of dispersion approaches infinity and converts to plug flow in cases where dispersion does not exist. Their equation is

$$\frac{C_0}{C_e} = \frac{4ae^{(1/2)d}}{(1 + a)^2 e^{a/2d} - (1 - a)^2 e^{-a/2d}} \tag{1-53}$$

where

$$a = \sqrt{1 + 4Ktd} \quad \text{and} \quad d = \frac{D}{vL} = \frac{Dt}{L^2} \tag{1-54}$$

in which d is the diffusivity constant or dispersion number (dimensionless) and has values of zero and infinity, respectively, for plug-flow and completely mixed systems; D is the axial dispersion coefficient (area per time); v is the fluid velocity (length per time); and L is the characteristic length of travel path of a typical particle in the tank (length).

The second term in the denominator of equation (1-53) is small and, as an approximation, may be neglected. The resulting simplified equation is

$$\frac{C_0}{C_e} = \frac{4ae^{(1-a)/2d}}{(1+a)^2} \tag{1-55}$$

The approximate formula should not be used when the value of d exceeds 2.0, but is generally satisfactory for lesser values.

A detailed discussion of the dispersion model will not be presented because most biological processes are designed on the basis of either complete mixing or plug flow. However, Thirumurthi (1969) has proposed it for the design of waste stabilization ponds, and it will be discussed more thoroughly in that section. The interested reader is also referred to either Levenspiel (1972) or Weber (1972) for an excellent treatment of this particular model.

PROBLEMS

1-1 The rate of kill of *E. coli* with chlorine was investigated and the surviving organisms were determined by a plate count. The following data were obtained:

Contact time (min)	*Number of organisms remaining*
0	40,800
2	36,000
4	23,200
6	16,000
8	12,800
10	10,400
15	5,600
20	3,000
25	1,500

Determine the reaction order of the disinfection process and the value of the associated reaction-rate constant, K.

1-2 Various forms of the Michaelis–Menten equation have been devised because equation 1-20 is the equation of a rectangular hyperbola and the constant terms, K_m and R_{max}, are not easily determined from the plot. The Lineweaver–Burk equation has the form

$$\frac{1}{r} = \frac{K_m}{R_{max}} \frac{1}{[S]} + \frac{1}{R_{max}}$$

Develop the Lineweaver–Burk equation from the Michaelis–Menten equation and then plot the following data according to this equation.

[S] (*moles*)	r (*moles/min*)
0.002	0.045
0.005	0.115
0.020	0.285
0.040	0.380
0.060	0.460
0.080	0.475
0.100	0.505

From the plot evaluate K_m and R_{max}.

1-3 Prove that $K_m = [S]$ when $r = \frac{1}{2}R_{max}$.

1-4 A certain reaction was found to have an activation energy of 20,000 cal/mole and a specific reaction-rate constant of 2.0 min^{-1} at 20°C. Calculate the specific reaction-rate constant at 0°C.

1-5 Compare the total volume requirements for the following reactor systems:
a/ Single CFSTR.
b/ Two CFSTRs connected in series.
c/ Four CFSTRs connected in series.
d/ PF reactor.
It is desired that the reactant concentration be reduced from 100 mg/ℓ to 20 mg/ℓ for a flow of 1 MGD. Assume that first-order kinetics are followed and the rate constant has a value of 0.8 day^{-1}.

REFERENCES

GOLDMAN, J. C., W. J. OSWALD, AND D. JENKINS, "The Kinetics of Inorganic Carbon Limited Algal Growth," *Journal of the Water Pollution Control Federation*, **46**, 554 (1974).

LEVENSPIEL, O., *Chemical Reaction Engineering*, John Wiley & Sons, Inc., New York, 1972.

METCALF & EDDY, INC., *Wastewater Engineering*, McGraw-Hill Book Company, New York, 1972.

SEGEL, I. H., *Biochemical Calculations*, John Wiley & Sons, Inc., New York, 1968.

THIRUMURTHI, D., "Design Principles of Waste Stabilization Ponds," *Journal of the Sanitary Engineering Division, ASCE*, **95**, 311 (1969).

WEBER, WALTER, J., JR., *Physicochemical Processes for Water Quality Control*, Wiley–Interscience, New York, 1972.

WEHNER, J. F., AND R. H. WILHEM, "Boundary Conditions of Flow Reactors," *Chemical Engineering Science*, **6**, 89 (1958).

Fundamentals of Microbiology

The objective of wastewater treatment is to remove pollutants from the water. Generally, the wastewater pollutants of primary concern are soluble and insoluble organics, various forms of nitrogen and phosphorus, and inert insoluble materials. In most cases both soluble and insoluble organic material as well as nitrogen can be effectively removed through biological action if a favorable environment is provided for the living microorganisms. Some phosphorus can also be removed, by incorporation into new cellular mass, but the percent removal is normally to a lesser degree than it is for nitrogen and organic material.

To effectively design any biological wastewater treatment process, it is necessary that the engineer have a basic understanding of (1) the nutritional requirements of microorganisms, (2) the environmental factors that affect microbial growth, (3) the metabolism of microorganisms, and (4) the relationship between microbial growth and substrate utilization. Such key factors will be presented and reviewed in this chapter.

2-1
Nutritional Requirements

All biological processes employed by the environmental engineer find their basis in the nutritional requirement of microorganisms. For example, in the activated sludge process a microbial suspension is aerated in the presence of wastewater containing dissolved and colloidal organic matter. During the period of aeration the organic material is removed by the microorganisms and used to support both life and growth functions. After a time the microorganisms are separated from the wastewater and a liquid stream relatively free of organic contamination is discharged.

For microorganisms, nutrients (1) provide the material required for synthesis of cytoplasmic material, (2) serve as an energy source for cell growth and biosynthetic reactions, and (3) serve as acceptors for the electrons released in the energy-yielding reactions. Classification of the nutrient requirements of microorganisms is shown in Table 2-1.

TABLE 2-1

CLASSIFICATION OF NUTRIENT REQUIREMENTS

Function	*Source*
Energy source	Organic compounds
	Inorganic compounds
	Sunlight
Electron acceptor	O_2
	Organic compounds
	Combined inorganic oxygen (NO_3^-, NO_2^-, SO_4^{2-})
Carbon source	CO_2, HCO_3^-
	Organic compounds
Trace elements and growth factors such as vitamins.	

Considering their nutritional requirements, microorganisms can be divided into specific classes. On the basis of the chemical form of carbon required, microorganisms are classified as (1) *autotrophic*, which are organisms (commonly called *autotrophs*) that use CO_2 or HCO_3^- as their sole source of carbon and construct from these all their carbon-containing biomolecules, or (2) *heterotrophic*, which are organisms (commonly called *heterotrophs*) that require carbon in the form of relatively complex, reduced organic compounds such as glucose.

On the basis of the energy source required, microorganisms are classified as (1) *phototrophs*, which are organisms that use light as their energy source, or (2) *chemotrophs*, which are organisms that employ oxidation–reduction reactions to provide their energy. Chemotrophs can be further classified on the basis of the type of chemical compounds oxidized (i.e., the electron donor). For example, *chemoorganotrophs* are organisms that use complex organic molecules as their electron donor, whereas *chemoautotrophs* use simple inorganic molecules such as hydrogen sulfide or ammonia. Some typical reactions for the various classifications of microorganisms are given in Table 2-2.

TABLE 2-2

TYPICAL MICROBIAL REACTIONS

Microbial reaction	*Nutritional classification*
$CO_2 + 2H_2O \xrightarrow{\text{light}} \underset{\text{new cells}}{(CH_2O)} + O_2 + H_2O$	Autotrophic, photosynthetic
$\underset{\text{cells}}{(CH_2O)} + O_2 \longrightarrow CO_2 + H_2O$	Cellular respiration, aerobic
$C_6H_{12}O_6 + 6O_2 \longrightarrow 6CO_2 + 6H_2O$	Heterotrophic (chemoorganotrophic), aerobic
$C_6H_{12}O_6 \longrightarrow 2C_2H_6O + 2CO_2$	Heterotrophic, anaerobic, fermentative
$C_2H_3O_2 \longrightarrow CH_4 + HCO_3$	Heterotrophic, anaerobic, fermentative
$C_6H_{12}O_6 + 12KNO_3 \longrightarrow 12KNO_2 + 6H_2O + 6CO_2$	Heterotrophic, anaerobic, intermolecular oxidative–reductive
$2NH_3 + 3O_2 \longrightarrow 2HNO_3 + 2H_2O$	Autotrophic, chemosynthetic (chemoautotrophic), aerobic
$5S + 2H_2O + 6HNO_3 \longrightarrow 5H_2SO_4 + 3N_2$	Chemoautotrophic, anaerobic

Microbial Enzymes

All the activities of a microbial cell depend upon food utilization and all the chemical reactions involved are controlled by enzymes. *Enzymes* are proteins produced by the living cell that act as catalysts to accelerate specific reactions in accordance with the rate equations discussed in Section 1-1. Enzymes are specific in that they will catalyze only certain kinds of reactions, and they will act on only one kind of substance. There is a momentary combination, usually only a few hundredths of a second, between the enzyme and the chemical undergoing change. During this combination the chemical reaction occurs and a new compound(s) is formed. There is relatively little attraction between the new compound and the enzyme, so the enzyme is immediately freed to combine with another molecule of the substance for which it has specificity.

Microbial enzymes catalyze three types of reactions: hydrolytic, oxidative, and synthetic. *Hydrolytic* enzymes are used to hydrolyze complex insoluble food substances into simple soluble components that can pass through the cell membranes into the cell by diffusion. These enzymes are generally excreted by the organism into the surrounding medium and are spoken of as *extracellular* enzymes to distinguish them from *intracellular* enzymes, which are liberated only when the cell has disintegrated. The hydrolysis reaction involves the addition of water to a complex compound and the splitting of the compound into simpler and more soluble products.

Energy-yielding reactions are catalyzed by intracellular enzymes, and these reactions provide energy for the maintenance and growth of microorganisms. All such reactions involve oxidations and reductions, wherein the addition or removal of oxygen or hydrogen are of primary importance. Most microorganisms oxidize their food by the enzymatic removal of hydrogen from the molecule. Enzymes known as *dehydrogenases* remove the hydrogen from the compound, an atom at a time, and pass it on from one enzyme system to another until it is used to reduce the final *hydrogen acceptor*. The final hydrogen acceptor is determined by the aerobic or anaerobic nature of the surrounding medium and the character of the cells performing the reaction. For aerobic reactions, oxygen is the final hydrogen acceptor and water is formed. Under anaerobic conditions, an oxidized compound accepts the hydrogen and a reduced compound is formed.

Energy is liberated during oxidation and energy is consumed during reduction. The net result is that more energy is liberated than is used up and the excess is available for use by the cell.

The synthesis of cellular material for both the maintenance of cells and for new cells is catalyzed by intracellular *synthetic* enzymes. Vast numbers of enzymes are required to synthesize the many types of complex compounds found in microbial cells. The great amount of energy required for the synthetic reactions is obtained from the oxidations that occur during energy metabolism.

The activity of enzymes is affected by environmental conditions, particularly temperature, pH, and the presence of certain ions, such as PO_4^{3-}, Mg^{2+}, or Ca^{2+}. The effect of temperature was shown in Figure 1-7. Every enzyme has a range of pH in which it can operate. Some are most active in an acid medium, others at neutrality, and still others in an alkaline medium. When the pH is increased or decreased beyond the optimum, the activity of the enzyme decreases until it disappears. The ions listed above may accelerate the action of some enzymes and are necessary for the activity of others. In addition, salts of heavy metals such as $HgCl_2$ and $CuSO_4$ will, in time, inactivate enzymes.

Since enzymes are affected by environmental conditions, environmental effects on microbial growth are very significant and must be considered in biological waste treatment design.

2-2
Environmental Effects on Microbial Growth

The physical environment in which microorganisms are contained influences to a large degree their growth processes. Thus, to ensure optimum treatment efficiency, a proper environment must be provided for any biological treatment process. In this regard temperature, oxygen requirement and pH are probably the most imp^rtant considerations.

Temperature Effects

All the processes of growth are dependent on chemical reactions, and the rates of these reactions are influenced by temperature. Thus, the rate of microbial growth as well as the total amount of growth can be affected by temperature. Figure 2-1 shows the effect of temperature on growth rate. This figure reflects a minimum temperature below which growth does not occur. As the temperture is increased, a point will be reached where the rate of growth is a maximum. The temperature corresponding to this point is called the *optimum temperature*. With a further increase in temperature the heat-sensitive cell components such as enzymes are denatured and the growth rate drops rapidly. A maximum temperature above which growth does not occur is generally noted after a small increase in temperature above the optimum value.

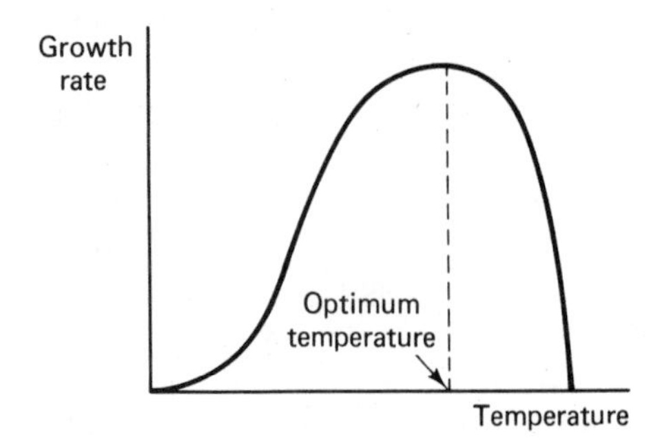

Figure 2-1. Effect of Temperature on Microbial Growth Rate.

Based on the temperature range within which they can proliferate, bacteria can be classified as psychrophilic, mesophilic, or thermophilic. Figure 2-2 reflects the acceptable temperature range for each class of bacteria. The hatched portion within each range indicates an approximate optimum temperature, the temperature that allows for the most rapid growth during a short period of time (12 to 24 h). In their respective classes, facultative thermophiles and facultative psychrophiles are the bacteria which have opti-

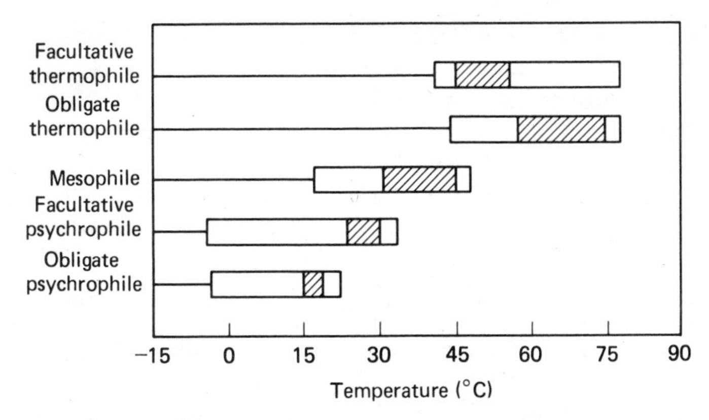

Figure 2-2. Temperature Ranges for Reproduction of Psychrophilic, Mesophilic, and Thermophilic Bacteria. (After Doetsch and Cook, 1973.)

mum temperatures that extend into the mesophilic range. Optimum temperatures for obligate thermophiles and obligate psychrophiles lie outside the mesophilic range.

Oxygen Requirements

The presence or absence of molecular oxygen divides organisms into three distinct classes. More specifically, microorganisms can be classified on the basis of their electron acceptor. Organisms that use molecular oxygen as their electron acceptor are known as *aerobes*, whereas organisms that use some molecule other than molecular oxygen are known as *anaerobes*. *Facultative* organisms can use either oxygen or some other chemical compound as their electron acceptor. However, growth is more efficient for these organisms under aerobic conditions.

Obligate aerobes are unable to grow in the absence of oxygen, and *obligate anaerobes* are poisoned by the presence of oxygen. A few microorganisms grow best at very low molecular oxygen concentrations. They are known as *microaerophiles*.

The principal significance of the electron acceptor used by the microorganism is the completeness of the resulting reaction and thus the amount of energy that becomes available for growth and life-support facilities. Among heterotrophs, aerobic and facultative microorganisms generally oxidize their food completely, whereas the anaerobic fermenters do not. The difference in energy yield can be seen by comparison of the following equations for glucose metabolism:

$$C_6H_{12}O_6 + 6O_2 \longrightarrow 6CO_2 + 6H_2O + 689{,}000 \text{ cal}$$

$$C_6H_{12}O_6 \longrightarrow 2C_2H_6O + 2CO_2 + 31{,}000 \text{ cal}$$

In the first equation, oxidation of the compound is complete, and therefore the maximum amount of energy is liberated. Complete oxidation can also occur through facultative intermolecular oxidation–reduction, as shown by the equation in Table 2-2, although a reduced compound remains and the energy yield is slightly less.

Some aerobic autotrophic bacteria can also completely oxidize the inorganic compounds they use for energy. *Thiobacillus thiooxidans* is a good example:

$$2S + 2H_2O + 3O_2 \longrightarrow 2H_2SO_4 + 237{,}000 \text{ cal}$$

Nitrosomonas, however, only partially oxidizes ammonia under aerobic conditions:

$$3NH_3 + 3O_2 \longrightarrow 2HNO_3 + 2H_2O + 66{,}500 \text{ cal}$$

Energy-transfer mechanisms are discussed in Section 2-3.

pH Effects

For most bacteria, and thus for most wastewater treatment processes, the extremes of the pH range for growth falls somewhere between 4 and 9. The optimum pH for growth generally lies between 6.5 and 7.5. Wilkinson (1975) suggests that bacteria grow best when the pH is slightly on the alkaline side, whereas algae and fungi grow best when the pH is slightly on the acid side. Biological wastewater treatment processes are seldom operated at optimum growth conditions, however, and full-scale experience has shown that both extended aeration activated sludge and aerated lagoon systems can be successfully operated when the pH is between 9 and 10.5. By contrast, both systems are very vulnerable to pH levels below 6.0.

Figure 2-3 illustrates the effect of pH on the growth rate of most microorganisms. Such a response is attributed to a change in enzyme activity with pH. The hydrogen-ion concentration is considered to be one of the most important factors that influence enzyme activity.

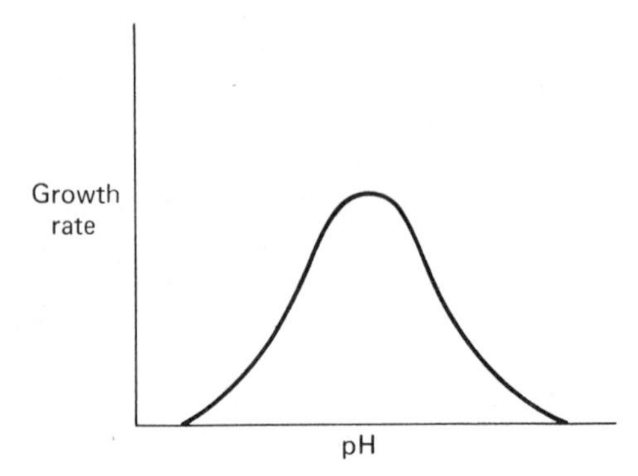

Figure 2-3. Effect of pH on Microbial Growth Rate.

Experiments by Randall et al. (1972) have shown that attached filamentous systems, principally fungi, can efficiently metabolize organic matter down to a pH of 2.65. It has also been shown that the same type of system can operate efficiently at a pH above 9.0 (Kato and Sekikawa, 1967). Thus, while most conventional processes are usually operated within narrow pH limits, it is possible to treat organic wastewaters over a wide pH range.

2-3 Metabolism of Microorganisms

The nutrient substances absorbed by microorganisms are subjected to numerous different biochemical reactions. Among these are the oxidative, exothermic, enzymatic degradation processes known as *catabolism*, and the reductive, endothermic, enzymatic synthesis processes known as *anabolism*. Catabolism results in the release of free energy inherent in the complex structure of large organic molecules. This energy is conserved by the organism in the form of the phosphate-bond energy of adenosine triphosphate (ATP). Reactions that release energy are termed *exergonic*. Anabolism, on the other hand, is a synthesis process that results in an increase in size and complexity of chemical structures. Thus, energy is required in this process and is obtained from the phosphate-bond energy of ATP. It can, therefore, be stated that microorganisms function by capturing the energy liberated by *exergonic* reactions and use this energy to drive *endergonic* reactions. The connecting link between energy-producing and energy-requiring reactions is the adenosine triphosphate (ATP)/adenosine diphosphate (ADP) system.

Free energy can be defined as the capacity to do useful work. More specifically, the Gibbs free energy, G, can be defined mathematically as

$$\Delta G = \Delta H - T\,\Delta S \qquad (2\text{-}1)$$

where G, H, T, and S represent free energy, enthalpy (heat energy), temperature, and entropy, respectively. The Δ quantities refer to the value of the term in the final state minus its value in the initial state when considering a process change. Thus, during the course of a chemical reaction, the amount of energy produced or absorbed is referred to as the free-energy change (ΔG) of the reaction. The sign of ΔG, negative or positive, predicts whether the reaction can proceed spontaneously. If ΔG has a positive value, the reaction will not proceed spontaneously. In this case energy is required to drive the reaction. Such a reaction is termed *endergonic*.

As a standard of reference, consider the overall reaction for the hydrolysis (breaking of a chemical bond through the addition of water) of ATP:

$$ATP + H_2O \longrightarrow ADP + H_3PO_4 \qquad (2\text{-}2)$$

At standard conditions of concentration (1.0 *M*), pH = 7.0, and a temperature of 30°C, the free-energy change for this reaction is −8400 cal/mole. Structural changes for this reaction are illustrated in Figure 2-4.

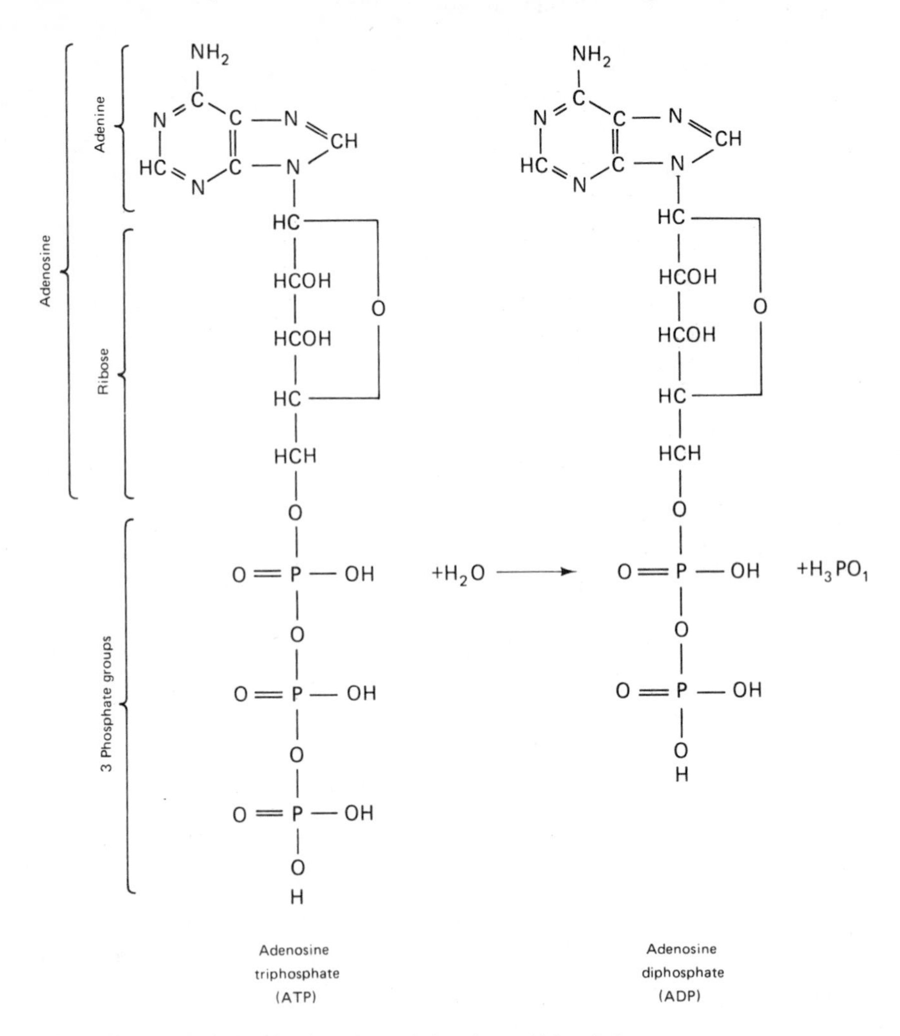

Figure 2-4. Hydrolysis of Adenosine Triphosphate. (After Pelczar and Reid, 1972.)

It is important to note that in microorganisms the energy released during a phosphate group transfer is not wasted in a hydrolysis reaction. It is utilized to effect synthesis of needed metabolites by coupling exergonic reactions with endergonic reactions. Thus, for each mole of phosphate released, 8.4 kcal of energy becomes available to the microorganisms for biosynthetic processes.

Energy Metabolism of Heterotrophs

Heterotrophs use the same substances as sources of carbon and energy. A portion of the absorbed material is oxidized to provide energy while the remaining portion is used as building blocks for cellular synthesis.

Generally speaking, where organic molecules are involved, oxidation reactions are actually *dehydrogenations* (i.e., reactions that involve the loss of hydrogen atoms). However, since a hydrogen atom contains an electron, the compound that loses this atom has also lost an electron and has been oxidized.

Heterotrophs recover the energy released during oxidative catabolism reactions in two ways. The first is by *substrate-level phosphorylation*, where a portion of the energy released during an oxidation reaction is used to drive the endergonic phosphorylation reaction of ADP to ATP. Thus, energy storage is provided in the form of phosphate-bond energy.

The second means of recovering the energy released during catabolism is by *oxidative phosphorylation*. In this case electrons are produced by the oxidation of an electron donor (DH_2). These electrons (usually in pairs) are then passed through an electron-transport system to a terminal electron acceptor (A). The electron-transport system is a series of electron carriers arranged so that the large amount of energy produced by the oxidation of the electron donor is released in small packets which are used to drive the endergonic phosphorylation reaction of ADP to ATP. Hence, once again energy storage is provided in the form of phosphate-bond energy.

Before proceeding to a study of the various pathways of energy metabolism, a brief discussion of the function of enzyme cofactors is warranted. Many enzymes depend for activity only on their structure as proteins. However, some enzymes require cofactors for activity. These *cofactors* can be either metal ions or complex organic molecules called *coenzymes*. These coenzymes usually function as carriers of electrons or of specific atoms or functional groups that are transferred during the enzymatic reaction. One of the most important coenzymes is nicotinamide adenine dinucleotide (NAD denotes oxidized form, NADH + H denotes reduced form). Figure 2-5 gives the oxidized and reduced structure of this coenzyme.

Three general methods exist by which heterotrophic microorganisms can obtain energy.

Fermentation. In this case the carbon and energy source is broken down by a series of enzyme-mediated reactions. A portion of the energy released during

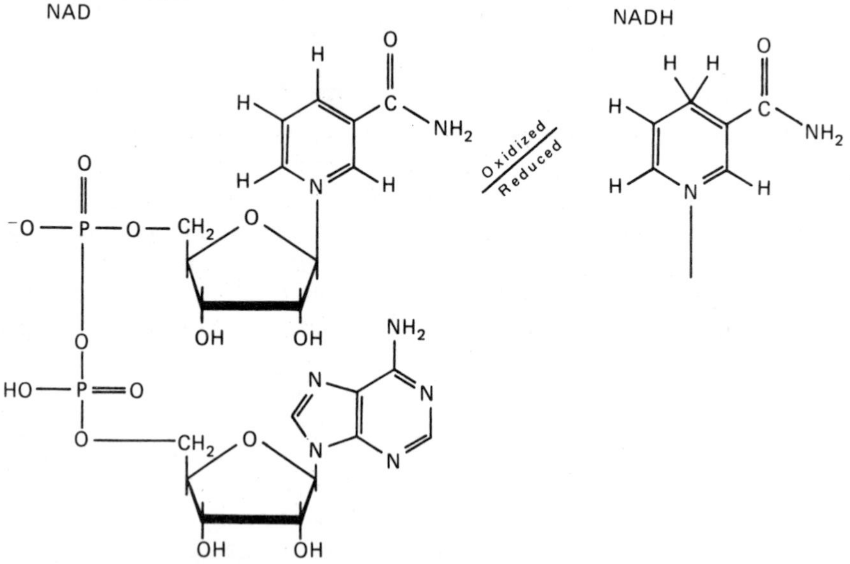

Figure 2-5. Structure of Oxidized and Reduced Forms of Nicotinamide Adenine Dinucleotide. (After Brock, 1970.)

the oxidation stages is retained by substrate-level phosphorylation. A unique feature of fermentation is that no external electron acceptor is required.

Consider the dissimilation of glucose as illustrated in Figure 2-6. This is actually a four-stage process. The first stage is endergonic and requires energy that is obtained from cleaving the terminal phosphate group from ATP molecules. The second stage involves the cleavage of one six-carbon sugar to form two interconvertible three-carbon compounds. The third stage is the energy-yielding stage, where substrate-level phosphorylation occurs. In the fourth stage, depending on the type of microorganisms present, different reactions occur which serve to regenerate NAD from the NADH produced during the oxidation of 3-phosphoglyceraldehyde.

Pyruvate is the pivotal compound in metabolism. If an external electron acceptor is present, pyruvate can be converted to acetyl CoA, which enters the Krebs cycle. However, when no external electron acceptors are present, pyruvate may undergo any of several alternative reactions which serve to regenerate NAD from NADH. One such reaction is shown in stage 4 of Figure 2-6. However, Figure 2-7 shows many other products which can be derived from pyruvic acid. The importance of acetic, butyric, and proprionic acid formation will become apparent during the study of anaerobic processes.

Fermentation is a name used to designate any anaerobic mechanism of energy production that does not involve an electron transport chain.

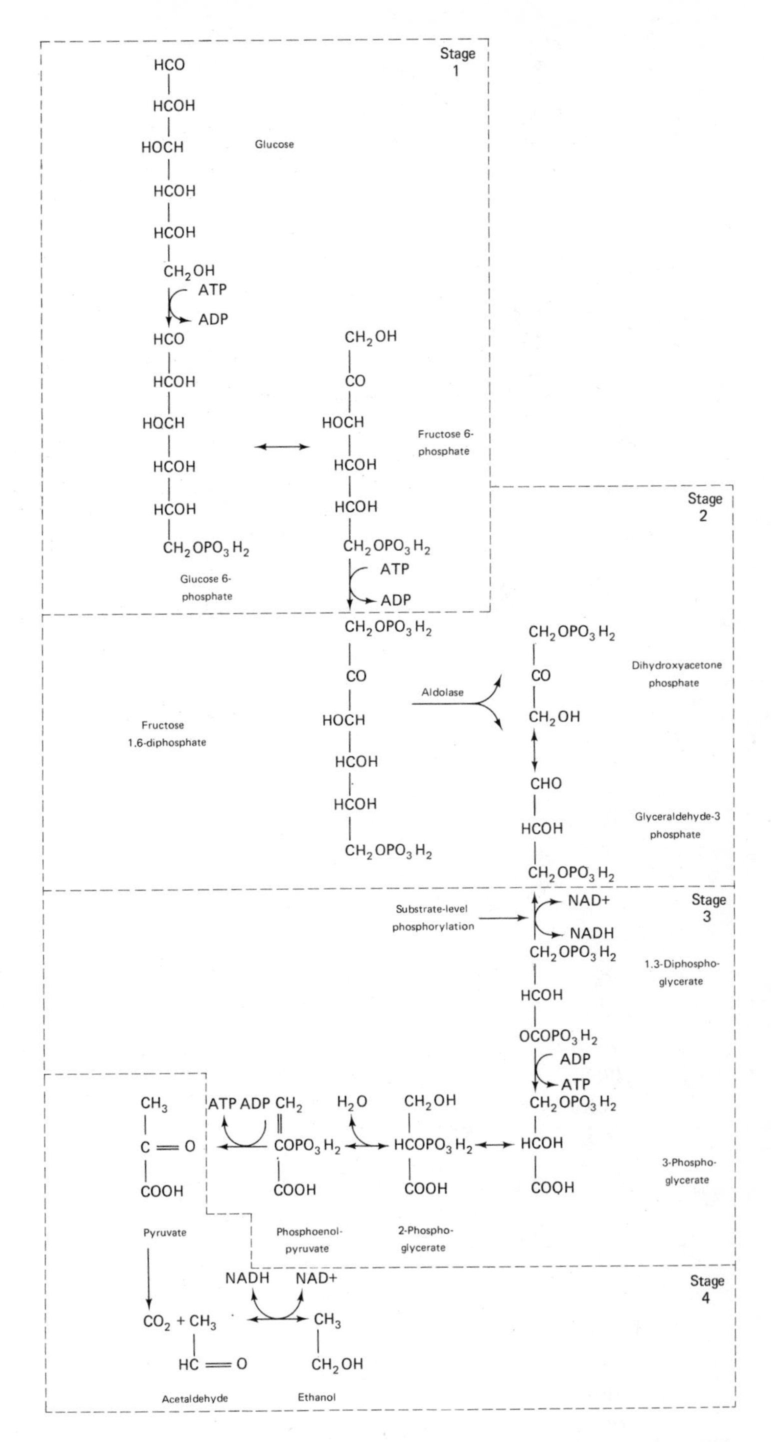

Figure 2-6. Fermentation Pathway. (After Brock, 1970.)

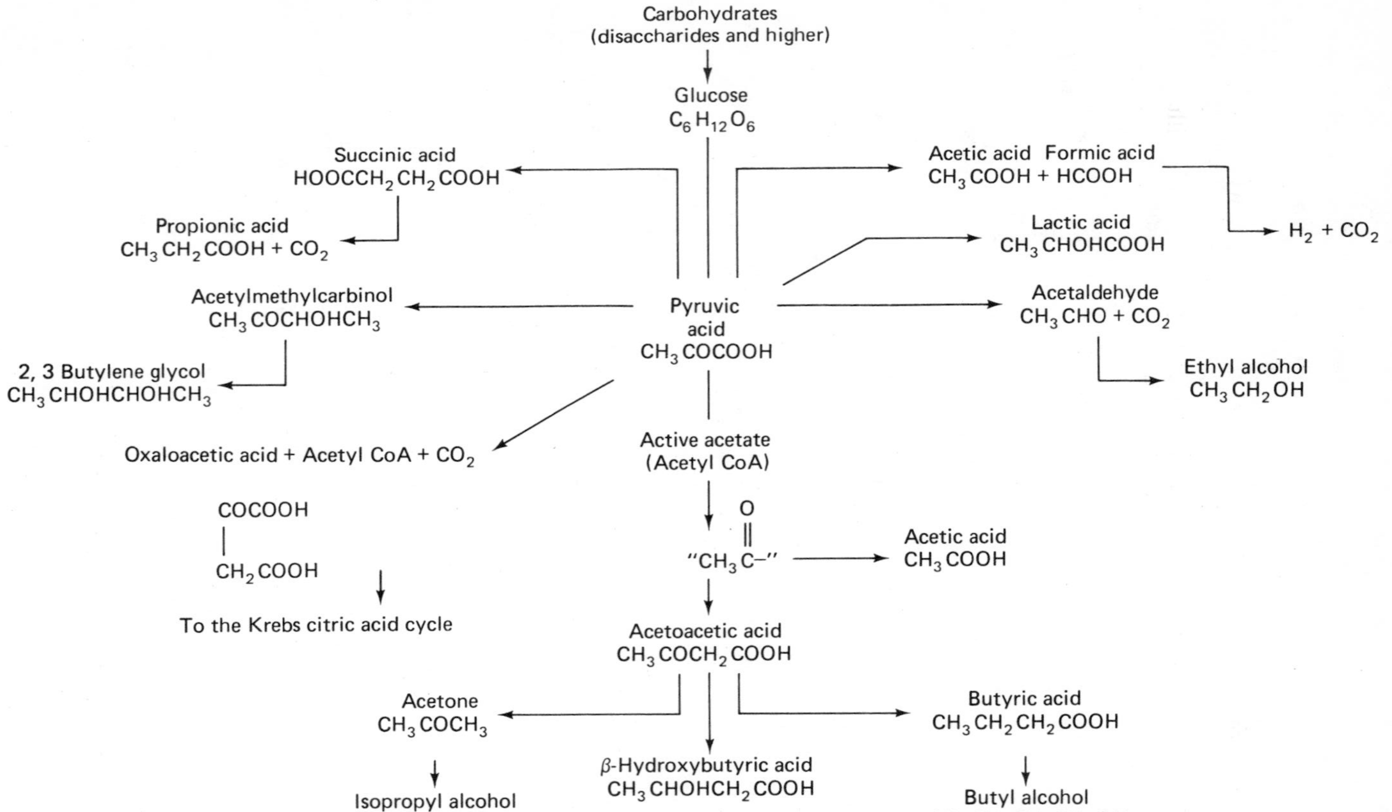

Figure 2-7. Representation of the Pivotal Nature of Pyruvic Acid. (After Pelczar and Reid, 1972.)

Aerobic Respiration. In aerobic respiration, the carbon and energy source is initially broken down by the series of enzyme-mediated reactions shown in Figure 2-6. However, in this case oxygen serves as an external electron acceptor so that pyruvic acid is converted to acetyl CoA, which enters the *Krebs cycle* (also known as the *tricarboxylic acid cycle* or the *citric acid cycle*), where it is oxidized to CO_2 and H_2O. The Krebs cycle is illustrated in Figure 2-8.

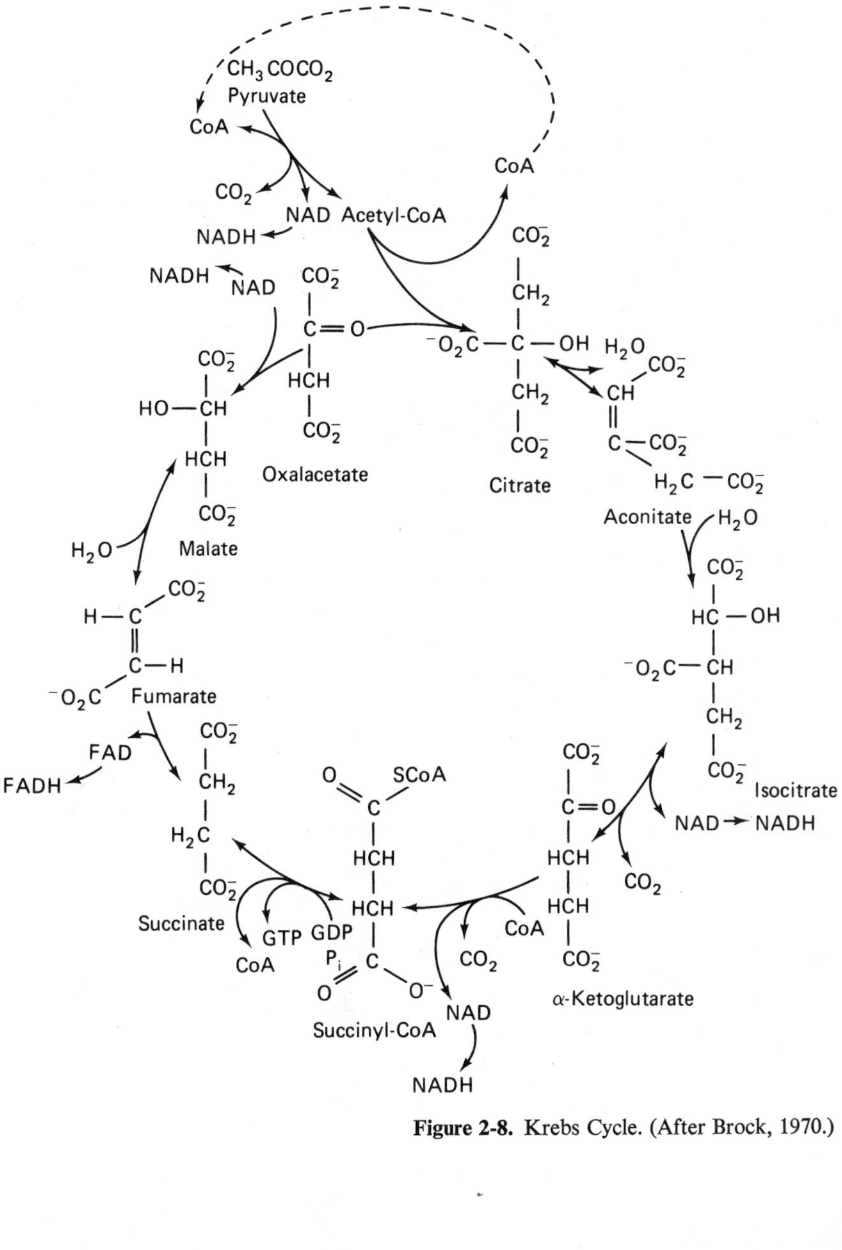

Figure 2-8. Krebs Cycle. (After Brock, 1970.)

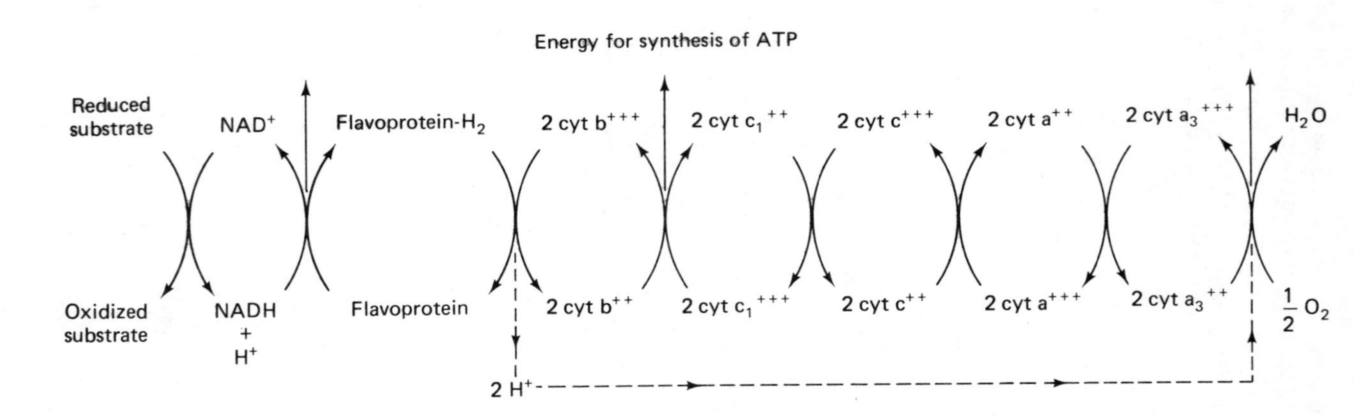

Figure 2-9. Electron Transport System Common to Most Bacteria. (After Pelczar and Reid, 1972.)

By far the greatest amount of useful energy produced during aerobic respiration comes from oxidative phosphorylation, which occurs as electrons are carried through an electron transport system. The electron transport system used by most bacteria is given in Figure 2-9. The primary purpose of this system is to reoxidize the reduced coenzymes formed as a result of oxidation reactions. There are a maximum of three phosphorylation sites located along the electron transport chain. If the electrons enter the chain at the level of NAD, three ATPs per pair of electrons transported will be realized. However, if the electrons enter the chain at the level of FAD, only two ATPs per pair of electrons transported will be provided.

Anaerobic Respiration. Although oxygen is the most efficient external electron acceptor, not all bacteria use it in this capacity. Certain bacteria utilize other inorganic electron acceptors by a process called *anaerobic respiration.* Compounds commonly used during anaerobic respiration are presented in Figure 2-10. This figure shows that sulfate is reduced to sulfide when sulfate is used as the terminal electron acceptor. Similarly, nitrate can be reduced to ammonia, nitrous oxide, or molecular nitrogen, whereas carbon dioxide is reduced to methane by some bacteria.

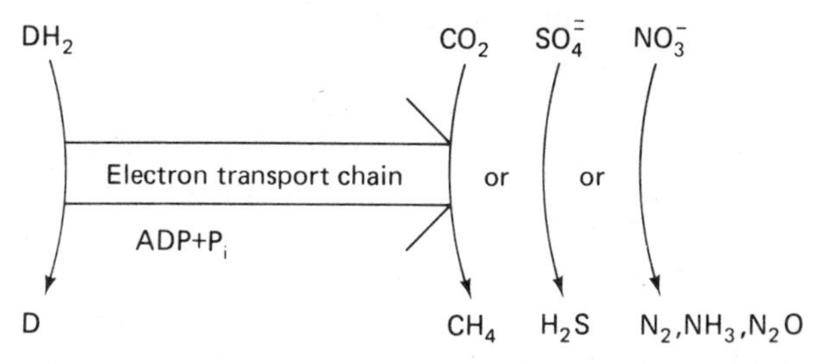

Figure 2-10. Electron Transport Chain for Anaerobic Respiration. (After Wilkinson, 1975.)

The metabolic pathways followed in the breakdown of the carbon and energy source are the same for both anaerobic and aerobic respiration. Yet, there are two basic differences between these two processes. These are (1) the ultimate fate of electrons produced in the oxidation reactions, and (2) the amount of ATP formed by oxidative phosphorylation. The amount of ATP formed when a pair of electrons are passed through the electron transport system depends on the difference in redox potential between the electron donor and acceptor. Oxygen will generally have a lower redox potential than other inorganic electron acceptors. Hence, more ATP will usually be realized from aerobic respiration.

Energy Budget

It is instructive to summarize the energy yield for the different types of energy metabolism presented for heterotrophic microorganisms. In fermentation, depending on the starting point (i.e., whether the dissimilation reaction begins at glucose or glycogen), the net yield is two ATPs or three ATPs per hexose unit introduced. However, in aerobic respiration, when anaerobic breakdown to pyruvate is followed by the Krebs cycle, 38 to 39 ATPs per hexose unit are realized. This means that the utilization of glucose from the standpoint of ATP production is much more efficient in the presence of oxygen. Hence, a much higher percentage of a unit amount of glucose will be assimilated to cell material in aerobic growth compared to anaerobic growth. Anaerobic biological treatment processes take advantage of this characteristic.

Energy Metabolism of Autotrophs

On the basis of their energy production, autotrophic microorganisms can be classified into two groups.

Chemoautotrophs. These organisms use inorganic carbon as their carbon source and obtain their energy by the oxidation of inorganic compounds. Generally, oxygen is used as the terminal electron acceptor, and ATP is formed by oxidative phosphorylation. A good example of chemoautotrophic microorganisms are bacteria from the genus *Nitrobacter.* Bacteria of this genus obtain their energy through the oxidation of nitrite ions to nitrate ions. A schematic presenting the level at which the electrons enter the electron transport chain is given in Figure 2-11. This figure shows that the electrons enter at the level of cytochrome c. Thus, only one ATP per NO_2^- unit oxidized will be realized from this system.

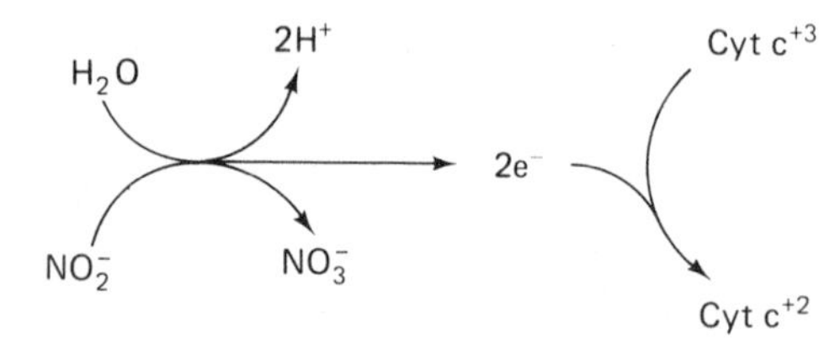

Figure 2-11. Entry Level for Electrons in Nitrite Oxidation. (After Pelczar and Reid, 1972.)

Table 2-3 presents the energy-yielding reactions of several chemoautotrophic bacteria. In all cases oxygen is the terminal electron acceptor.

Photoautotrophs. Generally, the organisms of interest in this class are certain algae that use free carbon dioxide as a carbon source and derive their energy from the sunlight. As algae are a principal source of oxygen in certain pond

TABLE 2-3

ENERGY-YIELDING REACTIONS FOR SOME CHEMOAUTOTROPHIC BACTERIA

Bacteria group	*Electron donor*		*Product*
Nitrifying bacteria	NH_3	$\longrightarrow$	NO_2^-
	NO_2^-	$\longrightarrow$	NO_3^-
Sulfur bacteria	H_2S	$\longrightarrow$	S
	S	$\longrightarrow$	SO_4^{2-}
Hydrogen bacteria	H_2	$\longrightarrow$	H_2O
Iron bacteria	Fe^{2+}	$\longrightarrow$	Fe^{3+}

Source: After Wilkinson (1975).

systems, the process of photosynthesis should be understood by those who engineer such treatment processes. Photosynthesis can be defined as the production of organic compounds from carbon dioxide and water using light energy. The overall result of photosynthesis is oxidation and reduction. Water is oxidized and the electrons released are used to reduce carbon dioxide to carbohydrates.

The photosynthetic process follows two basic reactions known as light and dark reactions. In the light reaction, light energy is captured and converted to ATP and the reduced form of nicotinamide adenine dinucleotide phosphate, NADPH. In the dark reaction the ATP and NADPH from the light reaction are consumed during the reduction of carbon dioxide to carbohydrate.

The small portion of the electromagnetic spectrum (see Figure 2-12) that can be seen is called the *visible region* (between 400 and 700 nm). The light energy for photosynthesis comes from this region. The light is absorbed by pigments of the photosynthetic system. Chlorophyll is the primary light-trapping pigment in green cells such as green algae. Most oxygen-producing

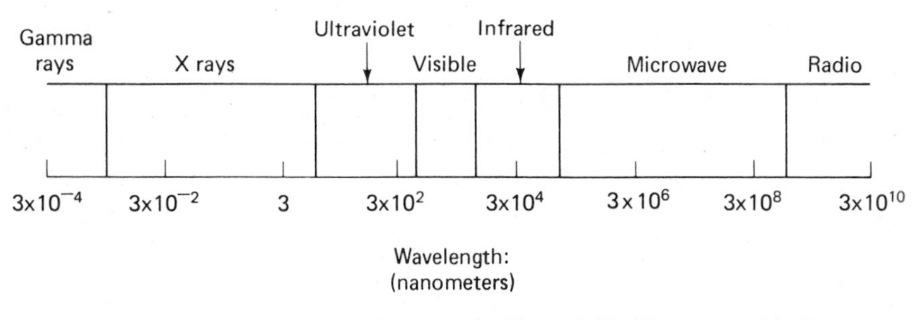

Figure 2-12. Electromagnetic Spectrum.

photosynthetic cells contain two kinds of chlorophyll, one of which is always chlorophyll a. Other forms may be chlorophyll b in green plants, chlorophyll c in brown algae, or chlorophyll d in red algae. Small amounts of a pigment with a light absorption maximum at 700 nm have been found in most photosynthetic cells. This pigment has been designated P700 (Lehninger, 1970). Initially, short-wavelength light is absorbed to supply energy for the oxidation of water, which results in the liberation of oxygen and the production of electrons of high potential energy. The energy level of the electrons is gradually reduced by passing them through an electron transport system. Energy liberated during this process is conserved in the form of ATP. The electrons are again energized to a high-potential-energy level by long-wavelength light absorbed by the pigment P700. These high-energy electrons are then used in the reduction of NADP to NADPH, which gives the reducing power for the dark reaction. A schematic representation of the light reaction is illustrated in Figure 2-13.

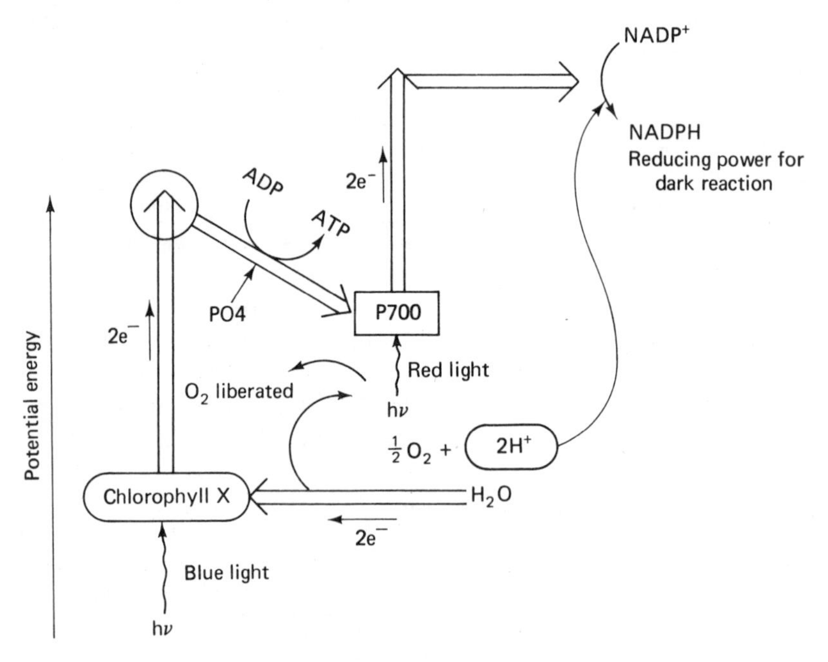

Figure 2-13. Electron Flow During Photosynthesis.

The overall dark reaction can be written as

$$6 \text{ ribulose 1,5-diphosphate} + 6CO_2 + 18ATP + 12NADPH + 12H^+ \longrightarrow$$

6 ribulose 1,5-diphosphate + hexose

$$+ 18P_i + 18ADP + 12NADP \qquad (2\text{-}3)$$

The dark reaction is a very complex cyclic mechanism. However, the end result is that a hexose molecule is ultimately formed from six molecules of CO_2. Ribulose 1,5-diphosphate is shown on both sides of equation 2-3, because it is a necessary component that is regenerated at the end of the cycle. All other cytoplasmic material can be obtained through metabolic conversion of the carbohydrate formed in the dark reaction. It should be stressed that the dark reaction does not occur during the hours of darkness, but it is so named because it does not require light energy as such but rather utilizes the products of the light reaction.

2-4 Kinetic Relationships from Microbiology

When a small number of viable bacterial cells are placed in a closed vessel containing excessive food supply in a suitable environment, conditions are established in which unrestricted growth takes place. Cell growth reflects the functioning of the enzyme system leading to the addition of macromolecular products to the cytoplasm. However, addition of cytoplasmic mass and growth of an organism do not go on indefinitely, and after a characteristic size is reached, the cell divides due to hereditary and internal limitations. With many species of bacteria, the rate of growth may follow a growth pattern similar to that shown in Figure 2-14 (Lamanna, 1965; Knaysi, 1951). The curve shown in this figure may be divided into six well-defined phases (Monod, 1949):

1/ *Lag phase:* adaptation to a new environment, long generation time, null growth rate, and cell size and rate of metabolic activity maximum.

2/ *Acceleration phase:* decreasing generation time and increasing growth rate.

3/ *Exponential phase:* minimal and constant generation time, maximal and constant specific growth rate, maximum rate of substrate conversion, and achievement of steady-state as indicated by a nearly constant ratio of DNA/cell, RNA/cell, protein/cell, constant cell density, and minimum cell size.

4/ *Declining growth phase:* increasing generation time and decreasing specific growth rate due to gradual decrease in substrate concentration and increased accumulation of toxic metabolites.

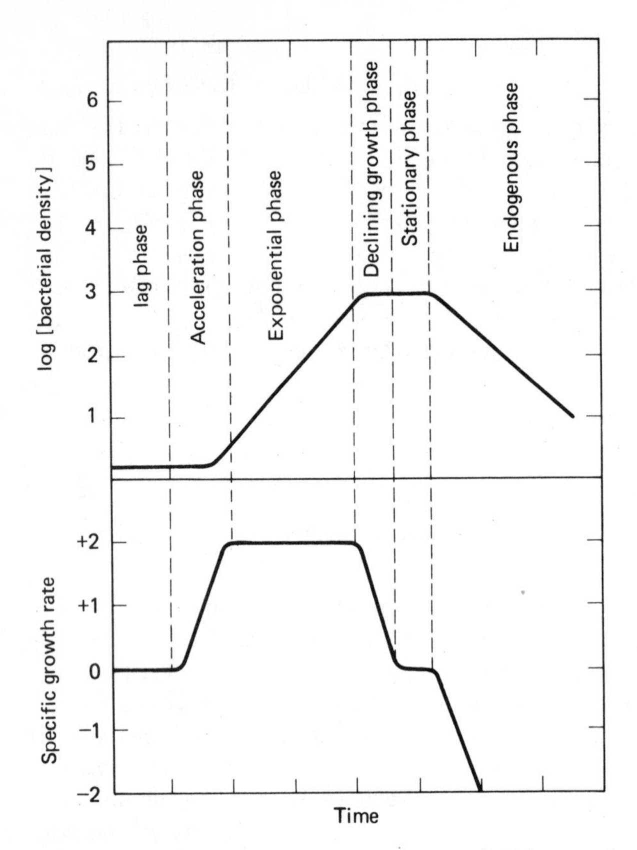

Figure 2-14. Characteristic Growth Curves of Cultures of Microorganisms. (After Monod, 1949.)

5/ *Stationary phase:* exhaustion of nutrients, high concentration of toxic metabolites, and maximum physical crowding. Wilkinson (1975) notes that the stationary phase can result from a balance between growth and death but normally is a result of the cells remaining in a state of suspended animation.

6/ *Endogenous phase:* endogenous metabolism, high death rate, and cell lysis.

It should be emphasized that the growth cycle described is not a basic property of bacterial cells but is a result of their interaction with the environment in a closed system. It is possible to maintain the cells in the exponential phase of growth for long periods of time when an open system such as a continuous flow process is used. Furthermore, exponential growth can be maintained at less than the maximum rate.

Biomass Growth Rate

Some of the more important prerequisites for growth of biomass in a bacterial culture are (1) an energy source, (2) a carbon source, (3) an external electron acceptor when required, and (4) a suitable physicochemical environment.

When all the requirements for growth are met, then, for some increment of time, Δt, the incremental increase in biomass concentration, Δx, is proportional to the concentration of biomass present, x. Thus, it is possible to write

$$\Delta x \propto x\,\Delta t \tag{2-4}$$

Equation 2-4 may be written as an equality by introducing the proportionality constant μ:

$$\Delta x = \mu x\,\Delta t \tag{2-5}$$

Dividing both sides of equation 2-5 by Δt and taking the limit as $\Delta t \longrightarrow 0$ gives the derivative

$$\left(\frac{dx}{dt}\right)_g = \mu x \tag{2-6}$$

where the derivative $(dx/dt)_g$ expresses the biomass growth rate and has the dimensions mass volume^{-1} time^{-1}.

Letting x_0 represent the biomass concentration at $t = 0$, integration of equation 2-6 gives

$$\ell\text{n}\, x = \ell\text{n}\, x_0 + \mu t \tag{2-7}$$

Then, if the logarithms are converted to base 10, equation 2-7 becomes

$$\log x = \log x_0 + \frac{\mu t}{2.3} \tag{2-8}$$

Equation 2-8 suggests that a plot of log x versus time will give a straight line of slope $\mu/2.3$.

Equation 2-7 can also be rearranged into the form

$$\ell\text{n}\left(\frac{x}{x_0}\right) = \mu t \tag{2-9}$$

or

$$x = x_0 e^{\mu t} \tag{2-10}$$

Growth that follows the rate law expressed by equation 2-10 is called *exponential growth.*

At this point it is instructive to discuss the significance of the proportionality constant μ. First, consider that equation 2-6 must be dimensionally correct to be meaningful. As x is given as concentration, it has the units mass volume^{-1} and since the dimensions of $(dx/dt)_g$ must be the same as μx, it follows that μ has the dimension of reciprocal time. Equation 2-6 also shows that the fractional growth rate, $x^{-1}(dx/dt)_g$, at any time is a constant and this constant is μ:

$$\frac{(dx/dt)_g}{x} = \mu \tag{2-11}$$

The parameter μ represents the rate of growth per unit amount of biomass and is termed the *specific growth rate*.

Pirt (1975) notes that the law of exponential growth will be followed as long as there is no change in the composition of the biomass and the environmental conditions remain constant. A change in environmental conditions is probably the most common cause of deviation from exponential growth. This will always occur at some point during the growth of a batch culture.

Monod (1949) extended the quantitative description of the classic growth curve shown in Figure 2-14 to include both the exponential and declining-growth-rate regions. From experimental studies he observed that the growth rate, $(dx/dt)_g$, was a function not only of organism concentration but also of some limiting nutrient concentration. He described the relationship between the residual concentration of the growth-limiting nutrient and the specific growth rate of biomass by the equation

$$\mu = \mu_m \frac{S}{K_s + S} \tag{2-12}$$

where μ = specific growth rate, time^{-1}
μ_m = maximum value of μ at saturation concentrations of growth-limiting substrate, time^{-1}
S = residual growth-limiting substrate concentration, mass volume^{-1}
K_s = saturation constant numerically equal to the substrate concentration at which $\mu = \mu_m/2$, mass volume^{-1}

This relationship is illustrated in Figure 2-15, which shows that the relationship between the specific growth rate and the growth-limiting nutrient concentration has the same form as the Michaelis–Menten equation (equation 1-20), which describes the saturation of an enzyme with its substrate.

From equation 2-12 it is seen that specific growth rate can have any value between zero and μ_m, provided that the substrate concentration can be held constant at a given value. Any system designed for the continuous cultivation of microorganisms meets this condition. The design of many biological wastewater treatment processes is based on this characteristic.

When applying the Monod relationship, the S term must be the growth-limiting nutrient concentration. The carbon and energy source, as measured by ultimate biochemical oxygen demand (BOD_u), chemical oxygen demand (COD), or total organic carbon (TOC), is usually considered to be the growth-limiting nutrient in biological wastewater treatment processes. However,

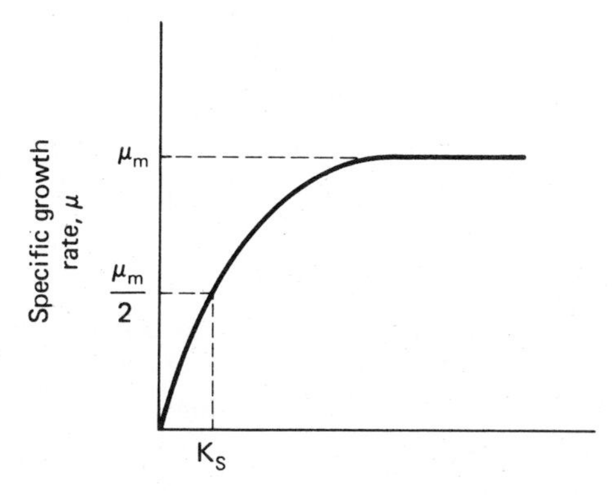

Figure 2-15. Relationship Between Specific Growth Rate and Growth-Limiting Nutrient Concentration.

it must be kept in mind that growth can be controlled by other substances, such as nitrogen and phosphorus.

In this text, only BOD_u and degradable COD will be used as relative measures of the organic content of the wastewater. This will be done to avoid mixing computations which use biokinetic constants based on BOD_5 measurements with BOD_u values, when the oxygen requirement for a particular activated sludge process is being evaluated.

Growth Yield

Growth yield Y is defined mathematically as

$$\frac{\Delta x}{\Delta S} = Y \tag{2-13}$$

where Δx is the incremental increase in biomass which results from the utilization of the incremental amount of substrate, ΔS. Taking the limit of $\Delta x/\Delta S$ as $\Delta S \rightarrow 0$ gives the derivative

$$\frac{dx}{dS} = Y \tag{2-14}$$

The quantitative nutrient requirement of an organism is given by the growth yield, as reflected in equation 2-14.

Monod (1949) observed that as long as there was no change in the composition of the biomass and the environmental conditions remained constant, the growth yield Y remained a constant quantity. Thus, designating the initial biomass and substrate concentration as x_0 and S_0, respectively, and letting x and S represent the corresponding concentrations during growth, Pirt (1975) has shown that the relationship between growth and substrate utilization can be expressed as

$$x - x_0 = Y(S_0 - S) \tag{2-15}$$

For a growth-limiting substrate, a culture reaches its maximum biomass concentration, x_m, near the end of the declining-growth phase. At this point it can be assumed that the concentration of the growth-limiting substrate is zero (i.e., $S \approx 0$). For this situation, equation 2-15 reduces to

$$x_m - x_0 = YS_0 \qquad (2\text{-}16)$$

or

$$x_m = x_0 + YS_0 \qquad (2\text{-}17)$$

Thus, for a growth-limiting substrate, equation 2-17 suggests that a plot of x_m versus S_0 will give a linear trace of slope Y. Such a construction is presented in Figure 2-16. If linearity breaks down at high substrate concentrations, factors other than substrate concentration are limiting growth in this region.

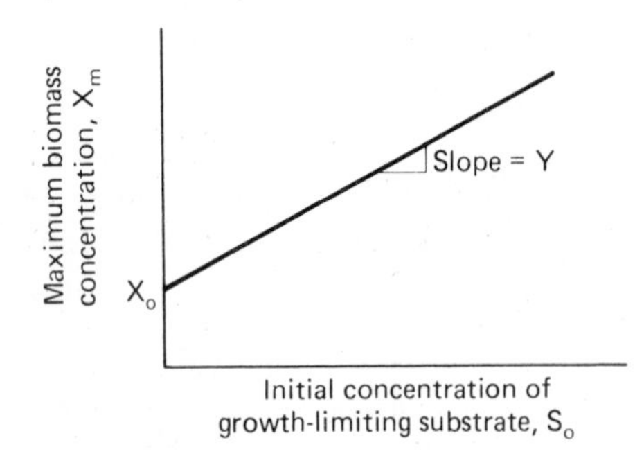

Figure 2-16. Effect of Growth-Limiting Substrate Concentration on the Growth of Biomass.

Substrate Utilization in Microbial Cultures

Pirt (1975) suggests that for some increment of time, Δt, the incremental change in substrate concentration, ΔS, is proportional to the concentration of biomass present, x, Such a relationship is expressed by

$$\Delta S \propto x\,\Delta t \qquad (2\text{-}18)$$

Equation 2-18 can be written as an equality by introducing the proportionality constant q:

$$\Delta S = qx\,\Delta t \qquad (2\text{-}19)$$

Dividing both sides of equation 2-19 by Δt and taking the limit as $\Delta t \to 0$ gives the derivative

$$\left(\frac{dS}{dt}\right)_u = qx \qquad (2\text{-}20)$$

where the derivative $(dS/dt)_u$ represents the substrate utilization rate and has the dimensions of mass volume^{-1} time^{-1}. Equation 2-20 suggests that the *fractional substrate utilization rate*, $x^{-1}(dS/dt)_u$, at any time is a constant, and this constant is q:

$$\frac{(dS/dt)_u}{x} = q \qquad (2\text{-}21)$$

The parameter q represents the substrate utilization rate per unit amount of biomass and is termed the *specific substrate utilization rate*. Parameter q has the dimension of time^{-1}.

Recalling that

$$\mu = \frac{(dx/dt)_g}{x} \tag{2-11}$$

and that

$$Y = \frac{dx}{dS} \tag{2-14}$$

it is possible to develop a relationship between specific substrate utilization rate, specific growth rate, and growth yield which has the form

$$\frac{(dS/dt)_u}{x} = q = \frac{(dx/dt)_g/x}{dx/dS} = \frac{\mu}{Y}$$

or

$$q = \frac{\mu}{Y} \tag{2-22}$$

Equation 2-22 can be used to estimate the substrate demand at various growth rates.

Effect of Substrate Concentration on Growth Rate

The exponential growth phase in batch cultures suggests that the rate of substrate utilization may be unaffected by substrate concentration over a wide range of substrate concentrations; that is, the rate of substrate utilization follows zero-order kinetics with respect to substrate concentration. Furthermore, because growth rate decreases in the declining-growth phase, it is not unreasonable to assume that for a growth-limiting substrate, the rate of substrate utilization would be directly proportional to substrate concentration at low substrate concentration. Thus, it might be expected that substrate utilization could be described by an equation similar to the Michaelis–Menten equation for enzyme kinetics. Such an equation might have the form

$$q = q_m \frac{S}{K_s + S} \tag{2-23}$$

where q_m is the *maximum specific substrate utilization rate* obtained when $S \gg K_s$. Substituting from equation 2-22 for q and using μ_m/Y for q_m, equation 2-23 reduces to the Monod equation (equation 2-12), which describes the relationship between specific growth rate and substrate concentration.

Energy and Carbon Source Requirements

For heterotrophic microorganisms it has been shown that the substrate acts both as a carbon and an energy source. For these organisms it is necessary to distinguish between that fraction of the substrate which is channeled into the synthesis function (i.e., provides the building blocks for organism growth) and that fraction of the substrate which is channeled into the energy function and subsequently oxidized to provide energy for all cellular functions. Such a distinction can be made by performing a substrate balance for substrate utilized during an increment of time, Δt.

$$\begin{bmatrix}\text{total substrate}\\ \text{utilized}\end{bmatrix} = \begin{bmatrix}\text{substrate utilized}\\ \text{for synthesis}\end{bmatrix} + \begin{bmatrix}\text{substrate oxidized}\\ \text{for energy}\end{bmatrix} \tag{2-24a}$$

Equation 2-24a can be expressed mathematically as

$$\Delta S = (\Delta S)_S + (\Delta S)_E \tag{2-24b}$$

If Δx represents the increase in biomass concentration during the time increment Δt and if both sides of equation 2-24b are divided by Δx, the resulting expression is

$$\frac{\Delta S}{\Delta x} = \frac{(\Delta S)_S}{\Delta x} + \frac{(\Delta S)_E}{\Delta x} \tag{2-25}$$

Recalling that $Y = \Delta x/\Delta S$, equation 2-25 can be rewritten as

$$\frac{1}{Y} = \frac{1}{Y_S} + \frac{1}{Y_E} \tag{2-26}$$

It is important to realize that Y_E is not a real value because no biomass is produced from the substrate associated with this term. Rather, it indicates that fraction of the substrate removed per unit of biomass produced which is channeled into energy metabolism. Thus, it represents substrate that is not synthesized to new cellular mass and its value is always negative. It should also be noted that $(\Delta S)_S$ actually equals the real Δx, and therefore $(\Delta S)_S/\Delta x$ is always 1, and $1/Y_S$ also equals unity.

Pirt (1975) has proposed that microorganisms require energy for growth (synthesis) as well as for maintenance functions such as turnover of cell materials, active transport, motility, and so on. Thus, the last term on the right side of equation 2-25 can be expressed as

$$\frac{(\Delta S)_E}{\Delta x} = \frac{(\Delta S)_{GE} + (\Delta S)_{ME}}{\Delta x} = \frac{1}{Y_E} \tag{2-27}$$

where $(\Delta S)_{GE}$ represents that portion of the substrate oxidized for energy which is utilized in the growth function and $(\Delta S)_{ME}$ represents that portion of the substrate oxidized for energy which is utilized in the maintenance function. $(\Delta S)_{ME}$ represents the total amount of substrate channeled into the energy of maintenance function, which includes energy for the biomass originally present in the system as well as energy for the biomass produced during the substrate utilization process. When the maintenance energy requirement is zero, equation 2-27 reduces to a minimum value, which is

$$\frac{(\Delta S)_E}{\Delta x} = \frac{(\Delta S)_{GE}}{\Delta x} = \frac{1}{Y_E} \tag{2-28}$$

This represents an optimum yield condition because a portion of the substrate that might have been oxidized to provide for energy of maintenance will now be assimilated into new biomass. Under such conditions the Y term in equation 2-26 will have a maximum value which is referred to as the "true" growth yield, Y_T. Equation 2-26 then becomes

$$\frac{1}{Y_T} = \frac{1}{Y_S} + \frac{1}{Y_E} \tag{2-29}$$

Equation 2-24b can be written in the form

$$\left(\frac{dS}{dt}\right)_u = \left(\frac{dS}{dt}\right)_{uS} + \left(\frac{dS}{dt}\right)_{uE} \tag{2-30}$$

where $(dS/dt)_u$ represents the overall substrate utilization rate, $(dS/dt)_{uS}$ represents the rate of substrate utilization for synthesis, and $(dS/dt)_{uE}$ represents the rate of substrate utilization for energy. The term representing the rate of substrate utilization for energy can be further separated such that

$$\left(\frac{dS}{dt}\right)_{uE} = \left(\frac{dS}{dt}\right)_{uGE} + \left(\frac{dS}{st}\right)_{uME} \tag{2-31}$$

where $(dS/dt)_{uGE}$ represents the rate of substrate utilization related to energy for growth and $(dS/dt)_{uME}$ represents the rate of substrate utilization related to energy for maintenance. Assuming that the relationship between the amount of substrate used as building blocks for new cellular material and energy required for growth is a constant, the $(dS/dt)_{uGE}$ and $(dS/dt)_{uS}$ terms can be combined into the form

$$\left(\frac{dS}{dt}\right)_{uS} + \left(\frac{dS}{dt}\right)_{uGE} = \left(\frac{dS}{dt}\right)_{uG} \tag{2-32}$$

The term $(dS/dt)_{uG}$ represents the rate of substrate utilization for the growth function and includes substrate used as building blocks for new cellular material as well as substrate oxidized to provide energy for synthesis.

It has been shown (Pirt, 1975) that the rate of substrate utilization for energy of maintenance is proportional to the biomass present. This relationship can be expressed as

$$\left(\frac{dS}{dt}\right)_{uME} = bx \tag{2-33}$$

where b is a proportionality constant that represents the substrate utilized for the energy function per unit of biomass per unit time, that is, the *specific substrate utilization rate for energy of maintenance*. The parameter b has the dimension time^{-1}.

Substituting into equation 2-30 from equations 2-31, 2-32, and 2-33 gives

$$\left(\frac{dS}{dt}\right)_u = \left(\frac{dS}{dt}\right)_{uG} + bx \tag{2-34}$$

Equations 2-11 and 2-14 imply that

$$\frac{dS}{dt} = \frac{\mu x}{Y} \qquad (2\text{-}35)$$

Thus, substituting for $(dS/dt)_u$ and $(dS/dt)_{uG}$ in equation 2-34 from equation 2-35 results in the following formulation:

$$\frac{\mu x}{Y} = \frac{\mu x}{Y_g} + bx \qquad (2\text{-}36)$$

where Y_g is a constant which represents that fraction of the substrate removed which is channeled into the growth function. This includes substrate used for synthesis as well as substrate oxidized to provide energy for synthesis.

Dividing both sides of equation 2-36 by x yields an expression of the form

$$\frac{\mu}{Y} = \frac{\mu}{Y_g} + b \qquad (2\text{-}37)$$

When q is substituted for μ/Y (see equation 2-22), equation 2-37 reduces to

$$q = \frac{1}{Y_g}\mu + b \qquad (2\text{-}38)$$

The Y_g term in equation 2-38 is, in fact, Y_T. This can be illustrated by considering that over some increment of time, Δt, there is a Δx incremental increase in biomass and a ΔS incremental decrease in substrate. The relationship between metabolism and substrate removed can be expressed mathematically as

$$(\Delta S)_{\text{removed}} = (\Delta S)_{\substack{\text{synthesis}\\ \text{and energy}\\ \text{for synthesis}}} + (\Delta S)_{\substack{\text{energy of}\\ \text{maintenance}}} \qquad (2\text{-}39)$$

Since

$$\Delta S = \frac{\Delta x}{Y} \qquad (2\text{-}13)$$

and

$$(\Delta S)_{\substack{\text{energy of}\\ \text{maintenance}}} = bx \qquad (2\text{-}33)$$

it is possible to write

$$\frac{\Delta x}{x}\frac{x}{Y_A} = \frac{\Delta x}{x}\frac{x}{Y_g} + bx \qquad (2\text{-}40)$$

or

$$\frac{\mu}{Y_A} = \frac{\mu}{Y_g} + b \qquad (2\text{-}41)$$

where Y_A = variable yield coefficient which describes the actual amount of biomass produced per unit of substrate removed

Solving this equation for Y_g gives

$$\frac{1}{Y_g} = \left(\frac{1}{Y_A} - \frac{b}{\mu}\right) \tag{2-42}$$

For the condition when all the substrate is channeled into the growth function, that is, when there is no energy of maintenance requirement $b = 0$ and $Y_A = Y_T$. Hence,

$$\frac{1}{Y_g} = \frac{1}{Y_T} \tag{2-43}$$

or

$$Y_g = Y_T \tag{2-44}$$

and therefore,

$$q = \frac{1}{Y_T}\mu + b \tag{2-45}$$

Equation 2-45 suggests that a plot of specific substrate utilization rate versus specific growth rate will yield a linear trace of slope $1/Y_T$ and intercept b. Figure 2-17 illustrates such a plot.

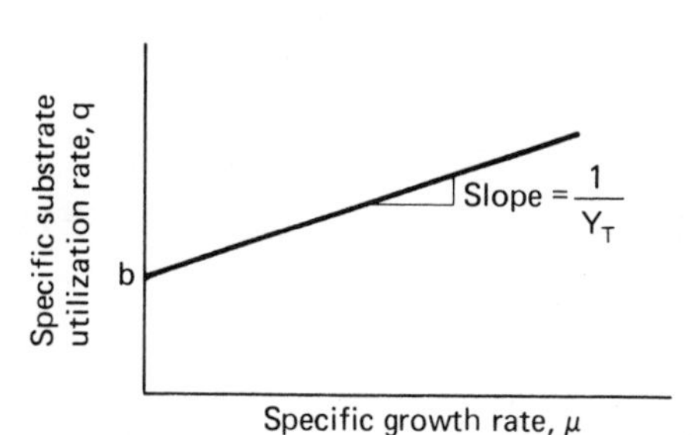

Figure 2-17. Plot of Specific Substrate Utilization Rate Versus Specific Growth Rate.

Maintenance as Endogenous Respiration

To account for the decrease in biomass production that is observed when the specific growth rate decreases, Herbert (1958) has suggested that the maintenance energy requirement is satisfied through endogenous metabolism; that is, cellular components are oxidized to satisfy the maintenance energy requirement rather than oxidizing a portion of the external substrate. For this situation a biomass balance can be written as

$$\begin{bmatrix}\text{net}\\ \text{growth}\end{bmatrix} = \begin{bmatrix}\text{total}\\ \text{growth}\end{bmatrix} - \begin{bmatrix}\text{biomass lost to}\\ \text{endogenous respiration}\end{bmatrix} \tag{2-46}$$

Equation 2-46 can be expressed on a rate basis as

$$\left(\frac{dx}{dt}\right)_g = \left(\frac{dx}{dt}\right)_T - \left(\frac{dx}{dt}\right)_E \tag{2-47}$$

The rate at which biomass is lost to endogenous respiration is proportional to the biomass present. Thus,

$$\left(\frac{dx}{dt}\right)_E = K_d x \qquad (2\text{-}48)$$

where K_d is a proportionality constant that represents the biomass lost to endogenous respiration per unit of biomass per unit of time. The constant K_d is termed the *microbial decay coefficient* and has the dimension time^{-1}.

Substituting for $(dx/dt)_E$ in equation 2-47 from equation 2-48 gives

$$\left(\frac{dx}{dt}\right)_g = \left(\frac{dx}{dt}\right)_T - K_d x \qquad (2\text{-}49)$$

The total growth $(dx/dt)_T$ can be expressed as

$$\left(\frac{dx}{dt}\right)_T = Y_T \left(\frac{dS}{dt}\right)_u \qquad (2\text{-}50)$$

since it is assumed that all the substrate utilized is channeled into the growth function, whereas the energy for maintenance comes from the oxidation of cellular components. Thus, substituting for $(dx/dt)_T$ in equation 2-49 from equation 2-50, the following expression is obtained:

$$\left(\frac{dx}{dt}\right)_g = Y_T \left(\frac{dS}{dt}\right)_u - K_d x \qquad (2\text{-}51)$$

Dividing both sides of equation 2-51 by X yields an expression of the form

$$\mu = Y_T q - K_d \qquad (2\text{-}52)$$

or

$$q = \frac{1}{Y_T}\mu + \frac{K_d}{Y_T} \qquad (2\text{-}53)$$

This equation shows that at a high rate of growth (when μ is much greater than K_d), the component μ/Y_T is much larger than the K_d/Y_T component. As growth slows down, the first component decreases, reaching zero at $\mu = 0$ for which $q = K_d/Y_T$. Thus, at low specific growth rates, the μ/Y_T term in equation 2-53 is small implying that the majority of the substrate utilized goes to the maintenance function rather than for growth (i.e., a much larger fraction of the substrate utilized per unit of biomass is used in the energy of maintenance function than is used in the growth function). Equation 2-53 also suggests that the b term in equation 2-45 is equal to K_d/Y_T.

It is interesting to note that the expression relating specific substrate utilization rate to specific growth rate is the same regardless of the mechanism proposed to describe how a microorganism satisfies its energy of maintenance requirement. However, it will be shown in Chapter 4 that a different relationship for specific oxygen utilization rate is obtained when the Pirt theory is accepted instead of the Herbert theory.

2-5
Kinetic Relationships Applied in Process Design

Design equations for many different types of biological treatment processes can be developed by applying material balances to the particular system of interest. In these material-balance equations, derivatives that indicate the rate of substrate utilization and rate of growth are found. The design equations, which are developed from these material-balance equations, will depend upon the particular relationships chosen to describe the derivative rate expressions for substrate utilization and growth.

Lawrence and McCarty (1970) relate the rate of substrate utilization to the concentration of microorganisms in the reactor and to the concentration of substrate surrounding the organisms. This relationship has the form

$$\left(\frac{dS}{dt}\right)_u = \frac{kxS}{K_s + S} \tag{2-54}$$

where $\left(\frac{dS}{dt}\right)_u$ = overall substrate utilization rate, mass volume^{-1} time^{-1}

k = maximum specific substrate utilization rate, that is, the maximum rate of substrate utilization per unit of biomass, time^{-1}

S = substrate concentration surrounding the biomass, mass volume^{-1}

K_s = saturation constant which has a value equal to the substrate concentration when $(dS/dt)_u/X = \frac{1}{2}k$, mass volume^{-1}

x = active biomass concentration, mass volume^{-1}

Equation 2-54 indicates that the relationship between the rate of substrate utilization and substrate concentration is continuous over the entire range of substrate concentrations. This relationship is illustrated in Figure 2-18.

Equation 2-54 is the same as equation 2-23, which was proposed to describe the effect of substrate concentration on growth rate and was shown

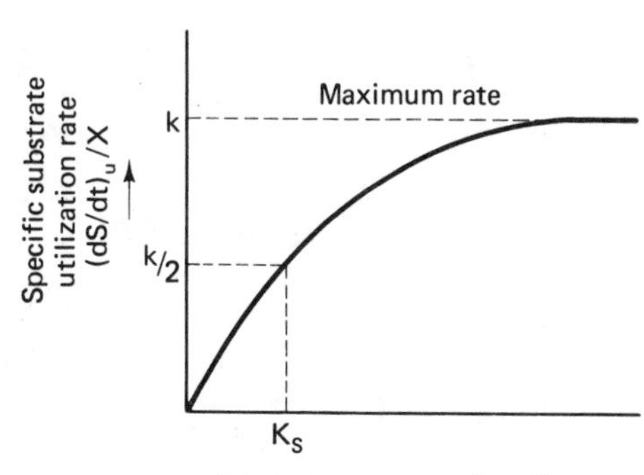

Figure 2-18. Relationship Between Substrate Utilization Rate and Substrate Concentration as Predicted by Equation 2-54.

to reduce to the Monod equation (equation 2-12). The parameter k is identical to the q_m term in equation 2-23 and will be used for all references to the maximum specific substrate utilization rate throughout the rest of this text.

For the limiting case, when S is much greater than K_s, K_s can be neglected in the sum term given in the denominator of equation 2-54. In this case equation 2-54 reduces to an expression that is zero-order with respect to substrate concentration,

$$\left(\frac{dS}{dt}\right)_u = kx \tag{2-55}$$

For the final limiting case, when S is much smaller than K_s, S can be neglected in the sum term given in the denominator. In this situation, equation 2-54 reduces to an expression that is first-order with respect to substrate concentration:

$$\left(\frac{dS}{dt}\right)_u = KxS \tag{2-56}$$

where

$$K = \frac{k}{K_s} = \text{specific substrate utilization rate constant, volume mass}^{-1}\text{ time}^{-1} \tag{2-57}$$

Equation 2-55 represents a zero-order reaction with respect to substrate concentration, while equation 2-56 represents a first-order reaction. Figure 2-19 illustrates the relationship between the rate of substrate utilization and substrate concentration given by the two limiting cases of equation 2-54, which together form what is sometimes referred to as the *discontinuous model for substrate utilization* (Garrett and Sawyer, 1960).

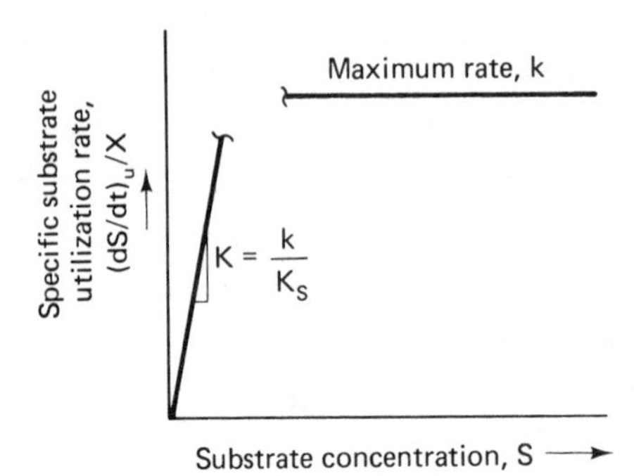

Figure 2-19. Relationsnip Between Substrate Utilization Rate and Substrate Concentration as Predicted by Equations 2-55 and 2-56.

Eckenfelder and Ford (1970) propose that the rate of substrate utilization in most biological wastewater treatment processes is adequately described by equation 2-56 and recommend its use in lieu of equation 2-54.

In recent work, Grady and Williams (1975) have presented data which suggest that neither equation 2-54 or 2-56 adequately describe the effects of a varying influent substrate concentration on the substrate utilization rate. For such a situation, the relationship proposed by Grau et al. (1975) appears to more accurately describe the rate of substrate utilization. Grau et al. proposed that

$$\left(\frac{dS}{dt}\right)_u = K_1 x \left(\frac{S}{S_0}\right)^n \tag{2-58}$$

where n = reaction order and is generally assumed to have a value of 1
S_0 = initial substrate concentration, mass volume^{-1}
S = substrate concentration surrounding the biomass at any time t, mass volume^{-1}
K_1 = specific substrate utilization rate constant, time^{-1}

An equation that describes the relationship between net growth rate of microorganisms and the rate of substrate utilization was developed empirically from waste treatment studies and reported by Heukelekian et al. (1951). This equation has the form

$$\left(\frac{dx}{dt}\right)_g = Y_T\left(\frac{dS}{dt}\right)_u - K_d x \tag{2-59}$$

where $\left(\frac{dx}{dt}\right)_g$ = net growth rate of the biomass, mass volume^{-1} time^{-1}

It is interesting to note that this equation is identical to equation 2-51, which was developed by proposing that the decrease in biomass production observed at low specific growth rates was due to oxidation of cellular components to provide energy for maintenance.

Sherrard and Schroeder (1973) have proposed that net growth rate is best described by the equation

$$\left(\frac{dx}{dt}\right)_g = Y_{\text{obs}}\left(\frac{dS}{dt}\right)_u \tag{2-60}$$

where Y_{obs} = variable observed yield coefficient

Equation 2-60 is basically the same as equation 2-59. The difference is equation 2-59 requires that the maintenance requirement be subtracted from the theoretical yield, whereas equation 2-60 describes the actual (observed) yield after the total energy requirement has been considered.

It is possible to develop an expression relating observed yield to specific growth rate. First, it is necessary to express equation 2-60 in the form

$$Y_{\text{obs}} = \frac{(dx/dt)_g}{(dS/dt)_u} \tag{2-61}$$

Then, multiplying the right side of equation 2-61 by x/x gives

$$Y_{\text{obs}} = \frac{(dx/dt)_g/x}{(dS/dt)_u/x}$$

or

$$Y_{obs} = \frac{\mu}{q} \tag{2-62}$$

Substituting from equation 2-53 for q in equation 2-62 results in an expression of the form

$$Y_{obs} = \frac{Y_T}{1 + K_d/\mu} \tag{2-63}$$

In this equation the dependence of the observed yield on the specific growth rate is shown.

Later chapters will show how design equations for various biological wastewater treatment processes can be developed by applying material balances and using the fundamental kinetic relationships given by equations 2-54, 2-56, 2-58, 2-59, 2-60, and 2-63.

PROBLEMS

2-1 An important problem in the operation of wastewater treatment plants is disposal of the waste sludge produced. One advantage of anaerobic (fermentation) processes over aerobic processes is that yield coefficients are lower, resulting in the production of a smaller quantity of waste sludge. Explain this characteristic on the basis of energy metabolism.

2-2 Chemoautotrophic bacteria are responsible for the oxidation of NH_4^+-N to NO_3^--N in many biological wastewater treatment processes. These bacteria, collectively called nitrifiers, consist of the genera *Nitrosomonas* and *Nitrobacter*. The oxidation of ammonia to nitrate is considered a two-step sequential reaction as follows:

$$2NH_4^+ + 3O_2 \xrightarrow{Nitrosomonas} 2NO_2^- + 4H^+ + 2H_2O$$

$$2NO_2^- + O_2 \xrightarrow{Nitrobacter} 2NO_3^-$$

$$\text{overall reaction} \quad NH_4^+ + 2O_2 \xrightarrow{\text{Nitrifiers}} NO_3^- + 2H^+ + H_2O$$

Considering the overall reaction, use stoichiometric calculations to show how many mg/ℓ of alkalinity as $CaCO_3$ are destroyed and how many mg/ℓ of O_2 are required for the oxidation of 1 mg/ℓ of NH_4^+-N to NO_3^--N.

2-3 In denitrification (an anaerobic respiration process), nitrates and nitrites serve as electron acceptors in energy metabolism and are reduced to nitrogen gas. A wide variety of facultative bacteria, including the genera *Pseudomonas* and *Archromobacter*, are capable of bringing about denitrification. Denitrification in biological wastewater treatment will be carried out only if the organisms are supplied with an organic energy source and if no dissolved oxygen is available. Methanol is typically used as the organic energy source, and in this situation denitrification can be considered a two-step process, as indicated by the following reactions:

$$NO_3^- + \tfrac{1}{3}CH_3OH \longrightarrow NO_2^- + \tfrac{1}{3}CO_2 + \tfrac{2}{3}H_2O$$

$$NO_2^- + \tfrac{1}{2}CH_3OH \longrightarrow \tfrac{1}{2}N_2 + \tfrac{1}{2}CO_2 + \tfrac{1}{2}H_2O + OH^-$$

overall reaction $$NO_3^- + \tfrac{5}{6}CH_3OH \longrightarrow \tfrac{1}{2}N_2 + \tfrac{5}{6}CO_2 + \tfrac{7}{6}H_2O + OH^-$$

Considering the overall reaction, use stoichiometric calculations to find the increase in alkalinity (mg/ℓ as $CaCO_3$) and the mg/ℓ of methanol required for the reduction of 1 mg/ℓ of NO_3^--N to nitrogen gas.

2-4 Given the following experimental data obtained from a batch growth experiment at 20°C:

x (mg/ℓ)	*Time (hr)*
3467	2.0
3700	3.4
4100	4.8
4400	6.1
4786	7.7

Make the appropriate plot on semilog paper and determine the specific growth rate of the culture under the conditions of the experiment. What was the biomass concentration at the beginning of the exponential growth phase?

2-5 Determine the growth yield for a particular bacterial culture which gave the following response when grown in batch at 20°C:

x_m *(mg/ℓ)*	S_0 *(mg/ℓ)*
3300	580
3700	1160
4100	1920
4400	2320
4600	3200

Was substrate concentration the growth-limiting factor over the entire range of substrate concentrations studied?

2-6 A continuous culture growth experiment was conducted in a completely mixed reactor. The following experimental data were obtained at 20°C:

μ *(h^{-1})*	S_e *(mg/ℓ)*
0.66	20.0
0.50	10.0
0.40	6.6
0.33	5.0
0.28	4.0

Determine the magnitude of the maximum specific growth rate, μ_m, and the saturation constant, K_s.

2-7 To illustrate how the observed yield coefficient varies with specific growth rate, construct a plot of Y_{obs} versus $1/\mu$ assuming that $Y_T = 0.5$ and $K_d = 0.1\ day^{-1}$. The following specific growth rate values are to be used in working this problem:

μ (*day*$^{-1}$)
0.10
0.12
0.17
0.25
0.50
1.00

REFERENCES

BROCK, T. D., *Biology of Microorganisms*, Prentice-Hall, Inc., Englewood Cliffs, N.J., 1970.

DOETSCH, R. N., AND T. M. COOK, *Introduction to Bacteria and Their Ecobiology*, University Park Press, Baltimore, Md., 1973.

ECKENFELDER, W. W., JR., AND D. L. FORD, *Water Pollution Control*, Pemberton Press, Austin, Tex., 1970.

GARRETT, M. T., AND C. N. SAWYER, "Kinetics of Removal of Soluble BOD by Activated Sludge," in *Proceedings, 7th Industrial Waste Conference*, Purdue University, West Lafayette, Ind., Vol. 36, 1960, p. 5.

GRADY, C. P. L., JR., AND D. R. WILLIAMS, "Effects of Influent Substrate Concentration on the Kinetics of Natural Microbial Populations in Continuous Culture," *Water Research*, **9**, 171 (1975).

GRAU, P., M. DOHAŃYOS, AND J. CHUDOBA, "Kinetics of Multicomponent Substrate Removal by Activated Sludge," *Water Research*, **9**, 637 (1975).

HERBÉRT, D., in *Recent Progress in Microbiology, VII International Congress for Microbiology*, ed. by G. Tunevall, Almquist, and Wiksell, Stockholm, 1958, p. 381.

HEUKELEKIAN, H., H. E. OXFORD, AND R. MANGANELLI, "Factors Affecting the Quantity of Sludge Production in the Activated Sludge Process," *Sewage and Industrial Wastes*, **23**, 945 (1951).

KATO, K., AND Y. SEKIKAWA, "FAS (Fixed Activated Sludge) Process for Industrial Waste Treatment," in *Proceedings, 22nd Industrial Waste Conference*, Purdue University, West Lafayette, Ind., 1967, pp. 129, 926.

KNAYSI, G., *Elements of Bacterial Cytology*, 2nd ed., Cornell University Press, Ithaca, N.Y., 1951.

LAMANNA, C., AND M. F. MALLETTE, *Basic Bacteriology*, The Williams & Wilkins Co., Baltimore, Md., 1965.

LAWRENCE, A. W., AND P. L. MCCARTY, "Unified Basis for Biological Treatment Design and Operation," *Journal of the Sanitary Engineering Division, ASCE*, **96**, SA3, 757 (1970).

LEHNINGER, A. L., *Biochemistry*, Worth Publishers, Inc., New York, 1970.

MONOD, J., "The Growth of Bacterial Cultures," *Annual Review of Microbiology*, **3**, 371 (1949).

Pelczar, M. J., JR., AND R. D. REID, *Microbiology*, McGraw-Hill Book Company, New York, 1972.

PIRT, S. J., *Principles of Microbe and Cell Cultivation*, Halsted Press, a division of John Wiley & Sons, Inc., New York, 1975.

RANDALL, C. W., H. R. EDWARDS, AND P. H. KING, "Microbial Process for Acidic Low-Nitrogen Wastes," *Journal of the Water Pollution Control Federation*, **44**, 401 (1972).

SHERRARD, J. H., AND E. D. SCHROEDER, "Cell Yield and Growth Rate in Activated Sludge," *Journal of the Water Pollution Control Federation*, **45**, 1889 (1973).

WILKINSON, J. F., *Introduction to Microbiology*, Halsted Press, a division of John Wiley & Sons, Inc., New York, 1975.

3 Wastewater Characteristics and Flows

Industrial wastewater is that which is discharged from manufacturing plants, whereas domestic wastewater is discharged from institutions, commercial establishments, and residences. Any combination of wastewaters that are collected in sanitary sewers for treatment in a municipal treatment facility is termed *municipal wastewater*.

Untreated municipal wastewater has many undesirable components, some of which will deplete the oxygen budget when discharged into a receiving stream, while others may stimulate the growth of certain microorganisms such as algae. These undesirable components are composed of both organic and inorganic matter as well as soluble and insoluble material. Thus, the characterization of a wastewater is an important consideration before process selection and design is begun.

3-1 Organic Materials

The organic components of municipal wastewater are composed of a mixture of many different carbonaceous materials. As a result, tests for the organic content of such waters are generally nonspecific. The three most common tests are the biochemical oxygen demand (BOD), the chemical oxygen demand (COD), and total organic carbon (TOC).

Biochemical Oxygen Demand

The *biochemical oxygen demand* (BOD) test gives a measure of the oxygen utilized by bacteria during the oxidation of organic material contained in a wastewater sample. This test is based on the premise that all the biodegradable organic material contained in the wastewater sample will be oxidized to CO_2 and H_2O, using molecular oxygen as the electron acceptor. Hence, it is a direct measure of oxygen requirements and an indirect measure of biodegradable organic matter pollution.

The dilution bottle method of BOD employs bottles of approximately 300-ml volume. The BOD of the sample is estimated so that approximately 4 mg/ℓ of dissolved oxygen (DO) in the bottle will be used during 5 days of incubation. The volume fraction of the sample to be used is equal to the 4 mg/ℓ divided by the estimated BOD. For a waste with an estimated BOD of 200 mg/ℓ, the volume fraction is 0.02 and 6 ml of sample would be used in the 300-ml bottles. Multiple dilutions of one-half and twice this value are used to assure valid results. Multiple bottles in each dilution provide better accuracy. Dilution water made up of deionized water with appropriate nutrients, phosphate buffer, $MgSO_4$, $CaCl_2$, $FeCl_3$, and seed organisms (usually settled wastewater effluent) is used to fill the bottles and they are capped to exclude all air. A blank is run on the dilution water so that the oxygen demand of the seed material can be subtracted from the results.

Table 3-1 gives the BOD measurable with various sample dilutions prepared either on a percentage basis or by direct pipetting into bottles of 300-ml capacity. These bottles are then incubated at 20°C for some period of time, generally 5 days. After the desired incubation period the calculation of BOD can be made by either of the following equations:

For direct pipetting:

$$\text{BOD (in mg/}\ell) = \left[(DO_b - DO_i)\frac{\text{volume of bottle}}{\text{ml of sample}}\right] - (DO_b - DO_s) \tag{3-1}$$

For percent mixture:

$$\text{BOD (in mg/}\ell) = \left[(DO_b - DO_i)\frac{100}{\%}\right] - (DO_b - DO_i) \tag{3-2}$$

In these formulations DO_b and DO_i represent, respectively, the dissolved oxygen concentration in the blank and diluted sample at the end of the incubation period. DO_s represents the dissolved oxygen concentration originally present in the undiluted sample. Sawyer and McCarty (1967) point out that correction for the dissolved oxygen concentration in the undiluted sample is not necessary when the BOD exceeds 200 mg/ℓ.

TABLE 3-1

BOD MEASURABLE WITH VARIOUS SAMPLE DILUTIONS

Using percent mixtures		*By direct pipetting into 300-ml bottles*	
% Mixture	*Range of BOD*	*ml*	*Range of BOD*
0.01	20,000–70,000	0.02	30,000–105,000
0.02	10,000–35,000	0.05	12,000–42,000
0.05	4000–14,000	0.10	6000–21,000
0.1	2000–7000	0.20	3000–10,500
0.2	1000–3500	0.50	1200–4200
0.5	400–1400	1.0	600–2100
1.0	200–700	2.0	300–1050
2.0	100–350	5.0	120–420
5.0	40–140	10.0	60–210
10.0	20–70	20.0	30–105
20.0	10–35	50.0	12–42
50.0	4–14	100	6–21
100	0–7	300	0–7

Source: After Sawyer and McCarty (1967).

For the results of a BOD test to be valid, at least 2 mg/ℓ of oxygen should remain after incubation and more than 1 mg/ℓ of dissolved oxygen should be used during the incubation period.

Example 3-1

To determine the BOD of a municipal wastewater containing 2.0 mg/ℓ of dissolved oxygen, a 5-ml sample and 2 ml of seed material were added to each of three 300-ml volume BOD bottles, and the bottles were then completely filled with dilution water with DO at saturation. At the same time three blanks were prepared which contained no sample, 2 ml of seed material and dilution water. What is the BOD_5 of the wastewater if the sample bottles contained an average DO concentration of 2.5 mg/ℓ and the blank bottles an average DO concentration of 6.8 mg/ℓ after 5 days of incubation at 20°C?

solution

Substituting into equation 3-1 gives

$$BOD_5 = \left[(6.8 - 2.5)\frac{300}{5}\right] - (6.8 - 2.0)$$
$$= 253 \text{ mg}/\ell$$

A typical BOD progression curve can be divided into five zones for discussion. These zones are illustrated in Figure 3-1.

1/ The lag and synthesis dip is due to a combination of retarded oxidation as the organisms acclimate to the substrate and reduced oxygen utilization during the high-synthesis phase. It can be noted that the

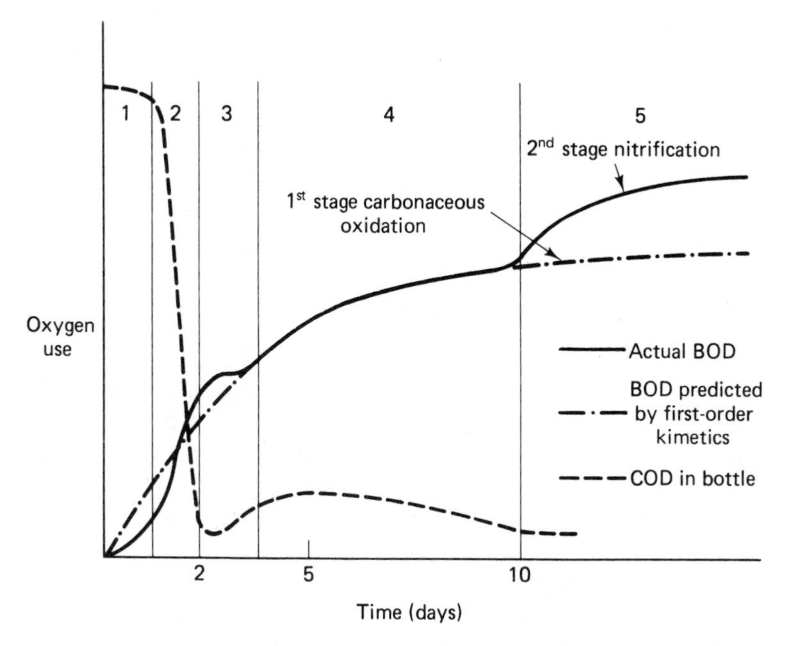

Figure 3-1. BOD Progression Curve.

BOD bottle becomes cloudy on the second day, indicating a large increase in microbial mass. Synthesis requires much less oxygen per unit of substrate utilized than does oxidation to carbon dioxide.

Synthesis:

$$\underset{\text{substrate}}{C_5H_{12}} + 2.25O_2 \longrightarrow \underset{\text{biomass}}{C_5H_7NO_2} + 2.5H_2O \qquad (3\text{-}3)$$

Oxidation:

$$C_5H_{12} + 8O_2 \longrightarrow 5CO_2 + 6H_2O \qquad (3\text{-}4)$$

2/ As the mass of organisms becomes large, the rate of synthesis decreases and the rate of the oxygen reaction (energy of maintenance requirement) increases. This causes a very rapid oxygen utilization in the second phase. As substrate becomes limiting, the rate of oxidation decreases at the end of this phase. Reduced oxidation rates at the end of the second phase may also be due to the fact that the more easily utilized materials have been assimilated and only the more difficult materials remain.

3/ During the second or third day, the microorganism concentration reaches a maximum and an oxidation plateau is reached. The reason for the plateau or third zone is not completely understood, but it has been speculated that it is due either to changeover from external substrate to cellular materials as a food source with a slight acclimation period or to the onset of rapid growth of predator organisms.

4/ The endogenous phase occurs after the plateau where cellular components are oxidized to provide energy for life-support functions.

5/ At about 10 days (sooner for treatment plant effluents), organisms that oxidize nitrogen compounds begin to predominate. The nitrifying organisms probably are present throughout the test, but because proteins are resistant to breakdown and much of the nitrogen is tied up in the protein, the nitrifiers do not predominate until nearly the end of the carbonaceous oxidation. This causes a second hump in the curve called the second-stage BOD or nitrification.

The practice in the United States is to measure the standard BOD at 5 days. This time was selected to minimize the effect of nitrification, and it assumes that settled sewage with a low population of nitrifiers was used as a seed. This approach is probably reasonable when the BOD results are to be used as a treatment plant efficiency index or design parameter. In treatment plant design, carbonaceous BOD removal is generally the major consideration. European practice favors the use of long-term BOD's, including nitrification. This is because stream pollution considerations are paramount, and all oxygen utilization potential must be considered.

The oxygen utilization in the BOD test is a biological reaction. For this reason, the bacterial environmental conditions, the initial mass of organisms, acclimation of the bacteria to the organics, and the food/microorganism ratio are important variables and possible sources of differences in results. A biochemical reaction cannot be expected to give the same precision as a pure chemical reaction because the mass and condition of the bacterial catalysts are of major importance. Biochemical reactions usually involve a complex series of intermediate reactions, and it is impossible to provide an exact mathematically based, theoretical explanation for the oxidation. This is particularly true in the BOD test, where a mixed and undefined group of organics is tested using a mixed and undefined group of bacteria. Nevertheless, it is generally assumed that BOD removal approximates first-order kinetics; that is, the rate of BOD removal (rate of oxidation of organic matter) is directly proportional to the amount of BOD remaining at any time. Mathematically, the expression for the time progression is

$$\frac{dL}{dt} = -KL \tag{3-5}$$

where $\frac{dL}{dt}$ = rate of BOD removal (rate at which organic matter is oxidized), mass volume^{-1} time^{-1}

L = concentration of BOD remaining, mass volume^{-1}

K = BOD reaction-rate constant, time^{-1}

Assuming that at $t = 0$, $L = L_u$, where L_u represents the ultimate BOD (the total amount of BOD present before any biological action has occurred), equation 3-5 can be integrated to give

$$L = L_u e^{-Kt} \tag{3-6}$$

If Y represents the amount of BOD satisfied at any time, t, then

$$L_u = L + Y \tag{3-7}$$

or

$$L = L_u - Y \tag{3-8}$$

Substituting for L in equation *3-6* from equation *3-8*, equation *3-6* can be expressed in the form

$$Y = L_u(1 - e^{-Kt}) \tag{3-9}$$

or in common logarithm form as

$$Y = L_u(1 - 10^{-K't}) \tag{3-10}$$

where

$$K' = \frac{K}{2.3} \tag{3-11}$$

The relationship among L_u, L, and Y is illustrated in the generalized BOD curves of Figures 3-2 and 3-3.

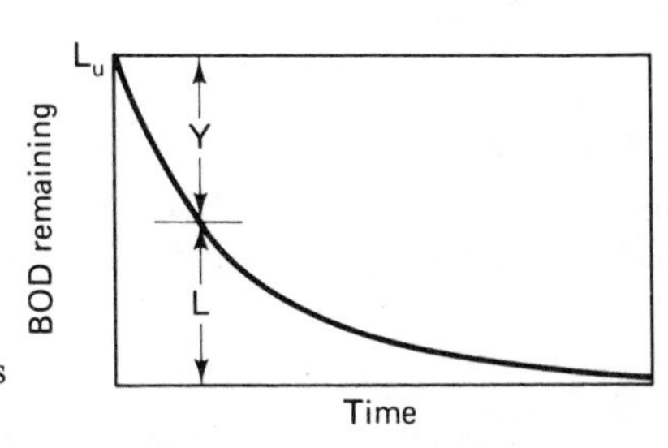

Figure 3-2 BOD Remaining Versus Reaction Time.

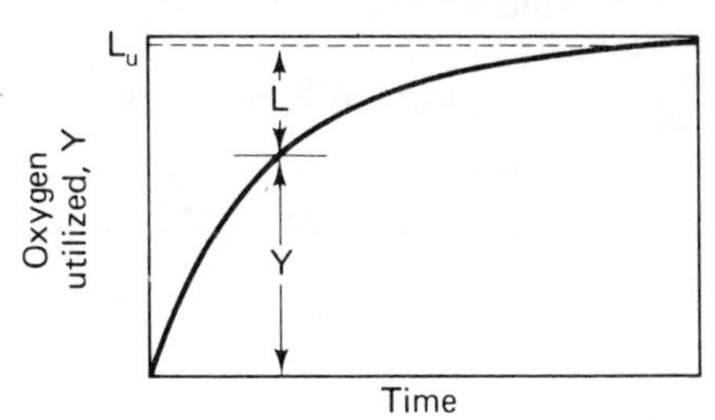

Figure 3-3. Oxygen Utilized Versus Reaction Time.

Example 3-2

The 5-day BOD of a wastewater is 250 mg/ℓ. Assuming K' to be 0.20 day^{-1}, what is the ultimate demand, L_u?

solution

Substituting into equation 3-10 and solving for L_u gives

$$250 = L_u[1 - 10^{-(0.20)(5)}]$$

or

$$L_u = 278 \text{ mg}/\ell$$

The rate at which organic matter is oxidized is dependent upon many factors, such as temperature, nutrients, biological population, and so on, and is reflected by the magnitude of the reaction-rate constant, K. For the results of the 5-day BOD test to be of value, the magnitude of the reaction-rate constant must be known. This can be illustrated by considering Figure 3-4. It is shown in the figure that two wastewaters may have nearly the same ultimate BOD, even though their 5-day BODs suggest a large difference in waste strengths. Thus, it is apparent that 5-day BOD values, without supplementary knowledge of the rate of oxidation, are of little practical value.

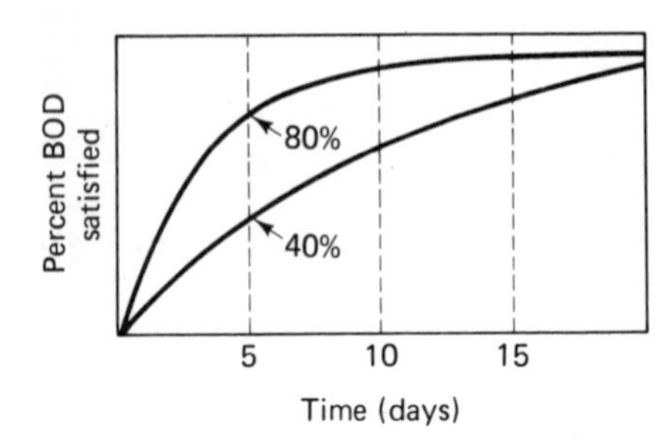

Figure 3-4. Effect of Reaction Rate Constant on Short-Term BOD.

Several procedures are available for determining the reaction-rate constant, K, and the ultimate BOD, L_u. However, the Thomas method is probably the simplest to use. It is based on the function

$$\left(\frac{t}{Y}\right)^{1/3} = (2.3K'L_u)^{-1/3} + \left[\frac{(K')^{2/3}}{3.43(L_u)^{1/3}}\right]t \qquad (3\text{-}12)$$

where Y = BOD satisfied during time, t, mass volume^{-1}
K' = base 10 reaction-rate constant, time^{-1}
L_u = ultimate BOD, mass volume^{-1}

Equation 3-12 suggests that a plot of $(t/Y)^{1/3}$ versus time will yield a linear trace of slope b and intercept a, where

$$K' = 2.61\,\frac{b}{a} \qquad (3\text{-}13)$$

and

$$L_u = \frac{1}{2.3K'(a)^3} \qquad (3\text{-}14)$$

Figure 3-5 illustrates the plot necessary to determine K' and L_u. To make such a plot, several observations of Y as a function of time are required. Mano-

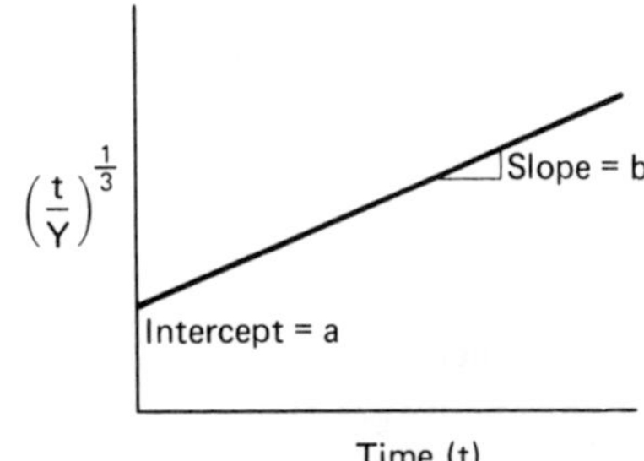

Figure 3-5. Determination of K' and L_u by the Thomas Graphical Method.

metric units or other devices that make it possible to read the oxygen utilization with time are the most convenient for this purpose. The data that are used should be limited to the first 10 days because of the probability of nitrification. Some commonly accepted K' and L_u values are presented in Table 3-2.

TABLE 3-2

SOME TYPICAL VALUES FOR K' AND L_u

Wastewater type	K' *(day*$^{-1}$*)*	L_u *(mg/ℓ)*
Weak domestic	0.152	150
Strong domestic	0.168	250
Primary effluent	0.152	75–150
Secondary effluent	0.052–0.100	10–75

Example 3-3

Using the data given below, determine the reaction-rate constant, K', and the ultimate BOD, L_u.

t (days)	0	1	2	3	4	5	6	7
BOD (mg/ℓ)	0	72	120	155	182	202	220	237

solution

1/ Arrange the data for plotting.

$(t/Y)^{1/3}$	0.240	0.255	0.268	0.280	0.291	0.301	0.309
t	1	2	3	4	5	6	7

2/ Plot the data given in step 1 and determine the slope and intercept of the linear trace.

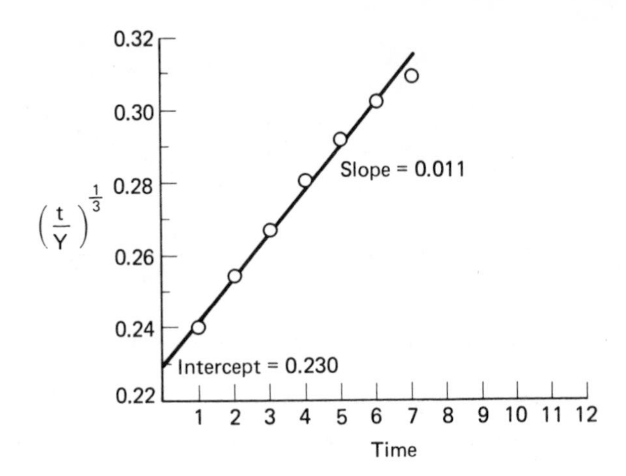

3/ Determine the reaction-rate constant, K', from equation 3-13.

$$K' = (2.61)\frac{0.011}{0.230}$$

$$= 0.125 \text{ day}^{-1}$$

4/ Determine the ultimate BOD, L_u, from equation 3-14.

$$L_u = \frac{1}{2.3(0.125)(0.23)^3}$$

$$= 286 \text{ mg}/\ell$$

A quick method for obtaining reaction constants from a minimum of data points is to measure the BOD at two evenly spaced times, such as 3 and 6 days. Equation 3-9 can be rearranged and solved simultaneously for the two points.

Example 3-4

If the 3-day BOD of a wastewater is 155 mg/ℓ and the 6-day BOD is 220mg/ℓ, determine L_u and K.

solution

1/ Rearrange equation 3-9 into the form

$$\ell n\left(\frac{L_u - Y}{L_u}\right) = -Kt$$

2/ Insert the known quantities into the expression given in step 1 and solve the equations simultaneously for L_u.

$$\ell n\left(\frac{L_u - 155}{L_u}\right) = -3K$$

$$\ell n\left(\frac{L_u - 220}{L_u}\right) = -6K$$

Therefore,

$$\ell n\left(\frac{L_u - 220}{L_u}\right) = 2\,\ell n\left(\frac{L_u - 155}{L_u}\right)$$

$$\frac{L_u - 220}{L_u} = \left(\frac{L_u - 155}{L_u}\right)^2$$

$$L_u^2 - 220L_u = L_u^2 - 310L_u + 24{,}025$$

$$90L_u = 24{,}025$$

$$L_u = 267 \text{ mg}/\ell$$

3/ Compute the reaction-rate constant.

$$\ell n\left(\frac{267 - 155}{267}\right) = -3K$$

$$K = 0.29 \text{ day}^{-1}$$

$$K' = \frac{0.29}{2.3} = 0.126 \text{ day}^{-1}$$

Although the standard BOD test calls for an incubation temperature of 20°C, field conditions often necessitate incubation at other temperatures. Eckenfelder (1970) suggests that the rate constant for carbonaceous demand may be adjusted for temperature effects by modifying equation 1-30 into the form

$$K'_T = K'_{20°C}\theta^{T-20} \qquad (3\text{-}15)$$

where $K'_{20°C}$ = BOD reaction-rate constant determined at 20°C, time^{-1}
K'_T = BOD reaction-rate constant at temperature of interest, time^{-1}
T = temperature of interest, °C
θ = 1.135 for the temperature range 4 to 20°C and 1.056 for the temperature range 20 to 30°C

Example 3-5

The ultimate and 5-day BOD values of a river water sample incubated at 20°C were found to be 38 and 27 mg/ℓ, respectively. What will be the river reaction-rate constant, K', if the river temperature is 10°C?

solution

1/ Substituting into equation 3-10 and solving for K' gives

$$27 = 38[1 - 10^{-K'(5)}]$$

or

$$K' = 0.107 \text{ day}^{-1}$$

2/ The reaction-rate constant is adjusted for temperature effects through the use of equation 3-15.

$$K' = (0.107)(1.135)^{10-20}$$

$$= 0.03 \text{ day}^{-1}$$

Special precautions should be made when determining the BOD of industrial waste. In such a case the use of nonacclimated biological seed is probably the most common factor responsible for erroneous BOD results. Furthermore, the presence of certain materials in the wastewater may have a toxic effect on the seed microorganisms. This effect is usually evidenced when the BOD values increase with increasing sample dilution.

Acclimated seeds can be developed in the laboratory by feeding the wastewater to be tested to an aerated flask of settled sewage organisms over a period of time. Two weeks is usually sufficient. If the waste has been discharged to a stream for a considerable period of time, a water sample from the stream some distance below the outfall will usually contain a population of acclimated organisms. Settled effluent from a plant treating the same wastewater can also be used, but it will typically contain a large population of nitrifiers and the onset of nitrification effects will occur early in the test.

Toxic materials can make it virtually impossible to measure the BOD of a wastewater using the standard static-bottle BOD test. Urban stormwater runoff is an excellent example of such a wastewater. In addition to the use of chelating agents to tie up toxic metals, toxic effects can be partially overcome by two techniques. One is to stir the contents of the bottle continuously. This has been demonstrated with TNT wastewater (Nay, 1971). Another method is to increase the organism concentration in the bottle. Microbial toxicity is commonly viewed as a toxic material concentration phenomenon. It is assumed that there is some concentration of the toxic material that is toxic, and all higher concentrations are increasingly toxic while lower concentrations are not toxic. This is not generally true, however, unless the concentration of microorganisms present is constant for all toxic material concentrations. In fact, toxicity is a function of the ratio of toxic material to microbial mass present in the system. Instead of a "threshold" concentration there is a "threshold" ratio. This has been demonstrated by both Fitzgerald (1964) and Randall and Lauderdale (1967).

If both continuous stirring and an increase in the concentration of microbial solids are used, toxic effects can generally be overcome enough to obtain reliable BOD values.

Biological oxidations are seldom as complete as chemical oxidations because of biologically resistant organic compounds and the fact that some potentially oxidizable biomass remains even at the end of long-term tests. The 5-day BOD is usually about 60 to 70% of the ultimate or long-term carbonaceous BOD and about 40 to 50% of the ultimate first- and second-stage BOD. This will vary considerably, however, with the type of waste being tested. It should also be recognized that a few wastewaters will not follow first-order BOD kinetics.

BODs were originated for analysis of stream pollution problems involving dilution and oxygen sag. These considerations are of less importance today, and there is a gradual shift to the use of CODs for treatment plant design and efficiency analyses because of the relative quickness and ease of testing.

Chemical Oxygen Demand

The *chemical oxygen demand* (COD) test is based on the principal that most organic compounds are oxidized to CO_2 and H_2O by strong oxidizing agents under acid conditions. The measurement represents the oxygen that would be needed for aerobic microbial oxidation to CO_2 and H_2O, assuming that all the organics are biodegradable.

The procedure consists of refluxing a sample containing organic material with sulfuric acid and an excess of standardized potassium dichromate. During the reflux period, the chemically oxidizable organic material reduces a stoichiometrically equivalent amount of dichromate, the remainder of which is measured by titration with standard ferrous ammonium sulfate. The amount of dichromate reduced is a measure of the amount of organic material oxidized.

Although it might be expected that the ultimate BOD of a wastewater would approximate the COD, in many cases this is not true. Eckenfelder and Ford (1970) suggest that the following factors may be responsible for differences noted between COD and ultimate BOD values:

1/ The COD test measures total oxidizable organics but is not capable of distinguishing between those that are biodegradable and those that are not. Furthermore, not all organic compounds are oxidizable by wet chemical methods. Some compounds, such as sugars, branched-chain aliphatics, and substituted benzene rings are completely oxidized, whereas compounds such as benzene, toluene, and pyridine are not oxidized. Some compounds such as straight-chain acids, alcohols, and amino acids may be only partially oxidized.

2/ Certain inorganic substances, such as sulfides, sulfites, thiosulfates, nitrites, and ferrous iron, are oxidized by dichromate, creating an inorganic COD that gives an error when measuring the organic content of a wastewater.

3/ COD results are independent of seed acclimation, a factor that may cause the BOD results to be low.

It is possible to calculate the theoretical chemical oxygen demand for organic compounds if the oxidation reaction is known. For example, consider the following reaction, which shows the oxidation of glucose to carbon dioxide and water:

$$\underset{180}{C_6H_{12}O_6} + \underset{(6 \times 32)}{6O_2} \longrightarrow 6CO_2 + 6H_2O \qquad (3\text{-}16)$$

$$\frac{6 \times 32}{180} = 1.066 \text{ g oxygen required per gram of glucose oxidized}$$

The molar ratio suggests that 1.066 g of oxygen is required per gram of glucose oxidized to CO_2 and H_2O.

For a given wastewater it is sometimes possible to establish a correlation between COD and BOD values. A typical correlation is presented in Figure 3-6. This figure also illustrates another interesting response, generally noted as wastewater moves through a biological treatment facility, which is a decrease in the BOD/COD ratio of the wastewater. The wastewater illustrated is shown to have approximately 100 mg/ℓ of nonbiodegradable material measured as COD.

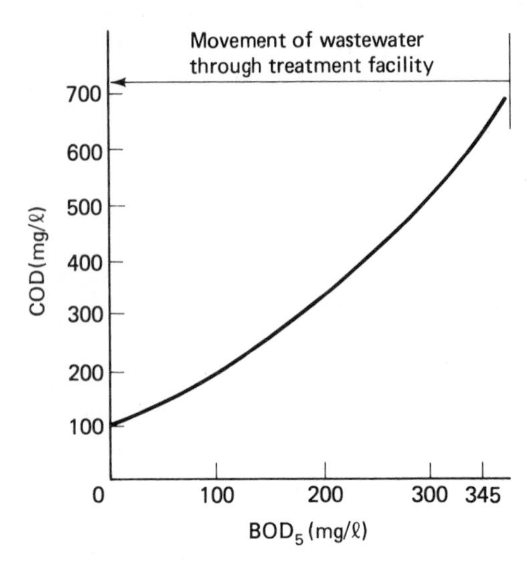

Figure 3-6. Typical BOD-COD Correlation as Wastewater Moves Through a Biological Treatment Plant. (After Eckenfelder and Ford, 1970).

Total Organic Carbon

The *total carbon analyzer* allows a total soluble carbon analysis to be made directly on an aqueous sample. The aqueous sample is injected directly into a combustion tube and heated to 950°C in a constant flow of carrier gas. Any organic matter is oxidized to CO_2 and water vapor on an asbestos packing impregnated with catalyst. These are carried from the combustion tube by the carrier gas. Outside the combustion tube, the water is condensed and the CO_2 is swept through a continuous-flow sample cell in a nondispersive infrared CO_2 analyzer. The amount of CO_2 measured is proportional to the initial sample concentration. Thus, a standard curve can be developed which can be used to determine total soluble carbon for any wastewater sample. Organic carbon can be determined by removing the inorganic carbon prior to analysis by acidification and sparging (which does not remove volatile acids such as formic and acetic acids) or by providing a dual-combustion-tube total carbon analyzer, where a low-temperature combustion tube is used for inorganic carbon analysis. Organic carbon is then taken as the difference between total carbon and inorganic carbon. A flow diagram for a dual combustion tube total carbon analyzer is presented in Figure 3-7.

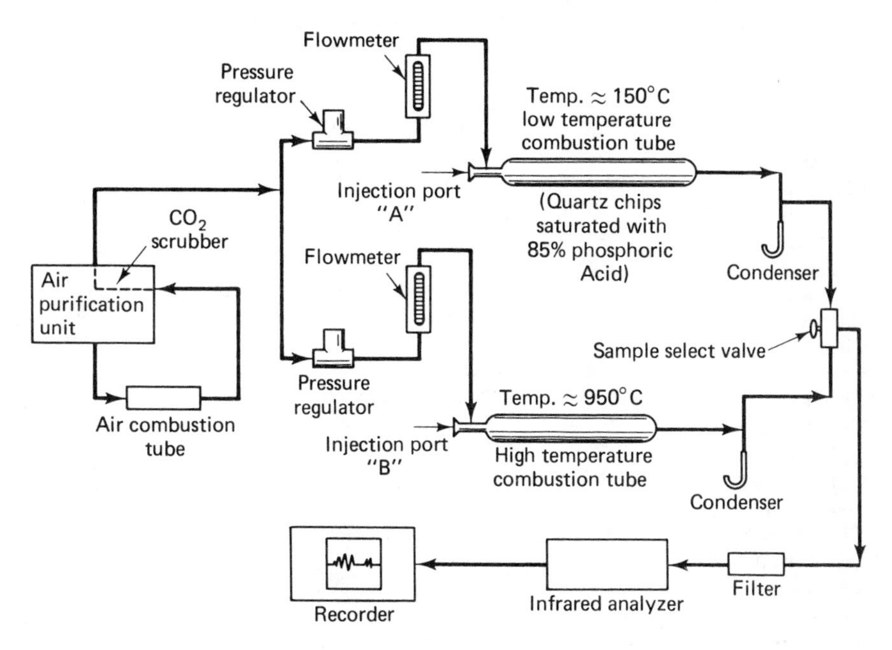

Figure 3-7. Flow Diagram of Dual-Combustion-Tube Total Carbon Analyzer. (After Eckenfelder and Ford, 1970.)

The greatest source of error in TOC analysis has been the inclusion of solids in the injected sample. The sample is typically very small (20 $\mu\ell$) and the presence or absence of one sizable organic particle can make a significant difference in the measured value. For this reason TOC has been used primarily for filtered samples containing soluble organics. However, equipment[1] has been developed which can overcome this limitation, and it is possible to measure the TOC of samples containing high solids concentrations, whether the solids are organic or inorganic.

In many cases TOC measurements can be correlated with COD and occasionally with BOD values. Because the time required for carbon analysis is generally short (a few minutes), such correlations are extremely helpful when monitoring treatment plant flows for efficiency control. Tables 3-3 and

[1]Dohrmann Division of Envirotech Corp., 3240 Scott Blvd., Santa Clara, Calif. 95050.

TABLE 3-3

OXYGEN DEMAND AND ORGANIC CARBON OF CERTAIN INDUSTRIAL WASTEWATERS

Type of waste	BOD_5 *(mg/ℓ)*	*COD (mg/ℓ)*	*TOC (mg/ℓ)*	BOD_5/TOC	*COD/TOC*
Chemical[a]	—	4,260	640	—	6.65
Chemical[a]	—	2,440	370	—	6.60
Chemical[a]	—	2,690	420	—	6.40
Chemical	—	576	122	—	4.72
Chemical	24,000	41,300	9,500	2.53	4.35
Chemical refinery	—	580	160	—	3.62
Petrochemical	—	3,340	900	—	3.32
Chemical	850	1,900	580	1.47	3.28
Chemical	700	1,400	450	1.56	3.12
Chemical	8,000	17,500	5,800	1.38	3.02
Chemical	60,700	78,000	26,000	2.34	3.00
Chemical	62,000	143,000	48,140	1.29	2.96
Chemical	—	165,000	58,000	—	2.84
Chemical	9,700	15,000	5,500	1.76	2.73
Nylon polymer	—	23,400	8,800	—	2.70
Petrochemical	—	—	—	—	2.70
Nylon polymer	—	112,600	44,000	—	2.50
Olefin processing	—	321	133	—	2.40
Butadiene processing	—	359	156	—	2.30
Chemical	—	350,000	160,000	—	2.19
Synthetic rubber	—	192	110	—	1.75

[a]High concentration of sulfides and thiosulfates.
Source: After Eckenfelder and Ford (1970).

3-4 contain BOD_5, COD, and TOC correlations observed for selected industrial wastewaters and municipal wastewaters.

As noted by Eckenfelder and Ford (1970), the COD/TOC ratio might be expected to approximate the molecular ratio of oxygen to carbon (i.e., $32/12 = 2.66$). However, because of variations in the chemical oxidizability of different organic compounds the COD/TOC ratio has been observed to fluctuate over a fairly wide range. These workers report a range of 1.75 to 6.65 for several industrial wastewaters.

Other tests for measuring the organic content of wastewaters include total oxygen demand (TOD), theoretical oxygen demand (ThOD), and total biological oxygen demand (T_bOD). These tests will not be presented here, because values derived from such measurements are not commonly used in the design of biological wastewater treatment processes. For those interested in further discussion of these tests, Metcalf and Eddy (1972) and Schroeder (1977) are recommended.

TABLE 3-4

OXYGEN DEMAND AND ORGANIC CARBON OF CERTAIN MUNICIPAL WASTEWATERS

Treatment	BOD_5 (*mg/ℓ*)	*COD* (*mg/ℓ*)	*TOC* (*mg/ℓ*)	BOD_5/TOC	*COD/TOC*
Raw	—	136	41	—	3.32
	—	—	—	1.39	—
	—	—	—	1.88	—
	—	230	54	—	4.26
	105	304	65	1.63	4.68
	92	264	70	1.32	3.76
	84	235	57	1.47	4.12
	76	227	49	1.55	4.64
	89	263	61	1.46	4.32
	72	228	55	1.31	4.14
Primary effluent	68	299	51	1.33	5.85
	66	220	61	1.08	3.60
	59	200	58	1.02	3.45
	50	161	44	1.14	3.66
	57	197	54	1.06	3.65
	46	146	46	1.00	3.19
Final effluent	16	85	34	0.47	2.50
	19	95	38	0.50	2.50
	20	85	33	0.61	2.58
	11	77	30	0.37	2.56
	14	81	40	0.35	2.02
	11	78	35	0.31	2.23
	—	—	—	0.20	—
	—	—	—	0.69	—

Source: After Eckenfelder and Ford (1970).

3-2
Inorganic Materials

Classically, wastewater treatment has been directed toward the removal of suspended solids, biochemical oxygen demand, and reduction of bacterial contaminants. Conversion and removal of nutrients has been incidental to treatment processes. However, in recent years much attention has been focused on nutrients (specifically nitrogen and phosphorus) contained in wastewaters. When considering nutrient removal rather than conversion (ammonia nitrogen oxidized to nitrate nitrogen), it must be remembered that nitrogen and phosphorus are essential cellular constituents. Thus, wastewater must contain these materials in the minimum amounts necessary for biomass production if the organic removal process is to be effective. In this regard, pH and alkalinity are also important parameters because of the need to maintain the operating pH of a biological treatment process within a relatively narrow range for optimum microbial activity.

pH and Alkalinity

The *pH scale* is a means of designating the concentration of H^+ ions in an aqueous solution in the acidity range between 1.0 *M* H^+ and 1.0 *M* OH^-. The term pH is defined as

$$\text{pH} = \log\left(\frac{1}{[\text{H}^+]}\right) \tag{3-17}$$

where $[H^+]$ represents the hydrogen-ion concentration in moles/ℓ. Since the ionization constant for water is approximately 10^{-14}, the pH scale has an approximate range of 0 to 14, with pH 7 representing neutrality.

It should be noted that pH is actually defined as

$$\text{pH} = \log\left(\frac{1}{a_{\text{H}^+}}\right)$$

where a_{H^+} is defined as the activity of H^+. However, in this text no distinction will be made between activities and concentrations.

pH is important in biological wastewater treatment because most microorganisms grow best at pH values near neutrality. Although many aspects of bacterial cell structure and function are influenced by pH, it is the catalytic activity of enzymes that is especially sensitive. The effect of pH on microbial growth rate has been presented in Figure 2-3.

Acids are generally considered to be substances that dissociate to yield hydrogen ions, and bases those substances that dissociate to yield hydroxyl ions. The Brönsted–Lowry theory defines acids as *proton donors* and bases as *proton acceptors*. An acid–base reaction always involves a *conjugate acid–base pair* made up of a proton donor and a proton acceptor. For example, a monoprotic acid of the form HA and its associated anion, A^-, form a conjugate acid–base pair.

The tendency of any acid to dissociate (ionize) is described by its equilibrium expression. As an example, consider the ionization of the monoprotic acid HA:

$$\text{HA} \rightleftharpoons \text{H}^+ + \text{A}^- \tag{3-18}$$

The resulting equilibrium expression has the form

$$K_a = \frac{[\text{H}^+][\text{A}^-]}{[\text{HA}]} \tag{3-19}$$

where the brackets indicate concentrations in moles/ℓ and K_a represents the thermodynamic equilibrium constant for the reaction of interest. It is standard practice in engineering to work with thermodynamic equilibrium constants, that is, constants which are not corrected for deviation of the system from ideal behavior caused by factors such as ionic strength and temperature.

Table 3-5 gives the thermodynamic equilibrium constants of some acids and bases. The pK_a values listed in the right-hand column of Table 3-5 are

TABLE 3-5

THERMODYNAMIC EQUILIBRIUM CONSTANTS OF VARIOUS ACIDS (25°C)

Acid	K_a	pK_a
H_3PO_4	7.52×10^{-3}	2.12
$Fe(H_2O)_6^{3+}$	6.3×10^{-3}	2.2
HNO_2	4.6×10^{-4}	3.4
$Al(H_2O)_6^{3+}$	1.3×10^{-5}	4.9
H_2S	1.1×10^{-7}	7.0
$H_2PO_4^-$	6.23×10^{-8}	7.21
NH_4^+	3.3×10^{-10}	9.5
HPO_4^{2-}	2.2×10^{-13}	12.7
HS^-	1.2×10^{-15}	14.9

simply logarithmic transformations of K_a which are given by the expression

$$pK_a = \log\left(\frac{1}{K_a}\right) \qquad (3\text{-}20)$$

Note that strong acids (acids that have a low affinity for protons and as such give them up easily) have low pK_a values, whereas weak acids have large pK_a values.

Plotting the titration curve of a weak acid is an easy way to visualize the concept of the acid ionization constant. To construct a titration curve, small increments of a strong base solution (e.g., NaOH) are added to a solution of a weak acid, and the resulting pH is determined after each addition. A plot is then made of pH versus equivalents of base added, which results in an S-shaped curve. Such a curve shape can be explained by considering that increasing the concentration of hydroxyl ions in the solution causes the acid to surrender its protons. The pH at which this begins to occur depends on several factors which are characteristic of each proton-releasing group. The protons released by the acid combine with the hydroxyl ions as illustrated by the reaction

$$H^+ + OH^- \rightleftharpoons H_2O \qquad (3\text{-}21)$$

Thus, the pH does not change by a very large amount during the ionization. When the acid has given up most of its protons, the top of the titration curve is reached. At this point, further addition of hydroxyl ions causes a rapid rise in pH.

A titration curve for a weak monoprotic acid is presented in Figure 3-8. The shape of this curve can be expressed by the *Henderson–Hasselbalch equation*, which has the form

$$pH = pK_a + \log\left(\frac{[A^-]}{[HA]}\right) \tag{3-22}$$

At pH $= pK_a$, equimolar concentrations of proton donor (HA) and proton acceptor (A^-) exist. This condition is reflected at the midpoint of the titration curve presented in Figure 3-8.

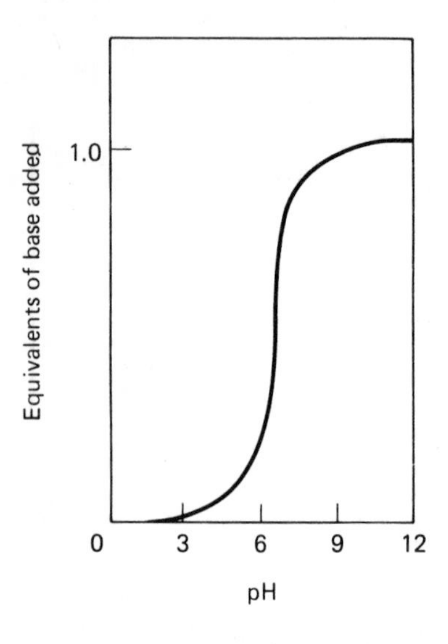

Figure 3-8. Titration Curve for a Weak Monoprotic Acid.

Example 3-6

A solution of ammonia in water is neutralized with dilute HCl to a pH of 8.5. What are the relative concentrations of free base and conjugate acid?

solution

1/ From Table 3-5 the pK_a value for ammonium is 9.25.

2/ Substituting into equation 3-22 and solving for the $[NH_3]/[NH_4^+]$ ratio,

$$8.5 = 9.25 + \log\left(\frac{[NH_3]}{[NH_4^+]}\right)$$

$$\log\left(\frac{[NH^3]}{[NH_4^+]}\right) = -0.75$$

$$\frac{[NH_3]}{[NH_4^+]} = 0.18$$

A solution is said to be buffered if it resists change of its hydrogen- or hydroxyl-ion concentration with the addition of acids or bases. Such a solution must contain a proton donor to react with a base and also a proton acceptor to react with an acid. Hence, a solution containing a weak acid and its salt or a weak base and its salt acts as a buffer. For example, assume that hydroxyl ions are added to a solution containing acetic acid and sodium acetate. The resulting reaction is illustrated by the equation

$$\begin{array}{c} HA\bar{c} \rightleftharpoons H^+ + A\bar{c} \\ + \\ OH^- \\ \upharpoonleft\downharpoonright \\ H_2O \end{array} \tag{3-23}$$

The hydroxyl ions react with hydrogen ions to form water. This causes a further ionization of HAc to replenish the hydrogen ions used in the water reaction. The overall effect is that there is little change in hydrogen-ion concentration until enough base has been added to almost stoichiometrically equal the acetic acid present. In using equation 3-22 to describe such systems, the $[A^-]$ is taken to be the initial salt concentration and [HA] the initial acid concentration.

A single compound having appropriate acidic and basic properties may also be a buffer. A good example is a solution of sodium bicarbonate because the bicarbonate ion, HCO_3^-, can either furnish protons or accept protons by reaction with either bases or acids.

$$HCO_3^- + OH^- \rightleftharpoons H_2O + CO_3^{2-} \tag{3-24}$$

$$HCO_3^- + H^+ \rightleftharpoons H_2CO_3 \tag{3-25}$$

Example 3-7

If 10 g of sodium bicarbonate is dissolved to give 1 ℓ of solution, will the final solution be acidic or basic? Assume that the solution temperature is 25°C.

solution

1/ There are two ways in which the bicarbonate ion can react in water.

a/ $$HCO_3^- + H_2O \rightleftharpoons H_3O^+ + CO_3^{2-}$$

Table 3-5 gives K_a for this reaction as 6.31×10^{-11}.

b/ $$HCO_3^- + H_2O \rightleftharpoons H_2CO_3 + OH^-$$

Since $K_a \times K_b = K_w$, the equilibrium constant for this reaction is

$$K_b = \frac{10^{-14}}{6.31 \times 10^{-11}} = 1.6 \times 10^{-4}$$

2/ If the first hydrolysis reaction predominates, the hydrogen-ion concentration will be greater and the solution will be acidic. However, if the second hydrolysis reaction predominates, the hydroxyl-ion concentration will be greater and the solution will be basic.

3/ In this case the equilibrium constant for the second reaction is much greater than that for the first reaction. It follows, therefore, that the solution will be basic.

The bicarbonate buffer system (H_2CO_3-HCO_3^-) has some interesting features. The proton donor species (carbonic acid) is in reversible equilibrium with dissolved CO_2:

$$H_2CO_3 \rightleftharpoons CO_{2(\text{dissolved})} + H_2O \tag{3-26}$$

In an open system such as that normally encountered in wastewater treatment, the dissolved CO_2 is in equilibrium with gaseous CO_2:

$$CO_{2(\text{dissolved})} = CO_{2(\text{gas})} \tag{3-27}$$

Henry's law states that the solubility of a gas in water is proportional to the partial pressure of the gas above the liquid surface. If the partial pressure were to be increased and all other variables held constant, the pH of the solution would decrease. This can be explained by noting that an increase in the dissolved CO_2 concentration would increase the H_2CO_3. According to the *Henderson–Hasselbalch equation*, pH is given by

$$\text{pH} = \text{p}K_a + \log\left(\frac{[HCO_3^-]}{[H_2CO_3]}\right) \tag{3-28}$$

Equation 3-28 then predicts a decrease in pH with an increase in H_2CO_3 concentration.

Alkalinity is defined as the capacity of a water to neutralize acids. For most wastewaters, pH values lie within the range 6.5 to 8.5. It is commonly assumed that this pH is controlled primarily by dissolved carbonate species (i.e., carbon dioxide, bicarbonate, and carbonate). Carbonate equilibrium in water can be represented by the reaction

$$CO_2 + H_2O \rightleftharpoons H_2CO_3 \rightleftharpoons HCO_3^- + H^+ \rightleftharpoons CO_3^{2-} + H^+ \tag{3-29}$$

Since H_2CO_3 is difficult to measure and since only a small fraction of the dissolved CO_2 is hydrolyzed to H_2CO_3, an apparent first ionization constant can be written incorporating the sum of the concentrations of dissolved CO_2 and H_2CO_3:

$$\frac{[H^+][HCO_3^-]}{[CO_2] + [H_2CO_3]} = \frac{[H^+][HCO_3^-]}{[H_2CO_3^*]} = K_1 = 4.45 \times 10^{-7} \text{ at } 25°\text{C} \tag{3-30}$$

The second ionization constant, K_2, for the ionization of HCO_3^- is

$$\frac{[H^+][CO_3^{2-}]}{[HCO_3^-]} = K_2 = 4.69 \times 10^{-11} \text{ at } 25°\text{C} \tag{3-31}$$

where [ ] denotes concentration in moles/ℓ. Loewenthal and Marais (1976) propose that the ionization constants K_1 and K_2 may be corrected for temperature changes by using the following equations:

$$pK_1 = \frac{17{,}052}{T} + 215.21 \log T - 0.12675\, T - 545.56 \qquad (3\text{-}32)$$

$$pK_2 = \frac{2902.39}{T} + 0.02379\, T - 6.498 \qquad (3\text{-}33)$$

where T represents temperature in °K. Equation 3-32 was determined for the temperature range 0 to 38°C, whereas equation 3-33 was determined for the temperature range 0 to 50°C.

Example 3-8

The pH of a 0.1 M solution of sodium bicarbonate is adjusted to 7.0 with strong acid. A 100-ml portion of this solution is obtained, and to this volume 100 ml of 0.08 N HCl are added. What are the final $[H_2CO_3^*]$ and $[HCO_3^-]$ concentrations and resulting pH?

solution

1/ Determine the initial distribution of $[H_2CO_3^*]$ and $[HCO_3^-]$ species.

$$7.0 = 6.35 + \log\left(\frac{[HCO_3^-]}{[H_2CO_3^*]}\right)$$

$$\frac{[HCO_3^-]}{[H_2CO_3^*]} = 4.47$$

Since

$$[HCO_3^-] + [H_2CO_3^*] = 0.1$$

the concentration of the $[H_2CO_3^*]$ specie is given by

$$4.47[H_2CO_3^*] + [H_2CO_3^*] = 0.1$$

$$[H_2CO_3^*] = 0.018 \text{ M}$$

It is then possible to determine the $[HCO_3^-]$ concentration

$$[HCO_3^-] + 0.018 = 0.1$$

$$[HCO_3^-] = 0.082 \text{ M}$$

Since only a one-proton shift is involved, the number of millequivalents (meq) per 100 ml is given by

$$\text{meq of } H_2CO_3^*/100 \text{ ml} = 100 \times 0.018 = 1.8 \text{ meq}$$

$$\text{meq of } HCO_3^-/100 \text{ ml} = 100 \times 0.082 = 8.2 \text{ meq}$$

2/ Compute the meq of hydrogen ions added to the original volume.

$$\text{meq of } [H^+] \text{ added} = 100 \times 0.08 = 8.0 \text{ meq}$$

3/ Determine the new distribution of $[H_2CO_3^*]$ and $[HCO_3^-]$ species.

$$HCO_3^- + H^+ \rightleftharpoons H_2CO_3^*$$

The bicarbonate ions are reduced by the amount of hydrogen ions added:

$$\text{meq of } H_2CO_3^*/200 \text{ ml} = 1.8 + 8.0 = 9.8 \text{ meq}$$

$$\text{meq of } HCO_3^-/200 \text{ ml} = 8.2 - 8.0 = 0.2 \text{ meq}$$

The new concentration of each species is

$$[H_2CO_3^*] = \frac{9.8}{200} = 0.049 \text{ M}$$

$$[HCO_3^-] = \frac{0.2}{200} = 0.001 \text{ M}$$

4/ Compute the final pH using the Henderson–Hasselbalch equation.

$$\text{pH} = 6.35 + \log \frac{0.001}{0.049}$$

$$= 4.66$$

For pH buffering, a buffer may be characterized as having (1) a neutralizing capacity, and (2) an intensity. From equation 3-22 it can be seen that only the molar ratio of the proton acceptor to proton donor determines the pH of the mixture. However, the buffer capacity depends upon the actual amounts of the buffer components and controls the amount of hydrogen ions or hydroxyl ions that may be added without consuming all the buffering component. For example, total alkalinity is defined as the acid-neutralizing capacity of a water sample to pH 4.3.

Buffer intensity is defined as the number of moles of hydrogen ions or hydroxyl ions required to change the pH of 1 ℓ of solution by 1 pH unit. However, considering the titration curve presented in Figure 3-8, it is seen that buffer intensity varies with pH; for example, more base is required to effect a 1-unit pH change near the midpoint of the titration curves than is required near the ends of each curve. For this reason it is best to represent buffer intensity by a differential such that

$$\beta = -\frac{dA}{d\text{pH}} = \frac{dB}{d\text{pH}} \tag{3-34}$$

where β = buffer intensity

$\frac{dA}{d\text{pH}}$ = number of moles of strong acid required to produce a change of dpH

$\frac{dB}{d\text{pH}}$ = number of moles of strong base required to produce a change of dpH

The negative sign in equation 3-34 implies that the addition of strong acid decreases the pH.

In the titration of a homogeneous aqueous carbonate system with a strong acid, Weber and Stumm have (1963) shown that buffer intensity can be calculated from the expression

$$\beta = 2.3\left(\frac{\alpha([\text{alk.}] - [\text{OH}^-] + [\text{H}^+])\{[\text{H}^+] + (K_1K_2/[\text{H}^+]) + 4K_2\}}{K_1(1 + 2K_2/[\text{H}^+])} + [\text{H}^+] + [\text{OH}^-]\right) \tag{3-35}$$

where
$$\alpha = \frac{K_1}{K_1 + [\text{H}^+] + (K_1K_2/[\text{H}^+])} \tag{3-36}$$

β = buffer intensity, equivalents/unit pH
$[OH^-]$ = hydroxyl-ion concentration, moles/ℓ
$[H^+]$ = hydrogen-ion concentration, moles/ℓ
[alk.] = total alkalinity, equivalents-ℓ

Equation 3-35 can be used to estimate the expected pH change when a known concentration of hydrogen ions is added to or produced in a biological treatment process such as during biological nitrification.

Example 3-9

The wastewater from an industrial plant is discharged into the outfall sewer to a municipal treatment facility, at the rate of 0.2 MGD. The average flow in the sewer is 2 MGD, while the wastewater alkalinity is 200 mg/ℓ as $CaCO_3$ and the pH is 7.5. What will be the pH of the wastewater after mixing with the industrial waste if the industrial flow has a pH of 3.5? The temperatures of the municipal and industrial wastewaters are 10°C and 30°C, respectively. Assume that buffering in the wastewater is due to the carbonate system.

solution

1/ Estimate the temperature of the wastewater after mixing.

$$T = \frac{(10°\text{C})(2.0) + (30°\text{C})(0.2)}{2.0 + 0.2}$$
$$\approx 12°\text{C}$$

2/ Determine the values of K_1 and K_2 from equations 3-32 and 3-33.

$$\text{pK}_1 = \frac{17{,}052}{285} + 215.21 \log(285) - 0.12675(285) - 545.56$$
$$= 6.46 \longrightarrow K_1 \approx 10^{-6.5}$$

$$\text{pK}_2 = \frac{2902.39}{285} + 0.02379(285) - 6.498$$
$$= 10.46 \longrightarrow K_2 \approx 10^{-10.5}$$

3/ Convert alkalinity concentration to equivalents per liter.

$$[\text{alk.}] = \frac{0.2\ \text{g}/\ell}{\text{eq. wt. of CaCO}_3} = \frac{0.2}{50} = 4 \times 10^{-3}\ \text{eq}/\ell$$

4/ Calculate the hydrogen-ion concentration.

$$[H^+] = 10^{-pH} = 10^{-7.5}$$

5/ Compute the hydroxyl-ion concentration.

$$[OH^-] = \frac{K_w}{[H^+]} = \frac{10^{-14}}{10^{-7.5}} = 10^{-6.5}$$

6/ Determine the value of α using equation 3-36.

$$\alpha = \frac{10^{-6.5}}{10^{-6.5} + 10^{-7.5} + (10^{-6.5} \times 10^{-10.5}/10^{-7.5})}$$

$$= 0.91$$

7/ Estimate the buffer intensity of the municipal wastewater using equation 3-35.

$$\beta = 2.3\left[\frac{(0.91)(4 \times 10^{-3} - 10^{-6.5} + 10^{-7.5})\{10^{7.5} + (10^{-6.5} \times 10^{-10.5}/10^{-7.5}) + 4 \times 10^{-10.5}\}}{10^{-6.5}[1 + (2 \times 10^{-10.5}/10^{-7.5})]} + 10^{-7.5} + 10^{-6.5}\right]$$

$$= 8.37 \times 10^{-4} \text{ equivalents}/\ell$$

or

$$\beta = 8.37 \times 10^{-4}\,\frac{\text{eq}}{\ell} \times 3.78\,\frac{\ell}{\text{gal}} \times 2 \times 10^{6}\,\frac{\text{gal}}{\text{day}}$$

$$= 6.3 \times 10^{3}\,\frac{\text{eq}}{\text{day}}$$

8/ Approximate the addition of hydrogen-ion concentration resulting from the industrial waste flow.

$$\Delta[H^+] = 3.2 \times 10^{-4}\,\frac{\text{eq}}{\ell} \times 3.78\,\frac{\ell}{\text{gal}} \times 0.2 \times 10^{6}\,\frac{\text{gal}}{\text{day}}$$

$$= 2.4 \times 10^{2}\,\frac{\text{eq}}{\text{day}}$$

9/ Compute the expected pH change using equation 3-34.

$$\beta = \frac{\Delta A}{\Delta pH} = \frac{\Delta[H^+]}{\Delta pH}$$

The expression above can be solved for ΔpH to give

$$\Delta pH = \frac{2.4 \times 10^{2} \text{ eq/day}}{6.3 \times 10^{3} \text{ eq/day}}$$

$$= 0.038$$

Thus, the final pH of the mixture is

$$pH = 7.5 - 0.038 = 7.46$$

Nitrogen

The principal chemical species containing nitrogen which are of importance in wastewater treatment are ammonia, organic, nitrite, and nitrate nitrogen. These forms are related by the nitrogen cycle illustrated in Figure 3-9. In the ammonia/ammonium ion form (NH_3/NH_4^+), nitrogen has a valence of -3, whereas in the molecular nitrogen (N_2) form the valence is 0. For the nitrite form (NO_2^-) the valence for nitrogen is $+3$, and for the nitrate form (NO_3^-) the valence for nitrogen is $+5$.

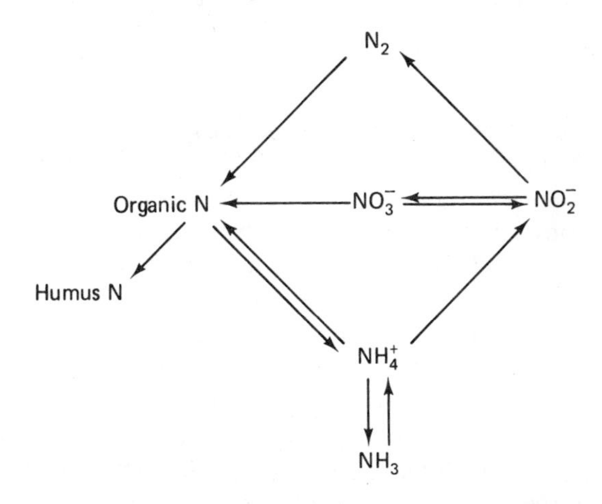

Figure 3-9. The Nitrogen Cycle.

Ammonia nitrogen exists in aqueous solution in the ammonia or ammonium-ion form. The predominant form depends upon the position of equilibrium of the following reaction:

$$NH_4^+ \rightleftharpoons NH_3 + H^+ \tag{3-37}$$

The equilibrium expression for this reaction has the form

$$K_a = \frac{[NH_3][H^+]}{[NH_4^+]} \tag{3-38}$$

or

$$\frac{K_a}{[H^+]} = \frac{[NH_3]}{[NH_4^+]} \tag{3-39}$$

where K_a = ionization constant
$[NH_3]$ = ammonia concentration, moles/ℓ
$[NH_4^+]$ = ammonium-ion concentration, moles/ℓ
$[H^+]$ = hydrogen-ion concentration, moles/ℓ

A mass balance for total ammonia nitrogen can be written as follows:

$$\text{total ammonia nitrogen concentration} = [NH_3] + [NH_4^+] \tag{3-40}$$

This mass-balance expression implies that the percentage distribution of NH_4^+ can be given by the formulation

$$\% \, NH_4^+ = \frac{[NH_4^+]}{[NH_4^+] + [NH_3]} \times 100 \qquad (3\text{-}41)$$

or

$$\% \, NH_4^+ = \frac{100}{1 + ([NH_3]/[NH_4])} \qquad (3\text{-}42)$$

Substituting for $[NH_3]/[NH_4^+]$ in equation 3-42 from equation 3-39 gives

$$\% \, NH_4^+ = \frac{100}{1 + (K_a/[H^+])} \qquad (3\text{-}43)$$

Equation 3-43 implies that the percentage distribution of ammonia and ammonium is pH-dependent, since $[H^+] = 10^{-pH}$. The ionization constant, K_a, has a value of 5.6×10^{-10} at 25°C. Thus, at a pH of 8.0, equation 3-43 shows that the percentage distribution of ammonium ion is 94.6%. Therefore, at pH values normally encountered in most biological treatment units, the ammonium ion is the predominant form.

Total Kjeldahl nitrogen is the analysis employed to determine the combined concentration of ammonia and organic nitrogen in wastewater, and the result is expressed as mg/ℓ of nitrogen. For raw municipal wastewaters TKN values have been found to range from 15 to 50 mg/ℓ. Table 3-6 summarizes the normal range of nitrogen compounds contained in domestic wastewater as well as typical removal efficiencies observed during primary and

TABLE 3-6

Nitrogen Concentration and Removals During Treatment of Municipal Wastewaters

	Raw wastewater	*Primary effluent*		*Secondary effluent assuming no nitrification*	
Nitrogen form	*mg/ℓ*	*mg/ℓ*	*% removal*	*mg/ℓ*	*% removal*
Organic nitrogen	10–25	7–20	10–40	3–6	50–80
Dissolved	4–15	4–15	0	1–3	50–80
Suspended	4–15	2–9	40–70	1–5	50–80
Ammonia nitrogen	10–30	10–30	0	10–30	0
Nitrite nitrogen	0–0.1	0–0.1	0	0–0.1	0
Nitrate nitrogen	0–0.5	0–0.5	0	0–0.5	0
Total nitrogen	15–50	15–40	5–25	15–40	25–55

Source: After McCarty (1970).

secondary treatment. Removal during primary treatment results mainly from sedimentation of insoluble organic nitrogen materials. Values given in Table 3-6 for nitrogen removal during secondary treatment assumes that nitrification does not occur. In this case nitrogen removal is primarily accomplished through bacterial assimilation. It must be stressed that nitrogen values in Table 3-6 cover the entire range of values that have been observed. Municipal wastewaters will generally have nitrogen contents in the range 15 to 25 mg/ℓ with 60% being contributed by the soluble fraction and 40% by the insoluble fraction.

Organic nitrogen must be considered as a potential source of ammonia nitrogen because deamination reactions that occur during the metabolism of organic material result in a release of ammonium ions. Proteins are high-molecular-weight organic compounds composed of amino acids joined in peptide linkage. Most amino acids can be described by the general formula

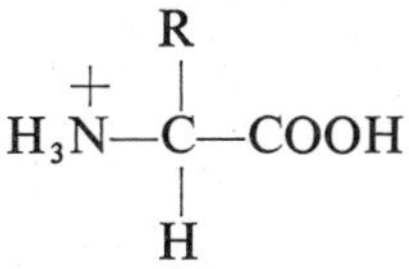

$$\mathrm{H_3\overset{+}{N}{-}\underset{\displaystyle H}{\overset{\displaystyle R}{C}}{-}COOH} \qquad (3\text{-}44)$$

where R represents a chemical group that may be hydrogen (giving the amino acid glycine), CH_2SH (cysteine), or 17 or so other groups. To form proteins they are joined by peptide bonds as follows:

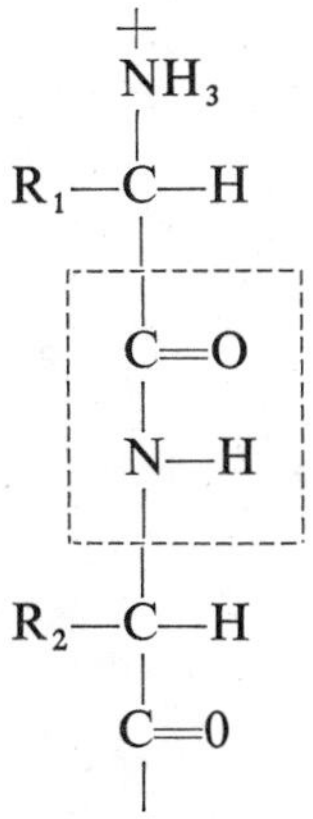

$$\begin{array}{c} \overset{+}{\mathrm{NH_3}} \\ | \\ \mathrm{R_1{-}C{-}H} \\ | \\ \boxed{\begin{array}{c} | \\ \mathrm{C{=}O} \\ | \\ \mathrm{N{-}H} \\ | \end{array}} \\ | \\ \mathrm{R_2{-}C{-}H} \\ | \\ \mathrm{C{=}0} \\ | \end{array} \quad \text{peptide bond} \qquad (3\text{-}45)$$

The peptide bonds are formed by the condensation of the α-carboxyl group of one amino acid with the α-amino group of another amino acid. Proteins can be used as an energy source by microorganisms. Under the influence of proteolytic enzymes (proteases), proteins are hydrolyzed to their constituent amino acids. The amino acids that are formed can enter the Krebs cycle (Figure 2-8) at a number of points after they have been deaminated (i.e., the NH_3 group removed). For example, consider the following reaction, which illustrates the deamination of aspartic acid:

$$\begin{array}{c}COOH\\ |\\ H\text{—}C\text{—}\overset{+}{N}H_3\\ |\\ CH_2\\ |\\ COOH\\ \text{aspartic acid}\end{array} \rightleftharpoons \begin{array}{c}COOH\\ |\\ C{=}O\\ |\\ CH_2\\ |\\ COOH\\ \text{oxaloacetic acid}\end{array} + NH_4^+ \tag{3-46}$$

Since oxaloacetic acid is a Krebs-cycle intermediate, the acid can enter at this point. A by-product of protein oxidation is the ammonium ion. Hence, biological stabilization of wastewater containing proteins will result in the presence of ammonium in the wastewater.

High animal forms (such as human beings) rid themselves of the waste product ammonia by releasing it in the form of urea, which is removed in the urine. Urea has the following chemical structure:

$$\begin{array}{c}NH_2\\ |\\ C{=}O\\ |\\ NH_2\end{array} \tag{3-47}$$

Outside the body urea is hydrolyzed rapidly in a reaction mediated by the enzyme urease to yield ammonium carbonate (Sawyer and McCarthy, 1967):

$$\begin{array}{c}NH_2\\ |\\ C{=}O\\ |\\ NH_2\end{array} + 2H_2O \xrightarrow{\text{urease}} (NH_4^+)_2 : CO_3^{2-} \tag{3-48}$$

This reaction accounts for the high percentage of ammonium and alkalinity in raw municipal wastewater.

It is highly desirable for the protection of certain receiving streams that biological wastewater treatment plants be designed to produce a nitrified effluent, that is, designed to establish a culture of nitrifying bacteria which will oxidize ammonium to nitrate while the wastewater is within the boundaries of the treatment facility. There are several disadvantages in discharging effluents containing ammonia nitrogen:

1/ Ammonia consumes dissolved oxygen in receiving waters. Under favorable environmental conditions it has been shown that organic nitrogen contained in wastewater will be converted to ammonium. Ammonium is the most reduced form of inorganic nitrogen and serves as the starting point for a two-stage biological process known as *nitrification.* There exists in the environment two groups of chemoautotrophic bacteria that can be associated with the process of nitrification. One group derives its energy through the oxidation of ammonium to nitrite, whereas the other group obtains energy through the oxidation of nitrite to nitrate. Both groups of bacteria obtain the carbon required for cell synthesis from carbon dioxide, carbonates, or bicarbonates (i.e., inorganic carbon forms).

The major nitrifying bacteria have been identified as belonging to the genera *Nitrosomonas* (responsible for the oxidation of ammonium to nitrite) and *Nitrobacter* (responsible for the oxidation of nitrite to nitrate). The oxidation of ammonium to nitrite ($N^{3-} \longrightarrow N^{3+}$) is a thermodynamically favorable reaction resulting in a free-energy release of between 66 and 84 kcal/mole at physiological concentrations (Painter, 1970). The amount of available energy actually used by the organisms depends on their energy utilization efficiency. For *Nitrosomonas*, energy utilization has been found to range from 5 to 14% (Alexander, 1961). It has been proposed that the oxidation of ammonium to nitrite occurs as a series of one-electron changes with intermediates as given by the following pathway (Doetsch and Cook, 1973):

$$NH_4^+ \longrightarrow (NH_2) \longrightarrow NH_2OH \longrightarrow (NHOH) \longrightarrow (NOH) \longrightarrow NO \longrightarrow NO_2^- \qquad (3\text{-}49)$$

However, it is customary to give only the overall reaction, which has the form

$$2NH_4^+ + 3O_2 \xrightarrow{Nitrosomonas} 2NO_2^- + 4H^+ + 2H_2O \qquad (3\text{-}50)$$

Since this is an oxidation reaction and, as such, electrons are released, an electron acceptor is required. Molecular oxygen serves this function. Without the availability of oxygen, nitrification will not occur.

The oxidation of nitrite to nitrate ($N^{3+} \longrightarrow N^{5+}$) is a single-step reaction by bacteria of the genus *Nitrobacter* and is also thermodynamically favorable, resulting in the release of approximately 17 kcal/mole of energy at physiological concentrations. The energy utilization efficiency of *Nitrobacter* has been estimated to be 5 to 10% (Alexander, 1961). The stoichiometric reaction for this oxidation is as follows:

$$2NO_2^- + O_2 \xrightarrow{Nitrobacter} 2NO_3^- \qquad (3\text{-}51)$$

The energy available to the nitrifiers per unit work is low, and therefore they are "slow growers" relative to the aerobic heterotrophs.

The overall reaction for the conversion of ammonium to nitrate can be expressed as

$$NH_4^+ + 2O_2 \longrightarrow NO_3^- + 2H^+ + H_2O \qquad (3\text{-}52)$$

Making the necessary calculation (Problem 2-2) it is seen that 4.57 mg of O_2 will be required per milligram of NH_4^+-N oxidized. Thus, if nitrification is allowed to occur in the receiving stream, second-stage BOD (nitrification) is exerted (see Figure 3-1) and a decrease in the oxygen resource will be experienced. Therefore, it is desirable to achieve nitrification in the wastewater treatment facility prior to discharge into the receiving stream so that the oxygen requirement for nitrification can be satisfied using an artificial means of aeration.

The reaction sequence illustrating the biological conversion of organic nitrogen to ammonia and ammonia to nitrate is presented in Figure 3-10. For streams, Figure 3-10 represents the steady-state spatial distribution of the nitrogen forms. In this case time represents travel time along a particular stream reach.

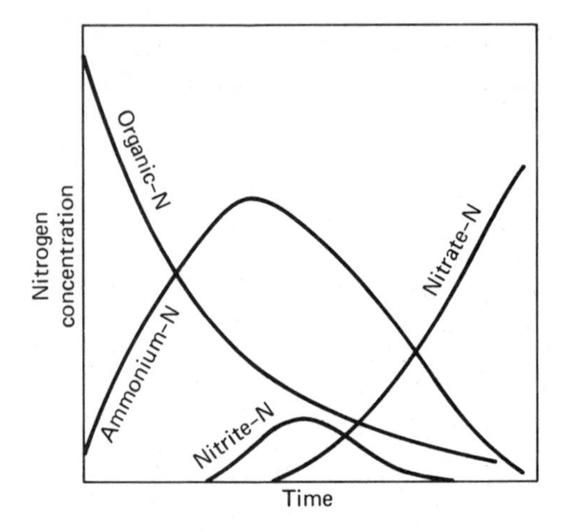

Figure 3-10. Nitrogen Conversion in a Polluted Stream. (After Sawyer and McCarty, 1967.

2/ Ammonia reacts with chlorine to form chloramines, which are less effective disinfectants than free available chlorine.

In aqueous solution, chlorine gas, Cl_2, hydrolyzes to yield Cl^- and Cl^+, a thermodynamically more stable system. This reaction is illustrated by the equation

$$Cl_2 + H_2O \rightleftharpoons HOCl + H^+ + Cl^- \qquad (3\text{-}53)$$

The equilibrium constant for this reaction has a value of 3×10^4 at 15°C, suggesting that the position of equilibrium is far to the right. In fact, in waters of pH greater than 3.0 and at chloride-ion concentrations less than 1000 mg/ℓ, free chlorine molecules (Cl_2) are virtually absent.

The hypochlorous acid (HOCl) formed in equation 3-53 is a weak electrolyte and will undergo partial ionization as follows:

$$HOCl \rightleftharpoons H^+ + OCl^- \qquad (3\text{-}54)$$

The extent of this reaction can be determined from the equilibrium expression which has the form

$$K_a = \frac{[H^+][OCl^-]}{[HOCl]} \qquad (3\text{-}55)$$

where $[H^+]$, $[OCl^-]$, and $[HOCl]$ represent the molar concentration of hydrogen ion, hypochlorite, and hypochlorous acid. Following the same line of reasoning used in the development of equation 3-43, an expression can be developed which gives the percentage distribution of hypochlorous acid with pH. The expression is as follows:

$$\%\ HOCl = \frac{100}{1 + (K_a/[H^+])} \tag{3-56}$$

The ionization constant, K_a, has a value of 3.3×10^{-8} at 20°C. Thus, using equation 3-56, the percentage distribution of hypochlorous acid between the pH values of 4 and 10 can be determined.

When chlorine is added to a wastewater containing ammonium, the initial reaction between chlorine and ammonia is almost entirely monochloramine.

$$NH_4^+ + HOCl \rightleftharpoons NH_2Cl + H^+ + H_2O \tag{3-57}$$

The time taken for this reaction is short (less than 1 min). From this substitution reaction it may be calculated (using molar ratios) that 1 part by weight of ammonium nitrogen requires 5 parts by weight of chlorine Cl_2/NH_4^+-N = 5:1). Thus, ignoring any loss of chlorine from other causes, all the chlorine will go toward producing monochloramines as long as the chlorine dose does not exceed five times the ammonium/nitrogen concentration.

If more chlorine is added than is required for equation 3-57, a continuing oxidation reaction occurs at a slower rate eventually producing mainly nitrogen.

$$2NH_2Cl + HOCl \rightleftharpoons N_2 + 3H^+ + 3Cl^- + H_2O \tag{3-58}$$

There is usually some appearance of dichloramine ($NHCl_2$) and nitrogen trichloride (NCl_3) when the chlorine dose exceeds the 5:1 ratio. However, the net result eventually corresponds to the overall equation

$$2NH_4^+ + 4HOCl \longrightarrow N_2 + 4H^+ + 4Cl^- + H_2O \tag{3-59}$$

The ammonium oxidation through monochloroamine to nitrogen gas stoichiometrically corresponds to a 7.6:1 weight ratio of Cl_2 to NH_4^+-N. However, in actual practice various side reactions with chlorine generally require that a ratio of 10:1 be provided if equation 3-59 is to be completed (White, 1972).

The reaction sequence described by equations 3-57 and 3-58 is illustrated by Figure 3-11. Point A on the curve represents the "breakpoint," or the point where continuing addition of chlorine to the wastewater gives a proportional increase in the residual.

It is important to note that in the disinfection process, before the breakpoint the residual available chlorine is present in the form of chloramines and related compounds. This is termed *combined chlorine*. However, after the breakpoint the residual available chlorine is present as *free chlorine*, which is

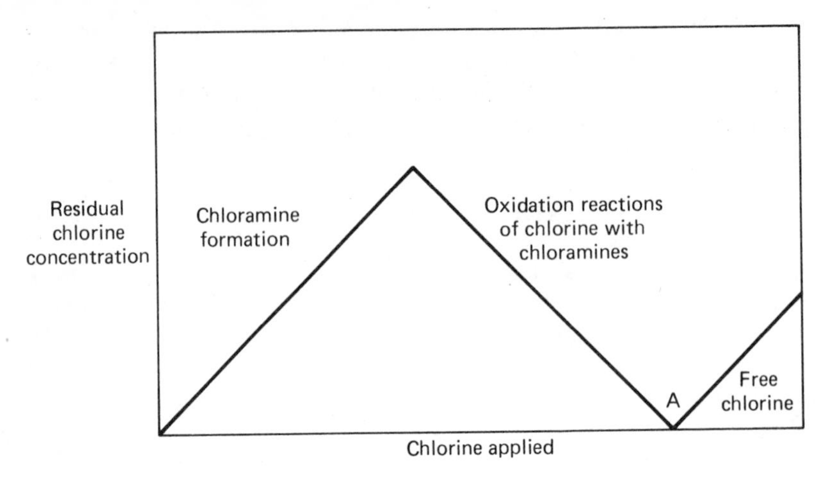

Figure 3-11. Idealized Breakpoint Curve.

a mixture at normal pH values of hypochlorous acid and hypochlorite ions. Since the bactericidal properties of free chlorine are superior to those of combined chlorine, it is often desirable to chlorinate to the point of establishing free chlorine. In treatment plants this will require a larger chlorine dose when ammonium is present.

3/ Ammonia is toxic to fish life. For certain receiving streams, it is important to control effluent ammonia concentration to prevent toxicity to fish life. It has been established that relatively low concentrations of ammonia in the un-ionized form will interfere with oxygen transport at the gills of the fish. A standard being used by the USEPA and the European Inland Fisheries Advisory Commission is 0.02 mg/ℓ un-ionized ammonia (NH_3-N) in the stream to prevent this problem.

The ionization of ammonia is a function of the pH of the receiving stream/effluent mixture as shown in equation 3-37. The percent un-ionized can be calculated for the pH and temperature of the stream mixture from

$$\% \, NH_3 = \frac{100}{1 + ([H^+]/K_a)} \tag{3-60}$$

Values of K_a at various temperatures are given in Table 3-7.

The critical conditions for analysis of a stream pollution problem is for the condition of the least dilution of the effluent with stream water, taking into account that higher temperature and pH mixtures may take precedence. Stream flow analysis is usually done on a statistical basis, as illustrated in the next section. Recorded stream flow data can often be fitted with a log-probability plot. In many areas, particularly Mountain and Western states, extreme low flows are encountered for short periods during the year. Under these conditions the 10% low flow established from the statistical plot is

TABLE 3-7

TEMPERATURE-INDUCED VARIATIONS IN THE IONIZATION CONSTANTS FOR AMMONIA AND WATER

Temperature (°C)	pK_w	pK_a
0	14.9435	10.0815
5	14.7338	9.9038
10	14.5346	9.7306
15	14.3463	9.5643
20	14.1669	9.3999
25	14.0000	9.2455
30	13.9965	9.0930
35	13.8330	8.9471

used for the minimum stream flow. The critical pH to use in computing the equilibrium ammonia concentration is usually taken as either the highest pH measured in the stream over the period of stream flow data recorded or the highest pH expected in the wastewater effluent. The decision to use either stream pH or effluent pH depends upon the magnitude of each flow and the difference between the two extreme pH values.

Example 3-10

It has been reported that un-ionized ammonia is toxic to fish in high concentration and that the allowable limit of 0.02 mg/ℓ un-ionized (NH_3-N) ammonia (as N) in a stream has been set. Determine the total amount of ammonia nitrogen (ionized and un-ionized as N) that can be in a waste effluent if the waste will be diluted in a receiving stream, with the following characteristics:

T (after mixing) = 15°C
flow = two times waste flow
ammonia (ionized and un-ionized as N) = 2.1 mg/ℓ
pH (after mixing) = 7.1

solution

1/ Compute the total ammonia (ionized and un-ionized as N) concentration allowable in the stream if the limit of 0.02 mg/ℓ un-ionized NH_3-N is not to be exceeded.

a/ Write appropriate equilibrium equations.

$$NH_4^+ \rightleftharpoons NH_3 + H^+$$

Thus,

$$K_a = \frac{[NH_3][H^+]}{[NH_4^+]}$$

and

$$H_2O \rightleftharpoons H^+ + OH^-$$

implying that

$$[H^+][OH^-] = K_w$$

or

$$[H^+] = \frac{K_w}{[OH^-]}$$

b/ Substitute for $[H^+]$ in the ammonium equilibrium expression from the equilibrium expression for water.

$$[NH_4^+] = \frac{[NH_3]K_w}{K_a[OH^-]}$$

c/ Determine the allowable ammonium concentration by taking the logarithm of both sides of the expression in step 1b.

$$\log [NH_4^+] = \log [NH_3] + \log K_w - \log K_a - \log [OH^-]$$

or

$$\log [NH_4^+] = \log [NH_3] + \text{pOH} + pK_a - pK_w$$

and since

$$\text{pH} + \text{pOH} = pK_w$$

it is possible to write

$$\log [NH_4^+] = \log [NH_3] + pK_a - \text{pH}$$

It is known that

$$NH_3\text{-N} = 0.02 \text{ mg}/\ell$$

or

$$[NH_3] = \frac{0.02 \times 10^{-3} \text{ g}/\ell}{14 \text{ g/mole N}}\left(\frac{17}{14}\right)$$

$$[NH_3] = 1.735 \times 10^{-6} \text{ mole}/\ell$$

Therefore,

$$\log [NH_4^+] = \log (1.735 \times 10^{-6}) + 9.56 - 7.1$$
$$= -3.3$$

Thus,

$$[NH_4^+] = 5.01 \times 10^{-4} \text{ mole}/\ell$$

or

$$NH_4^+\text{-N} = (5.01 \times 10^{-4} \text{ mole}/\ell \times 18 \text{ g/mole} \times 10^3)\left(\frac{14}{18}\right)$$
$$= 7.01 \text{ mg}/\ell$$

d/ The total allowable nitrogen is given by a mass-balance expression of the form

$$\text{total N} = (NH_4^+\text{-N}) + (NH_3\text{-N})$$
$$= 7.01 + 0.02 = 7.03 \text{ mg}/\ell$$

2/ Calculate the allowable nitrogen in the wastewater from a simple proportion equation.

$$\text{total nitrogen allowed in stream} = \frac{Q_w N_w + Q_s N_s}{Q_w + Q_s}$$

or

$$7.03 = \frac{(1)N_w + (2)(2.1)}{1 + 2}$$

$$N_w = 16.89 \text{ mg}/\ell \ (NH_4^+ + NH_3\text{-N})$$

Although a nitrified effluent is far preferable to one containing a large ammonia concentration, high nitrate levels may stimulate undesirable aquatic growths, and thereby contribute to the eutrophication problem. *Eutrophication* is the enrichment of water systems with plant nutrients, resulting in an increased concentration of photosynthetic organisms, mostly algae. Toerien (1975) lists several ways in which an overabundant aquatic growth reduces water quality.

1/ It increases cost of water treatment due to rapid filter clogging resulting in shorter filter runs.
2/ It interferes with water sports.
3/ Taste- and odor-causing compounds are produced as end products of algal metabolism.
4/ Loss of livestock and fish kills result from toxins produced by blue-green algae.
5/ Oxygen depletion occurs in the water body as a result of algal decay.

It has also been shown by Hoehn et al. (1978) and Thompson (1978) that, during growth, algae release organic compounds that become potentially carcinogenic trihalomethane chemicals upon chlorination, which greatly increases the significance of eutrophication in water supply reservoirs. The prevention of eutrophication problems in impoundments that are not eutrophied can only be effected through the prevention of nutrient addition. Hence, in many treatment situations it may be desirable to remove nitrogen from wastewater effluents. Care should be exercised, however, because nitrogen limitation when phosphorus is in abundance will generally stimulate the growth of blue-green algae instead of green algae and worsen the quality of the water for drinking purposes.

Biological treatment can be employed to remove nitrogen from wastewater. The applicable treatment process is called *denitrification*. This process requires that ammonium first be oxidized to nitrates (i.e., nitrification is essential). Certain facultative bacteria, which have the ability to derive energy by using nitrate as an electron acceptor (anaerobic respiration) in the absence of oxygen, reduce the nitrate to nitrogen gas, which is then stripped from the liquid, thereby decreasing the nitrogen content of the wastewater. Although nitrification is a necessary first step for nitrogen reduction, anoxic conditions and an easily degradable carbon source are also required. McCarty et al. (1969) recommend that methanol be used as the carbon source because of its low cost, but any readily available, inexpensive organic carbon source can be used.

McCarty (1970) proposes that denitrification may be viewed as a two-step process, as illustrated in the following reactions with methanol:

$$\text{step 1} \quad 3NO_3^- + CH_3OH \longrightarrow 3NO_2^- + CO_2 + 2H_2O \qquad (3\text{-}61)$$

$$\text{step 2} \quad 2NO_2^- + CH_3OH \longrightarrow N_2 + CO_2 + H_2O + 2OH^- \qquad (3\text{-}62)$$

$$\text{overall} \quad 6NO_3^- + 5CH_3OH \longrightarrow 3N_2 + 5CO_2 + 7H_2O + 6OH^- \qquad (3\text{-}63)$$

Stoichiometry considerations show that 1.9 mg/ℓ of methanol is required for each mg/ℓ of nitrate nitrogen reduced to molecular nitrogen.

Biological nitrification followed by denitrification is probably the most widely used method for removing nitrogen from wastewater. Interestingly, the primary rival techniques, ammonia stripping, ion exchange using clinoptilolite, and breakpoint chlorination, are all nonbiological and require that the nitrogen be in the ammonia(um) form for removal.

Phosphorus

Phosphorus is considered by most investigators to be the key nutrient in controlling eutrophication, since both nitrogen and carbon dioxide exist in the atmosphere and reach equilibrium concentrations in water. Thus, many treatment situations require that consideration be given to its removal.

Municipal wastewaters contain phosphorus in three different forms: (1) organic phosphorus, (2) orthophosphorus, and (3) condensed phosphorus. When treating wastewater biologically, bacteria assimilate orthophosphate (PO_4^{3-}) during their growth process. However, condensed phosphates (e.g., pyrophosphate, $P_2O_7^{4-}$, and tripolyphosphate, $P_3O_{10}^{5-}$) must first undergo enzymatic hydrolysis to the ortho form before they can be assimilated. Hurwitz et al. (1965) points out that 50 to 80% of the condensed phosphates present in municipal wastewater will be hydrolyzed during biological treatment. Normal ranges of phosphorus compounds contained in municipal wastewater and typical removals realized during both primary and secondary treatment are presented in Table 3-8.

In biological wastewater treatment, assimilation is the only means by which phosphorus is removed, except, possibly, when the water has an unusual chemical makeup and precipitation onto biological flocs occurs. Some treatment plants have reported biological phosphorus removal in quantities greater than that required for maximum growth. Such removal is generally explained on the basis of phosphorus storage by the cell and is

TABLE 3-8

PHOSPHORUS CONCENTRATION AND REMOVALS DURING TREATMENT OF MUNICIPAL WASTEWATERS

Phosphorus form as P	Raw wastewater (mg/ℓ)	Primary effluent		Secondary effluent	
		mg/ℓ	% Removal	mg/ℓ	% Removal
Organic phosphate	1–3	0.5–2	20–50	0.5–1	40–80
Orthophosphate	2–8	1–7	0–40	1–8	0–40
Condensed phosphate	2–8	2–8	0–20	1–3	40–80
Total phosphate	4–14	3–12	5–20	3–11	10–30

Source: After McCarty (1970).

referred to as "luxury uptake." This phenomenon has been observed in only a few plants and has not been duplicated at other plants by manipulation of operating conditions. Hence, when a high degree of phosphorus removal is required, process reliability is enhanced when chemical precipitation is used.

3-3 Solids Content

Total solids content is one of the most important physical characteristics of wastewater. Total solids include both the suspended solids and the dissolved solids which are obtained by separating the solid and liquid phase by evaporation.

Suspended solids are a combination of settleable solids and nonsettleable (colloidal) solids, which are usually determined by filtering a wastewater sample through a glass fiber filter contained in a Gooch crucible or through a membrane filter. *Settleable solids* are those which usually settle in sedimentation tanks during a normal detention period. This fraction is determined by measuring the volume of sludge in the bottom of an Imhoff cone after 1 h of settling.

Solids remaining after evaporation or filtration are dried, weighed, and then ignited. The loss of weight by ignition at 600°C is a measure of the *volatile solids*, which are classed as organic material. The remaining solids are the *fixed solids*, which are considered as inorganic (mineral) matter. The suspended solids associated with the volatile fraction are termed *volatile suspended solids* (VSS), and the suspended solids associated with the mineral fraction are termed *fixed suspended solids* (FSS) Table 3-9 presents a break-

TABLE 3-9

AVERAGE COMPOSITION OF MUNICIPAL WASTEWATER

State of solids	Solids			5-Day, 20°C BOD (mg/ℓ)	COD (mg/ℓ)
	Mineral (mg/ℓ)	Organic (mg/ℓ)	Total (mg/ℓ)		
Suspended	65	170	235	110	108
Settleable	40	100	140	50	42
Nonsettleable	25	70	95	60	66
Dissolved	210	210	420	30	42
Total	275	380	655	140	150

Source: After Fair (1971).

down of the solids content of a typical domestic wastewater. This table shows that a large fraction of the total organic content of wastewater is contributed by the suspended solids which are present.

The purpose of primary sedimentation is to reduce the velocity of the wastewater sufficiently to permit solids to settle. Primary sedimentation will remove most of the settleable solids, or about 40 to 70% of the suspended solids from domestic sewage. Since approximately 80% of the total BOD of municipal wastewater is contributed by suspended and colloidal solids, between 30 and 45% of the total BOD will be removed during this operation. Furthermore, Tables 3-6 and 3-8 show that between 5 and 25% of the total nitrogen and 5 to 20% of the total phosphorus will be removed at this point.

Shown in Figure 3-12 is the effect of clarifier surface loading rate on settleable solids removal, while the relationship between total suspended solids and settleable solids removal by primary sedimentation is given in Figure 3-13. The relationship between total suspended solids and BOD removal by primary sedimentation is shown in Figure 3-14. Figures 3-12, 3-13, and 3-14 reflect operational data from circular clarifiers located in Brazil and the United Kingdom. Rectangular clarifiers generally perform better than circular clarifiers.

The biomass wasted from an activated sludge system will represent from 0 to 40% of the ultimate BOD fed to the biomass, depending on the process modification and biological solids retention time (BSRT) used. For typical activated sludge operation the percent will vary from 25 to 30%. Since about 35% of the ultimate BOD in the plant influent is removed in the primary clarifier and 60 to 65% of the influent BOD to the plant is removed from wastewater in the form of solids that require further handling, only 35 to 40% of the BOD is actually destroyed.

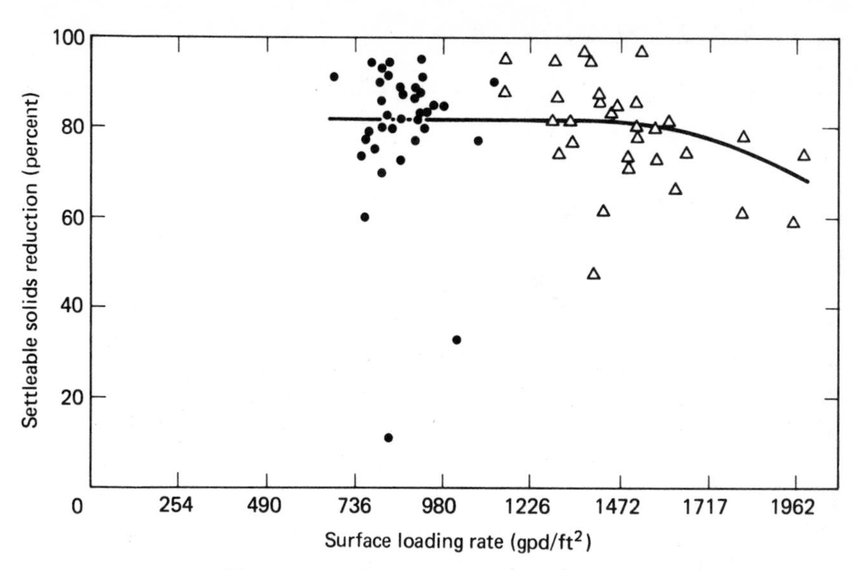

Figure 3-12. Effect of Surface Loading Rate on Settleable Solids Removal. (After Bradley, 1975.)

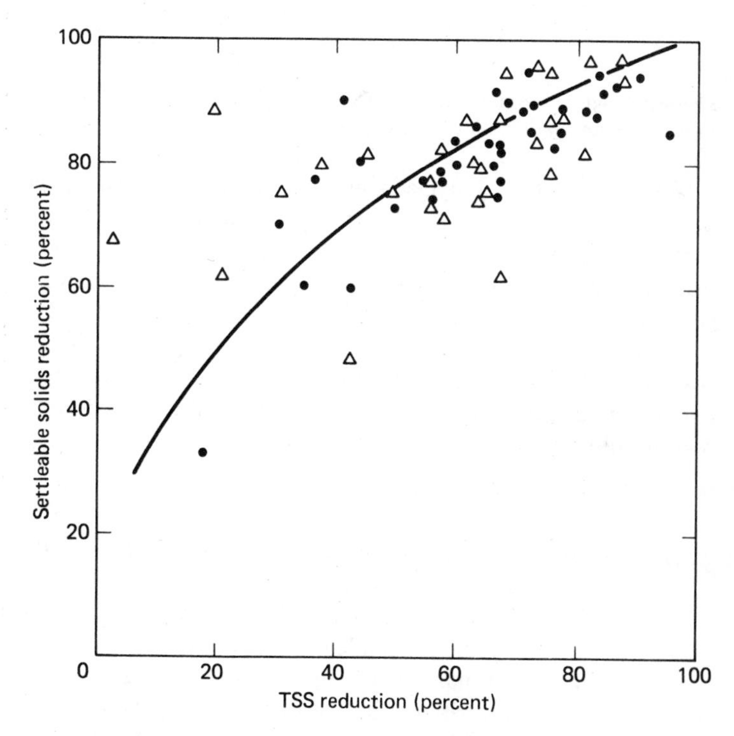

Figure 3-13. Relationship Between TSS and Settleable Solids Removal by Primary Sedimentation. (After Bradley, 1975).

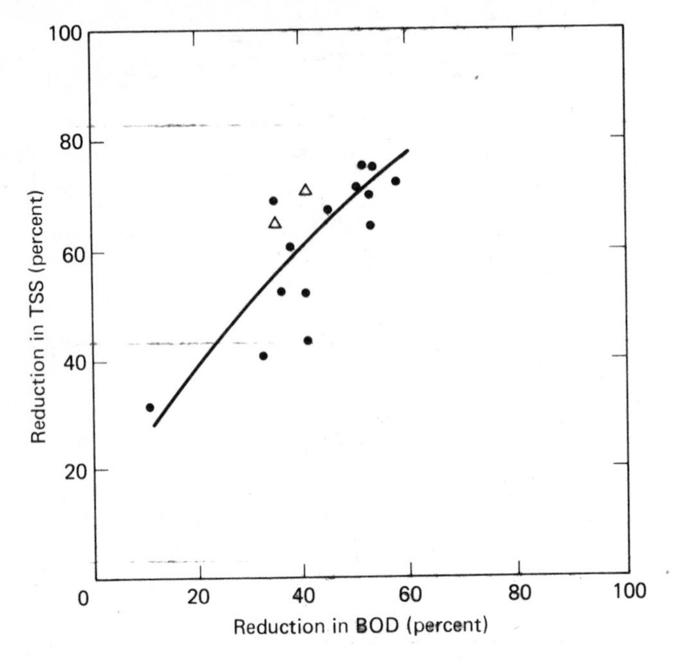

Figure 3-14. Relationship Between TSS and BOD Removal by Primary Sedimentation. (After Bradley, 1975.)

When computing sludge accumulation in the activated sludge process, not only is it necessary to consider the biomass production resulting from organic removal, but the accumulation of nonbiodegradable suspended solids originally present in the wastewater must be accounted for as well. It is generally assumed that total nonbiodegradable suspended solids in the influent are the sum of fixed suspended solids and nonbiodegradable volatile suspended solids. McKinney (1970) proposes than nonbiodegradable volatile suspended solids contained in raw wastewater be estimated from the relationship

$$\mathrm{NDVSS} = \left(\frac{\mathrm{COD}_0 - \mathrm{BOD}_{u0}}{\mathrm{COD}_0}\right)\mathrm{VSS} \qquad (3\text{-}64)$$

where NDVSS = nonbiodegradable volatile suspended solids in the raw wastewater, mg/ℓ

COD_0 = chemical oxygen demand of raw wastewater contributed by the suspended solids, mg/ℓ

BOD_{u0} = ultimate biochemical oxygen demand of raw wastewater contributed by the suspended solids, mg/ℓ

VSS = total volatile suspended solids concentration of raw wastewater, mg/ℓ

McKinney (1970) states that typically the volatile suspended solids fraction of domestic wastewater will be approximately 40% nonbiodegradable.

3-4 Wastewater Composition and Flow

Based on the concentration of its various components, wastewater may be classified as strong, medium, or weak. Typical compositions for domestic wastewater are presented in Table 3-10. In using this table it must be remembered that the composition and concentration of domestic wastewater is

TABLE 3-10

TYPICAL COMPOSITION OF DOMESTIC WASTEWATER

Constituent	*Concentration*[a]		
	Strong	*Medium*	*Weak*
Solids, total	1200	700	350
Dissolved, total	850	500	250
Fixed	525	300	145
Volatile	325	200	105
Suspended, total	350	200	100
Fixed	75	50	30
Volatile	275	150	70
Settleable solids (ml/ℓ)	20	10	5
Biochemical oxygen demand, 5-day, 20°C (BOD_5-20°)	300	200	100
Total organic carbon (TOC)	200	135	65
Chemical oxygen demand (COD)	1000	500	250
Nitrogen, (total as N)	85	40	20
Organic	35	15	8
Free ammonia	50	25	12
Nitrites	0	0	0
Nitrates	0	0	0
Phosphorus (total as P)	20	10	6
Organic	5	3	2
Inorganic	15	7	4
Chlorides[b]	100	50	30
Alkalinity (as $CaCO_3$)[b]	200	100	50
Grease	150	100	50

[a]All values except settleable solids are expressed in mg/ℓ.
[b]Values should be increased by amount in carriage water.
Source: After Metcalf and Eddy (1972).

highly variable, and because of this the data should serve only as a guide and not as a basis for design.

Wastewater flow from individual homes is generally estimated to be 45 gallons per capita per day (gpcd). When infiltration, business use, and schools are included, flows from residential areas may vary between 50 and 100 gpcd. It is generally accepted that each person contributes between 0.10

TABLE 3-11

APPROXIMATE WASTEWATER FLOWS AND PER CAPITA BOD CONTRIBUTIONS

Type	*Gallons per person per day*	*Pounds of BOD_5 per person per day*
Domestic waste from residential areas		
Large single-family houses	100	0.20
Typical single-family houses	75	0.17
Multiple-family dwellings (apartments)	60–75	0.17
Small dwellings or cottages	50	0.17
(If garbage grinders are installed multiply BOD by a factor of 1.5)		
Domestic waste from camps and motels		
Luxury resorts	100–150	0.20
Mobile home parks	50	0.17
Tourist camps or trailer parks	35	0.15
Hotels and motels	50	0.10
Schools		
Boarding schools	75	0.17
Day schools with cafeterias	20	0.06
Day schools without cafeterias	15	0.04
Restaurants		
Each employee	30	0.10
Each patron	7–10	0.04
Each meal served	4	0.03
Transportation terminals		
Each employee	15	0.05
Each passenger	5	0.02
Hospitals	150–300	0.30
Offices	15	0.05
Drive-in theaters, per stall	5	0.02
Movie theaters, per seat	3–5	0.02
Factories, exclusive of industrial and cafeteria wastes	15–30	0.05

Source: After Hammer (1975).

and 0.3 lb/day of BOD to the wastewater. The higher values are applicable in areas where garbage grinders are prevalent. Table 3-11 includes commonly accepted values for wastewater flows and per capita BOD contributions for various types of establishments and services. However, these data are subject to considerable variation. For strictly domestic wastewater, an extensive study documented by Chien and Jones (1975) indicated the average per capita loadings reported in Table 3-12. The wastewater flow given in this table does not include infiltration.

TABLE 3-12

PER CAPITAL LOADINGS FOR STRICTLY DOMESTIC WASTEWATER FLOWS

Parameter	*Per capita loading*
Wastewater flow	58 gal/day
BOD_5	0.1 lb/day
COD	0.2 lb/day
SS	0.08 lb/day

Source: After Chien (1975).

One method used to estimate the average domestic wastewater flow is to take 0.7–0.9 times the volume of water used for domestic consumption during the nonirrigation season. In most cases, however, the average wastewater flow from a community is taken as 100 gpcd, a value that includes flows from business establishments, residential areas, and infiltration, but does not include industrial wastes. As a guide in estimating industrial flows, data from various industries are reported in Tables 3-13 through 3-22. In these tables, the terms old, prevalent and new refer to the technology employed for plant processing. It must also be realized that even though most biological treatment processes are designed on the basis of average flow or 120% of average flow, the plant must be able to operate hydraulically at both minimum and maximum flows. Normally, minimum flows range from 20 to 50% of the average daily flow, whereas maximum flows are generally between 200 and 250% of the average daily flow, depending on the plant size.

When average wastewater flows are being estimated by totaling the various components of flow, it is necessary that extraneous flow due to infiltration be included. Values typically used for this purpose are given in Table 3-23.

TABLE 3-13

TYPICAL CANNING WASTES

Product	Waste volume (gal/case)	5-Day BOD		Suspended solids	
		mg/ℓ	lb/case	mg/ℓ	lb/case
Apples	29–46	1680–5530	0.64–1.31	300–600	0.10–0.20
Apricots	65–91	200–1020	0.15–0.56	200–400	0.14–0.25
Cherries	14–46	700–2100	0.16–0.50	200–600	0.05–0.14
Cranberries	11–23	500–2250	0.10–0.21	100–250	0.02–0.05
Peaches	51–69	1200–2800	0.69–1.20	450–750	0.24–0.34
Pineapples	74	26	0.002	—	—
Asparagus	80	16–100	0.01–0.07	30–180	0.02–0.12
Beans, baked	40	925–1440	0.31–0.48	225	0.07
Beans, green wax	30–51	160–600	0.15–0.67	60–150	0.02–0.04
Beans, kidney	20–23	1030–2500	0.19–0.45	140	0.02
Beans, lima, dried	20–23	1740–2880	0.30–0.60	160–600	0.05–0.10
Beans, lima, fresh	57–294	190–450	0.21–0.47	420	0.20–1.02
Beets	31–80	1580–7600	1.00–2.00	740–2220	0.50–1.00
Carrots	36	520–3030	0.11–0.67	1830	0.40
Corn, cream style	28–33	620–2900	0.17–0.66	300–675	0.07–0.17
Corn, whole kernel	29–80	1120–6300	0.74–1.50	300–4000	0.20–0.95
Mushrooms	—	76–850	4.77–53.38	50–240	3.14–15.07
Peas	16–86	380–4700	0.27–0.63	270–400	0.06–0.20
Potatoes, sweet	90	1500–5600	1.10–4.40	400–2500	0.31–1.95
Potatoes, white	—	200–2900	—	990–1180	—
Pumpkin	23–57	1500–6880	0.72–1.31	785–1960	0.38
Sauerkraut	3–20	1400–6300	0.10–0.30	60–630	0.01–0.10
Spinach	180	280–730	0.42–1.11	90–580	0.14–0.88
Squash	23	4000–11,000	0.76–2.09	3000	0.57
Tomatoes	3–114	180–4000	0.11–0.17	140–2000	0.06–0.13

Source: After Eckenfelder (1970).

TABLE 3-14
Wastewater Characteristics from Petroleum Refineries

	Flow (gal/bbl)		*BOD_5 (lb/bbl)*		*Phenol (lb/bbl)*		*Sulfide (lb/bbl)*	
Type of technology	*Average*	*Range*	*Average*	*Range*	*Average*	*Range*	*Average*	*Range*
Older	250	170–374	0.40	0.31–0.45	0.030	0.028–0.033	0.01	0.008
Typical	100	80–155	0.10	0.08–0.16	0.01	0.009–0.013	0.003	0.0028
Newer	50	20–60	0.05	0.02–0.06	0.005	0.001–0.006	0.003	0.0015

Source: After Eckenfelder (1970)

TABLE 3-15

WASTEWATER CHARACTERISTICS FROM THE PULP AND PAPER INDUSTRY

	Flow (gal/ton)		BOD_5 (lb/ton)		SS (lb/ton)	
Type of technology	*Average*	*Range*	*Average*	*Range*	*Average*	*Range*
Bleached kraft						
Older	110,000	75,000–140,000	200	50–350	200	80–370
Prevalent	45,000	39,000–54,000	120	30–220	170	50–200
Newer	25,000	—	90	—	90	—
Bleached sulfite						
Older	95,000	58,000–170,000	500	350–730	120	50–200
Prevalent	55,000	40,000–70,000	330	235–430	100	40–100
Newer	30,000	10,000–40,000	100	60–300	50	10–70
Bleached kraft						
Older	460	310–580	100	25–175	100	40–185
Prevalent	190	160–220	60	15–110	85	25–100
Newer	105	—	45	—	45	—
Bleached sulfite						
Older	390	240–710	250	175–365	60	25–100
Prevalent	230	170–290	165	118–215	50	20–50
Newer	125	415–170	50	30–150	25	5–35

Source: After Eckenfelder (1970).

TABLE 3-16

SLAUGHTERHOUSE WASTEWATER CHARACTERISTICS PER THOUSAND POUNDS OF LIVEWEIGHT KILLED[a]

Technology	*lb of BOD_5/1000 lb*	*gal/1000 lb*
Old	20.2	2112
Typical	14.4	1294
Advanced	11.3	1116

[a]Post catch basin.
Source: After Eckenfelder (1970).

TABLE 3-17

POULTRY PROCESSING WASTEWATER CHARACTERISTICS PER 1000 BIRDS

Plants	*lb of BOD_5/1000 birds*	*gal/1000 birds*
Older	31.7	4000
Newer	26.0	7300

Source: After Eckenfelder (1970).

TABLE 3-18

TANNERY WASTEWATER CHARACTERISTICS PER POUND OF LEATHER PROCESSED

Technology	*BOD_5 (lb/lb)*	*SS (lb/lb)*	*TDS (lb/lb)*	*Volume (gal/lb)*
Older	0.0916	0.260	0.380	10.5
Prevalent—newer	0.0883	0.250	0.350	9.5

Source: After Eckenfelder (1970).

TABLE 3-19
BREWERY WASTE CHARACTERISTICS

Brewery	Flow (gal/bbl)	BOD_5 (lb/bbl)	BOD_5 mg/ℓ	Suspended solids (lb/bbl)	Suspended solids mg/ℓ
A[a]	295[b]	2.1	1832[c]	1.3	1028[d]
New York City study	—	1.6	—	0.6	—
B	170	1.5	1040	1.8	
C	350	2.4	850	1.1	
Chicago breweries	320	3.1	1160	1.7	
C[a]	130	1.7	1500		470

[a]Separates spent grains and spent hops.
[b]Brewhouse, 130 gal/bbl; bottling, 165 gal/bbl.
[c]Brewhouse, 3580 mg/ℓ; bottling, 384 mg/ℓ.
[d]Brewhouse, 2220 mg/ℓ; bottling, 80 mg/ℓ; spent hops removed.
Source: After Eckenfelder (1970).

TABLE 3-20
DAIRY WASTEWATER

Product	BOD (lb/100 lb)	Volume (gal/100 lb)
Creamery butter	0.34–1.68	410–1350
Cheese	0.45–3.0	1290–2310
Condensed and evaporated milk	0.37–0.62	310–420
Ice cream[a]	0.15–0.73	620–1200
Milk	0.05–0.26	200–500

[a]Per 100 gal of product.
Source: After Eckenfelder (1970).

TABLE 3-21

WASTEWATER CHARACTERISTICS FROM THE TEXTILE-FINISHING INDUSTRY PER 1000 LB OF CLOTH PRODUCED

Process	*Technology*	*Vol (1000 gal)*	*BOD_5 (lb)*	*SS (lb)*	*TDS (lb)*
Wool	Old	73.7	450	—	—
	Prevalent	63.0	300	—	—
	New	62.0	50	—	—
Cotton	Old	50.0	170	80	245
	Prevalent	38.0	155	70	205
	New	35.0	140	62	187
Synthetic	Rayon	3–7	20–40	20–90	20–500
	Acetate	7–11	40–50	20–60	20–300
	Nylon	12–18	35–55	20–40	20–300
	Acrylic	21–29	100–150	25–150	25–400
	Polyester	8–16	120–250	30–150	30–600

Source: After Eckenfelder (1970).

TABLE 3-22

STEEL MILL WASTEWATER CHARACTERISTICS BASED ON QUANTITY PRODUCED PER INGOT TON PER DAY

	Old	*Prevalent*	*New*
Flow (gal/ton)	9860	10,000	13,750
Suspended solids (lb/ton)	103	125	184
Phenols (lb/ton)	0.069	0.064	0.064
Cyanides (lb/ton)	0.029	0.028	0.031
Fluorides (lb/ton)	0.033	0.031	0.031
Ammonia (lb/ton)	0.082	0.078	0.078
Lube oils (lb/ton)	3.08	2.72	2.37
Free acids (lb/ton)	3.03	3.54	3.40
Emulsions (lb/ton)	0.332	0.414	1.17
Soluble metals (lb/ton)	—	0.079	0.082

Source: After Eckenfelder (1970).

TABLE 3-23

SEWER INFILTRATION RATES[a]

Unit	*Minimum flow*	*Average flow*	*Maximum flow*
GPD/acre sewered area	500	2,000	5,000
GPD/mile of sewer pipe	5000	30,000	200,000
GPD/mile sewer pipe/inch diameter	500[c]	5,000	2,500
GPD/capita	25		200
GPD/manhole cover	75	100	150
GPD/mile 8-in. sewer[b]	1600	4,000[c]	12,000
GPD/mile 24-in. sewer[b]	4000	12,000[c]	36,000

[a]Typical values used in different engineering offices.
[b]Values indicated in some specifications on amount of infiltration permissible (new sewers).
[c]Ten-State Standards for new sewer construction.
Source: After Parker (1975).

Example 3-11

Estimate the average daily wastewater flow and BOD_5 concentration for the following community:

1/ Population = 15,000 persons.
2/ A 100-bed hospital.
3/ Four restaurants that serve an average of 200 meals/day each.
4/ A 700-student day school with cafeteria.
5/ A cotton mill that produces 100,000 lb of cloth per week.
6/ A slaughterhouse that processes 50,000 lb of live weight per day.
7/ Fifteen miles of sanitary sewer.

After the average daily wastewater flow has been computed, determine the maximum and minimum daily flows to be expected.

solution

1/ Compute the average daily domestic flow and BOD_5 loading using the data presented in Table 3-11.

$$\text{flow} = 15{,}000 \times 75 = 1{,}125{,}000 \text{ gal/day}$$

$$BOD_5 = 15{,}000 \times (0.2)(1.5) = 4500 \text{ lb/day}$$

2/ Calculate the hospital, restaurant, and school flows and BOD_5 loadings from Table 3-11.

$$\text{hospital flow} = 100 \times 225 = 22{,}500 \text{ gal/day}$$

$$\text{restaurant flow} = 4 \times 200 \times 4 = 3200 \text{ gal/day}$$

$$\text{school flow} = 700 \times 20 = 14{,}000 \text{ gal/day}$$

$$\text{hospital BOD}_5 = 100 \times 0.3 = 30 \text{ lb/day}$$

$$\text{restaurant BOD}_5 = 4 \times 200 \times 0.03 = 24 \text{ lb/day}$$

$$\text{school BOD}_5 = 700 \times 0.06 = 42 \text{ lb/day}$$

3/ Determine industrial flows and BOD_5 loadings from Tables 3-16 and 3-21.

$$\text{flow} = \frac{100{,}000}{5 \times 1000}(35) + \frac{50{,}000}{1000}(1116) = 56{,}500 \text{ gal/day}$$

$$\text{BOD}_5 = \frac{100{,}000}{5 \times 1000}(140) + \frac{50{,}000}{1000}(11.3) = 3365 \text{ lb/day}$$

4/ From Table 3-23 estimate the flow due to infiltration. It is assumed that no BOD is associated with infiltration.

$$\text{flow} = 30{,}000 \times 15 = 450{,}000 \text{ gal/day}$$

5/ Compute the total average flow and BOD_5 concentration.

$$\begin{aligned}
\text{total flow} &= 1{,}125{,}000 + 22{,}500 + 3200 + 14{,}000 \\
&\quad + 56{,}500 + 450{,}000 \\
&= 1{,}671{,}200 \text{ gal/day} \\
\text{total BOD}_5 \text{ load} &= 4500 + 30 + 24 + 42 + 3365 \\
&= 7961 \text{ lb/day} \\
\text{BOD}_5 &= 7961 \text{ lb/day} \times \frac{1}{1{,}671{,}200}\frac{\text{day}}{\text{gal}} \times \frac{1 \text{ gal}}{28.32\ \ell} \\
&\quad \times 454{,}000 \text{ mg/lb} \\
&= 76 \text{ mg}/\ell
\end{aligned}$$

6/ Estimate the expected maximum and minimum flows.

$$\text{minimum flow} = (0.3)(1.67 \text{ MGD}) = 0.50 \text{ MGD}$$

$$\text{maximum flow} = (2.25)(1.67 \text{ MGD}) = 3.8 \text{ MGD}$$

3-5 Wastewater Flow Gauging and Sampling

Many workers feel that for design purposes it is best to estimate wastewater flows and composition by monitoring flows in existing collection systems and then making the necessary corrections to predict future requirements. This method identifies storm inflows, infiltration, and industrial flows. However, this method of estimation is not without problems. The major concern here is that too much faith may be placed in 1 week or even 1 month of flow measurements, so that the data are interpreted as representing the maximum conditions that will occur. In most cases this is not sufficient. This can be understood by noting the cyclic nature of wastewater flow and composition over a 24-h period as illustrated in Figure 3-15, and then realizing that not only does wastewater flow and composition vary with the hours of the day but also with the week, with the month, and the year. It is not unusual for infiltration inflow to constitute 25 to 50% or more of domestic wastewater flow, and this is usually a seasonal response. Furthermore, municipal wastewater is not homogeneous since industrial discharges often occur in "slugs" of short duration.

Variations of municipal wastewater parameters can often be characterized as probabilistic. In such cases statistical methods can be used to quantify parameters such as BOD, SS, and wastewater flow. Ideally, data should be taken on a daily basis for a minimum of 1 year. However, analysis can be made with fewer data. The reliability of the analysis depends on the amount and quality of data taken. Hence, the importance of the sampling program cannot be overemphasized.

There are basically two types of samples which can be taken: (1) the grab sample, and (2) the composite sample. The *grab sample* shows the wastewater characteristics only at the time the sample is taken. It is generally used when (1) the wastewater flow and composition are relatively constant, (2) the wastewater flow is intermittent, or (3) composite samples obscure extreme conditions of the waste (e.g., pH and temperature). The minimum volume of the grab sample should be between 1 and 2 ℓ.

Composite samples are samples composed of mixtures of individual samples taken at different times. The amount of each individual sample that is added to the total mixture is proportional to the wastewater flow at the time the sample was taken:

$$\begin{array}{l}\text{portion of sample}\\ \text{needed per unit of}\\ \text{flow}\end{array} = \frac{\text{total volume of sample desired}}{[\text{average flow rate}] \times [\text{number of samples to be mixed}]} \tag{3-65}$$

The frequency of sampling depends on the variability of the wastewater flow and composition. For low variability, samples may only be required at intervals between 2 and 24 h, whereas for high variability, samples may be required as often as every 15 min. Individual portions of a composite sample should be between 25 and 100 ml, and the total composited volume should be between 2 and 4 ℓ.

Samples should be stored in a manner which ensures that the characteristics to be analyzed are not altered. Tables 3-24 and 3-25 report storage

TABLE 3-24

RECOMMENDED STORAGE PROCEDURE

Analysis	Sample storage	
	Refrigeration at 4°C	Frozen
Total solids	OK	OK
Suspended solids	Up to several days	No
Volatile suspended solids	Up to several days	No
COD	Up to several days	OK
BOD	Up to 1 day in composite sampling systems	Lag develops, must use fresh sewage seed

Source: EPA (1973).

TABLE 3-25

SAMPLE PRESERVATION

Parameter	Preservative	Maximum holding period
Acidity–alkalinity	Refrigeration at 4°C	24 h
Biochemical oxygen demand	Refrigeration at 4°C	6 h
Calcium	None required	7 days
Chemical oxygen demand	2 ml H_2SO_4/ℓ	7 days
Chloride	None required	7 days
Color	Refrigeration at 4°C	24 h
Cyanide	NaOH to pH 10	24 h
Dissolved oxygen	Determine on-site	No holding
Fluoride	None required	7 days
Hardness	None required	7 days
Metals, dissolved	5 ml HNO_3/ℓ	6 months
Metals, total	Filtrate: 3 ml 1:1 HNO_3/ℓ	6 months
Nitrogen, ammonia	40 mg $HgCl_2/\ell$, 4°C[a]	7 days
Nitrogen, Kjeldahl	40 mg $HgCl_2/\ell$, 4°C[a]	Unstable
Nitrogen, nitrate-nitrite	40 mg $HgCl_2/\ell$, 4°C[a]	7 days
Oil and grease	2 ml H_2SO_4/ℓ, 4°C	24 days
Organic carbon	2 ml H_2SO_4/ℓ (pH 2)	7 days
pH	Determine on-site	No holding
Phenolics	1.0 g $CuSO_4/\ell$ + H_3SO_4 to pH 4.0, 4°C	24 h
Phosphorus	40 mg $HgCl_2/\ell$, 4°C[a]	7 days
Solids	None available	7 days
Specific conductance	None required	7 days
Sulfate	Refrigeration at 4°C	7 days
Sulfide	2 ml Zn acetate/ℓ	7 days
Threshold odor	Refrigeration at 4°C	7 days
Turbidity	None available	7 days

[a]Disposal of mercury-containing samples is a recognized problem; research investigations are under way to replace it as a preservative.

Source: EPA (1973).

procedures and the applicability of refrigeration and freezing to several wastewater characteristics.

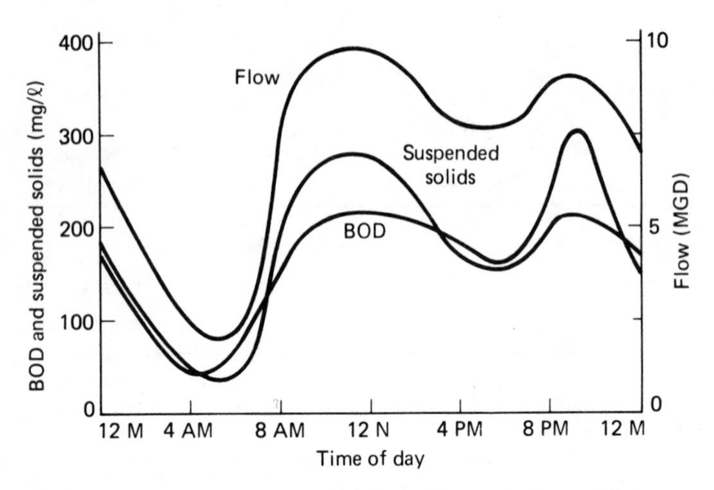

Figure 3-15. Typical Hourly Variations in Flow and Composition of Domestic Wastewater. (After Metcalf and Eddy, 1972.)

Example 3-12

The following samples were collected in the outfall sewer to a municipal treatment plant:

Sample number	*Flow (gpm)*
1	245
2	210
3	180
4	155
5	145
6	155
7	195
8	280
9	310
10	450
11	520
12	345
13	315
14	270
	Av. = 270

If a total composite volume of 2000 ml is required, how many milliliters of each sample is required?

solution

1/ Determine the portion of sample needed per unit of flow from equation 3-65.

$$\text{portion of sample needed per unit of flow} = \frac{2000}{270 \times 14} = 0.53 \text{ ml}$$

2/ Compute the total number of milliliters of each sample to be mixed.

Sample number	*Flow (gpm)*		*Amount (ml)*
1	245 × 0.53	=	130
2	210 × 0.53	=	111
3	180 × 0.53	=	95
4	155 × 0.53	=	82
5	145 × 0.53	=	77
6	155 × 0.53	=	82
7	195 × 0.53	=	103
8	280 × 0.53	=	148
9	310 × 0.53	=	164
10	450 × 0.53	=	239
11	520 × 0.53	=	276
12	345 × 0.53	=	183
13	315 × 0.53	=	167
14	270 × 0.53	=	143

Data Analysis

The purpose of a wastewater sampling program is to obtain sufficient data about the composition and flow of wastewater so that design values can be established. In obtaining data, the values for the various wastewater parameters will be observed to vary or fluctuate with time. The distribution of the values for each parameter can be examined for randomness. If data are randomly distributed, the data will produce a normally distributed (bell-shaped) curve. The data may also be skewed, since negative values are not possible or because a definite background value exists which is the minimum possible.

Probability plotting is an easy way to determine if certain sets of data fit a normal or log-normal distribution model. To use this method, the data of interest are generally plotted on both arithmetic and log probability paper, and then a determination is made as to whether a linear trace is given by either plot. If so, a normal or log-normal distribution, respectively, is followed. The probability plot can also be used to estimate the mean, median, and standard deviation of the data. The plots are particularly useful for obtaining mean values that are not distorted by a few excessively large or small values. The required procedure for probability plotting is as follows:

1/ Select the probability paper to be used—either arithmetic or logarithmic, or both.
2/ Arrange the data in increasing order of magnitude.
3/ Determine the plotting position of each value from the expression

$$\text{plotting position} = \frac{m - 0.5}{n}(100) \qquad (3\text{-}66)$$

where m = rank of value
n = number of observations

If the number of observations is very large (i.e., 50 to 100 or more), the data may be grouped by increments and the plotting position calculated from the formula

$$\text{plotting position} = \left(\frac{m}{n + 1}\right) 100 \qquad (3\text{-}66a)$$

With BOD, for example, the observations can be grouped by 50-mg/ℓ increments, and the midpoint of the increments used for plotting purposes. The following data would be typical when grouping is used:

Interval	*Number of samples in interval*	*m*	*Plotting position (%)*
200–249	5	5	5.2
250–299	4	9	9.3
.	.	.	.
.	.	.	.
.	.	.	.
1150–1199	4	96	99.0

4/ Trace the line of best fit. If the probability model chosen is correct, all except the extreme values should cluster around the trace. The probability scale represents percent equal to or less than values.
5/ When the data follow a normal distribution, the mean and median coincide. Thus, the 50 percentile value obtained from the probability paper represents both the mean and the median. This is not the typical case, however, and the purpose of the probability plot

is to define a reliable median that is the 50 percentile value for the line of best fit.

6/ For a normal distribution, the standard deviation is approximately equal to two-fifths of the difference between the 90 percentile value and the 10 percentile value. Actually, for a normal distribution, the standard deviation is equal to a value of plus or minus 34.13% from the mean or average value. This implies that 68.26% of all values fall within ±1 standard deviation from the mean.

Example 3-13

The raw wastewater flow from an industrial plant was monitored over a 19-day period of time, and the daily composite suspended solids (SS) concentration of the wastewater are reported in the following table:

Day	*SS (mg/ℓ)*
1	270
2	243
3	252
4	258
5	249
6	228
7	255
8	237
9	261
10	276
11	252
12	279
13	243
14	264
15	270
16	243
17	255
18	261
19	252

Determine the average suspended solids concentration and standard deviation by probability plotting.

solution

1/ Arrange the data in increasing order of magnitude and determine the plotting position of each value from equation 3-66.

SS in order of magnitude	*Rank of value*	*Plotting position*
228	1	2.6
237	2	7.9
243	3	13.2
243	4	18.4
243	5	23.7
249	6	29.0
250	7	34.2
252	8	39.5
252	9	44.7
255	10	50.0
255	11	55.3
258	12	60.5
261	13	65.8
261	14	71.0
264	15	76.3
270	16	81.6
270	17	86.8
276	18	92.1
279	19	97.4

2/ Using either arithmetic or logarithmic probability paper, trace the line of best fit. In this case the data follow an arithmetic normal distribution, as shown in Figure 3-16.

3/ From Figure 3-16 the mean suspended solids concentration of the wastewater is found to be 255 mg/ℓ.

4/ From Figure 3-16 the 10 percentile SS concentration is found to be 237 mg/ℓ, whereas the 90 percentile value is 272 mg/ℓ. The standard deviation is, therefore, approximately

$$\text{st. dev.} = \tfrac{2}{5}(272 - 237) = 14.0$$

Figures 3-17 through 3-20 present typical probability plots for the municipal wastewater parameters of BOD_5, BOD_5 loading, suspended solids, and flow, respectively. Note that the variations in BOD_5, BOD_5 loading, and flow follow a log-normal rather than an arithmetic normal distribution. Figure 3-20 also indicates that municipal wastewater flow can be divided into three separate groups, represented by the average flow for workday; the average flow for Sundays, and the wastewater flow, including the infiltration of rainfall.

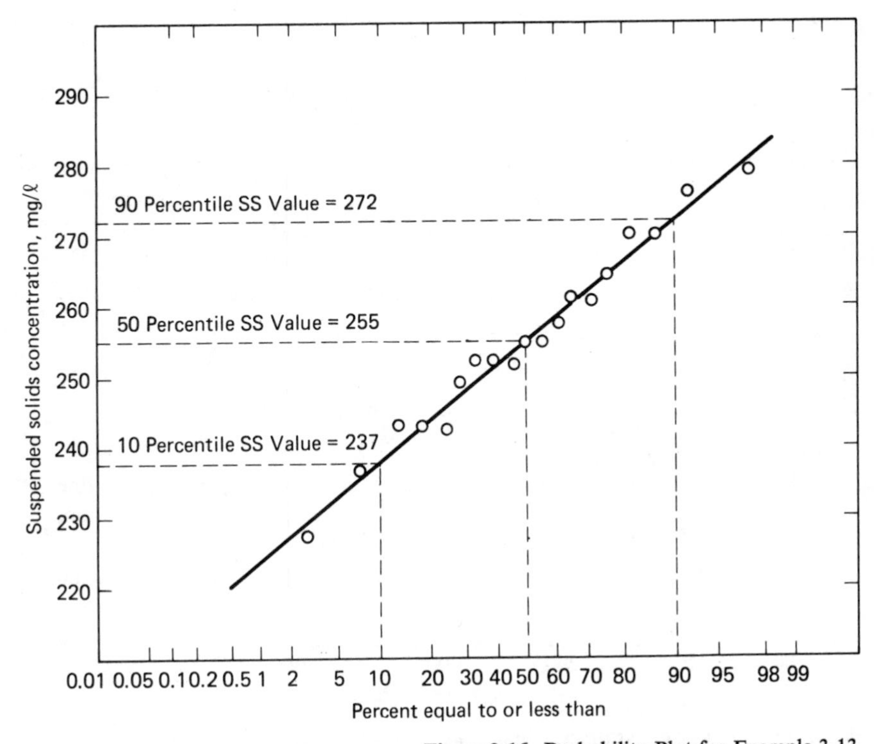

Figure 3-16. Probability Plot for Example 3-13.

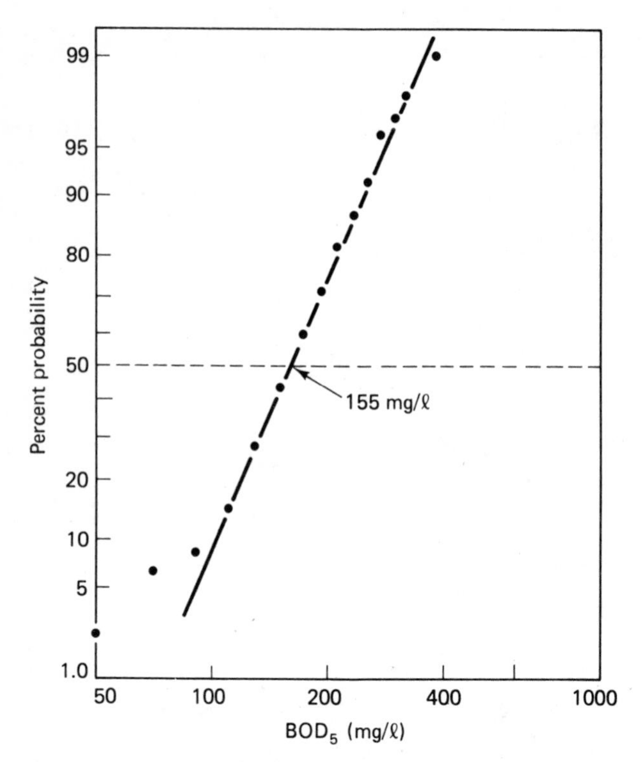

Figure 3-17. Probability of BOD_5 Influent. (After Malina et al., 1972.)

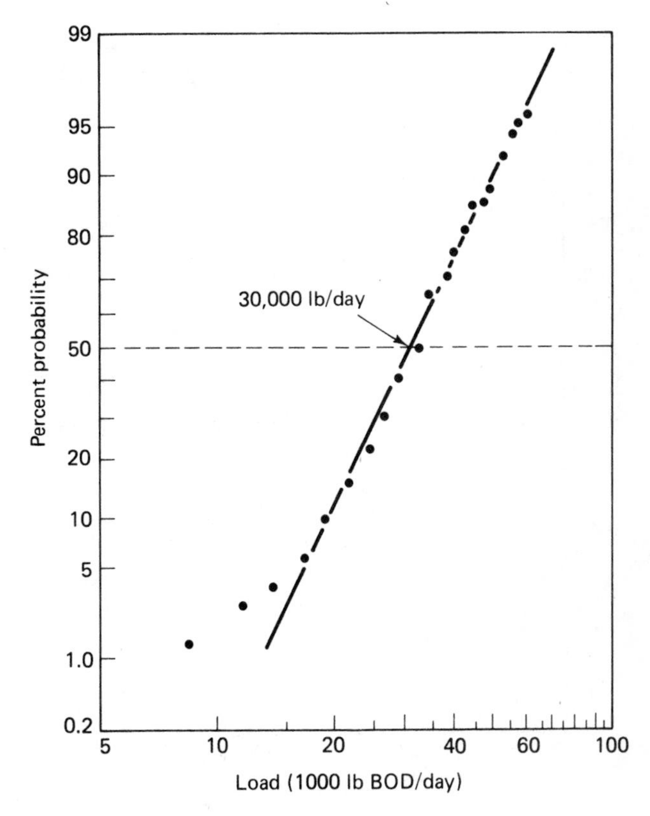

Figure 3-18. Probability of BOD_5 Loading. (After Malina et al., 1972.)

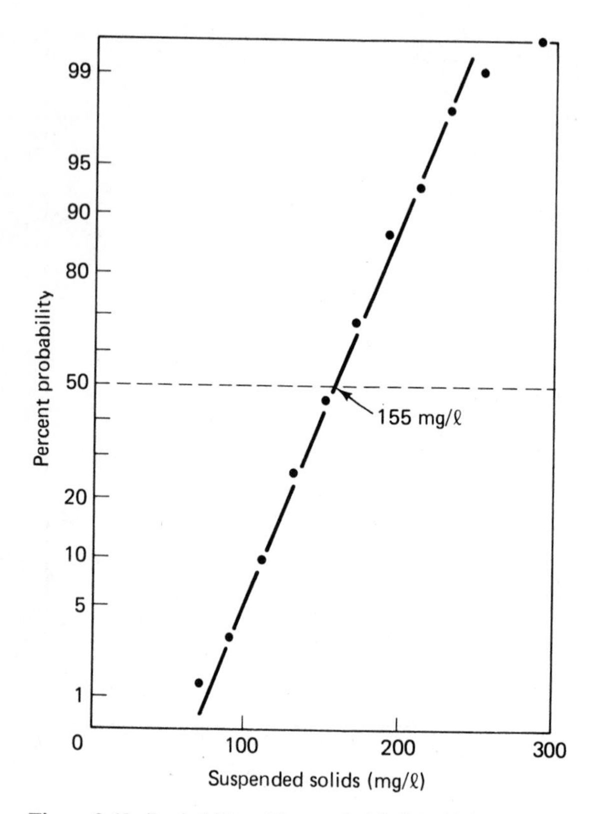

Figure 3-19. Probability of Suspended Solids. (After Malina et al., 1972.)

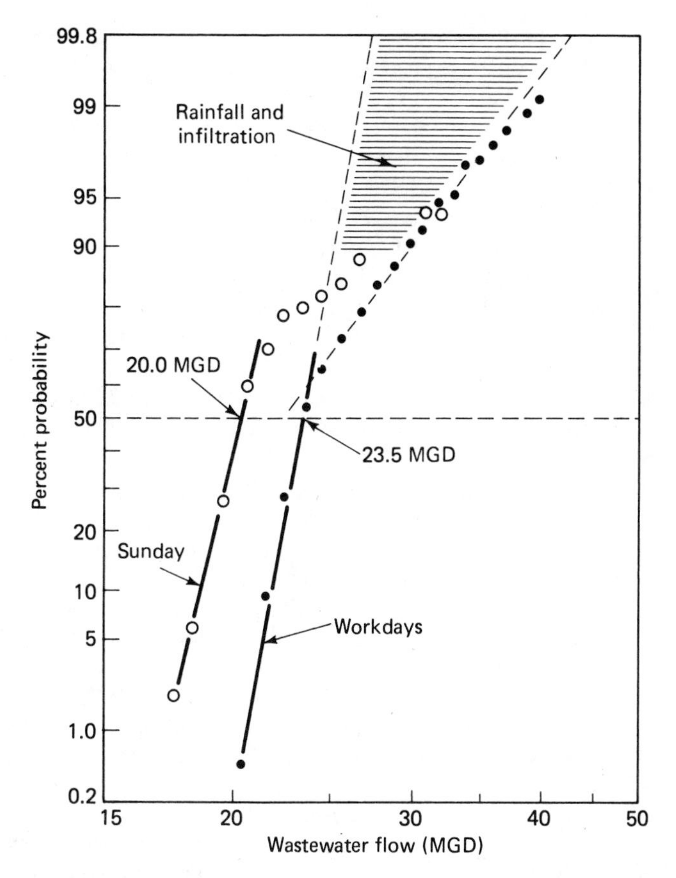

Figure 3-20. Probability of Wastewater Flow. (After Malina et al., 1972.)

PROBLEMS

3-1 The 5-day BOD of a municipal wastewater is 280 mg/ℓ. If the ultimate BOD is reported to be 350 mg/ℓ, what is the value of the BOD reaction-rate constant?

3-2 What percentage of the ultimate BOD of a wastewater has been satisfied after 3 days if the rate constant (base 10) is 0.10 day^{-1}?

3-3 The following data were determined in the laboratory by incubation at 20°C:

Time (days)	1	2	3	4	5	6	7	8	9	10
BOD (mg/ℓ)	4	9	18	28	50	64	70	71	73	78

Plot BOD versus time and determine if the data follow first-order kinetics; that is, does a lag period exist? If so, make the necessary adjustment by shifting the BOD axis and then determine K' and L_u.

3-4 Assuming that complete oxidation to carbon dioxide and water occurs, what is the theoretical chemical oxygen demand of the following compounds?
  a/ Phenol, C_6H_5OH.
  b/ Benzoic acid, C_6H_5COOH.
  c/ Ethanol, CH_3CH_2OH.
  d/ Glycerol, $CH_2OHCHOHCH_2OH$.

3-5 Given a liter of 0.2 M acetic acid solution at 25°C, how many grams of sodium acetate would have to be added to raise the pH to 4.0? Assume that no volume change occurs.

3-6 A buffer solution is 0.1 M in acetic acid and 0.1 M in sodium acetate (a 0.2 M acetate buffer). What is the pH after 4 ml of 0.05 *N* HCl is added to 10 ml of the buffer?

3-7 Each week an industry produces 2.0 MG of an acid waste (pH 1.5) during a 48-h production cycle. The plant engineer would like to discharge the waste into a river that passes near the plant. The river has a 7-day, 10-year low flow of 25 MGD, an alkalinity of 150 mg/ℓ as $CaCO_3$, and a pH of 7.7. If the temperature of the mixture is 25°C, will it be possible to consistently discharge all the waste into the river within a 5-day period if the river pH is to be maintained above 7.0?

3-8 A wastewater effluent containing 10 mg/ℓ of $NH_4^+ + NH_3$-N is discharged into a receiving stream containing no nitrogen in the reduced form. If the wastewater flow is 1 MGD and the critical stream flow is 10 MGD, what is the equilibrium concentration of NH_3-N in the mixture? Assume that the critical stream pH, as determined from 5 years of stream flow data, is 8.5 and the critical pH of the wastewater effluent is expected to be 8.0. The temperature of the wastewater is taken to be 30°C, and the temperature of the stream is 10°C. What is the maximum concentration of total ammonia that can be allowed in the effluent so that the standard of 0.02 mg/ℓ un-ionized ammonia will be met in the downstream?

REFERENCES

ALEXANDER, M., *Introduction to Soil Microbiology*, John Wiley & Sons, Inc., New York, 1961.

BRADLEY, R. M., "The Operating Efficiency of Circular Primary Sedimentation Tanks in Brazil and the United Kingdom," *The Public Health Engineer*, **13**, 5 (Jan. 1975).

BROWN and CALDWELL, Consulting Engineers, *Process Design Manual for Nitrogen Control*, EPA Technology Transfer, 1975.

CHIEN, J. S., AND J. D. JONES, "Pollutional Loadings Approach to Identifying I-I," *Water Pollution Control Federation Deeds and Data*, **12**, 1 (Oct. 1975).

Doetsch, R. N., and T. M. Cook, *Introduction to Bacteria and Their Ecobiology*, University Park Press, Baltimore, Md., 1973.

Eckenfelder, W. W., Jr., *Water Quality Engineering for Practicing Engineers*, Cahners Books, Boston, 1970.

Eckenfelder, W. W., Jr., and D. L. Ford, *Water Pollution Control*, Pemberton Press, Austin, Tex., 1970.

EPA, *Monitoring Industrial Wastewaters*. 1973.

Fair, G. M., J. C. Geyer, and D. A. Okun, *Elements of Water Supply and Wastewater Disposal*, 2nd ed., John Wiley & Sons, Inc., New York, 1971.

Fitzgerald, G. P., "Factors in the Testing and Application of Algicides," *Applied Microbiology*, **12**, 247 (1964).

Hammer, M. J., *Water and Waste-Water Technology*, John Wiley & Sons, Inc., New York, 1975.

Hoehn, R. C., R. P. Goode, C. W. Randall, and P. T. B. Shaffer, "Chlorination and Water Treatment for Minimizing Trihalomethanes in Drinking Water," in *Water Chlorination: Environmental Impact and Health Effects*, Vol. II, Ann Arbor Science Publishers, Inc., Ann Arbor, Mich., 1978, pp. 519–535.

Hurwitz, E. R., R. Beaudoim, and W. Walters, "Phosphates—Their 'Fate' in a Sewage Treatment Plant–Waterway System," *Water and Sewage Works*, **112**, 84 (1965).

Lehninger, A. L., *Biochemistry*, Worth Publishers, Inc., New York, 1970.

Loewenthal, R. E., and G. V. R. Marais, *Carbonate Chemistry of Aquatic Systems: Theory and Application*, Ann Arbor Science Publishers, Inc., Ann Arbor, Mich., 1976.

Malina, J. F., et al., "Design Guides for Biological Wastewater Treatment Processes," Center for Research in Water Resources Report 76, University of Texas, Austin, Tex., 1972.

McCarty, P. L., "Phosphorus and Nitrogen Removal by Biological Systems," in *Proceedings, Wastewater Reclamation and Reuse Workshop*, Lake Tahoe, Calif., June 25–27, 1970.

McCarty, P. L., L. Beck, and P. St. Amant, "Biological Denitrification of Wastewaters by Addition of Organic Materials," in *Proceedings, 24th Purdue Industrial Waste Conference*, West Lafayette, Ind., May 6, 1969.

McKinney, R. E., "Design and Operation of Complete Mixing Activated Sludge Systems," Environmental Protection Control Services Report, Vol. 1, No. 3, Lawrence, Kan., July 1970.

Metcalf and Eddy, Inc., *Wastewater Engineering*, McGraw-Hill Book Company, New York, 1972.

Nay, M. W., "A Biodegradable and Treatability Study of TNT Manufacturing Wastes with Activated Sludge Systems," Dissertation, Virginia Polytechnic Institute and State University, Blacksburg, Va., Dec. 1971.

Painter, H. A., "A Review of Literature on Inorganic Nitrogen Metabolism in Microorganisms," *Water Research*, **4**, 393 (1970).

Parker, H. W., *Wastewater Systems Engineering*. Prentice-Hall, Inc., Englewood Cliffs, N.J., 1975.

RANDALL, C. W., AND R. A. LAUDERDALE, "Biodegradation of Malathion," *Journal of the Sanitary Engineering Division, ASCE*, **93**, 145 (Dec. 1967).

SAWYER, C. N., AND P. L. MCCARTY, *Chemistry for Sanitary Engineers*, 2nd ed., McGraw-Hill Book Company, New York, 1967.

SCHROEDER, E. D., *Water and Wastewater Treatment*, McGraw-Hill Book Company, New York, 1977.

THOMPSON, B. C., "Trihalomethane Formation Potential of Algal Extracellular Products and Biomass," Master's thesis, Virginia Polytechnic Institute and State University, Blacksburg, Va., Mar. 1978.

TOERIEN, D. F., "South African Eutrophication Problems: A Perspective," *British Journal of Water Pollution Control*, **74**, 134 (1975).

WEBER, W. J., JR., AND W. STUMM, "Mechanism of Hydrogen Ion Buffering in Natural Waters," *Journal of the American Water Works Association*, **55**, 1553 (1963).

WHITE, G. C., *Handbook of Chlorination*, Van Nostrand Reinhold Company, New York, 1972.

Activated Sludge and Its Process Modifications

The *activated sludge process* is capable of converting most organic wastes to more stable inorganic forms or to cellular mass. In this process much of the soluble and colloidal organic material remaining after primary sedimentation is metabolized by a diverse group of microorganisms to carbon dioxide and water. At the same time, a sizable fraction is converted to a cellular mass that can be separated from the waste flow by gravity sedimentation.

Activated sludge is a heterogenous microbial culture composed mostly of bacteria, protozoa, rotifers, and fungi. However, it is the bacteria which are responsible for assimilating most of the organic material, whereas the protozoa and rotifers are important in removing the dispersed bacteria that otherwise would escape in the plant effluent.

The utilization of substrate (organic material) by a bacterial cell can be described as a three-step process: (1) the substrate molecule contacts the cell wall, (2) the substrate molecule is transported into the cell, and (3) metabolism of the substrate molecule by the cell takes place. However, as bacteria require substrate in the soluble form, colloidal or sterically incompatible molecules, which cannot be readily transported into a cell, have to be first adsorbed to the cell surface and then broken down or transformed externally to transportable fractions by exoenzymes or wall-bound enzymes.

To produce a high-quality effluent, the biomass (after removing the organic material from the wastewater) must be separated from the liquid stream. This is accomplished in the secondary clarifier and is effective only if the microbial species present readily agglomerate. Secondary clarification is nearly always the effluent quality-limiting step. The soluble BOD_5 of the effluent is generally below 5 mg/ℓ, but biomass solids carryover may produce an effluent BOD_5 of 20 mg/ℓ or greater.

Many studies have been presented which purport to describe the mechanism of biological flocculation, but the work of Pavoni (1972) is the most comprehensive. In this work biological flocculation was found to be governed by the physiological state of the microorganisms and did not occur until the microorganisms entered into the substrate depleted or endogenous growth phase. Biological flocculation is proposed to result from the interaction of exocellular polymers which accumulate at the cell surface during endogenous growth. Cells are bridged into three-dimensional matrices as a result of physical and electrostatic bonding of these polymers to the surface of the cells.

After separating the liquid phase from the solid phase, the biomass increase resulting from synthesis during substrate utilization is wasted and the remainder returned to the aeration tank. Thus, a relatively constant mass of microorganisms is maintained in the system, and performance of the process depends on the recycle of sufficient biomass. If biomass separation and concentration fails, the entire process fails.

The flow scheme for a typical activated sludge plant is presented in Figure 4-1. In general, this process may be considered to be one that involves (1) wastewater aeration in the presence of a microbial suspension, (2) solids–liquid separation following aeration, (3) discharge of a clarified effluent, and

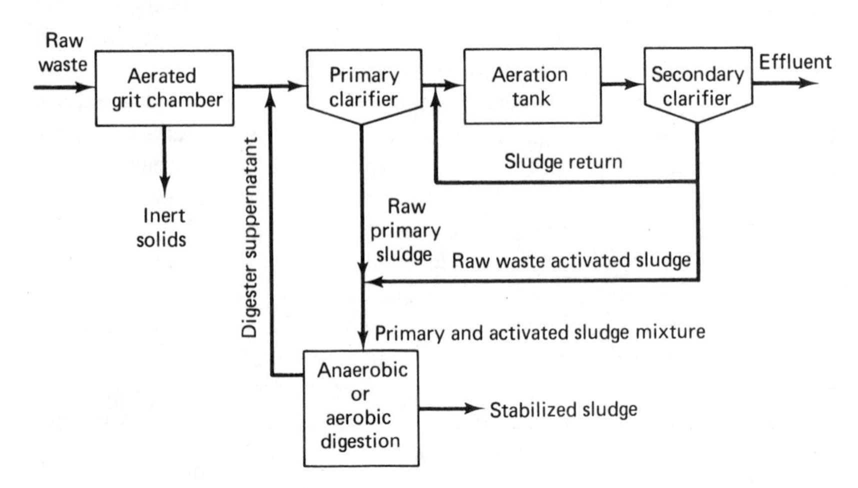

Figure 4-1. Typical Flow Diagram for an Activated Sludge Treatment Plant.

(4) wasting of excess biomass and return of the remaining biomass to the aeration tank (Eckenfelder et al., 1972).

4-1
Mixing Regime

One of the fundamental requirements in the design of an activated sludge process is to know what type of reactor (aeration tank) is best for a given problem. In this regard, actual reactor geometry is important because this determines the path of the liquid through the tank and fixes the gross mixing patterns.

In general, two types of mixing regimes are of major interest in the activated sludge process. The first is *plug flow.* As noted in Chapter 1, this type of mixing regime is characterized by orderly flow of mixed liquor through the aeration tank with no element of mixed liquor overtaking or mixing with any other element. There may be lateral mixing of mixed liquor, but, to conform with the definition, there must be no mixing or diffusion along the path of flow. In true plug flow, the retention time in the aeration tank is the same for all elements of mixed liquor.

The second type of mixing regime discussed in Chapter 1 was *complete mixing*. In this case the aeration tank contents are well stirred and uniform throughout. Thus, at steady-state, the effluent from the aeration tank has the same composition as the aeration tank contents.

It is recognized that true plug-flow or complete mixing conditions seldom occur during actual operation. However, if the system is properly designed the approximation will be sufficient for design and operation purposes.

The type of mixing regime is very important, as it affects (1) the oxygen transfer requirements in the aeration tank, (2) the susceptibility of the biomass to shock loads, (3) local environmental conditions (such as temperature) in the aeration tank, and (4) the kinetics governing the treatment process (Metcalf and Eddy, 1972). Ramifications of these factors will be discussed later in a section that will deal with activated sludge process modifications and considerations for their individual selection.

4-2
Kinetic Model Development

With few exceptions, kinetic models which have been proposed to describe the activated sludge process have been developed on the basis of steady-state conditions within the treatment system. In the discussion to follow, the more commonly used of these kinetic models are presented and the assumptions required in the development of each model are outlined. Such an approach is necessary if one is to understand the limitations of each model when used for plant design.

A typical flow scheme for a completely mixed activated sludge process is shown in Figure 4-2. In this figure, Q represents the rate of raw wastewater flow to the aeration tank; S_0, the substrate concentration in the raw wastewater; V_a, the volume of the aeration tank; X, the biomass concentration in the aeration tank and the effluent from the aeration tank; S_e, the steady-state substrate concentration after treatment; Q_r, the rate of sludge recycle; R, the sludge recycle ratio (i.e., Q_r/Q); Q_w, the rate of sludge wasting; and X_r; biomass concentration in the underflow from the secondary clarifier.

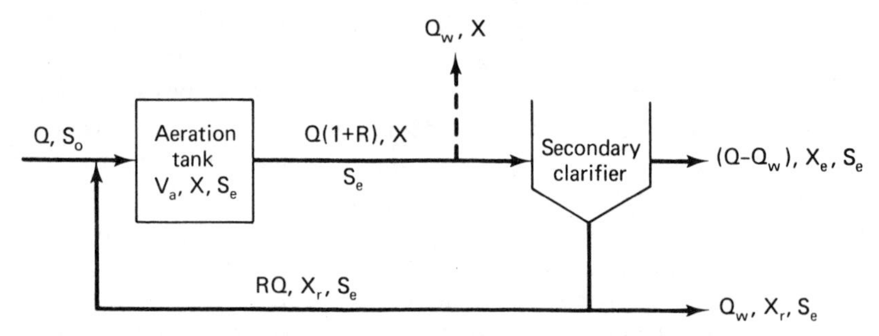

Figure 4-2. Typical Flow Scheme for a Completely Mixed Activated Sludge Plant.

Also, in Figure 4-2, two alternatives for *sludge wasting* are shown. Sludge wasting can be accomplished either from the sludge return line or directly from the aeration tank. Although sludge wasting from the sludge return line is traditional, sludge wasting from the aeration tank is more desirable, as this method offers an opportunity for better plant control and is also beneficial to subsequent sludge thickening operations, as it has been shown that higher solids concentrations can be achieved when dilute mixed liquor rather than concentrated return sludge is thickened (McCarty, 1966b). Thus, model development in this work will be based on the scheme of sludge wasting from the aeration tank.

Material-balance equations are derived from Figure 4-2. The development of the appropriate kinetic models from these equations is based on the following assumptions:

1/ Complete mixing is achieved in the aeration tank.
2/ Influent substrate concentration remains constant.
3/ No microbial solids are contained in the raw wastewater to the aeration tank.
4/ No microbial activity occurs in the secondary clarifier.

5/ No sludge accumulates in the secondary clarifier and a reasonable efficiency of solids–liquid separation is accomplished.
6/ All biodegradable substrate is in the soluble form.
7/ Steady-state conditions prevail throughout the system.

In their work, Lawrence and McCarty (1970) emphasize the importance of the operational parameter called *biological solids retention time* (BSRT), symbolized by θ_c, which is defined as the average time a unit of biomass remains in the treatment system, or

$$\theta_c \equiv \frac{(X)_T}{(\Delta X/\Delta t)_T} \tag{4-1}$$

where $(X)_T$ = total active biomass in treatment system, mass

$\left(\frac{\Delta X}{\Delta t}\right)_T$ = total quantity of active biomass withdrawn from the system daily, mass time^{-1}; this includes solids purposely wasted plus those lost in the effluent

For a completely mixed activated sludge process with solids recycle, following the assumptions previously outlined, equation 4-1 can be written as

$$\theta_c = \frac{XV_a}{Q_wX + (Q - Q_w)X_e} \tag{4-2}$$

Furthermore, for steady-state conditions, equations 2-11 and 4-1 suggest that

$$\theta_c = \frac{1}{\mu} \tag{4-3}$$

Hence, the importance of θ_c as a control parameter is apparent: by controlling θ_c one controls the specific growth rate and thus the physiological state of the organisms in the system.

A material-balance equation for biomass around the entire treatment system in Figure 4-2 gives

$$\begin{bmatrix}\text{net rate of change}\\ \text{in amount of biomass}\\ \text{within system}\end{bmatrix} = \begin{bmatrix}\text{rate at which}\\ \text{biomass appears}\\ \text{in system}\end{bmatrix} - \begin{bmatrix}\text{rate at which}\\ \text{biomass leaves}\\ \text{system}\end{bmatrix} \tag{4-4}$$

Since biomass appears in the system as a result of growth and leaves as a result of hydraulic action in both the sludge wastage stream and the effluent, equation 4-4 can be expressed as

$$\left(\frac{dX}{dt}\right)V_a = \left(\frac{dX}{dt}\right)_g V_a - [Q_wX + (Q - Q_w)X_e] \tag{4-5}$$

The V_a term appears in equation 4-5 because the biomass, X, is expressed as concentration. Multiplying by volume puts the rate term on a mass per unit time basis.

Substituting from equation 2-51 for $(dX/dt)_g$ and from equation 4-2 for $Q_wX + (Q - Q_w)X_e$ reduces equation 4-5 to

$$\left(\frac{dX}{dt}\right)V_a = \left[Y_T\left(\frac{dS}{dt}\right)_u - K_dX\right]V_a - \frac{XV_a}{\theta_c} \tag{4-6}$$

At steady-state,

$$\begin{bmatrix}\text{rate at which}\\ \text{biomass appears}\\ \text{in system}\end{bmatrix} = \begin{bmatrix}\text{rate at which}\\ \text{biomass leaves}\\ \text{system}\end{bmatrix}$$

which implies that

$$\left(\frac{dX}{dt}\right)V_a = 0$$

Thus, at steady-state, equation 4-6 can be written as

$$0 = \left[Y_T\left(\frac{dS}{dt}\right)_u - K_dX\right]V_a - \frac{XV_a}{\theta_c} \tag{4-7}$$

or

$$\frac{1}{\theta_c} = Y_T\frac{(dS/dt)_u}{X} - K_d \tag{4-8}$$

An expression for S_e can be obtained by substituting for $(dS/dt)_u$ from equation 2-54. This substitution into equation 4-8 gives

$$\frac{1}{\theta_c} = Y_T\frac{kS_e}{K_S + S_e} - K_d \tag{4-9}$$

or

$$\frac{1}{\theta_c} + K_d = \frac{Y_TkS_e}{K_S + S_e} \tag{4-10}$$

Expanding this expression,

$$\frac{K_S}{\theta_c} + K_SK_d + \frac{S_e}{\theta_c} + S_eK_d = Y_TkS_e \tag{4-11}$$

or

$$\frac{K_S(1/\theta_c + K_d)}{Y_Tk - (1/\theta_c + K_d)} = S_e \tag{4-12}$$

Multiplying the left side of this expression by θ_c/θ_c gives

$$\frac{K_S(1 + K_d\theta_c)}{Y_Tk\theta_c - (1 + K_d\theta_c)} = S_e \tag{4-13}$$

or

$$\frac{K_S(1 + K_d\theta_c)}{\theta_c(Y_Tk - K_d) - 1} = S_e \tag{4-14}$$

A material balance for substrate entering and leaving the aeration tank can be written as

$$\begin{bmatrix}\text{net rate of change} \\ \text{in amount of substrate} \\ \text{within aeration tank}\end{bmatrix} = \begin{bmatrix}\text{rate at which} \\ \text{substrate enters} \\ \text{aeration tank}\end{bmatrix} - \begin{bmatrix}\text{rate at which} \\ \text{substrate dis-} \\ \text{appears from} \\ \text{aeration tank}\end{bmatrix} \tag{4-15}$$

which can be expressed as

$$\left(\frac{dS}{dt}\right) V_a = QS_0 + RQS_e - \left(\frac{dS}{dt}\right)_u V_a - (1 + R)QS_e \tag{4-16}$$

Equation 4-16 implies that substrate is removed from the aeration tank through microbial utilization as well as by hydraulic action.

At steady-state,

$$\begin{bmatrix}\text{rate at which substrate} \\ \text{enters aeration tank}\end{bmatrix} = \begin{bmatrix}\text{rate at which substrate} \\ \text{disappears from aeration} \\ \text{tank}\end{bmatrix}$$

This implies that

$$\left(\frac{dS}{dt}\right) V_a = 0$$

Therefore, at steady-state, equation 4-16 reduces to

$$\left(\frac{dS}{dt}\right)_u = \frac{Q(S_0 - S_e)}{V_a} \tag{4-17}$$

Multiplying both sides by $1/X$, equation 4-17 becomes

$$\frac{(dS/dt)_u}{X} = \frac{Q(S_0 - S_e)}{XV_a} \tag{4-18}$$

Substituting from equation 4-18 for $(dS/dt)_u/X$ in equation 4-8 and solving for X,

$$X = \frac{Y_T Q(S_0 - S_e)}{V_a(1/\theta_c + K_d)} \tag{4-19}$$

Multiplying the right side of equation *4-19* by θ_c/θ_c gives

$$X = \frac{\theta_c Y_T Q(S_0 - S_e)}{V_a(1 + K_d\theta_c)} \tag{4-20}$$

Equations 4-8, 4-14, and 4-20 were developed by Lawrence and McCarty (1970) and have been widely accepted by the environmental engineering community. However, if equation 2-56 is substituted for $(dS/dt)_u$ in equation 4-8, the resulting expression is

$$\frac{1}{\theta_c} = Y_T K S_e - K_d \tag{4-21}$$

or

$$S_e = \frac{1/\theta_c + K_d}{Y_T K} \tag{4-22}$$

Multiplying the right side of this expression by θ_c/θ_c gives an equation for S_e which has the form

$$S_e = \frac{1 + K_d\theta_c}{Y_T K\theta_c} \tag{4-23}$$

Another popular relationship for S_e can be developed by substituting KXS_e from equation 2-56 for $(dS/dt)_u$ in equation 4-17:

$$\frac{Q(S_0 - S_e)}{V_a} = KXS_e \tag{4-24}$$

or

$$\frac{Q(S_0 - S_e)}{XV_a} = q = KS_e \tag{4-25}$$

where q represents the specific substrate utilization rate and has the dimension time^{-1}. This expression will give a straight line of slope K when $Q(S_0 - S_e)/XV_a$ versus S_e (measured as BOD_u) is plotted as shown in Figure 4-3. As previously mentioned, in this text substrate concentration will be measured as either BOD_u or degradable COD.

Equation 4-25 may also be rearranged into the form

$$\frac{S_e}{S_0} = \frac{1}{1 + KX(V_a/Q)} \tag{4-26}$$

Equations 4-25 and 4-26 result when substrate utilization is assumed to follow first-order kinetics. However, steady-state biomass concentration is independent of the substrate utilization-rate relationship. Thus, equation 4-20 describes X regardless of whether $(dS/dt)_u$ is described by the continuous hyperbolic relationship or the first-order approximation of this relationship.

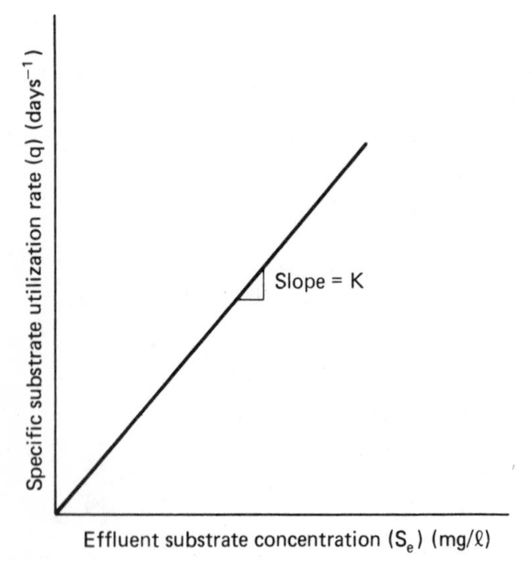

Figure 4-3. Plot of Specific Substrate Utilization Rate Versus Effluent Substrate Concentration.

A relationship between the recycle ratio, R, and θ_c can be developed from a material-balance equation for biomass entering and leaving the aeration tank. From Figure 4-2, such a mass balance can be written as

$$\begin{bmatrix}\text{net rate of change}\\ \text{in amount of biomass}\\ \text{within aeration tank}\end{bmatrix} = \begin{bmatrix}\text{rate at which}\\ \text{biomass appears}\\ \text{in aeration tank}\end{bmatrix} - \begin{bmatrix}\text{rate at which}\\ \text{biomass leaves}\\ \text{aeration tank}\end{bmatrix} \qquad (4\text{-}27)$$

or

$$\left(\frac{dX}{dt}\right)V_a = RQX_r + \left(\frac{dX}{dt}\right)_g V_a - Q(1+R)X \qquad (4\text{-}28)$$

Substituting for $(dX/dt)_g$ from equation 2-59 gives

$$\left(\frac{dX}{dt}\right)V_a = RQX_r + \left[Y_T\left(\frac{dS}{dt}\right)_u - K_dX\right]V_a - QX - RQX \qquad (4\text{-}29)$$

At steady-state

$$\begin{bmatrix}\text{rate at which}\\ \text{biomass appears}\\ \text{in aeration tank}\end{bmatrix} = \begin{bmatrix}\text{rate at which}\\ \text{biomass leaves}\\ \text{aeration tank}\end{bmatrix}$$

This implies that

$$\left(\frac{dX}{dt}\right)V_a = 0$$

Thus, assuming steady-state conditions and substituting for $(dS/dt)_u$ from equation 2-56, equation 4-29 reduces to

$$0 = RQX_r + (Y_TKXS_e - K_dX)V_a - QX - RQX \qquad (4\text{-}30)$$

A further substitution for S_e from equation 4-23 gives

$$0 = RQX_r + \left[Y_TKX\left(\frac{1+K_d\theta_c}{Y_TK\theta_c}\right) - K_dX\right]V_a - QX - RQX \qquad (4\text{-}31)$$

Rearranging equation 4-31 and solving for θ_c,

$$\frac{1}{\theta_c} = \frac{Q}{V_a}\left(1 + R - R\frac{X_r}{X}\right) \qquad (4\text{-}32)$$

Equation 4-32 shows θ_c to be a function of the ratio X_r/X and the recycle ratio, R. The ratio X_r/X is a function of the settling characteristics of the biomass and of the efficiency of the secondary clarifier.

When secondary clarifiers are operating properly, solids capture should approach 100%. For this situation the maximum solids concentration in the sludge return line can be estimated from

$$(X_r)_{\max} = \frac{10^6}{\text{SVI}} \qquad (4\text{-}33)$$

where SVI represents sludge volume index. An important point to remember is that X_r values calculated on the basis of SVI are in terms of total sus-

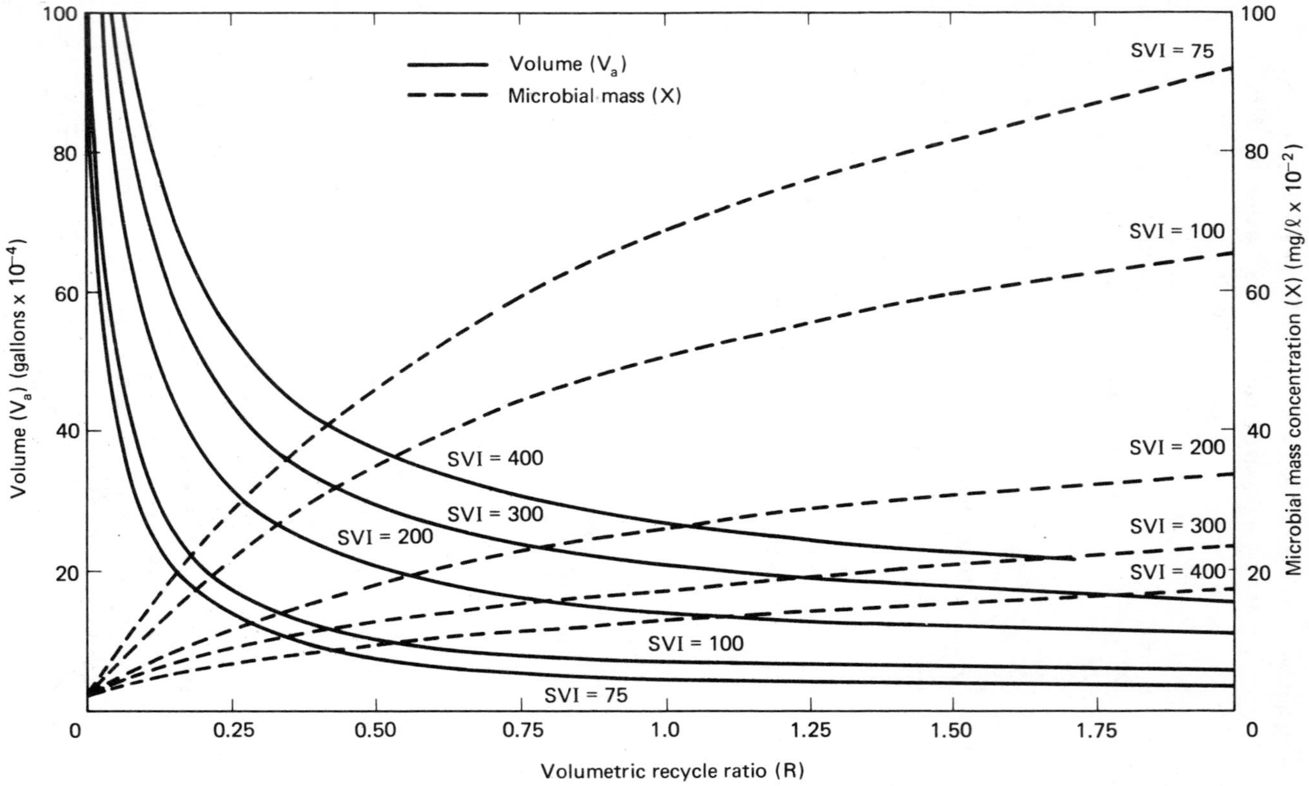

Figure 4-4. Relationships Between Volumetric Recycle Ratio and Reactor Volume and Microbial Mass Concentration for Various Values of SVI. (After Lawrence and McCarty, 1970.)

pended solids. These values should be converted to volatile suspended solids before X_r is used in equation 4-32.

A more accurate determination of X_r can be made using the flux plot method and equation 4-117, as discussed later in this chapter, and that technique should be used if sufficient data are available. If it is not available, however, equation 4-33 yields a usable first estimate. It should be understood that, although equation 4-33 will be used in subsequent examples, more accurate values of X_r should be used when possible.

Once the desired operating θ_c and treatment efficiency are determined, the total weight of biomass in the aeration tank (XV_a) can be computed from equation 4-20 for given values of Q, S_0, k, K_S, Y_T, and K_d. Equation 4-32 can then be used to calculate the steady-state biomass concentration and the aeration tank volume for assumed values of R and X_r. Figure 4-4 illustrates the relationship among X, V_a, R, and X_r for a specific set of operating conditions. In this figure, X_r values were estimated from equation 4-33.

Using equation 4-14 or equation 4-23 and equation 4-20, and knowing the values of the constants k, K_S, Y_T, and K_d for a specified wastewater, biological culture and a particular set of environmental conditions, the steady-state concentration of biomass and substrate in the aeration tank can be predicted for any value of BSRT, aeration tank volume, and raw wastewater substrate concentration. Figure 4-5 illustrates the effect of influent substrate concentra-

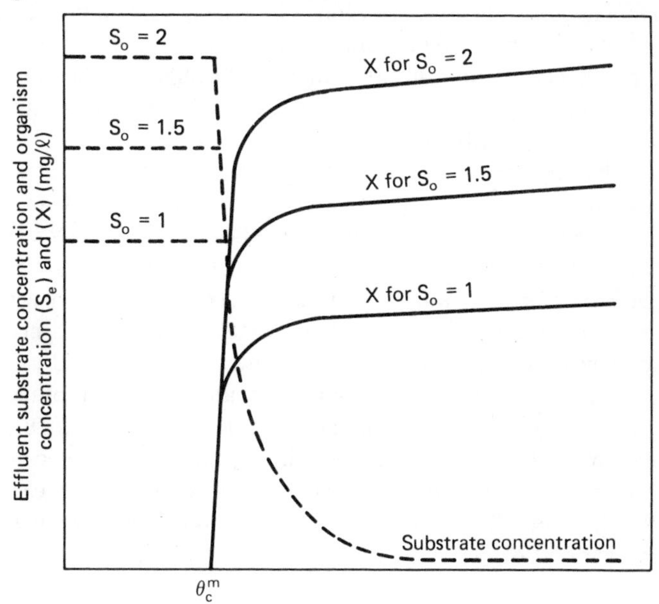

Figure 4-5. Effect of Influent Substrate Concentration on a Completely Mixed Activated Sludge Process. (After Andrews, 1971.)

tion on a completely mixed activated sludge process. From this figure it can be seen that there is a certain value of θ_c below which no substrate is removed. This value is called the *minimum biological solids retention time* and is designated as θ_c^m. This is the BSRT at which biomass is removed from the system faster than it is being produced. Therefore, if a process is operating at a BSRT below this minimum value, washout will occur (i.e., all the biomass will be lost from the system). After washout has occurred, the effluent substrate concentration will be the same as the influent substrate concentration, as no microorganisms will be present to utilize the organic material. Hence, for these conditions, assuming that the rate of substrate utilization follows the relationship given by equation 2-54, it is possible to develop an expression for θ_c^m which has the form

$$\frac{1}{\theta_c^m} = Y_T \frac{kS_0}{K_S + S_0} - K_d \tag{4-34}$$

However, if it is assumed that the rate of substrate utilization follows first-order kinetics, the expression for θ_c^m becomes

$$\frac{1}{\theta_c^m} = Y_T K S_0 - K_d \tag{4-35}$$

Figure 4-5 also suggests that for the kinetic models previously developed, effluent substrate concentration is independent of influent substrate concentration. As can be seen from this figure, for a given value of θ_c, higher substrate concentrations in the influent will result in higher steady-state biomass concentrations in the aeration tank while the effluent substrate concentration remains essentially unchanged.

The independence between effluent substrate concentration and influent substrate concentration, as implied by equations 4-14 and 4-23, is of great importance in the design and control of wastewater treatment plants. For example, when θ_c is used for plant control, there is no requirement to evaluate the *mixed liquor volatile suspended solids* (MLVSS), which is generally taken to be a measure of the biomass in the system, nor is there a need to monitor the influent and effluent substrate concentration (measured as BOD_5, COD, or TOC). As long as θ_c is held constant, any change in influent substrate concentration will result only in a change in the steady-state biomass concentration while the effluent quality will remain constant. Because of this, it has been proposed that θ_c can be controlled by hydraulic means (Walker, 1971; Burchett and Tchobanoglous, 1974). Furthermore, because of its simplicity, the method of hydraulic control has become popular in actual plant operation.

The independence between influent and effluent substrate concentration also suggests that when determining the kinetic coefficients required for process design, it is not necessary to use a wastewater of the same concentration expected under field conditions (Grady and Williams, 1975).

At this point it should be noted that no one has been able to prove the superiority of either equation 2-54 or equation 2-56 for describing the actual rate of substrate utilization in the activated sludge process. However, in plant design, equation 2-56 offers two advantages: (1) it is a mathematical convenience which makes analysis easier, and (2) it does not require that accurate values for the biokinetic coefficients k and K_S be determined, a task which in most cases is quite difficult. It should also be stressed that the kinetic models thus far developed have been based on several assumptions. It is, therefore, worthwhile to discuss the effects of these assumptions on the validity of the models.

Example 4-1

A completely mixed activated sludge process is to be used to treat a wastewater flow of 1 MGD having a BOD_u of 200 mg/ℓ. Design criteria are as follows:

$$X = 2000 \text{ mg/}\ell \text{ as MLVSS}$$

$$S_e = 10 \text{ mg/}\ell \text{ BOD}_u$$

$$R = 30 \text{ to } 40\% \text{ of wastewater flow}$$

$$Y_T = 0.5$$

$$K = 0.1 \ \ell\text{/mg-day}$$

$$K_d = 0.1 \text{ day}^{-1}$$

$$\text{MLVSS} = 0.8 \text{ MLSS}$$

Compute the required aeration tank volume and operating BSRT. Determine the effects on process efficiency when the SVI changes from 80 to 160 and no adjustment is made in operating BSRT.

solution

1/ From equation 4-25, calculate the required aeration tank volume.

$$q = KS_e = (0.1)(10) = 1.0 \text{ day}^{-1}$$

Furthermore,

$$q = \frac{Q(S_0 - S_e)}{XV_a}$$

Therefore,

$$V_a = \frac{1(200 - 10)}{(2000)(1.0)} = 0.095 \text{ MG}$$

2/ Compute the operating BSRT from equation 4-21.

$$\frac{1}{\theta_c} = (0.5)(1.0) - 0.1$$

or

$$\theta_c = 2.5 \text{ days}$$

3/ Determine the effect of changing SVI on process efficiency by first applying equation 4-32 to obtain the variation in X with recycle and X_r for a θ_c of 2.5 days. These data are presented in the following tabulation:

SVI	*R*	X_r *(mg/ℓ)*	*X (mg/ℓ)*
80	0.3	12,500 × 0.8 = 10,000	2342
80	0.4	12,500 × 0.8 = 10,000	2897
160	0.3	6250 × 0.8 = 5,000	1171
160	0.4	6250 × 0.8 = 5,000	1448

With the new steady-state biomass concentration, the actual effluent substrate concentration for each operating condition can be calculated from a modification of equation 4-25, where q/K is substituted for S_e in the basic formulation.

$$q = \frac{Q[S_0 - (q/K)]}{XV_a}$$

or

$$q = \frac{1[200 - (q/0.1)]}{(2342)(0.095)} = 0.86 \text{ day}^{-1}$$

Similar calculations using the remaining data yield the following S_e values:

X (mg/ℓ)	S_e *(mg/ℓ)*
2342	8.6
2897	7.0
1171	16.5
1448	13.6

Thus, unless the operating BSRT is adjusted when the SVI changes from 80 to 160, the desired effluent quality will be exceeded.

It was first assumed that complete mixing would be achieved throughout the aeration tank. However, if plug flow is the type of mixing regime established in the aeration tank, the kinetic models developed from the material-balance expressions will differ considerably from those previously developed.

For an aeration tank in which plug flow describes the mixing regime, the substrate concentration will decrease and the biomass concentration will increase as the waste stream passes through the tank. Therefore, since steady state is not established within the reactor, it is very difficult to develop mathematical models that adequately describe treatment kinetics when this type of mixing exists. However, in Chapter 1 it was noted that in a CFSTR, the initial high driving force (substrate concentration) was instantaneously reduced to the final low driving force present at the reactor exit. As a consequence, the reaction time required to achieve a desired removal efficiency would be greater for a CFSTR than for a PF reactor. That is to say, for a fixed volume, the efficiency of the PF reactor is greater than that for a CFSTR. Lawrence and McCarty (1970) have compared the effluent substrate concentration and treatment efficiency for both completely mixed and plug-flow mixing regimes. This comparison is presented in Figure 4-6 and shows the plug-flow system to be more efficient than the completely mixed system. Since there is evidence to indicate that ideal plug-flow conditions do not actually exist in practice but rather that conditions in a plug-flow tank actually approach complete mixing (Grieves et al., 1964; Milbury et al., 1965), kinetic models developed where complete mixing is assumed can be used for the design of plug-flow systems. Such a design procedure is conservative and any deviation toward plug flow under actual operating conditions may result in a treatment efficiency higher than that predicted in design. This is not to say, however, that all aeration tanks should be designed for complete mixing and then actually constructed to produce plug flow. Each type of mixing regime has certain advantages which makes it desirable under specific circumstances.

In assumption 2 it was stated that the influent substrate concentration would remain at some constant value with no variation. This is a valid assumption when designing activated sludge plants for some industrial processes or when equalization basins are to be provided to dampen fluctuations in waste strength. However, when designing a plant to treat a wastewater with a substrate concentration that is variable, experimental evidence indicates that equation 2-54 or 2-56 no longer accurately describes the rate of substrate utilization.

Pure culture work has shown that single substrates are transported into a cell according to the relationship described by equation 2-54, and work by Riesing (1971) under such conditions shows that effluent substrate concentration is independent of influent substrate concentration. However, Grau et al. (1975) and Grady and Williams (1975) have recently presented data which indicate that when mixed microbial cultures are grown on substrates of multicomponent mixtures, the influent and effluent substrate concentrations are no longer independent. Since such experimental conditions in many cases more closely describe the complex systems of interest to the design engineer, these results will have a significant effect on current design and control procedures.

Figure 4-7 shows the results of the study by Grady and Williams (1975) fitted to the first-order substrate utilization-rate relationship given by equation 4-25, which is only a special case of the Monod-type relationship used by Lawrence and McCarty (1970). This figure shows that the specific substrate utilization-rate constant, K, is significantly affected by the influent substrate concentration. Hence, one must assume that neither of the substrate utilization-rate relationships given by equation 2-54 and 2-56 accurately de-

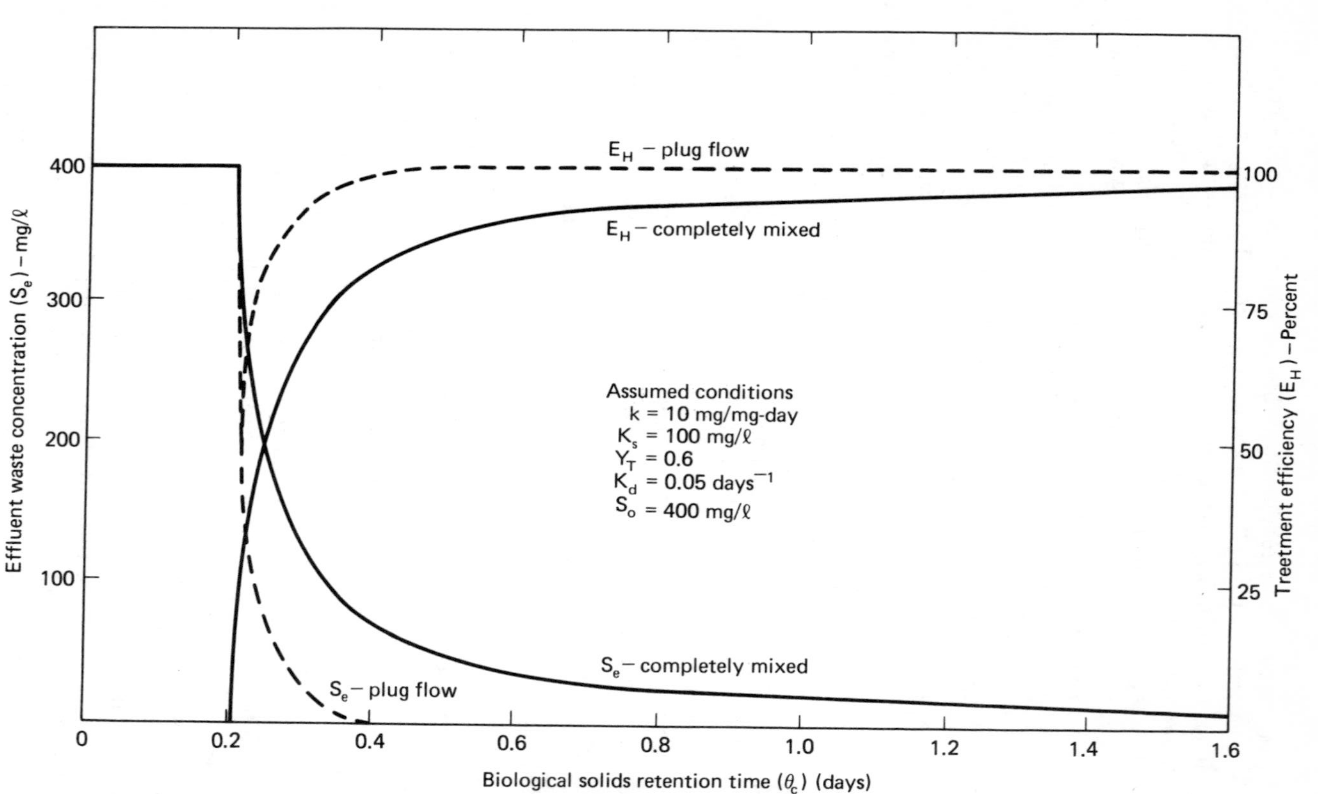

Figure 4-6. Comparison Between Steady-State Effluent Waste Concentrations and Waste Treatment Efficiencies for Plug-Flow and Completely Mixed Activated Sludge Processes. (After Lawrence and McCarty, 1970.)

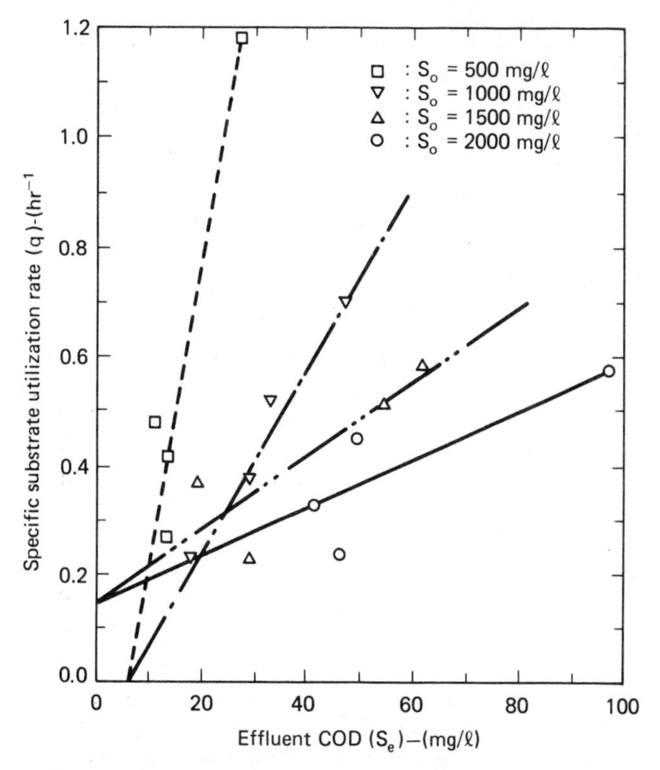

Figure 4-7. Data from Completely Mixed Activated Sludge System Without Cell Recycle, Treating a Variable-Strength Multicomponent Substrate, Fitted to a First-Order Substrate Utilization Rate Model. (After Grady, 1975.)

scribes the rate of substrate utilization in the activated sludge process when mixed microbial cultures and variable strength, multicomponent organic substrates are involved. For such a situation, Grau et al. (1975) have proposed equation 2-58, where n is taken as 1.

If one substitutes for $(dS/dt)_u$ in equation 4-17 from equation 2-58, the expression becomes

$$\frac{Q(S_0 - S_e)}{V_a} = K_1 X \frac{S_e}{S_0} \tag{4-36}$$

or

$$\frac{Q(S_0 - S_e)}{XV_a} = K_1 \frac{S_e}{S_0} \tag{4-37}$$

Figure 4-8 shows the data of Grady and Williams (1975) fitted to equation 4-37. Although there is some scatter to the data, the fit is much better than that obtained if equation 4-25, which is based on equation 2-56, is used to describe the experimental results. Thus, equation 2-58 more accurately de-

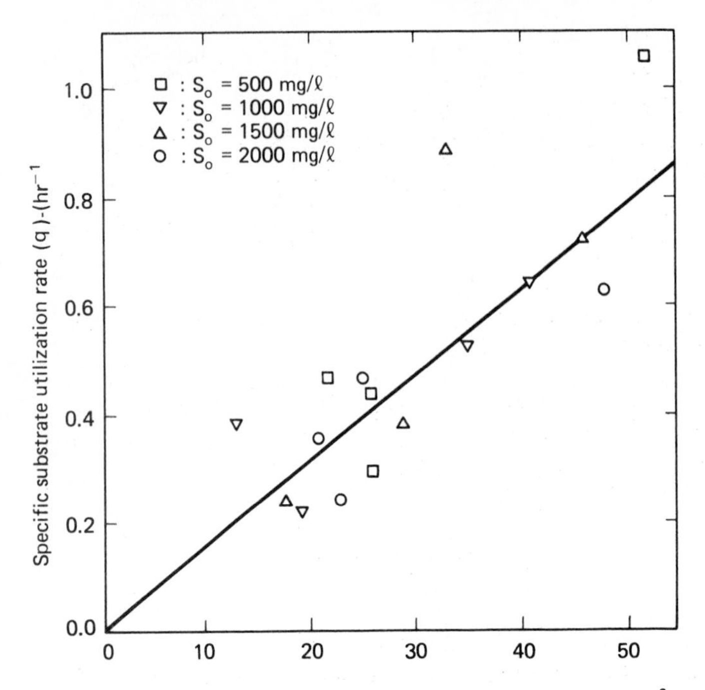

Figure 4-8. Data of Grady Fitted to Grau Model. (After Benefield, 1977.)

scribes the rate of substrate utilization in the activated sludge process when a mixed microbial culture uses a variable strength, multicomponent substrate whose availability is measured as BOD_5, COD, or TOC, and n is assumed to be 1.

If equation 2-58 is accepted as describing the rate of substrate utilization, one may substitute $K_1X(S_e/S_0)$ for $(dS/dt)_u$ in equation 4-8 and obtain the expression

$$\frac{1}{\theta_c} = Y_T\left(K_1\frac{S_e}{S_0}\right) - K_d \tag{4-38}$$

Solving equation 4-38 for S_e gives

$$S_e = \frac{S_0(1 + K_d\theta_c)}{Y_TK_1\theta_c} \tag{4-39}$$

However, the expression for the biomass concentration will not change from that given by equation 4-20.

Equation 4-39 can be used to determine θ_c^m, by recalling that at θ_c^m, $S_e = S_0$. This implies that

$$Y_TK_1\theta_c^m = 1 + K_d\theta_c^m \tag{4-40}$$

or

$$\theta_c^m = \frac{1}{Y_T K_1 - K_d} \tag{4-41}$$

Figure 4-9 illustrates the effect of influent substrate concentration on the effluent concentration of a completely mixed activated sludge process when equation 4-39 is used to determine S_e. From this figure it can be seen that to maintain the same effluent quality when influent substrate concentration increases, the BSRT must be increased.

The relationship among θ_c, R, and X_r is not affected by the rate of substrate utilization. Hence, equation 4-32 remains valid even when the Grau relationship is used.

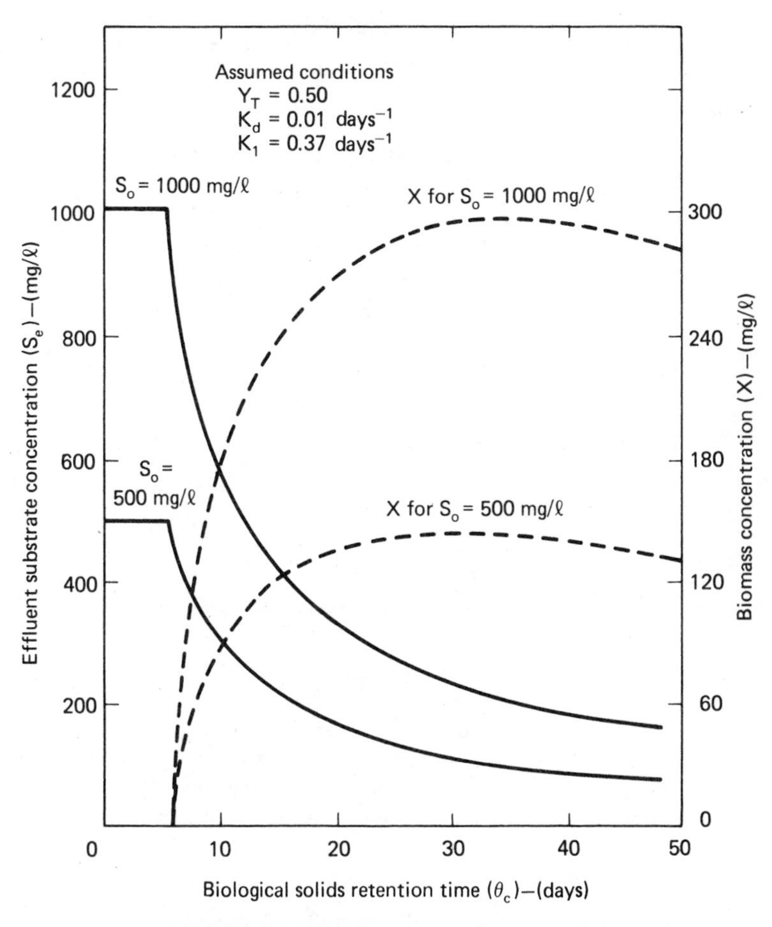

Figure 4-9. Effect of Influent Substrate Concentration on a Completely Mixed Activated Sludge System as Predicted by the Grau Substrate Utilization Rate Model. (After Benefield, 1977.)

Example 4-2

The BOD_u of the wastewater flow from an industrial plant was observed to vary between 100 and 400 mg/ℓ. If a BOD_u of 200 mg/ℓ is selected for design purposes, what size aeration tank is required to treat 1 MGD of this waste? Design criteria are as follows:

$$X = 2000 \text{ mg}/\ell$$

$$S_e = 10 \text{ mg}/\ell$$

$$Y_T = 0.5$$

$$K_1 = 17.0 \text{ day}^{-1}$$

$$K_d = 0.1 \text{ day}^{-1}$$

complete mix flow regime

Also determine θ_c^m for this process and compare it to θ_c^m for Example 4-1.

solution

1/ From equation 4-37 calculate the required aeration tank volume.

$$q = K_1 \frac{S_e}{S_0} = (17.0)\left(\frac{10}{200}\right) = 0.85 \text{ day}^{-1}$$

Furthermore,

$$q = \frac{Q(S_0 - S_e)}{XV_a}$$

or

$$V_a = \frac{1(200 - 10)}{(2000)(0.85)} = 0.112 \text{ MG}$$

2/ Compute θ_c^m from equation 4-41 for the Grau model.

$$\theta_c^m = \frac{1}{(0.5)(17) - 0.1} = 0.12 \text{ day}$$

3/ For Example 4-1, θ_c^m will be determined from equation 4-35.

$$\frac{1}{\theta_c^m} = (0.5)(0.1)(200) - 0.1$$

or

$$\theta_c^m = 0.10 \text{ day}$$

Comparing the minimum θ_c values calculated in steps 2 and 3 shows that for the kinetic constants selected for design, washout in both systems will occur at approximately the same time. However, when the Grau model is used, the time of washout is independent of the influent substrate concentration. This is not true for the other substrate utilization-rate models.

It can reasonably be asked why the influent substrate concentration affects the effluent concentration. It seems unreasonable to expect bacteria to respond to substrate concentrations they do not contact and, according to complete mixing theory, they are surrounded by the effluent concentration, not the influent concentration. Perhaps the failure to achieve complete mixing conditions is partially responsible. This is unlikely since the data come from laboratory systems that are notoriously overmixed. Daigger and Grady (1977) have pointed out that the soluble organic matter in the effluent is not the original substrate but is, in fact, a product of microbial metabolism. Then, using the concept that the activated sludge process is a producer of organic compounds as well as a user of them, they proposed the following model to predict effluent organic matter concentration:

$$z = Y_T S_0 \left(\frac{K_0}{\mu} + K_2 + K_3\right) + S_e \tag{4-42}$$

or, since S_e is very small,

$$z \simeq Y_T S_0 \left(\frac{K_0}{\mu} + K_2 + K_3\right) \tag{4-43}$$

where z is the total effluent organic concentration, Y_T, S_0, and S_e are the true growth yield, the influent substrate concentration, and the effluent substrate concentration, respectively, μ is the specific growth rate, and K_0, K_2, and K_3 are end-product formation constants.

Thus, the total quantity of metabolic end products formed is directly related to the influent substrate concentration and the effluent quality is related to the influent concentration. Since most of the effluent BOD is generally in the form of unsettled biomass solids, however, this phenomenon would appear to be of limited significance.

Assumption three states that no microbial solids are contained in the raw wastewater. This is a valid assumption that will hold true in the majority of cases. Even when the wastewater contains biomass, the amount is generally so small compared to the quantity of biomass in the return sludge that it can be neglected.

In assumption 4 it was stated that no microbial activity occurs in the secondary clarifier. This assumption is not strictly valid, as operational data have shown that there is activity in the clarifier. A good example is the rising sludge problem associated with denitrification. However, operational data also indicate that nearly all the organic materials removed during the treatment process are removed in the aeration tank. Thus, by assuming that no microbial activity occurs in the secondary clarifier, model development is simplified while sacrificing little in terms of accuracy.

Assumption 5 states that no sludge will accumulate in the secondary clarifier. This assumption led to the basic equation for biological solids retention time for a completely mixed activated sludge process,

$$\theta_c = \frac{XV_a}{Q_w X + (Q - Q_w)X_e} \tag{4-2}$$

Again, this assumption is not totally valid. Observations during plant operation will show that, indeed, biomass does accumulate in the secondary clarifier. As a result, there are individuals who advocate using the total amount of biomass in the entire system when computing BSRT,

$$(\theta_c)_T = \frac{(X)_T}{Q_w X + (Q - Q_w)X_e} \tag{4-44}$$

where $(\theta_c)_T$ = total system BSRT, time
$(X)_T$ = total amount of biomass in the treatment system, mass

Stall and Sherrard (1978) have compared many of the parameters currently used to design and control the activated sludge process. In their study it was noted that removal efficiency was independent of the particular BSRT used to control the process. This observation is presented in Figure 4-10. These investigators also noted that correlations can be made between all the control parameters. Figure 4-11 illustrates the correlation found between the BSRT based only on biomass in the aeration tank and the BSRT based on the total biomass in the system (i.e., biomass in the aeration tank, the secondary clarifier, and the sludge return line). Thus, as long as the BSRT to be used is specified, the ultimate results in terms of treatment efficiency should be the same. However, Stall and Sherrard do suggest the use of θ_c rather than $(\theta_c)_T$ because of the difficulty in evaluating the solids concentration in the secondary clarifier and the sludge return line.

Assumption 6 states that all biodegradable substrate is in the soluble form. Although this is true for certain industrial wastewaters, all municipal wastewaters contain insoluble organics. For example, a typical domestic wastewater may have a BOD between 100 and 300 mg/ℓ and a suspended solids

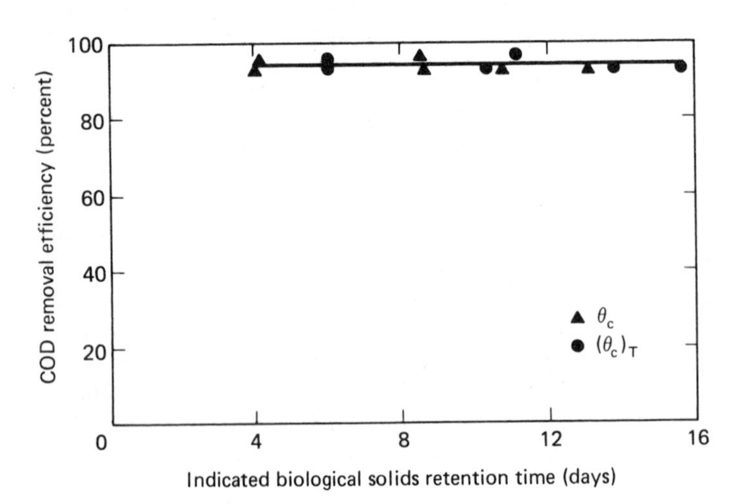

Figure 4-10. COD Removal Efficiency Versus Various Measurements of Biological Solids Retention Time. (After Stall and Sherrard, 1978.)

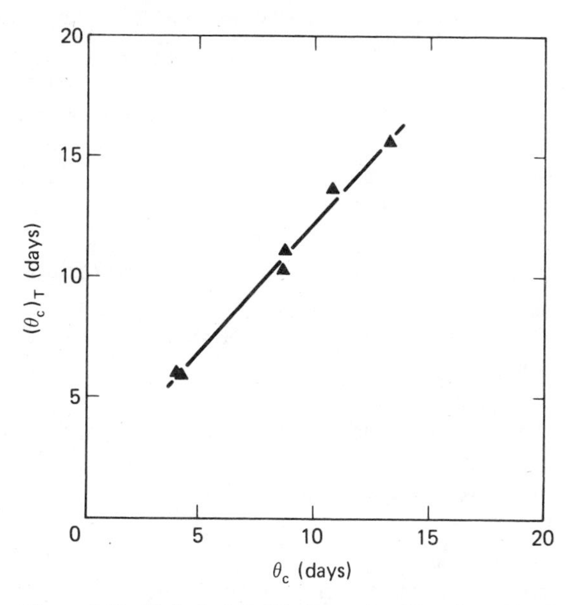

Figure 4-11. Biological Solids Retention Based on Total Biomass Versus Biological Solids Retention Time Based Only on Biomass in Aeration Tank. (After Stall and Sherrard, 1978.)

concentration of 100 and 300 mg/ℓ, which may be responsible for as much as 80% of the BOD.

Since suspended matter may represent a large portion of the organic material in the wastewater, it is important to include this material when a calculation is made for substrate concentration. This is accomplished during the laboratory investigation to determine the kinetic coefficients that describe rates of growth and substrate utilization. Here it is assumed that the substrate removed by the activated sludge process is given by the difference between influent total BOD or COD and effluent soluble BOD or COD. A word of caution when using COD: this parameter does not distinguish between organics which are amenable to biological degradation and those which are not. In fact, Jenkins and Garrison (1968) found that effluent COD from an activated sludge process treating domestic wastewater may contain a nondegradable fraction as large as 15% of the influent COD. Lawrence (1975) states that 30 mg/ℓ is a reasonable approximation for refractory COD in municipal effluents. Industrial wastewaters may contain nonbiodegradable COD concentrations of 200 mg/ℓ or more. To circumvent this problem and still maintain the flexibility inherent in the COD test, Gaudy and Gaudy (1972) propose the use of ΔCOD, which is a measure of only that portion of the total COD that is available for biological metabolism.

Even though it was assumed that all biodegradable substrate was in the soluble form, the kinetic models developed on this basis can be applied to substrates containing suspended organic material when a proper procedure is followed in the determination of the biokinetic coefficients describing substrate utilization. It should be realized, however, that the models developed for effluent substrate concentration apply only to the soluble biodegradable organic portion of the effluent.

The final assumption made in developing the various kinetic models was that flow was constant and steady-state conditions prevailed throughout the system. Although this assumption greatly simplifies the problem of developing kinetic models, it is not valid for most situations. For example, it is well known that municipal wastewater flows vary with time in volume and strength. One may then ask why, if steady-state is the exception rather than the rule, most kinetic models are based on this condition. An obvious answer is the mathematical complexity of describing the experimental results obtained from non-steady-state studies. Although considerable effort has been directed in recent years toward development of dynamic models of the activated sludge process, at the present time the use of such models is normally restricted to research activities. Present design practice generally considers only steady-state conditions at projected minimum, average, and maximum hydraulic and organic loadings. This is not to say that at some future time the use of dynamic models in design will not be common practice.

4-3 Process Modifications

The objective of an activated sludge process is to remove soluble and insoluble organics from a wastewater stream and to convert this material into a flocculant microbial suspension. Classically, this was accomplished by mixing the wastewater with a biological culture in a long, narrow aeration basin with a volume sufficient to provide a 6- to 8-h contact period between the two components. The biomass was then separated from the liquid stream in a secondary clarifier. A portion of this biological sludge was wasted and the remainder returned to the head of the aeration tank. Such a process arrangement has been termed *conventional activated sludge treatment.*

Conventional Activated Sludge Treatment

A typical flow pattern for conventional activated sludge treatment is shown in Figure 4-12. As the tank geometry is long and narrow, the mixing regime approaches plug flow. The air diffusers are generally located along one

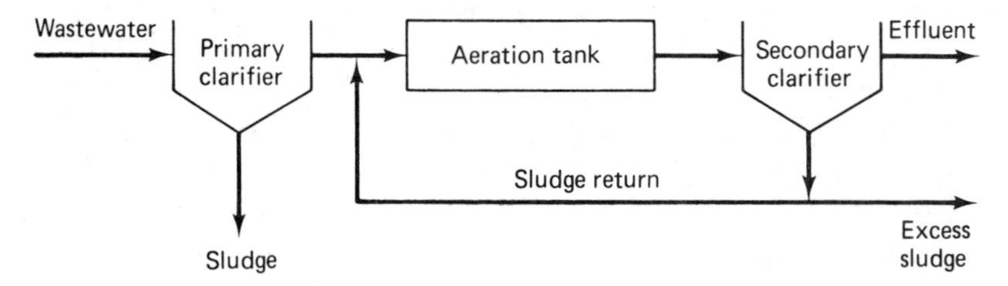

Figure 4-12. Conventional Activated Sludge Treatment.

side of the tank at depths of 8 to 12 ft below the surface. Such an arrangement produces a spiral flow pattern through the aeration tank (see Figure 4-13). It should be noted that many engineers today feel that spiral flow is undesirable because of coring and that diffusers should be evenly spaced across the bottom of the aeration tank.

Operating experience soon revealed a number of problems with this design. For example, as the biomass was recycled back to the head of the aeration tank and there mixed with the incoming wastewater, it was observed that the oxygen requirements at this point often exceeded the capability of the aeration system. Furthermore, it was observed that the oxygen requirements at the

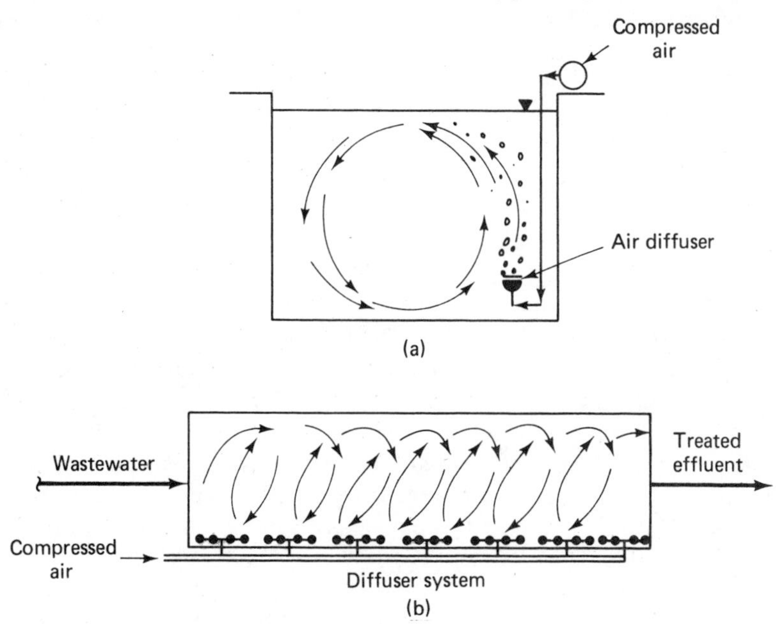

Figure 4-13. Conventional Activated Sludge Process Showing Plug-Flow and Spiral-Flow Diffused Aeration: (a) End View; (b) Top View. (After Lipták, 1974.)

exit were much less than at the head of the tank. It was also found that such a flow arrangement increased the probability of process failure due to shock loads of toxic or high-strength waste because these loads were not distributed throughout the aeration tank but rather were concentrated at the entrance.

Because of inherent deficiencies such as those just mentioned, numerous modifications to the original conventional process have been proposed. *Tapered aeration* is one such modification.

Tapered Aeration

A typical flow pattern for this process modification is that given by Figure 4-12. It is evident that the flow pattern of the tapered aeration modification is identical to the conventional process. The actual difference between the two processes is found in the diffuser arrangement. In tapered aeration the diffusers are spaced so that more air is supplied at the head of the tank, where the oxygen demand is greatest, and is then decreased along the tank length as the demand decreases. Such an arrangement is more economical than supplying a constant amount of air along the entire tank.

Step-Aeration

Figure 4-14 illustrates the flow pattern for a typical step-aeration process. In this modification, return sludge is mixed with a portion of the wastewater and enters the head of the aeration tank. Wastewater is also fed into the tank at different points along its length. Advantages of this process modification are (1) better equalization of waste load, (2) lower peak oxygen demand, (3) better distribution of oxygen demand over tank length, and (4) smaller overall aeration tank volume. Andrews et al. (1974) point out that this process will also provide an additional degree of operational control when one or more stages are used to add or remove sludge with variations in the wastewater.

This process is normally designed as a series of completely mixed reactors such as those shown in Figure 4-15. Such an approach, as pointed out earlier, is a conservative one. The kinetic models developed earlier, although valid, are not directly applicable to process design without modification.

In the development of design equations using the flow pattern given in Figure 4-15, it will be assumed that (1) the inflow is divided equally among the stages, (2) the volume of each stage is equal, and (3) no significant error will be incurred if the average active biomass concentration in the system is used for design, that is, $X_1 \approx X_a$, $X_2 \approx X_a$, and $X_3 \approx X_a$, where $X_a = (X_1 + X_2 + X_3)/3$.

For the flow scheme given in Figure 4-15, BSRT is given by the expression

$$\theta_c = \frac{3X_aV}{Q_wX_a + (Q_0 - Q_w)X_e} \tag{4-45}$$

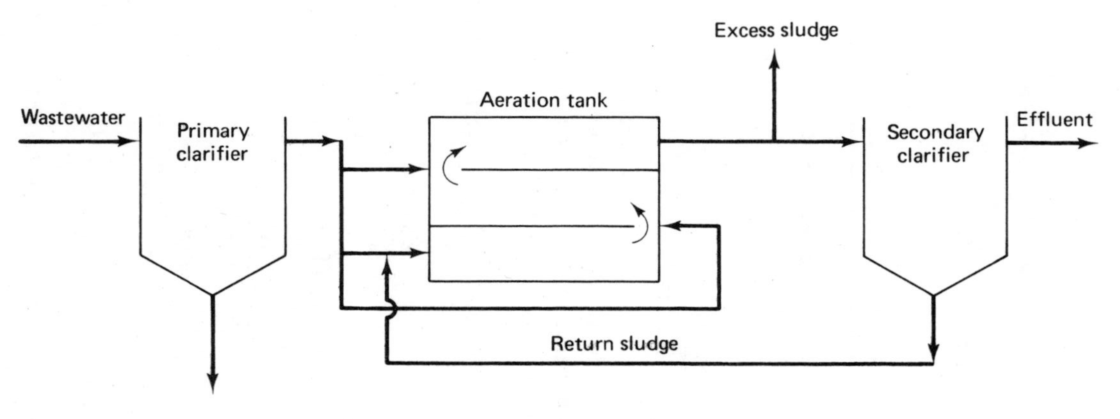

Figure 4-14. Step-Aeration Activated Sludge Process.

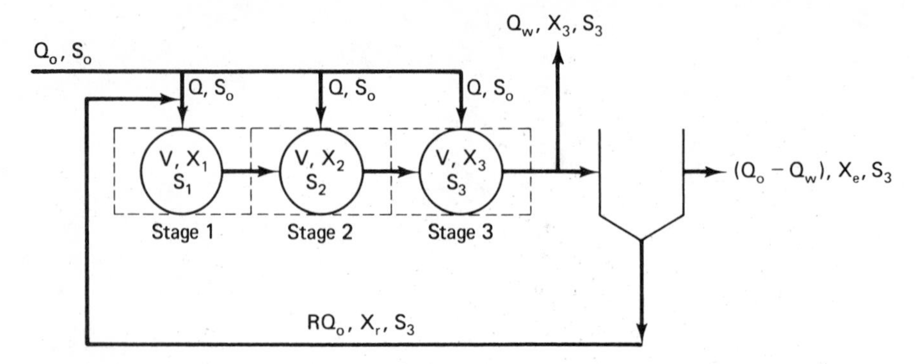

Figure 4-15. Completely Mixed Multistage Reactor System Used to Approximate Step-Aeration.

where

$$V = \text{volume of each stage}$$

Although a three-stage process is used in this development, a similar approach could be employed in developing design equations for more than three stages.

Steady-state material-balance expressions for substrate utilization rate and microbial growth rate for each individual stage have the form:

$$QS_0 + RQ_0S_3 - \left(\frac{dS}{dt}\right)_{u1} V - Q_0(\tfrac{1}{3} + R)S_1 = 0 \tag{4-46}$$

$$QS_0 + Q_0(\tfrac{1}{3} + R)S_1 - \left(\frac{dS}{dt}\right)_{u2} V - Q_0(\tfrac{2}{3} + R)S_2 = 0 \tag{4-47}$$

$$QS_0 + Q_0(\tfrac{2}{3} + R)S_2 - \left(\frac{dS}{dt}\right)_{u3} V - Q_0(1 + R)S_3 = 0 \tag{4-48}$$

$$RQ_0X_r + \left(\frac{dX}{dt}\right)_{g1} V - Q_0(\tfrac{1}{3} + R)X_a = 0 \tag{4-49}$$

$$Q_0(\tfrac{1}{3} + R)X_a + \left(\frac{dX}{dt}\right)_{g2} V - Q_0(\tfrac{2}{3} + R)X_a = 0 \tag{4-50}$$

$$Q_0(\tfrac{2}{3} + R)X_a + \left(\frac{dX}{dt}\right)_{g3} V - Q_0(1 + R)X_a = 0 \tag{4-51}$$

where stages 1, 2, and 3 are denoted by the subscripts 1, 2, and 3, respectively.

A material-balance equation for microbial mass around the entire system under steady-state conditions may be expressed as

$$Y_T\left[\left(\frac{dS}{dt}\right)_{u1} V + \left(\frac{dS}{dt}\right)_{u2} V + \left(\frac{dS}{dt}\right)_{u3} V\right] - K_d3X_aV - [Q_wX_a + (Q_0 - Q_w)X_e] = 0 \tag{4-52}$$

Equations 4-46, 4-47, and 4-48 can be rewritten as

$$\left(\frac{dS}{dt}\right)_{u_1} V = QS_0 + RQ_0S_3 - Q_0(\tfrac{1}{3} + R)S_1 \quad (4\text{-}53)$$

$$\left(\frac{dS}{dt}\right)_{u_2} V = QS_0 + Q_0(\tfrac{1}{3} + R)S_1 - Q_0(\tfrac{2}{3} + R)S_2 \quad (4\text{-}54)$$

$$\left(\frac{dS}{dt}\right)_{u_3} V = QS_0 + Q_0(\tfrac{2}{3} + R)S_2 - Q_0(1 + R)S_3 \quad (4\text{-}55)$$

Substituting for the appropriate $(dS/dt)_u$ terms in equation 4-52 from equations 4-53, 4-54, and 4-55 and for the $[Q_wX_a + (Q_0 - Q_w)X_e]$ term from equation 4-45 while recalling that $Q = Q_0/3$, the resulting expression is

$$\frac{1}{\theta_c} = \frac{Y_T[Q_0(S_0 - S_3)]}{3X_aV} - K_d \quad (4\text{-}56)$$

To develop an expression for the recycle ratio, R, consideration must be given to the fact that

$$\left(\frac{dS}{dt}\right)_{u(\text{overall})} = \left(\frac{dS}{dt}\right)_{u_1} + \left(\frac{dS}{dt}\right)_{u_2} + \left(\frac{dS}{dt}\right)_{u_3} \quad (4\text{-}57)$$

and

$$\left(\frac{dX}{dt}\right)_{g(\text{overall})} = \left(\frac{dX}{dt}\right)_{g_1} + \left(\frac{dX}{dt}\right)_{g_2} + \left(\frac{dX}{dt}\right)_{g_3} \quad (4\text{-}58)$$

The relationship between these two expressions has the form

$$\left(\frac{dX}{dt}\right)_{g(\text{overall})} = Y_{\text{obs}}\left(\frac{dS}{dt}\right)_{u(\text{overall})} \quad (4\text{-}59)$$

When proper substitutions are made from equations 4-46 through 4-51, this relationship will reduce to

$$Q_0(X_a + RX_a - RX_r) = Y_{\text{obs}}Q_0(S_0 - S_3) \quad (4\text{-}60)$$

Solving this expression for R gives

$$R = \frac{Y_{\text{obs}}(S_0 - S_3) - X_a}{X_a - X_r} \quad (4\text{-}61)$$

For plant design two additional relationships are required. The first of these relationships is developed by substituting KX_aS_1 into equation 4-53 for $(dS/dt)_u$ and solving for S_1:

$$S_1 = \frac{QS_0 + RQ_0S_3}{KX_aV + Q_0(\tfrac{1}{3} + R)} \quad (4\text{-}62)$$

The second relationship is developed by substituting KX_aS_2 into equation 4-54 for $(dS/dt)_u$ and solving for S_2:

$$S_2 = \frac{QS_0 + Q_0(\tfrac{1}{3} + R)S_1}{KX_aV + Q_0(\tfrac{2}{3} + R)} \quad (4\text{-}63)$$

In the development of these equations it has been assumed that the substrate utilization-rate constant, K, will be the same in each reactor. Under actual field conditions this may not be true, since slowly degradable compounds will accumulate. However, since it is assumed that each reactor receives an equal amount of raw wastewater, the variation in the value of K between reactors should be very small, so that little error is incurred by assuming a constant value for K.

Example 4-3

Determine the aeration volume required for a three-stage step-aeration activated sludge process if the design flow is 10 MGD and the influent BOD_u is 273 mg/ℓ. Pertinent design criteria are:

$$X_a = 2200 \text{ mg/ℓ as MLVSS}$$

$$S_e = 8 \text{ mg/ℓ } BOD_u$$

$$SVI = 98$$

$$MLVSS = 0.78 \text{ MLSS}$$

$$Y_T = 0.5$$

$$K_d = 0.05 \text{ day}^{-1}$$

$$K = 0.032 \text{ ℓ/mg-day}$$

solution

1/ Assume an operating θ_c value and calculate the volume of one stage by rearranging equation 4-56 into the form

$$V = \frac{Y_T[Q_0(S_0 - S_3)]}{3X_a[(1/\theta_c) + K_d]}$$

Assuming that θ_c is equal to 6 days, the required volume is

$$V = \frac{0.5[(10)(273 - 8)]}{3(2200)(\frac{1}{6} - 0.05)}$$

$$= 0.93 \text{ MG}$$

2/ Compute the observed yield coefficient corresponding to the operating θ_c assumed in step 1 from equation 2-63.

$$Y_{obs} = \frac{Y_T}{1 + K_d\theta_c}$$

$$= \frac{0.5}{1 + (0.05)(6)}$$

$$= 0.39$$

3/ Calculate the recycle ratio, R, from equation 4-61.

$$R = \frac{(0.39)(273 - 8) - 2200}{2200 - [(10^6/SVI)(0.78)]}$$

$$= 0.37$$

4/ Estimate the effluent substrate concentration from the first stage by applying equation 4-62.

$$S_1 = \frac{(\frac{10}{3})(273) + (0.37)(10)(8)}{[(0.032)(2200)(0.93) + (10)(\frac{1}{3} + 0.37)]}$$
$$= 13 \text{ mg}/\ell$$

5/ Approximate the effluent substrate concentration from the second stage by applying equation 4-63.

$$S_2 = \frac{(\frac{10}{3})(273) + (10)(\frac{1}{3} + 0.37)(13)}{[(0.032)(2200)(0.93) + (10)(\frac{2}{3} + 0.37)]}$$
$$= 13.1 \text{ mg}/\ell$$

6/ Determine the required specific substrate utilization rate for the final stage.

$$q = KS_3$$
$$= (0.032)(8) = 0.26 \text{ day}^{-1}$$

7/ Compute the influent substrate concentration for the final stage by simple proportions.

$$S_{0(\text{final})} = \frac{Q_0(\frac{2}{3} + R)S_2 + (Q_0/3)S_0}{Q_0(\frac{2}{3} + R) + (Q_0/3)}$$
$$= \frac{10(\frac{2}{3} + 0.37)(13.1) + (\frac{10}{3})(273)}{10(\frac{2}{3} + 0.37) + (\frac{10}{3})}$$
$$= 76.7 \text{ mg}/\ell$$

8/ Calculate the actual specific substrate utilization rate for the final stage.

$$q = \frac{(1 + R)Q_0[S_{0(\text{final})} - S_3]}{X_aV}$$
$$= \frac{(1 + 0.37)(10)(76.7 - 8)}{(2200)(0.93)}$$
$$= 0.46 \text{ day}^{-1}$$

9/ Compare the specific substrate utilization rates calculated in steps 6 and 8. If the difference between the values is less than 0.01 day^{-1}, the assumed θ_c can be taken as correct. If the difference between the values is greater than 0.01 day^{-1}, a new value for θ_c must be assumed and the procedure repeated.

In this problem an assumed θ_c of 11.5 days will give the correct solution.

High-Rate Activated Sludge

For this particular process modification a low mixed liquor suspended solids concentration is maintained in the aeration tank. This enables the process to be operated so that the specific growth rate and specific substrate utilization rate are high (i.e., at a low BSRT). Equation 2-63 shows that under such a condition, the process yield is maximized, which implies that the total oxygen requirement per unit of substrate removed per unit time is minimized even though the specific oxygen utilization rate is high. This occurs because a smaller fraction of the substrate removed is oxidized for energy.

Although the organic removal rate is high on a unit biomass basis, the overall removal is not. The substrate removal efficiency of the high-rate process is typically 60 to 75%, mainly because the plant effluent generally contains a high solids concentration. This high solids concentration is a result of the physiological state of the organisms in the aeration tank. It was noted earlier that for efficient solids–liquid separation, it is necessary that bioflocculation occur. Since the biopolymers responsible for agglomeration are only produced at low specific growth rates, the problem of dispersed growth exists. Because of this characteristic, the high-rate process cannot be used where a high-quality effluent is required.

Complete Mixing

To obtain complete mixing in the aeration tank requires the proper choice of tank geometry, feeding arrangement, and aeration equipment. Two possible arrangements that will produce such a mixing regime are shown in Figures 4-16 and 4-17. Figure 4-16 illustrates the use of diffused air aeration, which requires feeding and discharge along the entire tank length. Figure 4-17 shows the use of mechanical aerators, which would probably necessitate the use of either a round or square tank.

Through the use of complete mixing it is possible to establish a constant oxygen demand as well as a uniform solids concentration throughout the tank volume. Furthermore, this process is highly resistant to upset from shock

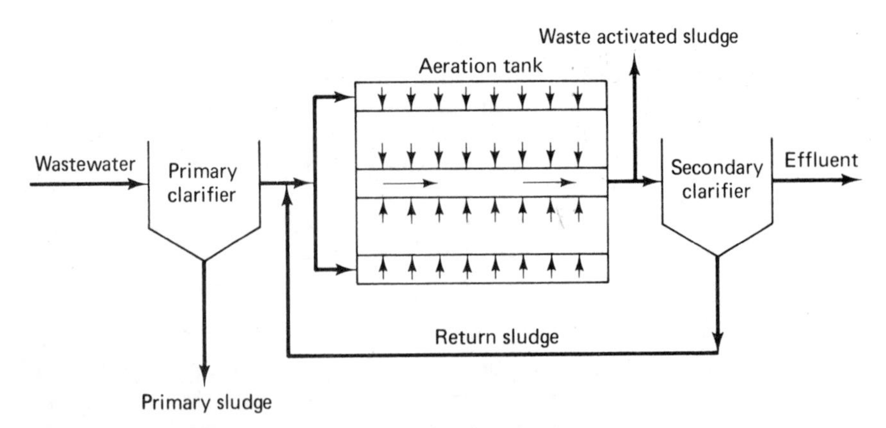

Figure 4-16. Completely Mixed Activated Sludge Process Employing Diffused Air Aeration.

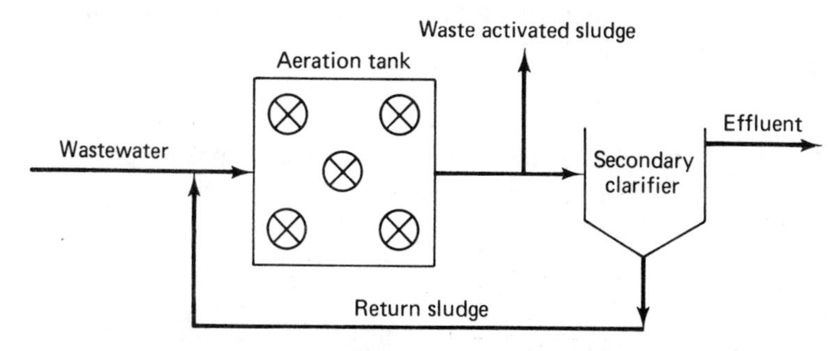

Figure 4-17. Completely Mixed Activated Sludge Process Employing Mechanical Aerators.

loadings because of the rapid blending of feed and tank contents. As a result of these advantantages, complete mixing has become very popular in the choice of mixing regimes.

It is not uncommon to omit primary clarification from completely mixed activated sludge plants, and there is no biological reason not to do so.

Extended Aeration

Extended aeration plants are generally small (applicable to flows less than 1 MGD) because of the large aeration tank volumes required and they almost invariably employ complete mixing. Figure 4-17 is typical of the flow pattern that may be used. However, primary clarification is generally omitted, and should be, as it defeats the principal objective of this modification, the minimization of sludge handling.

Theoretically, the extended air process is designed such that all substrate removed is channeled into energy metabolism and oxidized so that no excess biomass is produced and sludge handling is eliminated. Yet, it should be realized that in practice there is a buildup of nondegradable material which must be periodically wasted or the solids in the effluent will increase. Nevertheless, acceptable design practice is based on an absolute growth rate of zero,

$$\left(\frac{dX}{dt}\right)_g = 0 \qquad (4\text{-}64)$$

This expression implies that the amount of biomass produced during organic removal is equal to the amount oxidized to provide for energy requirements. This can be represented as

$$Y_T\left(\frac{dS}{dt}\right)_u V_a = K_d X V_a \qquad (4\text{-}65)$$

or

$$q = \frac{K_d}{Y_T} \qquad (4\text{-}66)$$

According to equation 4-17, a material balance for substrate around the aeration tank yields

$$\left(\frac{dS}{dt}\right)_u = \frac{Q(S_0 - S_e)}{V_a} \qquad (4\text{-}17)$$

Dividing both sides of this expression by X and substituting for $(dS/dt)_u/X = q$ from equation 4-66, an expression for V_a can be developed which has the form

$$V_a = \frac{Y_T Q(S_0 - S_e)}{XK_d} \qquad (4\text{-}67)$$

In this design approach it is assumed there is no sludge produced, which implies there is no sludge wasting. Under such circumstances BSRT is no longer a useful design parameter. Thus, for the extended aeration process, the food/microorganism (F:M) ratio will be used as the loading criterion. This parameter will be discussed in some detail in a later section.

It is important to realize that even though the organisms in the extended air process are maintained in the endogenous growth phase, the biomass generally agglomerates poorly and tends to remain dispersed in the secondary clarifier, forming what is referred to as "pin floc." This may result in a considerable deterioration in effluent quality.

Since all the substrate that is removed is channeled into energy metabolism and oxidized, the extended aeration process maximizes the total oxygen requirement per unit of substrate removed per unit time and thus the associated energy costs. This is another important factor that should be considered during process selection.

Example 4-4

What aeration tank volume is required to treat a wastewater flow of 1 MGD and BOD_u concentration of 150 mg/ℓ? An extended aeration process is to be used and pertinent design criteria are as follows:

$$X_a = 4000 \text{ mg/}\ell \text{ as MLVSS}$$

$$Y_T = 0.5$$

$$K_d = 0.02 \text{ day}^{-1}$$

$$K = 0.1\ \ell\text{/mg-day}$$

solution

1/ Compute the expected soluble effluent substrate concentration by combining equations 4-66 and 4-25.

$$q = \frac{K_d}{Y_T} = KS_e$$

or

$$S_e = \frac{K_d}{Y_T K} = \frac{0.02}{(0.5)(0.1)}$$

$$= 0.4 \text{ mg/}\ell$$

2/ Calculate the required aeration tank volume from equation 4-67.

$$V_a = \frac{(0.5)(1)(150 - 0.4)}{(4000)(0.02)}$$

$$= 0.94 \text{ MG}$$

Contact Stabilization

When a complex mixed colloidal/soluble substrate is added to an endogenous activated sludge culture contained in a batch reactor, it is not unusual to observe a BOD removal response similar to that presented in Figure 4-18 or an oxygen utilization rate (OUR) response similar to that given in Figure 4-19. On the other hand, if only a readily biodegradable soluble substrate is added to the activated sludge culture, BOD removal and OUR responses would probably be similar to those illustrated by Figures 4-20 and 4-21, respectively, although an instantaneous removal of a portion of the soluble substrate would occur in both reactors. The different responses given by the two substrates can be attributed to the fact that the soluble organic material present in both substrates is removed from solution at a relatively constant rate by the microorganisms present. However, the insoluble organic material, present only in the mixed substrate, is rapidly removed by adsorption onto the floc surfaces and enmeshment in the floc structure. Then, after some short aeration period the insoluble material is solubilized, resulting in an increase in filtrate BOD_u. As the aeration period is continued, the previously

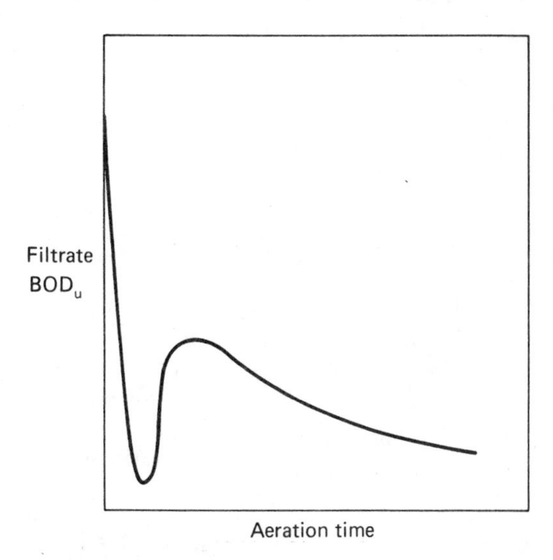

Figure 4-18. Variation in Filtrate BOD_u with Aeration Time for a Mixed Colloidal-Soluble Substrate and Activated Sludge Mixture.

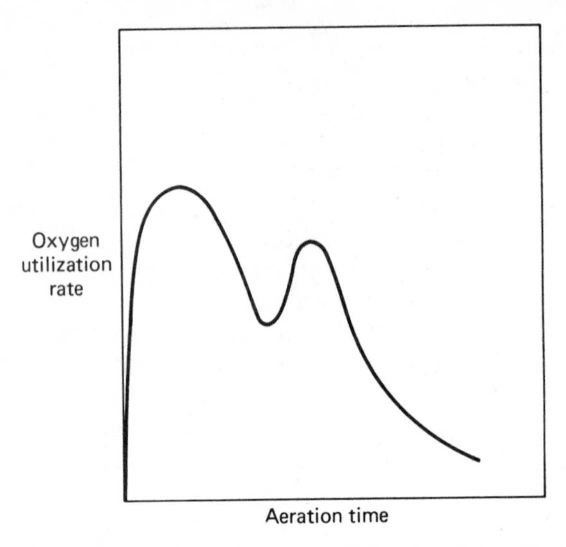

Figure 4-19. Variation in Oxygen Utilization Rate with Aeration Time for a Mixed Colloidal-Soluble Substrate and Activated Sludge Mixture.

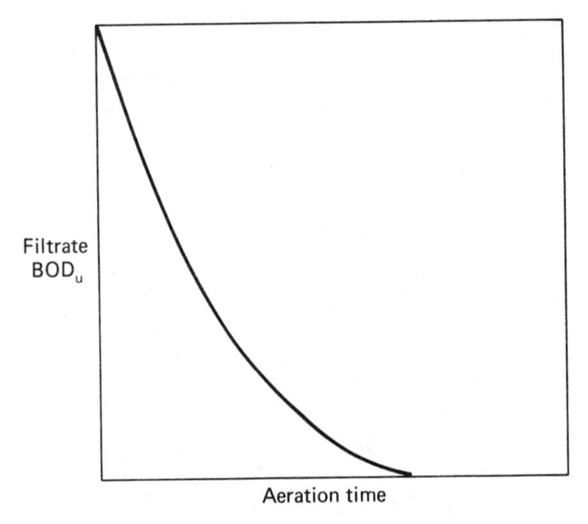

Figure 4-20. Variation in Filtrate BOD_u with Aeration Time for a Soluble Substrate and Activated Sludge Mixture.

insoluble material is metabolized. This is reflected by a decrease in filtrate BOD_u and OUR. A significant BOD bleed-back phenomenon usually occurs when a large percentage of the substrate is in the insoluble form. Although

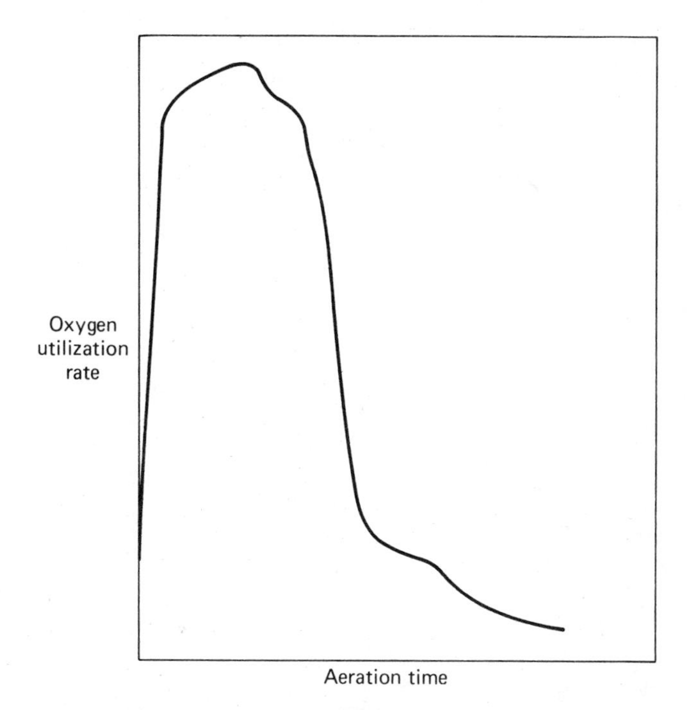

Figure 4-21. Variation in Oxygen Utilization Rate with Aeration Time for a Soluble Substrate and Activated Sludge Mixture.

transfer of the substrate from the liquid phase to the solid phase (biomass) is rapid, subsequent metabolism of the adsorbed substrate is necessary before the biomass is ready to treat additional wastewater. Such stabilization is also necessary to deplete internally stored substrate if the process is to continue removing substrate at peak efficiency. This has led to the utilization of the contact-stabilization flow scheme, as shown in Figure 4-22. To reduce the overall plant volume required to treat a specific wastewater, two aeration tanks are provided. The first tank provides contact between the biomass and the wastewater. It operates at a short retention time, sufficient only for the transfer of substrate from the liquid to the solid phase. The biomass is then separated from the wastewater in a secondary clarifier and the biological sludge channeled to the second aeration tank, where the organic material adsorbed to the biomass surface is metabolized or "stabilized."

The very brief retention time of the main wastewater stream in the contact tank offers considerable saving in total tank volume, because the further detention of the settled sludge for stabilization occupies much less volume than would be the case if the entire waste stream was contained in this tank. A comparison with conventional operation is given in Table 4-1.

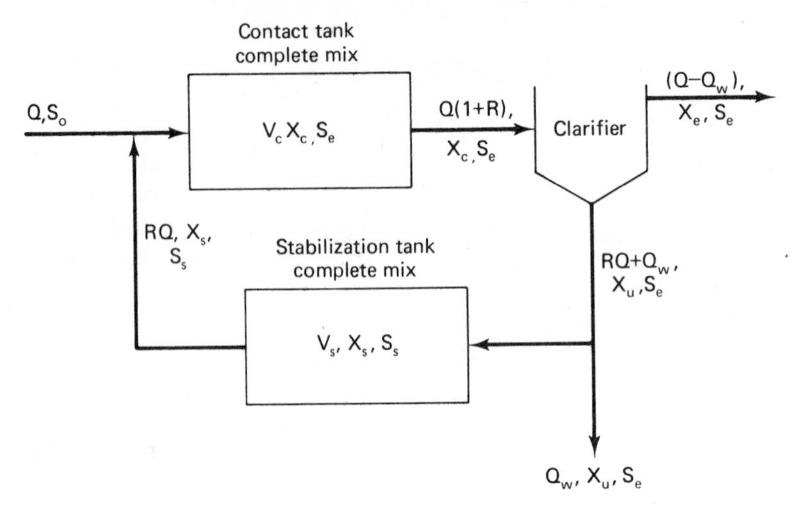

Figure 4-22. Flow Scheme for Contact-Stabilization Activated Sludge Process.

The table indicates that the sludge in the contact-stabilization process will be aerated 1/2 hr longer each cycle than conventional sludge. However, if the plant flow is Q and recycle is $0.5Q$, the volume of tankage in the respective plants would be $12Q$ and $5.25Q$ for the conventional and contact-stabilization process, respectively.

In Figure 4-22, Q represents the rate of raw wastewater flow to the contact tank; S_0, the total BOD_u of the raw wastewater which includes both the soluble and insoluble fraction; V_c, the volume of the contact tank; X_c, the biomass concentration in the contact tank as measured by the MLVSS concentration; S_e, the soluble effluent substrate concentration from the contact tank; X_u, the biomass concentration in the underflow from the secondary

TABLE 4-1

Reduction of Tank Volume by Contact-Stabilization Process

	Conventional		*Contact-Stabilization*	
Unit	*Flow retention (h)*	*Volume*	*Flow retention (h)*	*Volume*
Primary settling	1.5	$1.5Q$	None	—
Aeration tank	6	$6\,(1.5Q)$	0.5	$0.5\,(1.5Q)$
Sludge stabilization	None	—	6	$6\,(0.5Q)$
Final settling	1.5	$1.5Q$	1.5	$1.5Q$
Total	9	$12Q$	8	$5.25Q$

clarifier; V_s, the volume of the stabilization tank; and X_s, the biomass concentration in the stabilization tank.

When considering the contact-stabilization process as a treatment alternative, it is important to realize that it is not simply the colloidal nature of the waste, but rather the complexity, from a biodegradation standpoint, of the molecules being hydrolyzed that determines whether or not they are amenable to contact-stabilization treatment. If the solids are hydrolyzed at about the same rate as the soluble material is utilized, the BOD bleed-back phenomenon will be masked and the contact-stabilization flow scheme will not be applicable. A significant degree of storage of substrate for subsequent metabolism, however, favors contact-stabilization design. It is, therefore, important that the engineer ensure that contact-stabilization kinetics are applicable before considering this process as an alternative in wastewater treatment. A simple way to ascertain this is to simultaneously observe substrate removal and oxygen utilization in a batch system using realistic concentrations of biomass and substrate as well as a realistic initial food-to-microorganism ratio. If oxygen utilization continues at a rate considerably higher than endogenous for two or more times as long as the time required for substrate removal, contact-stabilization kinetics are indicated. If oxygen utilization returns to endogenous shortly after substrate removal is complete, contact-stabilization kinetics are not indicated. The two cases are illustrated by Figure 4-23. The time required for substrate removal, t_c, is related to the necessary contact time, whereas the time required for the oxygen utilization rate to return to the endogenous level, t_s, is related to the total aeration, or "stabilization" time, required.

Few mathematical models are found in the literature which are directly applicable to the design of a contact-stabilization activated sludge process. Benefield and Randall (1976) have proposed a series of such relationships which they derived from material-balance expressions derived from Figure 4-22. However, at the present time it is very difficult to apply the relationships to actual design because typical values for many of the biokinetic constants are not available.

Using the results of laboratory studies to determine the required retention in the contact tank and the settling properties of the biomass from this unit, a material-balance expression derived from Figure 4-22 can be used to develop an expression for stabilization tank volume. In developing this expression, it is assumed that (1) all insoluble material is adsorbed in the contact tank and metabolized in the stabilization tank, (2) all substrate entering the stabilization tank is metabolized, and (3) the yield coefficient for both the soluble and insoluble organic material is approximately the same. At steady-state, a material-balance expression for biomass entering and leaving the stabilization tank is given by

$$0 = RQX_u + Y_T QfS_0 + Y_T RQS_e - K_d X_s V_s - RQX_s \qquad (4\text{-}68)$$

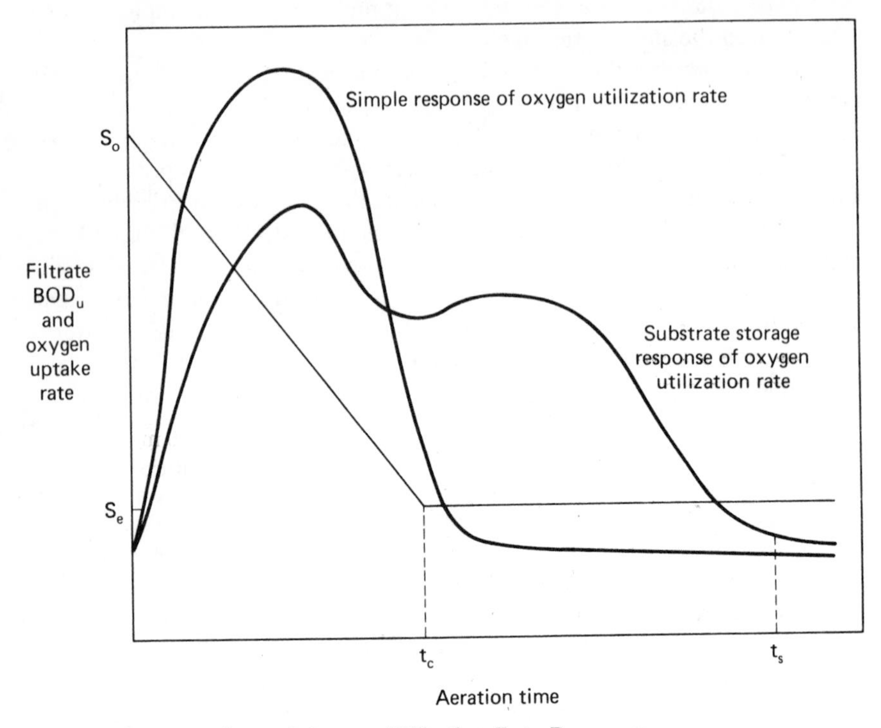

Figure 4-23. Comparison of Oxygen Utilization Rate Responses for Simple and Complex Substrates.

In this expression f represents the fraction of the influent total BOD_u which is insoluble and S_e represents the soluble effluent BOD_u from the contact tank. The term $Y_T Q f S_0$ represents growth in the stabilization tank due to insoluble BOD_u removal in the contact tank, while the term $Y_T R Q S_e$ represents growth due to soluble substrate removal in the stabilization tank. Solving equation 4-68 for V_s yields the following formulation:

$$V_s = \frac{RQ(X_u - X_s + Y_T S_e) + Y_T Q f S_0}{K_d X_s} \tag{4-69}$$

An expression for the recycle ratio, R, can be developed by considering that at steady-state, neglecting the increase in biomass due to synthesis in the contact tank, the quantities of biomass entering and leaving the contact tank are equal,

$$QX_0 + RQX_s = (1 + R)QX_c \tag{4-70}$$

The X_0 term represents the volatile suspended solids in the influent to the contact tank and is required in the basic expression because in the initial assumptions, it was assumed that no insoluble material was metabolized in the aeration tank. Solving equation 4-70 for R gives

$$R = \frac{X_c - X_0}{X_s - X_c} \tag{4-71}$$

Since X_0 is generally much less than X_c, X_0 can usually be neglected so that equation 4-71 reduces to

$$R = \frac{X_c}{X_s - X_c} \tag{4-72}$$

The critical factor in determining the stabilization tank volume is the quantity of biomass production desired. This value will be established by the operating θ_c. An extreme condition would be established if the stabilization tank volume were selected such that X_u and X_s were equal. For this case the process would be contact-stabilization operating in an extended aeration mode; that is, the system would be designed so that $\Delta X = 0$.

Example 4-5

Laboratory studies on a wastewater having a total BOD_u of 150 mg/ℓ have shown that after 45 min of contact with an activated sludge culture initially containing 2000 mg/ℓ MLVSS, the filtrate BOD_u is reduced to 15 mg/ℓ. Determine the aeration volumes for the contact and stabilization tanks using the following design criteria:

$$X_c = 2000 \text{ mg/ℓ as MLVSS}$$
$$\theta_c = 8 \text{ days}$$
$$f = 0.8$$
$$\text{SVI} = 110$$
$$\text{MLVSS} = 0.8 \text{ MLSS}$$
$$S_e = 15 \text{ mg/ℓ BOD}_u$$
$$Q = 2 \text{ MGD}$$
$$Y_T = 0.5$$
$$K_d = 0.1 \text{ day}^{-1}$$

solution

1/ Compute the contact tank volume on the basis of the contact time required to reduce the soluble BOD_u to the desired level.

$$V_c = t_c Q$$
$$= 45 \text{ min} \times \frac{1 \text{ day}}{1440 \text{ min}} \times 2 \text{ MG/day}$$
$$= 0.063 \text{ MG}$$

2/ Compute the observed yield coefficient from equation 2-63:

$$Y_{obs} = \frac{Y_T}{1 + K_d\theta_c}$$
$$= \frac{0.5}{1 + (0.1)(8)}$$
$$= 0.28$$

3/ Calculate the expected biomass production:

$$\Delta X = Q(3.78)\,Y_{obs}(S_0 - S_e)$$
$$= 2{,}000{,}000(3.78)(0.28)(150 - 15)$$
$$= 285{,}768{,}000 \text{ mg/day}$$

4/ Assuming that synthesis of all biomass occurs in the *stabilization tank*, develop an expression for biomass concentration in the stabilization tank in terms of R. A material balance for biomass entering and leaving the stabilization tank gives

$$X_s = X_u + \Delta X$$

or

$$X_s = \frac{10^6}{\text{SVI}}(0.8) + \Delta X$$

Therefore,

$$X_s = \frac{10^6}{\text{SVI}}(0.8) + \frac{285{,}768{,}000}{2{,}000{,}000(3.78)(R)}$$
$$= 7273 + \frac{38}{R}$$

5/ Compute R from equation 4-72:

$$R = \frac{X_c}{X_s - X_c}$$
$$= \frac{2000}{(7273 + 38/R) - 2000}$$
$$\simeq 0.38$$

6/ Using the value of R computed in step 5, determine X_s from the equation developed in step 4.

$$X_s = 7273 + \frac{38}{0.38}$$
$$= 7373 \text{ mg}/\ell$$

7/ Determine the required stabilization tank volume from equation 4-69:

$$V_s = \frac{(0.38)(2)[(10^6/\text{SVI})0.8 - 7373 + (0.5)(15)]}{(0.1)(7373)} + \frac{(0.5)(2)(0.8)(150)}{(0.1)(7373)}$$
$$= 0.06 \text{ MG}$$

Sludge Reaeration

In past years it was not uncommon to find activated sludge plants where the aeration capacity was inadequate or the aeration equipment was inefficient. Such deficiencies resulted in an insufficient oxygen supply to the biomass, which therefore reduced the organic removal efficiency. The problem was further aggravated by the time the biomass spent without oxygen in the secondary clarifier, causing the return sludge to create an abnormally high immediate oxygen demand upon entering the aeration tank. To alleviate this problem, many plants were designed to provide aeration of the return sludge. Figure 4-24 presents a typical sludge reaeration flowsheet.

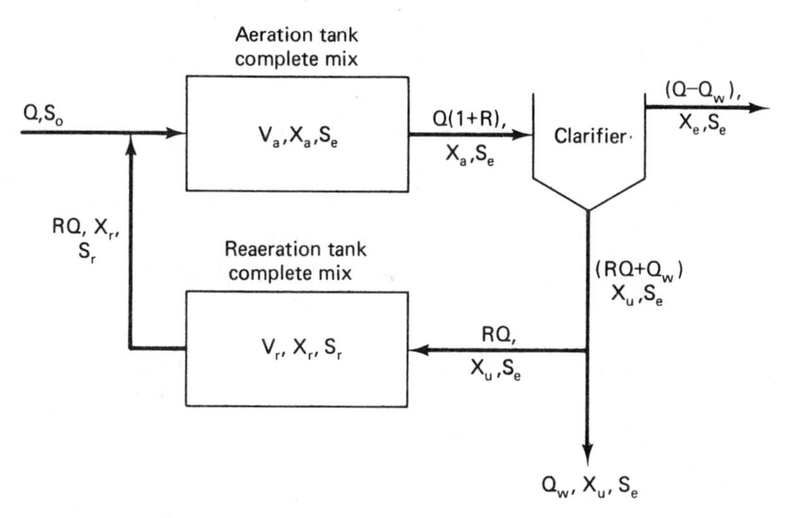

Figure 4-24. Flow Scheme for Sludge Reaeration Activated Sludge Process.

With the aeration equipment and design know-how available today, the original reason for using the sludge reaeration process is no longer valid. However, there are engineers who still feel that sludge reaeration conditions the biomass, resulting in increased BOD removal efficiency because of improved settling. Furthermore, it is not uncommon to find engineers who confuse the sludge reaeration and contact-stabilization processes.

In deriving design equations for the sludge reaeration process, it will be assumed that all substrate entering the reaeration tank is removed. Thus, no substrate will be present in the recycle from the reaeration tank to the aeration tank. For this situation, a material-balance equation for substrate entering and leaving the aeration tank has the form

$$\begin{bmatrix}\text{net rate of change} \\ \text{in amount of sub-} \\ \text{strate within aera-} \\ \text{tion tank}\end{bmatrix} = \begin{bmatrix}\text{rate at which sub-} \\ \text{strate enters} \\ \text{aeration tank}\end{bmatrix} - \begin{bmatrix}\text{rate at which} \\ \text{substrate dis-} \\ \text{appears from} \\ \text{aeration tank}\end{bmatrix} \qquad (4\text{-}73)$$

or

$$\left(\frac{dS}{dt}\right)V_a = QS_0 - \left(\frac{dS}{dt}\right)_u V_a - (1 + R)QS_e \qquad (4\text{-}74)$$

Substituting for $(dS/dt)_u$ from equation 2-56, the steady-state expression for aeration tank volume is given by

$$V_a = \frac{Q[S_0 - S_e(1 + R)]}{KX_aS_e} \tag{4-75}$$

A steady-state material balance expression for biomass entering and leaving the reaeration tank can be written as

$$0 = RQX_u + Y_TRQS_e - K_dX_rV_r - RQX_r \tag{4-76}$$

or

$$V_r = \frac{RQ[(X_u - X_r) + Y_TS_e]}{K_dX_r} \tag{4-77}$$

The term Y_TRQS_e represents the growth due to substrate removal, whereas $K_dX_rV_r$ represents the biomass lost to endogenous respiration. As in contact stabilization, the critical factor in the reaeration tank volume determination is the degree of solids destruction desired, which is controlled by the operating BSRT.

An expression for the required recycle ratio, R, can be developed by considering a steady-state material-balance expression for biomass entering and leaving the clarifier. In this case it is assumed that the effluent solids concentration is zero. Thus,

$$RQX_u = (1 + R)QX_a \tag{4-78}$$

or

$$R = \frac{X_a}{X_u - X_a} \tag{4-79}$$

Equation 4-79 is not exact, but it does provide a reasonable approximation which is accurate enough for design purposes.

Example 4-6

A sludge reaeration process is to be designed to treat a wastewater flow of 1 MGD which has a BOD_u concentration of 200 mg/ℓ. Determine the aeration volumes required if the following design criteria are applicable:

$$X_a = 2000 \text{ mg/ℓ as MLVSS}$$
$$\text{MLVSS} = 0.8 \text{ MLSS}$$
$$\text{SVI} = 100$$
$$S_e = 10 \text{ mg/ℓ BOD}_u$$
$$K = 0.05 \text{ ℓ/mg-day}$$
$$Y_T = 0.5$$
$$K_d = 0.1 \text{ day}^{-1}$$

solution

1/ Compute the specific substrate utilization rate from equation 4-25.

$$q = (0.05)(10)$$
$$= 0.5 \text{ day}^{-1}$$

2/ Calculate the operating θ_c from equation 4-21.

$$\frac{1}{\theta_c} = (0.5)(0.5) - 0.1$$

$$\theta_c = 6.7 \text{ days}$$

3/ Determine the observed yield cofficient from equation 2-63.

$$Y_{\text{obs}} = \frac{0.5}{1 + (0.1)(6.7)}$$

$$= 0.3$$

4/ Calculate the amount of sludge produced in mg/day.

$$\Delta X = 1{,}000{,}000(3.78)(0.3)(200 - 10)$$

$$= 215{,}460{,}000 \text{ mg/day}$$

5/ Assuming that all biomass production occurs in the aeration tank (i.e., neglecting any synthesis in the reaeration tank), a material-balance expression for biomass entering and leaving the aeration tank can be expressed as

$$X_a = \frac{RQX_r}{Q(1 + R)} + \frac{\Delta X}{Q(1 + R)(3.78)}$$

This expression can be solved for X_r to give

$$X_r = \frac{1 + R}{R}\left[X_a - \frac{\Delta X}{Q(1 + R)(3.78)}\right]$$

6/ Compute R from equation 4-79.

$$R = \frac{2000}{(10^6/100)(0.8) - 2000}$$

$$= 0.33$$

7/ From the expression developed for X_r in step 5, determine X_r.

$$X_r = \frac{1 + 0.33}{0.33}\left[2000 - \frac{215{,}460{,}000}{1{,}000{,}000(1 + 0.33)(3.78)}\right]$$

$$= 7887 \text{ mg}/\ell$$

8/ Compute the required aeration tank valume from equation 4-75.

$$V_a = \frac{1[200 - (10)(1 + 0.33)]}{(0.05)(2000)(10)}$$

$$= 0.19 \text{ MG}$$

9/ Determine the required reaeration tank volume from equation 4-77.

$$V_r = \frac{(0.33)(1)[(10^6/100)0.8 - 7887 + (0.5)(10)]}{(0.1)(7887)}$$

$$= 0.05 \text{ MG}$$

Pure Oxygen Aeration

Because of improved commercial availability, the use of high-purity oxygen for activated sludge aeration has increased substantially in the past few years, and it has been accepted as a standard process modification. The first full-scale study comparing the difference in growth and substrate utilization kinetics between an air and oxygen system was conducted at Batavia, New York, by the Linde Division of Union Carbide (Albertson et al., 1970). As a result of that study it was concluded that, in addition to permitting higher solids concentrations in the aeration unit because of increased oxygen transfer efficiency, the pure oxygen system produced less biomass and exhibited a higher substrate utilization rate than the air system. These phenomena were explained by suggesting that the high dissolved oxygen (DO) levels created by the use of oxygen caused metabolic changes within the microorganisms indigenous to the activated sludge process. Such an explanation is contrary to classical microbiological principles, and as a result, the validity of the Batavia study has been questioned by other researchers (Sherrard and Schroeder, 1973; Ball et al., 1972). However, additional researchers have supported their observed effects (Jewell and Mackenzie, 1973; Boon and Burgess, 1974).

Parker and Merrill (1976) suggest that as long as a *minimum* DO level of 2 mg/ℓ is maintained, the performance of an air system will be equal to that of an oxygen system. Benefield et al. (1977) found, though, that there generally will be a difference in the values of the kinetic coefficients relating to substrate utilization and cellular growth between an air system and an oxygen system. However, the difference between the kinetics of the two systems results from a difference in operational characteristics between small-volume high-solids systems and large-volume lesser-solids systems, and the inability of the VSS test to distinguish between proliferating, active but nonproliferating, and inactive cellular material. The flocs in the small-volume high-solids system are recycled much more frequently than those in a large-volume lesser-solids system operated at the same BSRT. During the interval of recycle, the flocs utilize the substrate within the floc lattice and, upon return, have a higher rate of substrate diffusion into the floc. Thus, the rate at which substrate is available to the individual cells is greater, and the more frequent rate of recycle produces a higher rate of substrate utilization.

The development of filamentous microorganisms has been a widespread problem with pure oxygen systems, particularly during startup, and this, too, has been addressed by Benefield et al. (1975). However, sludge settleability is typically superior in an oxygen system if filaments are not numerous. Figure 4-25 illustrates the difference in settleability as reported by Boon (1975). Related Studies also verified that there is less net sludge production in an oxygen unit when compared to a similarly operated air unit. These results are given in Table 4-2.

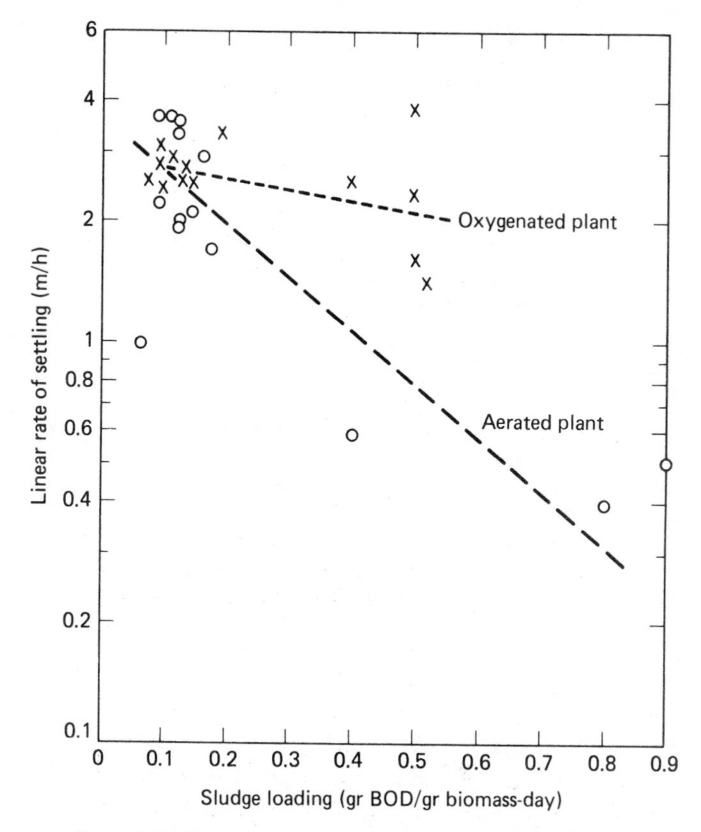

Figure 4-25. Variation in Linear Rate of Settling of Activated Sludge (Measured at Solids Concentration of 3500 mg/ℓ) with Sludge Loading of the Treatment Plant. (After Boon, 1975.)

The results show that for the same BSRT, the observed yield in the pure oxygen system was only 60% of that in the air system. It may be argued that the total biomass, at a specific BSRT, is not the same in a small-volume high-solids system as in a large-volume lesser-solids system. Therefore, comparing yield values between the two at a specific BSRT would be invalid. However, it has been shown that for a specific substrate utilization rate, q, the total biomass (measured as VSS) in a large-volume reactor will be the same as that in a small-volume reactor, regardless of the method of aeration or operation (see Figure 4-26). Thus, for the same specific substrate utilization rate, the steady-state specific growth rate must be greater in the small-volume high-solids unit. As noted by Benefield et al. (1977), this is most likely an artifact resulting from the inability of the VSS test to distinguish between proliferating, active but nonproliferating, and inactive cells. The benefit, nevertheless, is real.

TABLE 4-2

COMPARISON OF SLUDGE PRODUCTION IN OXYGEN AND AIR-ACTIVATED SLUDGE SYSTEMS

Type of treatment	*Wastewater treated*	*Sludge loading (g BOD/g sludge-day)*	*MLVSS (mg/ℓ)*	*BSRT (days)*	*Sludge production (g/g BOD removed)*	*Temperature of mixed liquor (°C)*
Oxygenated	Domestic	0.85	3600	2.6	0.60	13–14
Aerated		0.89	3450	2.1	0.98	
Oxygenated	Domestic and industrial	0.47	5600	2.8	0.96	10–12
Aerated		1.00	6950	11.0	0.95	
Oxygenated	Domestic	0.82	6000	2.2	0.68	11–20
Aerated		0.14	4900	11.0	0.70	
Oxygenated	Industrial and domestic	0.8	3610	3	0.60	Field installation
Aerated		0.9	3450	2	1.00	
Oxygenated	Industrial and domestic	0.6	4085	3	0.60	Field installation
Aerated		0.4	3670	3	0.90	

Source: After Boon and Burgess (1974).

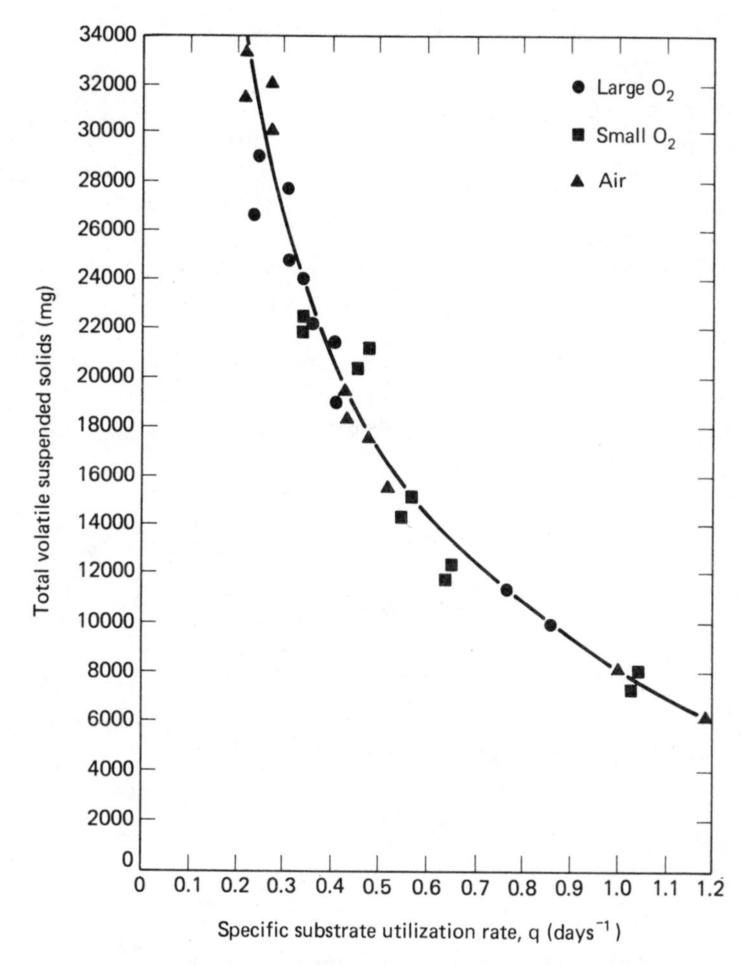

Figure 4-26. Total Volatile Suspended Solids Versus Specific Substrate Utilization Rate for a Small-Volume High-Solids and Two Large-Volume Low-Solids Activated Sludge Systems. (After Benefield, 1977.)

Thus, it is generally accepted that the pure oxygen process has the following advantages over an air system:

1/ Capability to meet higher oxygen demands.
2/ Ability to maintain a higher MLVSS concentration in the aeration tank and thus provide equivalent treatment in a smaller-volume aeration tank.
3/ Better sludge settling and thickening.
4/ Lower net sludge production per unit BOD removed.

5/ Can transfer more oxygen per horsepower ($\simeq$ 4 to 1).
6/ More stable treatment.

Advantages 3, 4, and 6 are the results of advantage 2.

With respect to the high MLVSS concentrations, it must be realized that there is an upper limit on the solids level that can be economically carried in the aeration tank. Speece and Humenick (1974) have found that, for municipal wastewaters, the solids-thickening limitation controls the economics at MLSS levels around 5500 mg/ℓ. Above this figure the money saved on a reduced aeration tank volume will be spent on the clarifier if adequate thickening is to be achieved.

Both covered and uncovered aeration tanks have been used with pure oxygen aeration, but generally the covered tanks have been used in full-scale plants. A typical flow pattern for a covered pure oxygen activated sludge process is shown in Figure 4-27.

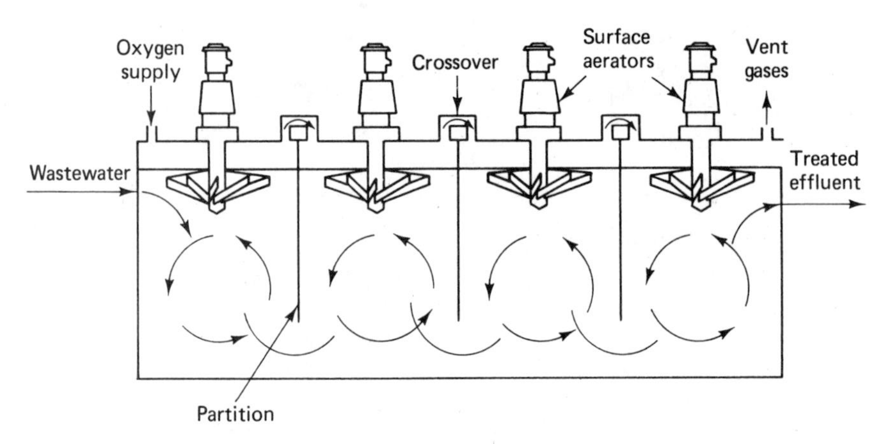

Figure 4-27. Typical Flow Pattern for a Covered Pure Oxygen Activated Sludge Process. (After Lipták, 1974.)

The Environmental Protection Agency (1977) has reviewed typical design and operation of covered and staged pure oxygen processes in the United States. A summary of typical design criteria is given in Table 4-3, and actual operating results are given in Table 4-4.

A principal disadvantage of the covered oxygen process is the accumulation of CO_2 in the wastewater, which results in a decrease in pH which is detrimental to nitrification. To achieve nitrification it is necessary to either strip the CO_2 from the liquid or neutralize it by chemical addition. This difficulty does not occur if a properly designed uncovered oxygen process is used.

Other disadvantages of the staged-covered process is that it (1) requires prescreening of the wastewater, (2) requires substantial shear from the

TABLE 4-3

TYPICAL DESIGN CRITERIA FOR UNOX PROCESS

Influent BOD_5 (mg/ℓ)	Aeration time (hr)	F:M (lb BOD_5/day)/ lb MLVSS	Aeration MLSS (mg/ℓ)	Recycle MLSS (%)	Clarifier overflow (gpd/ft^2)	Effluent BOD_5 (mg/ℓ)
188	1.6	0.62	5500	1.9	560	20
140	1.42	0.47	6250	—	1600	25
200	3.33	0.36	4700	1.9	630	30
200	1.9	0.60	5600	2.8	500	20
350	3.5	0.53	6000	3.0	800	27
200	1.56	0.68	5600	2.0	525	20
110	1.48	0.51	4200	2.2	750	15
370	3.5	0.85	5000	2.0	600	50
133	1.74	0.46	5000	2.2	630	25
1600	13.0	0.42	8000	2.8	540	160

Source: After EPA (1977).

TABLE 4-4

ACTUAL OPERATING DATA FROM UNOX PROCESS

Influent BOD_5 (mg/ℓ)	Aeration time (hr)	F:M (lb BOD_5/day)/lb MLVSS	Aeration MLSS (mg/ℓ)	Recycle MLSS (%)	Clarifier overflow (gpd/ft²)	Effluent quality	
						BOD_5 (mg/ℓ)	TSS (mg/ℓ)
157	1.97	1.10	2700	0.55	456	9	22
101	1.42	1.05	2750	0.85	1595	17	31
125	3.33	0.37	3000	1.20	630	12	50
160	2.16	0.60	4000	2.40	440	12	24
357	4.70	0.33	6400	1.60	590	32	79
245	2.40	0.68	4500	1.50	541	< 10	< 10
114	1.29	0.70	4760	1.66	645	13	18
425	4.20	0.75	4500	1.00	502	25	70
150	1.59	0.62	5100	1.60	704	17	13
1500	9.75	0.45	9600	3.0	360	90	60

Source: After EPA (1977).

impellers to produce small bubbles, and (3) is not economical for small plants.

A significant development for pure oxygen systems is the submerged rotating active diffuser (RAD) for uncovered oxygen systems (EPA, 1977). The RAD is designed to produce minute oxygen bubbles in the liquid without creating excessive turbulence or shear, which would tend to fragment the flocs. Oxygen utilization efficiencies of greater than 90% in conventional depth uncovered aeration tanks have been reported using this equipment. A cutaway view and a typical installation are shown in Figures 4-28 and 4-29.

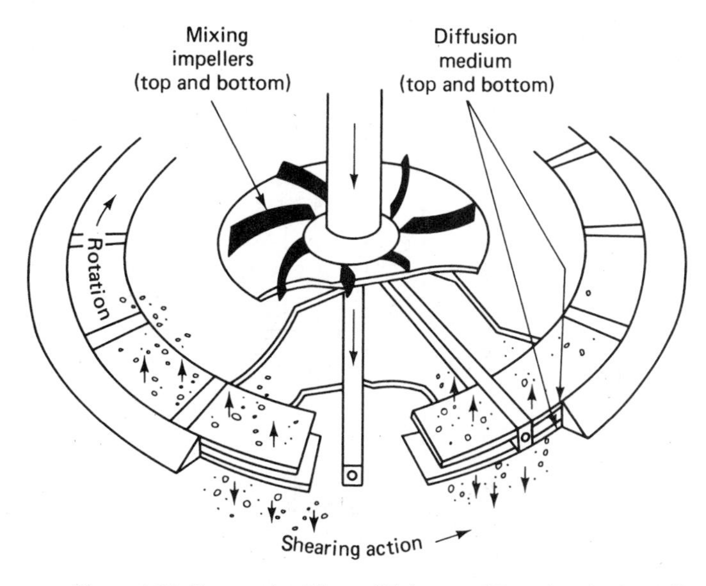

Figure 4-28. Perspective View of Submerged Rotating Active Diffuser Showing Gas Flow and Bubble Formation. (After EPA, 1977.)

Oxidation Ditches

The oxidation ditch is generally operated as an extended aeration process and employed for the treatment of small wastewater flows. A schematic of this process is presented in Figure 4-30. It consists of an oval or "racetrack"-shaped channel about 3 ft deep. A brush or caged rotor aerator is placed across the ditch to aerate the mixed liquor as well as impart unidirectional flow to the passing liquid. The mixed liquor circulates in the ditch at a rate of 1 to 2 ft/s (Metcalf and Eddy, 1972).

In the winter of 1975–1976, EPA Region VII (Iowa, Missouri, Kansas, and Nebraska) conducted a survey of secondary treatment facilities to determine if secondary treatment requirements (85% removal of BOD and suspended solids) were being met during the harsh winter months. Thirty facilities were

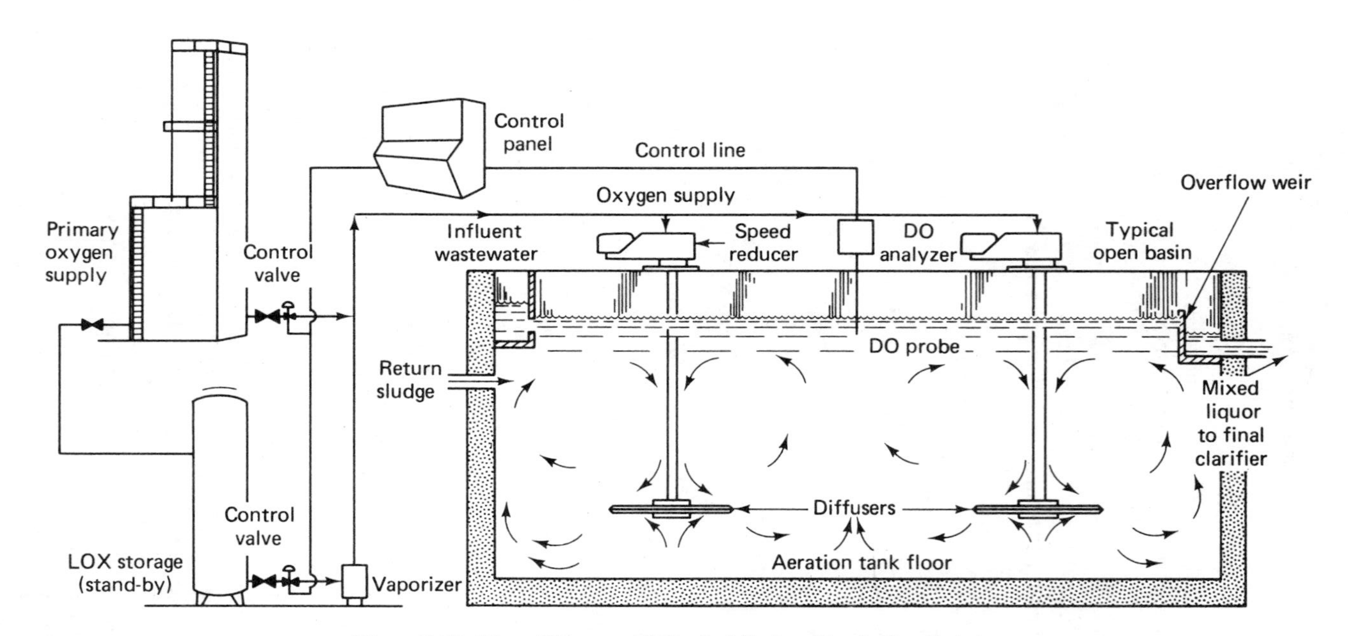

Figure 4-29. Flow Diagram of Typical System Employing Rotating Active Diffusers. (After EPA, 1977.)

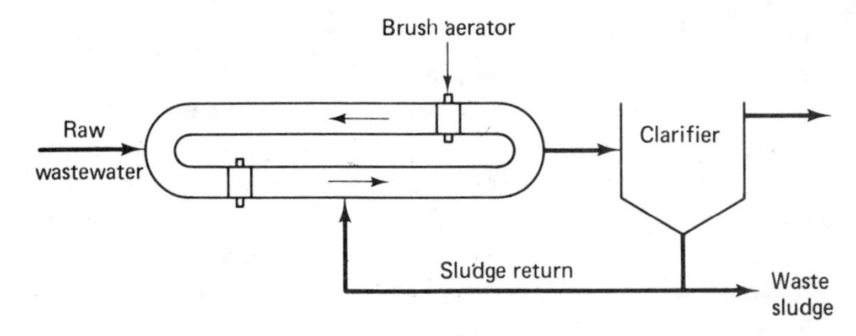

Figure 4-30. Schematic of Oxidation Ditch.

surveyed: 7 oxidation ditches, 7 activated sludge plants, and 16 trickling filters (covered and uncovered). Only 10 facilities were found to meet secondary treatment requirements, and 5 of these were oxidation ditches. Hoffmeier (1974) also found that oxidation ditches were effective treatment processes. He reported that as a class, oxidation ditches show the best performance of any type of secondary treatment facility in Colorado and presented data showing the relationship between effluent BOD and BOD loading for this type of treatment system (see Figure 4-31). These data indicate that as long as the BOD loading does not exceed 10 lb $BOD_5/1000$ ft^3-day, a good quality effluent can be expected. For the six plants shown in Figure 4-31 with a loading less than 10 lb $BOD_5/1000\ ft^3$-day, the average effluent suspended solids concentration was 25 mg/ℓ.

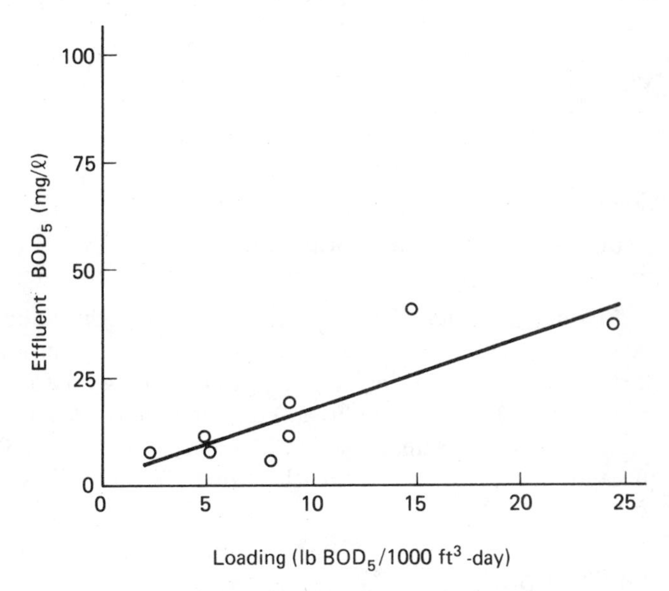

Figure 4-31. Oxidation Ditch Performance for Eight Colorado Plants. (After Hoffmeier, 1974.)

Of all the activated sludge process modifications, the oxidation ditch seems to be the least affected by operator skill. This should always be a consideration when selecting the particular modification to be used.

A useful modification of the oxidation ditch has been developed by Reid Engineering Company of Fredericksburg, Va,, and installed at a Holly Farms plant in New Market, Va. To avoid the aerosol, splashing, and icing problems that occur with brush aerators, they directed the entire flow through an underground, U-shaped passage and installed a turbine aerator at the entrance to the passage (see Figure 4-32).

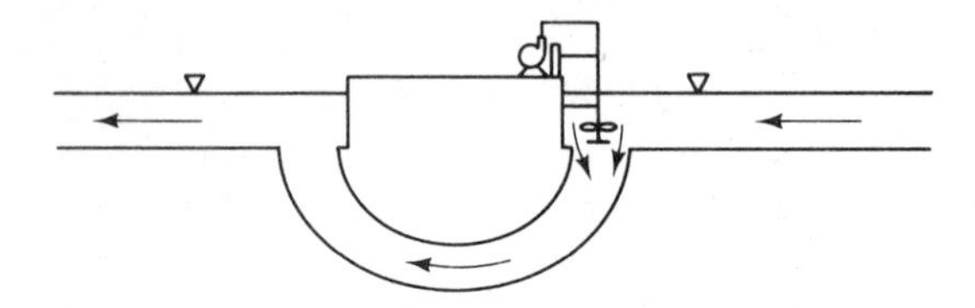

Figure 4-32. Subsurface Aeration of Oxidation Ditch.

Other Process Modifications

Although biological nitrification, biological denitrification, and the anaerobic-contact processes can be considered as valid modifications of the conventional activated sludge process, a discussion of these processes will be postponed until all factors that affect design for aerobic carbon removal have been covered. It is felt that such an approach will provide for a better understanding of the many factors that must be considered when designing the aforementioned process modifications.

4-4 Process Design Considerations

Mixing regime and the appropriate process modification are not the only factors that must be considered in activated sludge process design. Other factors that may be of importance are (1) loading criteria, (2) excess sludge production, (3) sludge viability, (4) oxygen requirement, (5) nutrient requirements, (6) temperature, (7) solids–liquid separation, and (8) effluent quality. Because of their importance, each of these factors will be discussed in some detail.

Loading Criteria

The food/microorganism (F:M) ratio has commonly been the design criterion for organic loading and is defined as the substrate load applied to the process per unit of biomass in the aeration tank per unit of time. Only the biomass in the aeration tank is used to compute the F:M ratio because it is assumed that substrate utilization occurs only in the aeration tank. Hence, the F:M ratio can be expressed mathematically as

$$\text{F:M} = \frac{QS_0}{XV_a} \tag{4-80}$$

An equation can be developed relating F:M ratio to θ_c by substituting for $(dS/dt)_u$ in equation 4-8 from equation 4-17.

$$\frac{1}{\theta_c} = Y_T \frac{Q(S_0 - S_e)}{XV_a} - K_d \tag{4-81}$$

Substituting for QS_0/XV_a from equation 4-80 then gives

$$\frac{1}{\theta_c} = Y_T\left(\text{F:M} - \frac{QS_e}{XV_a}\right) - K_d \tag{4-82}$$

Since the QS_e/XV_a term is normally very small compared to the value for the F:M ratio, equation 4-82 can be approximated as

$$\frac{1}{\theta_c} \approx Y_T(\text{F:M}) - K_d \tag{4-83}$$

Basically, what equation 4-83 shows is that large F:M ratios result in small values for θ_c, whereas small F:M ratios result in large θ_c values (i.e., there is an inverse relationship between F:M and θ_c).

Since $\theta_c^{-1} = \mu$ for a completely mixed activated sludge process operating at steady-state and because the physiological state of the microbial culture can be related directly to the magnitude of the specific growth rate, μ, the authors feel that θ_c is more informative than F:M as a design parameter. Thus, the procedure followed here will be to use θ_c as the design criterion for organic loading rather than the F:M ratio when a choice between the two parameters is possible.

The organic loading on the activated sludge process affects (1) process efficiency, (2) sludge production, (3) oxygen requirement, and (4) solids–liquid separation. For example, at very high organic loadings (i.e., short θ_c values), the substrate removal efficiency of the process is low (60 to 75%); sludge production is maximized, which also means that maximum nutrient removal will occur as nitrogen and phosphorus are required in the synthesis of new cytoplasmic material; total oxygen requirement is minimized because a larger fraction of the substrate removed is used for synthesis rather than oxidized for energy; and because a short BSRT implies a high specific growth rate, the biomass will be dispersed rather than agglomerated, causing problems with solids–liquid separation and producing a high-solids concentration in the plant effluent. Such a situation is typified by the high-rate activated sludge process. On the other hand, at very low organic loading (i.e., long θ_c

values), process efficiency is moderately high (75 to 95%); sludge production is minimal (meaning also that nitrogen and phosphorus removal as a result of cell synthesis is minimized); total oxygen requirement is maximized as a large fraction of the substrate removed is oxidized for energy rather than synthesized to new cellular material; and, even though a long BSRT implies a low specific growth rate, under extreme conditions, activated sludge may deflocculate, causing poor solids–liquid separation and resulting in a high-solids concentration in the plant effluent. Such a situation is typified by extended aeration.

The selection of the proper organic loading or BSRT for design is therefore a problem that requires the engineer to account for process efficiency, excess sludge production, oxygen requirement, and sludge settling characteristics, integrated so that the optimum design is produced. Typical values for various design parameters and process modifications are given in Table 4-5.

This discussion would not be complete without stressing the importance of recognizing the limitations of θ_c as the design criterion for organic loading. The mathematical models developed to describe treatment kinetics do not reflect the physical limitations of the system. These models place no limit on the organic loading or MLVSS concentration which may be imposed on the process, whereas under actual field conditions the loading and/or concentration may be limited by (1) the ability to maintain the biomass in suspension (i.e., provide adequate mixing), (2) the ability to efficiently separate the solid from the liquid phase in the secondary clarifier, and (3) the ability to maintain an adequate DO level in the aeration tank. Recognizing these limitations, the concept of BSRT provides the engineer with a very powerful tool for process design and control.

The design engineer should always have in mind the fundamental design constraints of all activated sludge processes. These are:

1/ Maximum microbial growth rate,
2/ Maximum oxygen transfer rate,
3/ Minimum hydraulic retention time,
4/ Mixed liquor suspended solids settleability.

Excess Sludge Production

Before the engineer can properly design sludge handling and disposal facilities, he must be able to estimate the total quantity of sludge produced during treatment of the wastewater. Earlier it was reported that during the period of aeration in the activated sludge process, organic material that is removed is either channeled into the growth function and used for synthesis of new cellular material or is channeled into energy metabolism and oxidized

TABLE 4-5

DESIGN CRITERIA FOR ACTIVATED SLUDGE PROCESSES

Process modification	θ_c *(days)*	*q (lb BOD_5 removed/ lb MLVSS-day)*	*MLSS (mg/ℓ)*	*V/Q (h)*	Q_r/Q	*BOD removal efficiency (%)*
Conventional	5–15	0.2–0.4	1500–3000	4–8	0.25–0.5	85–95
Completely mixed	5–15	0.2–0.6	3000–6000	3–5	0.25–1.0	85–95
Step aeration	5–15	0.2–0.4	2000–3500	3–5	0.25–0.75	85–95
High rate	0.2–0.5	1.5–5.0	600–1000	1.5–3	0.05–0.15	60–75
Contact stabilization	5–15	0.2–0.6	(1000–3000)[a] (4000–10,000)[b]	(0.5–1.0)[a] (3–6)[b]	0.25–1.0	80–90
Extended aeration	Not applicable	0.05–0.15	3000–6000	18–36	0.75–1.50	75–95
Pure-oxygen system	8–20	0.25–1.0	3000–5000	1–3	0.25–0.5	85–95
Sludge reaeration	5–15	0.2–0.4	(1500–3000)[c] (5000–10,000)[d]	(3–6)[c] (2–4)[d]	0.2 –0.6	80–90

[a]Contact tank.
[b]Stabilization tank.
[c]Aeration tank.
[d]Reaeration tank.
Source: After Metcalf and Eddy (1972).

to carbon dioxide and water. The absolute or net microbial growth rate resulting from substrate utilization is given by either equation 2-59 or 2-60.

$$\left(\frac{dX}{dt}\right)_g = Y_T\left(\frac{dS}{dt}\right)_u - K_d X \tag{2-59}$$

$$\left(\frac{dX}{dt}\right)_g = Y_{\text{obs}}\left(\frac{dS}{dt}\right)_u \tag{2-60}$$

Furthermore, it was shown that equation *2-63*

$$Y_{\text{obs}} = \frac{Y_T}{1 + K_d/\mu} \tag{2-63}$$

predicts a greater value of Y_{obs} for high specific growth rates than for low ones. Since $\theta_c^{-1} = \mu$ for the completely-mixed activated sludge process operating at steady-state, there exists an inverse relationship between Y_{obs} and θ_c. This relationship is shown graphically in Figure 4-33.

On a finite-time basis (generally 1 day), equation 2-60 can be expressed as

$$\Delta X = Y_{\text{obs}} \Delta S \tag{4-84}$$

or

$$\Delta X = Y_{\text{obs}} Q(8.34)(S_0 - S_e) \tag{4-85}$$

where ΔX = lb of biomass (dry-weight basis) produced in 1 day

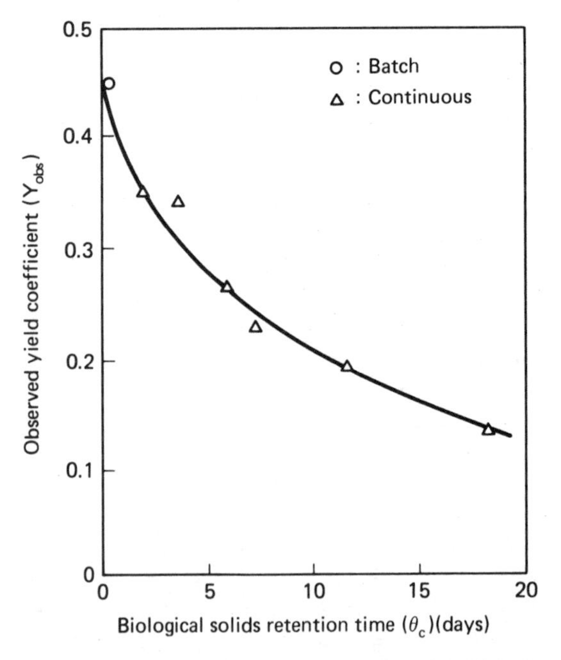

Figure 4-33. Observed Yield as a Function of Biological Solids Retention Time. (After Sherrard and Schroeder, 1973.)

Equation 4-85 can then be used to predict the increase in biomass as a result of biological metabolism. However, the engineer is also interested in knowing the total quantity of sludge generated during the treatment process. This includes not only the biomass produced as a result of organic removal but also the buildup of nonbiodegradable suspended solids originally present in the wastewater. Thus, equation 4-85 can be modified to account for the total sludge production in the process when primary clarification is not used.

$$SP = 8.34Q[Y_{obs}(S_0 - S_e) + \text{NDVSS} + \text{fSS}] \tag{4-86}$$

where SP = lb sludge (dry-weight basis) produced in 1 day
NDVSS = nonbiodegradable volatile suspended solids in the raw wastewater, mg/ℓ
fSS = fixed suspended solids in the raw wastewater, mg/ℓ

When primary clarification is used, the concentration of the NDVSS and the fSS will be reduced by some fractional amount (see Figures 3-13 and 3-14).

Lawrence (1975) proposes that for biological processes employing recycle, the concentration of nonbiological suspended solids in the mixed liquor can be determined from the equation

$$X_{NB} = \left(\frac{\theta_c}{t}\right)(\text{NDVSS} + \text{fSS}) \tag{4-87}$$

where X_{NB} represents the steady-state concentration of nonbiological mixed liquor suspended solids and t represents the hydraulic retention time in the aeration tank.

Sludge Viability

In the development up to this point, the active biomass concentration has been represented by X. Although such a representation is valid, a problem does result when it becomes necessary to actually measure the active biomass concentration in the system. Generally, it is assumed that volatile suspended solids will adequately measure this parameter. Yet, treatment of a soluble waste will produce a sludge floc that is composed of an active-proliferating fraction, an active-nonproliferating fraction, and an inert organic fraction. This situation is further complicated when the wastewater contains an insoluble, nonbiodegradable organic fraction. As an example, McKinney (1974) suggests that for an activated sludge process treating domestic wastewater, the active biomass represents only 30 to 50% of the MLSS. This figure drops to less than 10% for the extended aeration process. Weddle and Jenkins (1971) propose that the active heterotrophic microorganism content of activated sludge is only 10 to 20% of the MLVSS. In either case, it is no wonder that the VSS measurement for active biomass is suspect.

Inconsistencies do arise when VSS are used as a measure of active biomass, especially when the raw wastewater contains a large fraction of nonbiological suspended organic material. No doubt the actual percent of MLVSS which is active biomass is a function of BSRT, and this percent would be lowered by the inclusion of unmetabolized organics. Nevertheless, while many comparisons of more direct measurements such as cellular DNA (Agardy et al., 1963), organic nitrogen (Hartmann and Laubenberger, 1968), ATP content (Patterson et al., 1969), and dehydrogenase activity (Bucksteeg, 1966) have been made, a thorough evaluation led to the conclusion that VSS is just as accurate a measurement as any of them and it is much easier to measure (Weddle and Jenkins, 1971).

Oxygen Requirement

In Chapter 2 it was reported that oxygen is required as an electron acceptor in the energy metabolism of the aerobic heterotrophic organisms indigenous to the activated sludge process. In other words, a portion of the organic material removed is oxidized to provide energy for the maintenance function (nongrowth) and the synthesis function (growth). Any oxidation must be coupled with reduction (i.e., something must accept the electrons released during the oxidation process). Oxygen satisfies this requirement in aerobic microorganisms.

If the organisms in the aeration tank are deprived of an adequate supply of oxygen, process failure will occur, as noted by a rapid deterioration in effluent quality. Therefore, it is the engineer's responsibility to determine how much oxygen will be required by the organisms for a particular treatment situation and to design an aeration system capable of providing this amount of oxygen in addition to maintaining a dissolved oxygen residual of 2 mg/ℓ or greater throughout the aeration tank, while ensuring adequate mixing of the tank contents.

The *Ten-State Standards* require that aeration equipment must be capable of maintaining a DO level of 2 mg/ℓ in the aeration tank while providing thorough mixing of the solid and liquid phases. The air requirements for various process modifications which are commonly used are listed in Table 4-6.

There are also several methods available for calculating the oxygen requirement of an activated sludge process. One method is based on knowing the ultimate BOD of the waste and the amount of biomass wasted from the system each day (Metcalf and Eddy, 1972). To illustrate this, let the substrate concentration be represented by the nonspecific parameter, ultimate biochemical oxygen demand (BOD_u). Then, if all the substrate removed by the microorganisms was totally oxidized for energy purposes, the total oxygen requirement per day would be

$$\text{total } O_2 \text{ requirement per day} = 8.34Q(BOD_{u_{\text{influent}}} - BOD_{u_{\text{effluent}}}) \qquad (4\text{-}88)$$

TABLE 4-6

NORMAL AIR REQUIREMENTS FOR ACTIVATED SLUDGE PROCESSES

Process modification	*Cubic feet of air supplied/lb BOD_5 applied to aeration tank*
Conventional	1500
Step aeration	1500
Contact stabilization	1500
High rate	400–1500
Extended aeration	2000

Source: Great Lakes–Upper Mississippi River Board of State Sanitary Engineers (1971).

However, not all the substrate is oxidized; only a fraction of the total substrate removed is used for energy. The remainder is synthesized to new biomass, and at steady-state the amount of biomass produced is equal to the biomass wasted. Thus, the substrate synthesized to new biomass is not oxidized in the system and therefore exerts no requirement for oxygen.

If it is assumed that the biomass can adequately be described by the chemical formula $C_5H_7NO_2$ (Sykes, 1975), the oxygen required to oxidize a unit of biomass can be calculated in the following manner:

$$\underset{113}{C_5H_7NO_2} + \underset{(5 \times 32)}{5O_2} \longrightarrow 5CO_2 + 2H_2O + NH_3 \qquad (4\text{-}89)$$

$$\frac{5 \times 32}{113} = 1.42 \text{ units } O_2/\text{unit biomass oxidized}$$

It is therefore necessary to subtract that amount of oxygen which would be required to oxidize the biomass produced as a result of substrate utilization from the theoretical oxygen requirement given by equation 4-88. The resulting value is the actual oxygen requirement.

$$\begin{bmatrix} O_2 \text{ required} \\ \text{per day} \end{bmatrix} = \begin{bmatrix} \text{total amount of} \\ \text{substrate removed} \\ \text{per day given as} \\ \text{ultimate BOD} \end{bmatrix} - 1.42 \begin{bmatrix} \text{total active mass} \\ \text{of organisms} \\ \text{wasted each day} \end{bmatrix} \qquad (4\text{-}90)$$

Lawrence (1975) proposes that the oxygen requirement for carbon removal may be estimated from the following equation:

$$\Delta O_2 = 8.34Q[(1 - 1.42Y_T)(S_0 - S_e)] + 8.34(1.42)K_d X V_a \qquad (4\text{-}91)$$

where S_0 and S_e are given in terms of ultimate BOD, and the aeration tank volume is expressed in millions of gallons. The oxygen requirement in pounds per day is given by ΔO_2. Neither of the equations above account for nitrification oxygen requirements.

Additional equations for calculating oxygen requirement can be developed from the knowledge that the total oxygen requirement of the organisms in a biological system is related to the oxygen requirement for synthesis and the oxygen requirement for energy of maintenance. At this point, however, the reader must decide whether to accept the theory proposed by Pirt (1965)—that microorganisms, in the presence of an external substrate source, oxidize a portion of the substrate to satisfy maintenance energy requirements—or to accept the theory proposed by Herbert (1958)—that all the substrate is synthesized to new cellular material, and then a portion of this cellular material is oxidized to satisfy the energy of maintenance requirement.

If one accepts the theory of Herbert (1958), the rate of oxygen utilization can be expressed by the familiar equation (Goodman and Englande, 1974)

$$\frac{dO_2}{dt} = a\left(\frac{dS}{dt}\right)_u + bX \tag{4-92}$$

where $\frac{dO_2}{dt}$ = oxygen utilization rate, mass volume^{-1} time^{-1}

a = oxygen-use coefficient for synthesis

b = oxygen-use coefficient for energy of maintenance, time^{-1}

On a finite-time basis, this expression can be written as

$$\Delta O_2 = 8.34Qa(S_0 - S_e) + 8.34bXV_a \tag{4-93}$$

where ΔO_2 represents the oxygen requirement in pounds per day.

On the other hand, if the theory of Pirt (1965) is accepted, the rate of oxygen utilization would be expressed by

$$\frac{dO_2}{dt} = a_1\left(\frac{dX}{dt}\right)_g + b_1X \tag{4-94}$$

where a_1 = oxygen required to form a unit of biomass

b_1 = oxygen required per unit of biomass per unit time for life-support functions, time^{-1}

On a finite-time basis, this becomes

$$\Delta O_2 = a_1\,\Delta X + 8.34b_1XV_a \tag{4-95}$$

where ΔX is the pounds of biomass produced per day as calculated from equation 4-85. It is equation 4-95 which the authors feel more accurately predicts the oxygen requirement for the activated sludge process. This is largely a matter of opinion, for no one has proven beyond doubt which of the theories actually describes the mechanism of energy of maintenance, although it would seem that the process of synthesis before oxidation is somewhat wasteful in terms of energy utilization—and microorganisms do have a reputation for being efficient.

If it is assumed that organic nitrogen contained in the wastewater is ultimately transformed to ammonium as a result of microbial activity and the ammonium concentration is such that nitrification imposes a significant oxygen demand, nitrification should be considered in calculating the total oxygen requirement. Since 4.57 lb of oxygen are utilized in the oxidation of a pound of NH_4^+-N, the nitrogenous oxygen demand (NOD) is conservatively approximated by

$$\text{NOD} = 8.34Q(TKN_0)(4.57) \tag{4-96}$$

where NOD = nitrogenous oxygen demand, lb/day
TKN_0 = total Kjeldahl nitrogen in influent, mg/ℓ as nitrogen

Equation 4-90 then becomes

$$\Delta O_2 = 8.34Q(S_0 - S_e) - 1.42\,\Delta X + \text{NOD} \tag{4-97}$$

If the values for the coefficients in equations 4-93 and 4-95 were determined using a system that was accomplishing nitrification of the waste being treated, no adjustment for NOD would be necessary. If not, the following equations could be used:

$$\Delta O_2 = 8.34Qa(S_0 - S_e) + 8.34bXV_a + \text{NOD} \tag{4-98}$$

and

$$\Delta O_2 = a_1\,\Delta X + 8.34b_1XV_a + \text{NOD} \tag{4-99}$$

Example 4-7

A completely mixed activated sludge process is to be used to treat 10 MGD of wastewater containing 200 mg/ℓ of BOD_u. Compute the oxygen requirement for this process using the following design criteria:

$$X = 2500 \text{ mg/ℓ as MLVSS}$$
$$S_e = 5 \text{ mg/ℓ as } BOD_u$$
$$Y_T = 0.5$$
$$K_d = 0.1 \text{ day}^{-1}$$
$$K = 0.10 \text{ ℓ/mg-day}$$
$$\text{influent TKN} = 20 \text{ mg/ℓ as N}$$

solution

1/ Compute the required specific substrate utilization rate from equation 4-25

$$q = KS_e$$
$$= (0.10)(5) = 0.5 \text{ day}^{-1}$$

2/ Determine the aeration tank volume requirement from equation 4-25.

$$V_a = \frac{Q(S_0 - S_e)}{Xq}$$
$$= \frac{10(200 - 5)}{2500(0.5)} = 1.56 \text{ MG}$$

3/ Calculate the operating BSRT using equation 4-8.

$$\frac{1}{\theta_c} = Y_T q - K_d$$

or

$$\theta_c = 6.7 \text{ days}$$

4/ Determine the observed yield coefficient value from equation 2-63.

$$Y_{obs} = \frac{Y_T}{1 + K_d\theta_c}$$

$$= \frac{0.5}{1 + (0.1)(6.7)} = 0.3$$

5/ Estimate the biomass production by applying equation 4-85.

$$\Delta X = Y_{obs}Q(8.34)(S_0 - S_e)$$

$$= (0.3)(10)(8.34)(200 - 5) = 4879 \text{ lb/day}$$

6/ Approximate the total oxygen requirement by using equation 4-97, assuming that the nitrification efficiency is 100%.

$$\Delta O_2 = 8.34Q(S_0 - S_e) - 1.42\,\Delta X + \text{NOD}$$

ΔX can be substituted for the quantity of biomass wasted each day because at steady-state, the quantity of biomass produced each day equals the amount wasted each day.

$$\Delta O_2 = 8.34(10)(200 - 5) - 1.42(4879) + 38.1(10)(20)$$

$$= 16{,}955 \text{ lb/day}$$

Example 4-8

For the problem outlined in Example 4-5, compute the oxygen requirement in both the contact tank and the stabilization tank. Additional design criteria that may be required in this determination are as follows:

$$\text{influent TKN} = 20 \text{ mg}/\ell \text{ as N}$$

The total Kjeldahl nitrogen is composed of 60% ammonium nitrogen and 40% insoluble organic nitrogen.

solution

In computing the oxygen requirement and distribution in the contact-stabilization process, it will be assumed that (1) carbonaceous oxygen demand in the contact tank is created only by the soluble substrate removed in this tank and the energy of maintenance requirement of the microorganisms; (2) the nitrogenous oxygen demand in the contact tank is created only by the oxidation of the ammonium fraction of the TKN; (3) carbonaceous oxygen demand in the stabilization tank is created by the oxidation of the insoluble organic material removed in the contact tank, by the oxidation of the soluble organic material contained in the flow to the stabilization tank and by the energy of maintenance requirement of the microorganisms; and (4) the nitrogenous oxygen demand in the stabilization tank is due to the oxidation of ammonium created by deamination reactions involving organic compounds containing nitrogen.

1/ Compute the oxygen requirement for the contact tank by applying equation 4-91.

$$\begin{aligned}\Delta O_2 &= 8.34Q\{(1 - 1.42Y_T)[(1 - f)(S_0) - S_e]\} \\ &\quad + 8.34(1.42)K_d V_c X_c + 0.6\text{NOD} \\ &= (8.34)(2)\{[1 - (1.42)(0.5)][(1 - 0.8)(150) - 15)]\} \\ &\quad + (8.34)(1.42)(0.1)(0.063)(2000) + (0.6)(38.1)(2)(20) \\ &= 1136 \text{ lb/day}\end{aligned}$$

2/ Determine the oxygen requirement in the stabilization tank using the following modification to equation 4-91.

$$\begin{aligned}\Delta O_2 &= (1 - 1.42Y_T)[(8.34)(Q)(f)(S_0) + (8.34)(R)(Q)(S_e)] \\ &\quad + 8.34K_d X_s V_s + 0.4\text{NOD} \\ &= [1 - (1.42)(0.5)][(8.34)(2)(0.8)(150) \\ &\quad + (8.34)(0.38)(2)(15)] + 8.34(0.1)(7373)(0.06) \\ &\quad + (0.4)(38.1)(2)(20) \\ &= 1587 \text{ lb/day}\end{aligned}$$

These calculations suggest that 42% of the total oxygen requirement is created in the contact tank, whereas 58% is in the stabilization tank. This is very near the 40–60 distribution typically quoted in the literature. However, it should be noted that the size of the stabilization tank will significantly affect this distribution.

Nutrient Requirement

Most of the dry weight of microorganisms consists of carbon, hydrogen, nitrogen, and oxygen which are released as gaseous compounds on combustion of the cell. The remaining ash elements constitute from 2 to 14% of the dry weight of the cell. On the average, carbon accounts for about 50% of the dry weight, oxygen for about 30%, nitrogen from 7 to 14%, and hydrogen for approximately 7%. Phosphorus accounts for about 50% of the total weight of the ash. Most of the remaining ash is composed of potassium, sodium, magnesium, sulfur, calcium, chlorine, and iron, with trace amounts of manganese, zinc, molybdenum, boron, chromium, and cobalt.

Organic removal in the activated sludge process is accomplished by aerobic heterotrophic mircoorganisms which utilize a portion of the organic material as a carbon source for synthesis of new biomass and oxidize the remainder to provide energy for the synthesis function as well as for other nongrowth functions. In order for the synthesis function to proceed, the wastewater must contain an adequate supply of all the elements which are found in the cytoplasmic material of a cell. Generally, this requirement is easily met with municipal wastewater, but some industrial wastewaters are found to be deficient in nitrogen and/or phosphorus. When this occurs, these elements must be added if the treatment process is to function properly.

The nitrogen and phosphorus requirements are commonly based on the BOD_5 of the wastewater, where a $BOD_5/N/P$ ratio of 100: 5: 1 is considered acceptable. Although experience has shown that such an approach will certainly fulfill the nitrogen and phosphorus requirements, the economics of it must be questioned when one considers that the amount of biomass which is produced will decrease as the BSRT is increased. Since nitrogen and phosphorus are required in the synthesis of new cellular material, it follows that the requirement for these elements will decrease as biomass production decreases.

A widely used molecular formula to describe the composition of the biomass (which also includes the phosphorus content of the cells) is given by (McCarty, 1970)

$$C_{60}H_{87}O_{23}N_{12}P$$

This structure has a formula weight of 1374. The unit fraction represented by nitrogen is 168/1374, or 0.122, while phosphorus comprises 31/1374, or 0.023, of the unit weight. It is therefore possible to approximate the nitrogen and phosphorus requirement for a particular set of treatment conditions by the equations

$$\begin{bmatrix}\text{nitrogen requirement}\\ \text{(lb/day)}\end{bmatrix} = 0.122\,\Delta X \qquad (4\text{-}100)$$

where ΔX = biomass produced per day as calculated from equation 4-85 and

$$\begin{bmatrix}\text{phosphorus requirement}\\ \text{(lb/day)}\end{bmatrix} = 0.023\,\Delta X \qquad (4\text{-}101)$$

Such an approach reflects the decreasing nutrient requirement with increasing BSRT which, although not precise, does offer a simple method of estimating this requirement with the expectation of some reduction in chemical cost (where required) at higher values of θ_c. Figure 4-34 illustrates the change in nutrient requirement with increasing BSRT.

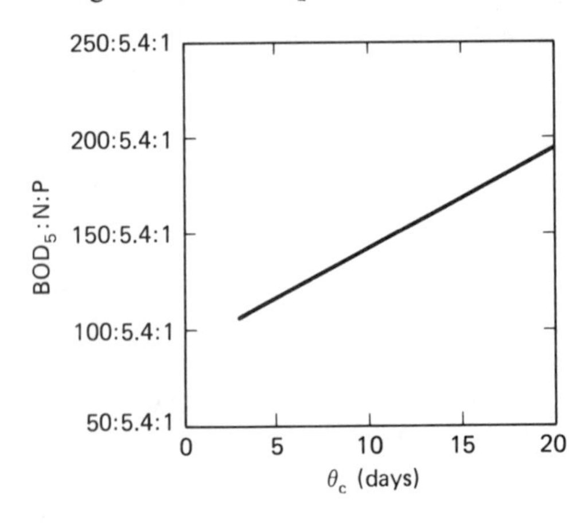

Figure 4-34. Effect of θ_c on Activated Sludge Nutrient Requirements. (After Sherrard, 1975.)

Example 4-9

Considering the problem given in Example 4-7, if the total phosphorus concentration of the raw wastewater is 7 mg/ℓ, are the nutrients nitrogen and phosphorus available in the required amounts so that neither would limit growth?

solution

1/ Compute the nitrogen and phosphorus loading to the plant.

a/ Nitrogen:

$$\text{N(lb/day)} = 8.34(10)(20) = 1668 \text{ lb/day}$$

b/ Phosphorus:

$$\text{P(lb/day)} = 8.34(10)(7) = 584 \text{ lb/day}$$

2/ Determine the N and P required to satisfy cell synthesis requirements.

a/ Nitrogen:

$$\text{N(lb/day)} = 0.122(4879) = 595 \text{ lb/day}$$

b/ Phosphorus:

$$\text{P(lb/day)} = 0.023(4879) = 112 \text{ lb/day}$$

These calculations show that nitrogen and phosphorus are available in the required amounts for growth to occur unrestricted relative to nutrient concentrations.

Temperature Effects

Temperature is an important consideration in the design of an activated sludge process because of the effect it has on microbial activity. The rate of biochemical reactions in the cell increases with temperature up to a maximum; with a further increase in temperature, the rate of activity will decline as enzyme denaturation occurs. The effect on bacterial growth rate of such a response is shown in Figure 4-35. Here it becomes evident that the physiological state of bacteria can be directly affected by temperature changes in their

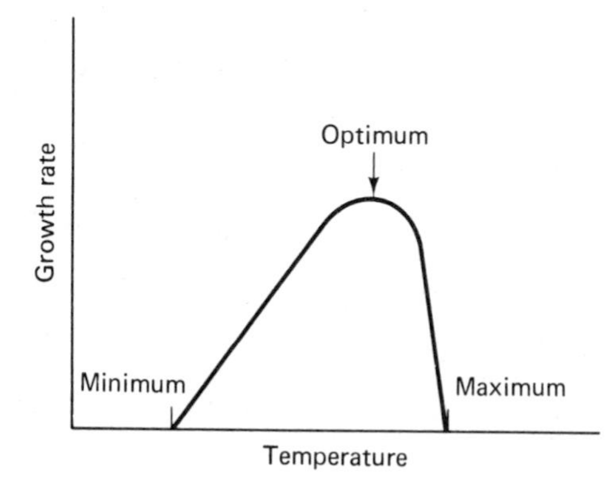

Figure 4-35. Effect of Temperature on Bacterial Growth Rate. (After Brock, 1970.)

environment. Because of this it is necessary that the engineer account for seasonal changes in the operating temperature (when significant) of an activated sludge process and know what effect such changes will have on the values of the biokinetic coefficients used in process design as well as the effect on the settling characteristics of the biomass.

The expression most commonly used in biological wastewater treatment to adjust reaction rates for temperature variation is the modified Arrhenius relationship developed in Chapter 1.

$$C_2 = C_1\theta^{T_2-T_1} \qquad (1\text{-}30)$$

where C represents the kinetic constant of interest. Experimental evidence indicates that the substrate utilization-rate constant, K, does indeed vary according to the Arrhenius relationship over the normal range of operating temperatures. Figure 4-36 illustrates such a response when the utilization rate for glucose by a mixed culture was studied under batch conditions. To correct the substrate utilization reaction-rate constant, K, for temperature, equation 1-30 is represented in the following form:

$$K_2 = K_1\theta^{T_2-T_1} \qquad (4\text{-}102)$$

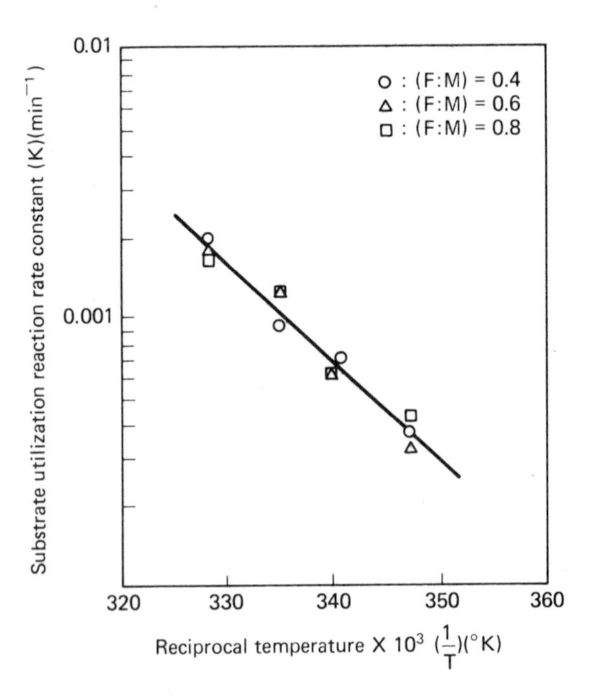

Figure 4-36. Variation in the Substrate Utilization Reaction Rate Constant with Temperature. (After Benefield et al., 1975.)

where temperature is expressed in degrees Celsius. Because of the high solids levels maintained in activated sludge systems, the process has been found to be fairly insensitive to temperature, with typical θ values of 1.01 to 1.04 reported.

Research indicates that the yield coefficient, Y_T, and the microbial decay coefficient, K_d, probably will not vary according to the modified Arrhenius relationship (Muck and Grady, 1974; Benefield et al., 1975; Topiwala and Sinclair, 1971). Furthermore, there is some disagreement as to what a particular response will be to variations in temperature. For example, Figure 4-37 illustrates the steady-state response of Y_T to temperature changes as noted by Muck and Grady (1974). On the other hand, Benefield et al. (1975) found that between 15 and 25°C, temperature variations have a minimal effect on the yield coefficient (see Figure 4-38). Experimental work by Sayigh and Malina (1976) support the findings of Benefield et al. (1975). These workers found that the value of Y_T was basically independent of temperature within the range 4 to 20°C.

Benefield et al. (1975) also found that the microbial decay coefficient varied according to the Arrhenius relationship between 15 and 25°C but deviated appreciably at 32°C. However, as 32°C is outside the operating range of most activated sludge plants, it is reasonable to assume that variations in K_d with

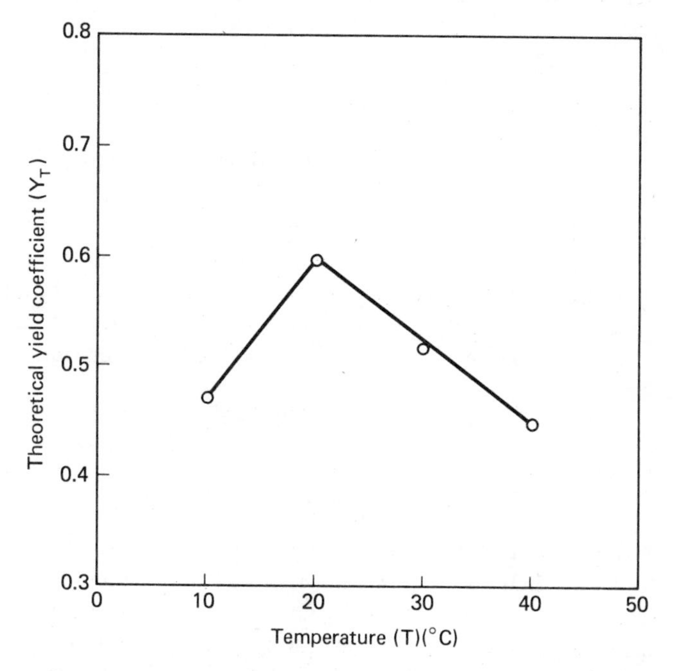

Figure 4-37. Relationship Between Theoretical Yield and Temperature. (After Muck and Grady, 1974.)

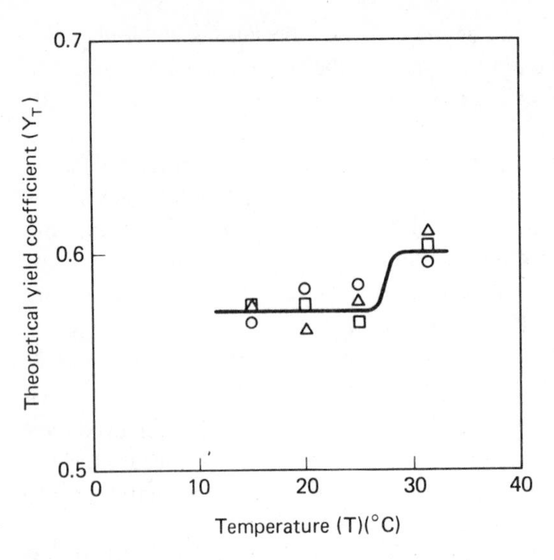

Figure 4-38. Relationship Between Theoretical Yield and Temperature. (After Benefield et al., 1975.)

temperature can be estimated by equation 1-30. In this case the following formulation is generally employed:

$$(K_d)_2 = (K_d)_1 \theta^{T_2 - T_1} \qquad (4\text{-}103)$$

where θ varies between 1.02 and 1.06.

Although equations 4-102 and 4-103 can be used to approximate the variation in reaction rates with temperature, it is recommended that, when possible, all biokinetic constants be evaluated at the aeration tank temperature expected under both critical summer and winter conditions.

The effect of temperature change on the settleability of the biomass is also important. Benedict and Carlson (1973), studying the effect of temperature acclimation between 4 and 32°C, found the following sludge volume indexes:

4°C culture SVI = 110

19°C culture SVI = 98

32°C culture SVI = 45

The values were obtained from cultures subjected to organic loadings of 0.25 to 0.4 mg BOD/day/mg MLSS. On the basis of these data it was concluded that settleability was not adversely affected by operating temperatures within the range of 4 to 32°C. However, it must be stressed that the validity of using SVI as a measure of sludge settleability is highly suspect, and caution must be used when comparing the settleability of different sludges using SVI (Dick and Vesilind, 1969).

Solids–Liquid Separation

If the activated sludge process is to operate efficiently, the secondary clarifier must provide (1) effective separation of the biological solids from the liquid phase, and (2) concentration of these solids before their return to the aeration tank. In many cases these clarifiers are designed to ensure efficient solids separation (i.e., clarification).

This is accomplished by dividing the expected flow by an acceptable overflow rate. Commonly used overflow rates for various MLSS concentrations are given in Table 4-7. These values are based on peak wastewater flow because the return flow is removed from the clarifier bottom and is therefore assumed not to contribute to the upward flow velocity. Typical clarifier depths range between 7 and 14 ft.

Designing a secondary clarifier to ensure that clarification is achieved does not, however, mean that adequate solids concentration will occur. Indeed, thickening considerations usually result in a larger cross-sectional area than that required for clarification.

The solids flux theory provides the basis for determining the required thickening capacity of a clarifier. Solids flux is defined as the mass of solids passing a unit area of a particular plane in a unit time. In a batch test the downward movement of the sludge solids is due only to the settlement velocity. Thus, the batch solids flux can be denoted as

$$G_B = UC \tag{4-104}$$

where G_B = batch solids flux, mass area^{-1} time^{-1}
U = sludge settling velocity taken to be equal to the zone settling velocity (ZSV) of a suspension having an initial suspended solids concentration of C, distance time^{-1}
C = initial suspended solids concentration, mass volume^{-1}

TABLE 4-7

RECOMMENDED PEAK FLOW OVERFLOW RATES (gpd/ft²) FOR VARIOUS MLSS CONCENTRATIONS

MLSS	*Percent recycle*	
(mg/ℓ)	*25*	*50*
500	1400	1400
1000	1400	1400
1500	1200	1200
2000	1200	1200
2500	1150	960
3000	960	800
3500	823	685

Source: After Metcalf and Eddy (1972).

In a continuous unit such as the secondary clarifier, two factors contribute to the downward movement of the sludge solids: (1) downward movement due to the removal of sludge in the underflow, and (2) downward movement due to settlement or the action of gravity. Here the total solids flux can be written as

$$G_T = G_u + G_B \tag{4-105}$$

where G_T = total solids flux, mass area^{-1} time^{-1}
G_u = solids flux due to underflow, mass area^{-1} time^{-1}

The solids flux due to the underflow can be expressed as

$$G_u = VC \tag{4-106}$$

where V = downward velocity due to sludge removal, distance time^{-1}

Furthermore, assuming that solids and downward velocities are uniformly distributed throughout the cross-sectional area of the clarifier,

$$V = \frac{Q_u}{A} \tag{4-107}$$

where Q_u = underflow rate, volume time^{-1}
A = cross-sectional area of clarifier, area

Substituting from equations 4-104 and 4-106 into equation 4-105 gives

$$G_T = VC + UC \tag{4-108}$$

A unique zone settling velocity, U, can be determined experimentally for designated solids concentrations within a specified range. By selecting a value for V, the magnitude of G_T for all solids concentrations in the specified range can be obtained from equation 4-108. Plotting these data will generally yield a total flux curve much like that presented in Figure 4-39. The limiting flux,

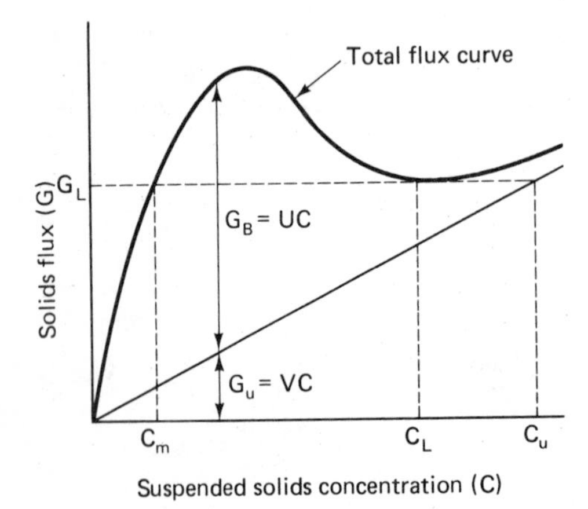

Figure 4-39. Total Possible Flux for Sludge Concentrations That Could Exist in a Continuous Thickner. (After Dick and Suidan, 1975.)

G_L, shown in this figure is equal to the minimum value of the total flux curve to the right of the maximum and is defined as that solids flux which requires the largest area to transmit a given quantity of solids. Theoretically, a design based on the limiting flux rate will always provide the maximum suspended solids concentration in the underflow as long as the influent suspended solids concentration is between C_m and C_L.

Dick and Young (1972) found that the relationship between zone settling velocity and solids concentration could be expressed as

$$U = gC^{-h} \tag{4-109}$$

where g and h are empirical constants.

Substituting for U in equation 4-108 gives

$$G_T = VC + gC^{-h}(C) \tag{4-110}$$

or

$$G_T = VC + gC^{(1-h)} \tag{4-111}$$

Differentiating equation 4-111 with respect to C gives

$$\frac{\partial G_T}{\partial C} = g(1 - h)C^{-h} + V \tag{4-112}$$

For the situation where sludge wasting is from the aeration tank,

$$Q_u = RQ \tag{4-113}$$

Thus, substituting for V in equation 4-112 from equation 4-107 gives

$$\frac{\partial G_T}{\partial C} = g(1 - h)C^{-h} + \frac{RQ}{A} \tag{4-114}$$

Figure 4-39 shows that the limiting flux occurs when $\partial G_T/\partial C = 0$ (i.e., when the tangent to the total flux curve is horizontal). Hence, the solids concentration obtained by setting equation 4-114 equal to zero is the solids concentration exhibiting the limiting flux.

$$C_L = \left[\frac{g(h - 1)A}{RQ}\right]^{1/h} \tag{4-115}$$

To confirm that the limiting flux does occur at $\partial G_T/\partial C = 0$, it must be shown that $\partial^2 G_T/\partial C^2$ is positive (i.e., greater than zero).

$$\frac{\partial^2 G_T}{\partial C^2} = \frac{gh(h - 1)}{C^{1+h}} \tag{4-116}$$

Thus, equation 4-116 is positive when g is positive and h has a value greater than 1. Values for g and h obtained experimentally show that this is indeed the case.

Substituting C_L for C and G_L for G_T in equation 4-111 and then substituting for C_L from equation 4-115, the limiting flux can be expressed as

$$G_L = [g(h-1)]^{1/h} \frac{h}{h-1}\left(\frac{RQ}{A}\right)^{h-1/h} \tag{4-117}$$

Assuming 100% solids capture and neglecting the sludge wastage flow, a material-balance equation for solids entering and leaving the clarifier is given by

$$Q(1+R)C_0 = G_L A = RQC_u \tag{4-118}$$

where C_0 represents the solids concentration in the effluent from the aeration tank and C_u represents the solids concentration in the underflow from the clarifier. The C_0 term represents the total suspended solids concentration and can be calculated by combining the biomass concentration with equation 4-87.

$$C_0 = X + (\text{NDVSS} + f\text{SS})\frac{\theta_c}{t} \tag{4-119}$$

Equation 4-118 can be rearranged to give an equation for C_u.

$$C_u = \frac{G_L A}{RQ} \tag{4-120}$$

Equation 4-118 also shows that if thickening is to be successful, the solids loading must not exceed the limiting flux. Therefore,

$$\frac{Q(1+R)C_0}{A} = [g(h-1)]^{1/h} \frac{h}{h-1}\left(\frac{RQ}{A}\right)^{h-1/h} \tag{4-121}$$

where the term on the left side of the equation represents the solids loading rate. Equation 4-121 can be rearranged into the form

$$\frac{Q}{A} = \frac{g(h-1)\left(\frac{h}{h-1}\right)^h (R)^{h-1}}{(C_0)^h(1+R)^h} \tag{4-122}$$

This equation can then be used to determine the required clarifier area to achieve thickening (Mynhier and Grady, 1975).

To apply equation 4-122, it is necessary to determine values for the constants g and h. This should be accomplished in the laboratory. However, as an aid to design engineers, Mynhier and Grady (1975) have tabulated representative values for these constants, which are given in Table 4-8.

TABLE 4-8

TYPICAL VALUES FOR g AND h

g (ft/min)[a]	h	System type	Process loading factor: Value	Process loading factor: Basis
0.43×10^{-8}	2.62	Conventional activated sludge	0.5	Pounds BOD/pounds MLVSS
3.95×10^{-8}	2.50	Conventional activated sludge	—	—
13.5×10^{-8}	2.58	Laboratory	—	—
2.9×10^{-8}–27×10^{-8}	2.25	11 conventional activated sludge plants	—	—
2.62×10^{-6}	1.70	Conventional activated sludge	0.24	Pounds BOD/pounds MLSS
27×10^{-6}	1.55	Conventional activated sludge	—	—
29.6×10^{-6}	1.63	Conventional activated sludge	0.06	Pounds BOD/pounds sludge

[a]1 ft/min = 5.08×10^{-3} m/s.
Source: After Mynhier and Grady (1975).

Example 4-10

A completely mixed activated sludge process is to be used to treat a wastewater flow of 10 MGD with a BOD_u of 200 mg/ℓ. Determine the required clarifier area (assuming thickening controls the design) and the expected underflow solids concentration. The following design criteria are applicable:

$$C_0 = 3000 \text{ mg/}\ell \text{ as MLSS}$$

$$R = 0.2 \text{ to } 0.3$$

$$g = 2.9 \times 10^{-8} \text{ ft/min}$$

$$h = 2.65$$

solution

1/ Compute the allowable solids loading by applying equations 4-122.

$$\frac{Q}{A} = \frac{(2.9 \times 10^{-8})(2.65 - 1)\left(\dfrac{2.65}{2.65 - 1}\right)^{2.65}(0.2)^{(2.65-1)}}{\left(3.0 \text{ g/}\ell \times \dfrac{1 \text{ lb}}{454 \text{ g}} \times 3.8 \dfrac{\ell}{\text{gal}} \times \dfrac{1 \text{ gal}}{8.34 \text{ lb}}\right)^{2.65}(1 + 0.2)^{2.65}}$$

$$= 0.0347 \text{ ft/min}$$

$$= 0.0347 \times 10{,}770 = 374 \text{ gpd/ft}^2$$

Note that the feed solids concentration is expressed as pounds of solids per pound of water (i.e., in its dimensionless form).

$$A = \frac{Q}{\text{solids loading}} = \frac{10{,}000{,}000}{374}$$

$$= 26{,}738 \text{ ft}^2$$

Similar calculations for $R = 0.3$ yield the values listed in the following table:

C_0 *(mg/ℓ)*	*R*	*Q/A (gpd/ft²)*	*A (ft²)*
3000	0.2	374	26,738
3000	0.3	592	16,892

These data show that 26,738 ft² is the limiting area.

2/ Determine the limiting flux from equation 4-118 using the limiting area calculated in step 1.

$$G_L = \frac{Q(1 + R)C_0}{A}$$

$$= \frac{(10)(8.34)(1 + 0.2)(3000)}{26{,}738} = 11.2 \text{ lb/day/ft}^2$$

C_0 *(mg/ℓ)*	*R*	G_L *(lb/day/ft²)*
3000	0.2	11.2
3000	0.3	12.1

3/ Estimate the underflow solids concentration using equation 4-120 and the critical solids loading computed in step 1.

$$C_u = \frac{G_L A}{RQ}$$

$$= \frac{(11.2 \text{ lb/day-ft}^2)\left(454 \text{ g/lb} \times \dfrac{1 \text{ gal}}{3.8\,\ell} \times 1000 \text{ mg/g}\right)}{(0.2)(374 \text{ gal/day-ft}^2)}$$

$$= 17{,}852 \text{ mg}/\ell$$

C_0 *(mg/ℓ)*	*R*	C_u *(mg/ℓ)*
3000	0.2	17,889
3000	0.3	12,884

Retention time is also a very important consideration in clarifier design, especially when nitrification occurs in the aeration tank. Because anoxic conditions prevail in the clarifier, certain bacteria will use nitrate as an electron acceptor during endogenous respiration. The nitrate is reduced to molecular nitrogen, which is a gas and as such will rise to the surface carrying with it sludge solids. These solids will ultimately be removed in the effluent, resulting in a deterioration in plant effluent quality. Schroeder (1977) recommends that the maximum retention time be less than $1\frac{1}{2}$ h and, where possible, less than 1 h.

Lawrence and Milness (1971) point out that the settling characteristics of activated sludge are dependent upon the physiological state of the organisms present (which is controlled by regulating the BSRT), and Bisogni and Lawrence (1971) have presented experimental data which support this view (see Figure 4-40). In this regard, significant amounts of surface-active

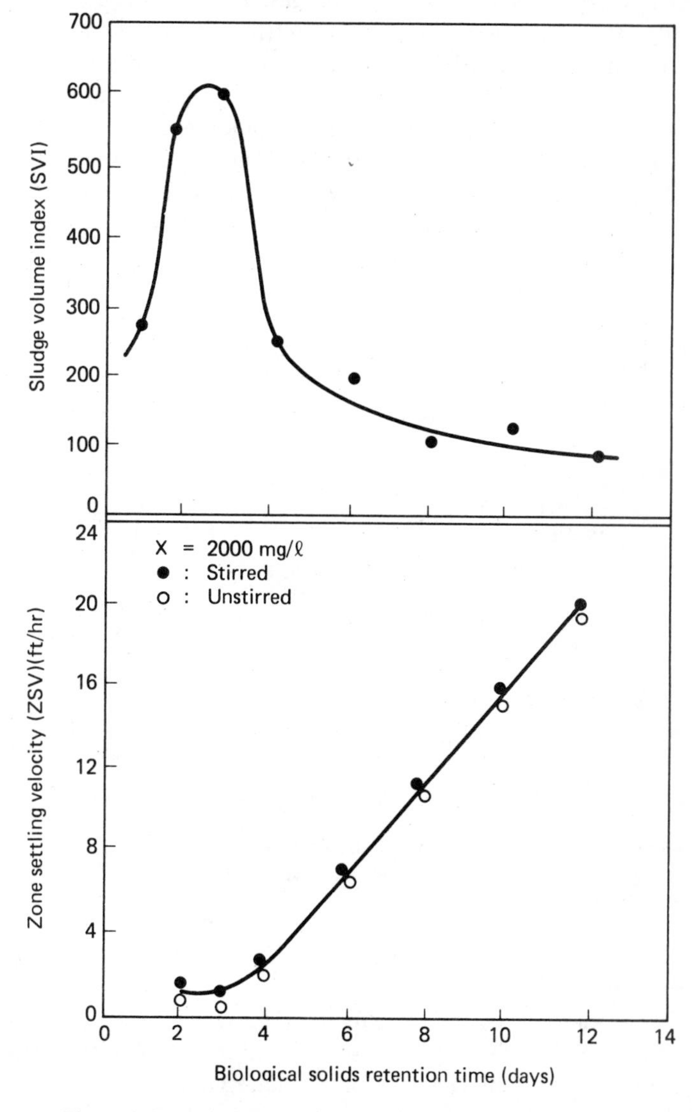

Figure 4-40. Relationship Among Zone Settling Velocity, Sludge Volume Index, and BSRT. (After Bisogni and Lawrence, 1971.)

materials such as proteins are released into the mixed liquor at long BSRTs, and this may create a foaming problem at the plant.

Figure 4-40 reflects the variation in sludge settling characteristics with BSRT as predicted by the zone settling velocity and the sludge volume index. However, Dick and Vesilind (1969) suggest that the SVI is not a true measure of the sludge settling characteristics since it measures only one position on the settling curve and therefore cannot be used to compare the settling characteristics of two different sludges. This point can be illustrated by considering two sludges with different settling characteristics but with the same SVI, such as those shown in Figure 4-41. Furthermore, the suspended solids concentration also affects the SVI. For example, a sludge with a suspended solids concentration of 10,000 mg/ℓ would have an SVI of 100 if it did not settle at all. Figure 4-42 shows the maximum SVI for various suspended solids concentrations. Thus, as proposed by Vesilind (1974), it should be standard practice to report the suspended solids concentration with SVI.

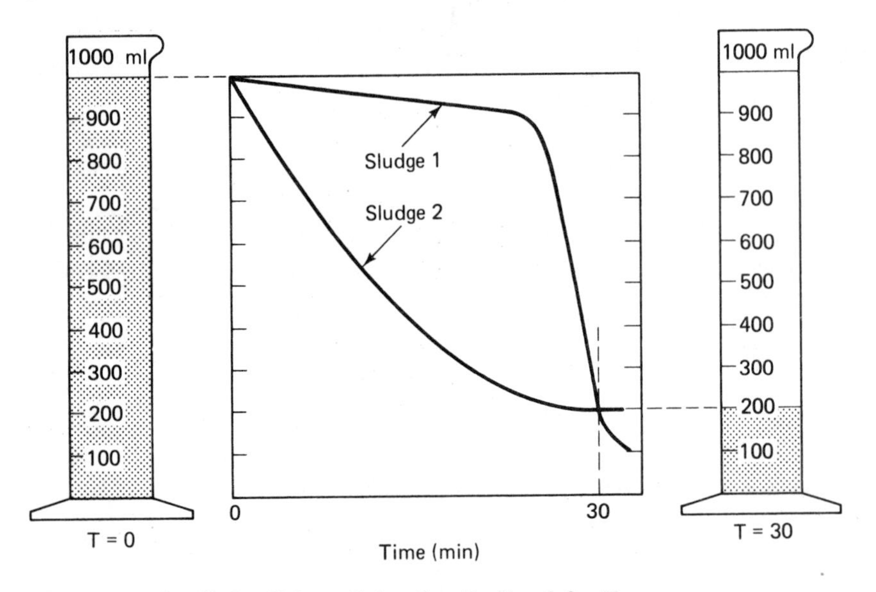

Figure 4-41. The Sludge Volume Index Can Be Equal for Two Sludges Having Very Different Settling Characteristics. (After Vesilind, 1974.)

Effluent Quality

The present secondary effluent requirement specifies a 30-day average effluent BOD_5 of 30 mg/ℓ. As the effluent from biological treatment processes contains both soluble and insoluble material, the 30 mg/ℓ value is the sum of the soluble effluent BOD_5 and the insoluble effluent BOD_5. The soluble

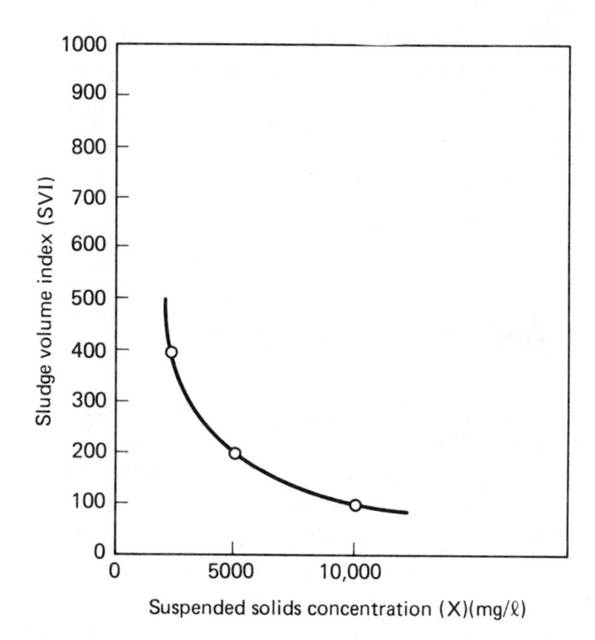

Figure 4-42. Maximum Possible Sludge Volume Index. Calculations Based on Sludge Interface Height of 1000 ml. (After Vesilind, 1974.)

BOD_5 is mainly a result of soluble organics that escape treatment, whereas the majority of the insoluble BOD_5 is due to biomass that escapes separation in the secondary clarifier.

In well-operating activated sludge plants the effluent suspended solids concentration will generally be less than 20 mg/ℓ. However, because of floc deterioration at very long θ_c values, the suspended solids concentration in the effluent from an extended aeration process may exceed 70 mg/ℓ. In either case, degradation of the effluent suspended solids creates a significant portion of the total effluent BOD_5 from an activated sludge process. As the kinetic equations previously developed apply only to the soluble portion of the effluent BOD_5, another relationship is required if the insoluble BOD_5 is to be considered.

Eckenfelder et al. (1972) proposes that the 5-day oxygen demand created by the degradation of effluent suspended solids can be estimated from the expression

$$\text{insoluble BOD}_5 = 5(1.42 K_d x_a C_e) \qquad (4\text{-}123)$$

where K_d represents the microbial decay coefficient in day^{-1}; x_a represents that portion of the effluent suspended solids which is active biomass (x_a ranges from about 0.8 for high-rate systems to about 0.1 for extended aerations systems); C_e represents the effluent suspended solids concentration in mg/ℓ; the constant 5 accounts for the 5-day BOD incubation period, while 1.42 represents the oxygen consumed in the oxidation of 1 mg of biomass. Considering equation 4-123, total effluent BOD_5 can then be approximated by the relationship

$$\text{total effluent BOD}_5 = S_e + 5(1.42K_d x_a C_e) \tag{4-124}$$

4-5
Evaluation of Biokinetic Constants

Before a particular mathematical model can be used in process design, the biokinetic constants K or K_1, Y_T, K_d, a and b, or a_1 and b_1 should be evaluated. The particular value for each of these constants depends on the type of waste to be treated and the microbial population established.

A laboratory study is required to evaluate the constants. In such a study, a lab-scale activated sludge unit much like that shown in Figure 4-43 is

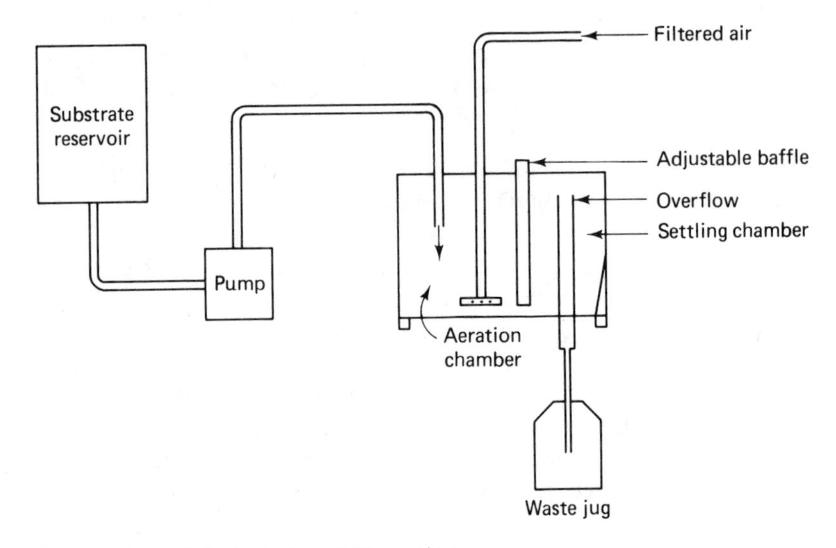

Figure 4-43. Lab-Scale Activated Sludge Unit.

employed. The system should be operated over a wide θ_c range such as 3 to 20 days. System control is achieved by sludge wasting. For a system such as that shown in Figure 4-43, the required wasting rate, when wasting directly from the aeration tank, can be approximated from

$$Q_w = \frac{VX_a}{\theta_c X_a} = \frac{V}{\theta_c} \tag{4-125}$$

where V = volume under aeration when baffle is pulled and entire tank contents are mixed

X_a = average active biomass concentration (MLVSS) during a 24-h period, measured when entire tank contents are completely mixed

However, when determining the biokinetic coefficients, a more accurate value of θ_c is needed than is given by V/Q_w. For this situation θ_c is calculated from

$$\theta_c = \frac{X_a V}{Q_w X_a + (Q - Q_w) X_e} \tag{4-126}$$

where X_e represents the volatile suspended solids concentration in the waste jug, measured at the end of a 24-h period.

If the rate of substrate utilization follows the relationship given by equation 2-56, a plot of $Q(S_0 - S_e)/X_a V$ versus S_e will give a graph similar to that shown in Figure 4-44 when BOD_u is used as a measure of the substrate concentration. The slope of the line gives the substrate utilization-rate constant, K. However, when COD is used as a measure of the substrate concentration, the plot will be similar to that shown in Figure 4-45. In this case the intercept on the x-axis represents the soluble nondegradable fraction of the substrate measured as COD. Keep in mind here that the kinetic models which have been developed predict only the soluble degradable organic fraction contained in the effluent.

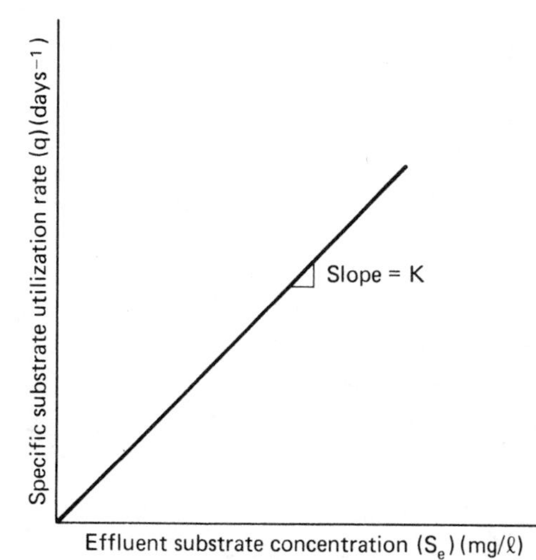

Figure 4-44. Plot of Specific Substrate Utilization Rate Versus Effluent Substrate Concentration (measured as BOD_u).

If the rate of substrate utilization follows the relationship given by equation 2-58, then a plot of $Q(S_0 - S_e)/X_a V$ versus S_e/S_0 will be similar to that shown in Figure 4-46. The slope of the line gives the substrate utilization-rate constant, K_1.

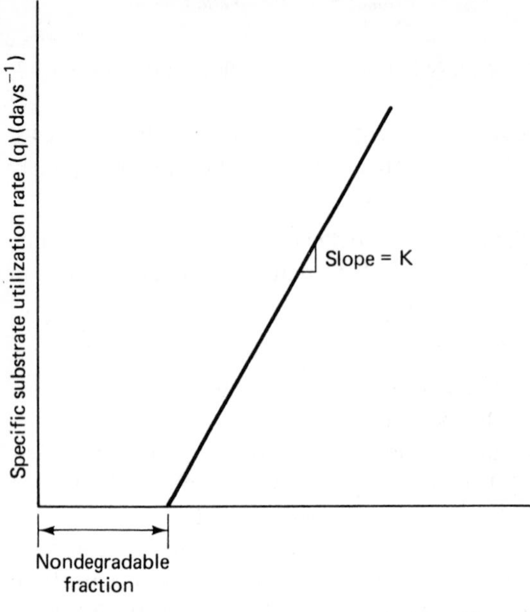

Figure 4-45. Plot of Specific Substrate Utilization Rate Versus Effluent Substrate Concentration (measured as COD).

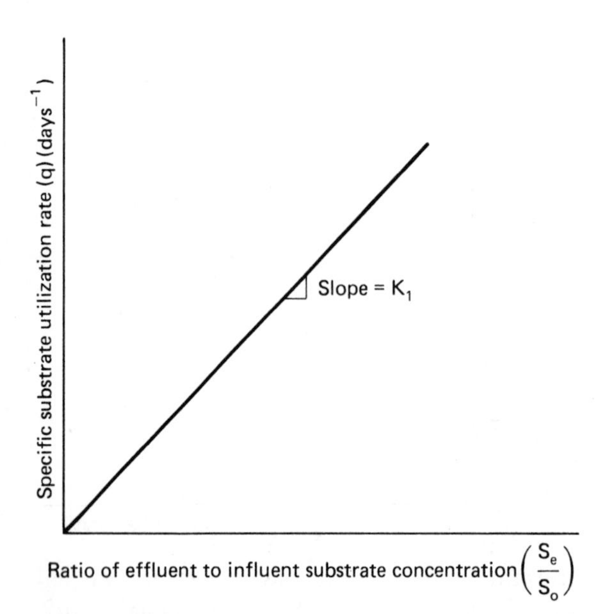

Figure 4-46. Plot of Specific Substrate Utilization Rate Versus Ratio of Effluent to Influent Substrate Concentration.

A plot of $Q(S_0 - S_e)/X_aV$ versus μ (i.e., θ_c^{-1}) will give a trace such as that shown in Figure 4-47. The intercept of the trace has a value equal to K_d/Y_T, while the slope has a value equal to $1/Y_T$.

The oxygen use coefficients a and b can be determined by plotting specific oxygen utilization rate versus $Q(S_0 - S_e)/X_aV$. Such a plot is illustrated in Figure 4-48, where the slope of the line represents a and the intercept on the

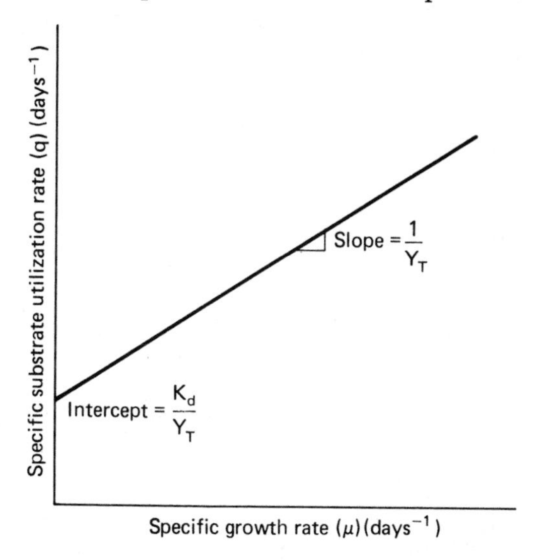

Figure 4-47. Plot of Specific Substrate Utilization Rate Versus Specific Growth Rate.

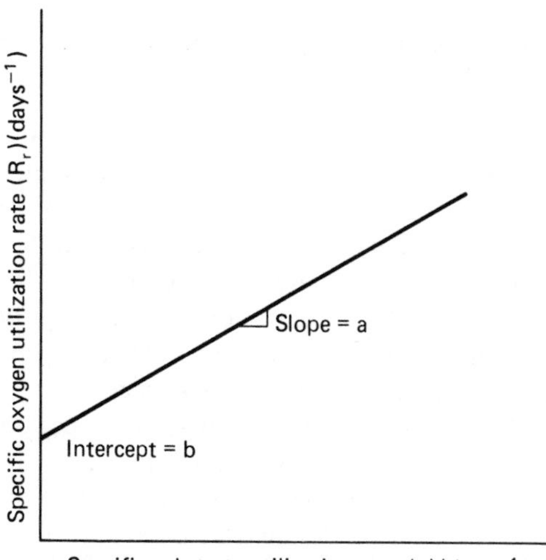

Figure 4-48. Plot of Specific Oxygen Utilization Rate Versus Specific Substrate Utilization Rate.

y-axis represents b. On the other hand, the oxygen use coefficients a_1 and b_1 are found from a graph of specific oxygen utilization rate versus specific growth rate, μ. In this case, the slope of the line gives a_1 while the intercept gives b_1 (see Figure 4-49). Eckenfelder and O'Conner (1961) have reported that the coefficient, a, varies from 0.35 to 0.55 for a wide variety of industrial wastes. The value of b appears to be in the range 0.07 to 0.2 day^{-1}. However, not enough data have been collected on the use of a_1 and b_1 to report typical values for these constants.

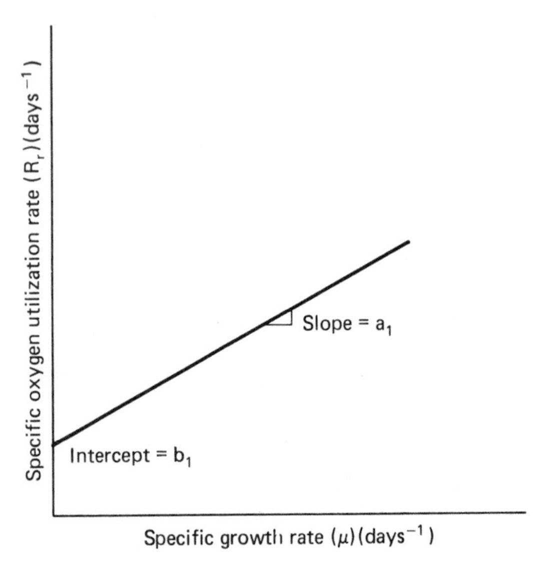

Figure 4-49. Plot of Specific Oxygen Utilization Rate Versus Specific Growth Rate.

During the course of the laboratory study, the relationship between SVI and θ_c should be graphed in a manner similar to that shown in Figure 4-50. It is also necessary to determine the empirical constants g and h which are required in determining the clarifier area to achieve thickening. These constants can be evaluated by making a log-log plot of the interfacial settling velocity versus concentration over the range of θ_c's covered during the laboratory investigation (see Figure 4-51). The slope of the line of best fit gives $-h$ and the intercept along the y-axis gives g.

It should be understood that the values obtained for the biokinetic coefficients represent mean values and that any shift in the mixed microbial population composition would alter the value of these constants.

For effective design the biokinetic coefficients should be established through laboratory or pilot-plant studies. However, during facilities planning when numerous treatment alternatives are under consideration, it is desirable

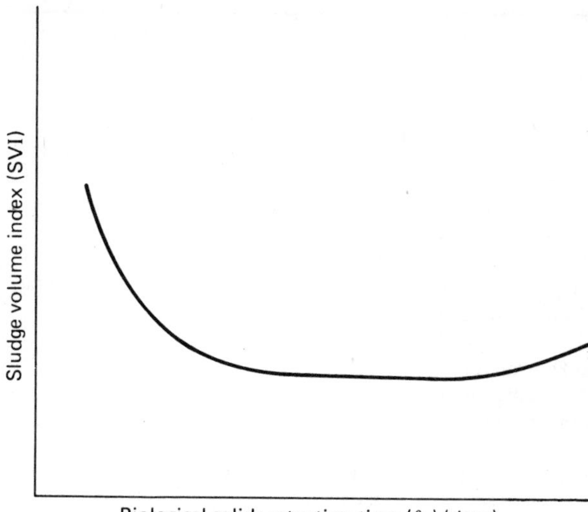

Figure 4-50. Relationship Between Sludge Volume Index and Biological Solids Retention Time.

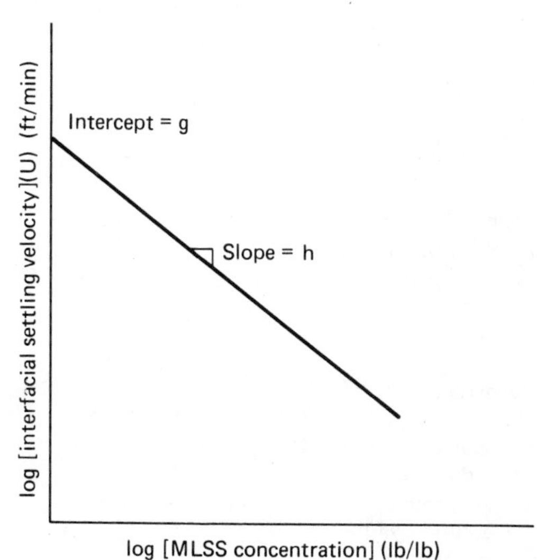

Figure 4-51. Relationship Between Interfacial Settling Velocity and Solids Concentration.

sometimes simply to select constant values representative of the wastewater to be treated. This allows general comparisons to be made between process alternatives without the cost of extensive laboratory experimentation. Table 4-9 reflects a partial list of biokinetic constants which have been compiled and

TABLE 4-9

BIOKINETIC COEFFICIENTS FOR VARIOUS WASTEWATERS

	Kinetic coefficients					
Wastewater type	μ_{max} *per hour*	K_s *(mg/ℓ)*	K_d *per hour*	Y_T *(mg/mg)*	K *(ℓ/mg-h)*	*Coefficient basis*
Ammonia base, semichemical					4.6×10^{-4}	BOD_5
Brewery					2.2×10^{-4}	BOD_5
				0.44		
Chemical Industry					1.4×10^{-4}	BOD_5
					2.0×10^{-4}	
Coke plant, ammonia liquor					11×10^{-4}	COD
Domestic sewage					10.8×10^{-4}-13.7×10^{-4}	BOD_5
	0.40	60				COD
	0.46	55				COD
	0.16	22	0.0029	0.67		COD
			0.0023	0.5		BOD_5
			0.0020	0.67		BOD_5
	0.55	120	0.0025	0.5		BOD_5
Organic chemical					0.73×10^{-4}	BOD_5
					0.50×10^{-4}	BOD_5
Petrochemical					2.4×10^{-4}-2.8×10^{-4}	BOD_5
Pharmaceutical					2.1×10^{-4}	BOD_5
					5.7×10^{-4}	BOD_5
Phenolic					0.92×10^{-4}	BOD_5
Poultry processing	3.0	500	0.030	1.32		BOD_5
Pulp and paper mill			0.0083	0.47	4.17×10^{-4}	BOD_5
Pulp and paper, kraft			0.0015			
Refinery					3.5×10^{-4}	BOD_5
			0.0104	0.53	10.0×10^{-4}	BOD_5
Rendering					15×10^{-4}	BOD_5
Shrimp processing	0.77	85.5	0.0667	0.50		BOD_5
Soybean	0.50	355	0.006	0.74		BOD_5
Tetraethyllead					7.1×10^{-4}	BOD_5
Textile, combined cotton, synthetic, and wool finishing and dyeing			0.030	0.38		COD
Textile, polyester dyeing			0.050	0.32		COD
Textile, wool dyeing, carbonizing, and washing			0.040	0.69		COD

TABLE 4-9 (Cont.)

Wastewater type	*Kinetic coefficients* μ_{max} *per hour*	K_s *(mg/ℓ)*	K_d *per hour*	Y_T *(mg/mg)*	*K (ℓ/mg-h)*	*Coefficient basis*
Textile, nylon scouring and dyeing			0.0014	0.25	1.5×10^{-4}	BOD_5
Thiosulfate			0.000833	0.029[a]	1.1×10^{-4}	COD
			0.000417	0.035[a]	2.1×10^{-4}	COD
Vegetable and fruit processing						
Apples			0·00479	0.57		
Carrots			0.00313	0.38		COD
Citrus			0.00633	0.465		
Corn			0.00271	0.41		COD
Pea			0.00117	0.49		
			0.00345	0.32		COD
Peach			0.00479	0.46		
Pear			0.00479			
Potato			0.00792	0.88		
Vegetable oil					3.1×10^{-4}	BOD_5
Whey, cottage cheese manufacturing			0.00229	0.40		BOD_5

[a]Thiosulfate basis.
Source: After Mynhier and Grady (1975).

presented by Mynhier and Grady (1975). Biokinetic coefficient values reported in Table 4-10 are those proposed by Lawrence (1975) for activated sludge systems treating municipal wastewaters.

TABLE 4-10

BIOKINETIC COEFFICIENTS FOR ACTIVATED SLUDGE SYSTEMS TREATING MUNICIPAL WASTEWATER AT 20°C

Coefficient	*Unit*	*Value range*
Y_T	mg VSS/mg COD	0.35–0.45
K_d	day^{-1}	0.05–0,10
k	mg COD/mg VSS-day	6–8
K_s	mg/ℓ COD	25–100

Source: After Lawrence (1975).

4-6
Nitrification

In Chapter 3 the advantages of producing a nitrified effluent were discussed in detail. In this chapter it was shown that chemoautotrophic bacteria are responsible for the oxidation of NH_4^+-N to NO_3^--N in many biological wastewater treatment processes. These bacteria, collectively called nitrifiers, consist of the genera *Nitrosomonas* and *Nitrobacter*. The oxidation of ammonia to nitrate is considered as a two-step sequential reaction as follows:

$$2NH_4^+ + 3O_2 \xrightarrow{Nitrosomonas} 2NO_2^- + 4H^+ + 2H_2O$$

$$2NO_2^- + O_2 \xrightarrow{Nitrobacter} 2NO_3^-$$

$$NH_4^+ + 2O_2 \xrightarrow{\text{Nitrifiers}} NO_3^- + 2H^+ + H_2O$$

In the overall reaction, hydrogen ions are released. If insufficient alkalinity is present to buffer the system, the pH will decrease. Since the nitrifiers consume CO_3^{2-}, insufficient alkalinity is likely to occur. As the nitrifiers are very sensitive to pH changes, it is important that the engineer know what pH changes to expect during nitrification so that chemical addition may be provided when required. When insufficient alkalinity is present, the system is also carbon-limited for nitrifiers and carbon must be added.

During nitrification, inorganic carbon is utilized for cell synthesis. If the inorganic carbon and the nitrogen incorporated into cell synthesis of the nitrifiers is neglected, equations 3-52, 3-34, and 3-35 can be used to estimate the pH change expected to occur as a result of biological nitrification. To illustrate how this may be accomplished, consider a treatment system operating at a temperature of 25°C and a pH of 7.4 and with a total alkalinity of 300 mg/ℓ as $CaCO_3$ before the onset of nitrification. To estimate the pH after complete nitrification, if the TKN concentration of the wastewater is 10 mg/ℓ as nitrogen, after consideration has been given to the nitrogen incorporated into heterotrophic cell synthesis, the procedure is as follows:

1/ Convert total Kjeldahl nitrogen concentration to concentration as as ammonium.

$$10\left(\frac{18}{14}\right) = 12.85 \text{ mg}/\ell$$

2/ From equation 3-52 calculate mg/ℓ of hydrogen ions formed during complete nitrification of 12.85 mg/ℓ ammonium.

$$\underset{18}{NH_4^+} + \underset{(2 \times 32)}{2O_2} \longrightarrow NO_3^- + \underset{(2 \times 1)}{2H^+} + H_2O$$

$$\left(\frac{2 \times 1}{18}\right) 12.85 = 1.43 \text{ mg}/\ell \text{ hydrogen ions formed}$$

3/ Convert concentration of hydrogen ions in mg/ℓ to concentration in moles/ℓ.

$$\text{moles}/\ell = \frac{0.00143}{1}\ \text{g}/\ell = 0.00143$$

4/ Convert alkalinity in mg/ℓ as $CaCO_3$ to equivalents/ℓ.

$$\frac{0.3\ \text{g}/\ell}{100/2} = 0.006 = 6 \times 10^{-3}\ \text{eq}/\ell$$

5/ Calculate α_1.

$$\alpha_1 = \frac{10^{-6.4}}{10^{-6.4} + 10^{-7.4} + [(10^{-6.4})(10^{-10.4})/10^{-7.4}]}$$

$$= 0.91$$

6/ Compute buffer intensity, β, from equation 3-35.

$$\beta = 2.3\left[\frac{(0.91)(6 \times 10^{-3})(10^{-7.4})}{10^{-6.4}}\right]$$

$$= 12.55 \times 10^{-4}\ \text{eq/pH unit}$$

7/ Estimate expected change in pH from equation 3-34.

$$\beta = \frac{\Delta A}{\Delta pH}$$

or

$$\Delta pH = \frac{\Delta A}{\beta} = \frac{1.43 \times 10^{-3}}{12.55 \times 10^{-4}}$$

$$= 1.14$$

8/ Calculate the resulting pH.

$$pH = 7.4 - 1.14 = 6.26$$

To make such a calculation, it is necessary that the steady-state pH established during carbon oxidation be determined. This can be accomplished by following the procedure outlined by Brown and Caldwell (1975), which requires the use of the following three equations and the assumption that for a typical municipal wastewater, 1.38 lb of CO_2 are produced for each pound of O_2 consumed:

1/
$$pH = pK_1 - \log\left(\frac{[H_2CO_3^*]}{[HCO_3^-]}\right) \qquad (3\text{-}28)$$

where pK_1 = 6.38 at 20°C

$[H_2CO_3^*]$ = total concentration of dissolved CO_2 plus H_2CO_3, moles/ℓ; however, as the amount of H_2CO_3 is generally very small, this value can be approximated by considering only CO_2

$[HCO_3^-]$ = alkalinity expressed as HCO_3^- concentration,

2/ $$C_{eq} = HP_g \quad (4\text{-}127)$$

where C_{eq} = concentration of dissolved carbon dioxide, mg/ℓ
H = Henry's law constant (for CO_2, H has a value of 2352 mg/ℓ-atm at 10°C, 1688 mg/ℓ-atm at 20°C and 1315 mg/ℓ-atm at 30°C)
P_g = partial pressure of gas in equilibrium with the liquid, atm.

3/ $$P_g V = nRT \quad (4\text{-}128)$$

where V = off-gas volume, cm^3
R = gas constant, 82.6 cm^3-atm/mole-deg
T = temperature, °K
n = moles of gas

To illustrate this method, consider Example 4-11.

Example 4-11

A 10-MGD wastewater flow containing 200 mg/ℓ alkalinity as $CaCO_3$ is to be treated in an activated sludge plant operating at 20°C. To achieve the desired degree of treatment, 40,000 lb oxygen/day must be supplied. What will be the operating pH if the aeration system has a transfer efficiency of 10% and it is assumed no nitrification has occurred?

solution

1/ Convert the alkalinity concentration to moles/ℓ of HCO_3^-.

$$\text{mg/ℓ } HCO_3^- = 200\left(\frac{61}{50}\right)$$
$$= 244$$

Then

$$\text{moles/ℓ} = \frac{0.244 \text{ g/ℓ}}{61} = 4 \times 10^{-3}$$

2/ Assume a steady-state pH and calculate the equilibrium CO_2 concentration from equation 3-28 (assume that pH = 7.4 for this problem).

$$7.4 = 6.38 - \log\left(\frac{[CO_2]}{[4 \times 10^{-3}]}\right)$$

$$[CO_2] = 3.82 \times 10^{-4} \text{ mole/ℓ}$$

3/ Convert $[CO_2]$ to pounds per gallon of wastewater treated.

$$\frac{(3.82 \times 10^{-4})(44 \text{ g/mole})(3.78 \text{ ℓ/gal})}{(454 \text{ g/lb})} = 1.4 \times 10^{-4} \text{ lb/gal}$$

4/ Convert $[CO_2]$ to mg/ℓ.

$$(3.82 \times 10^{-4})(44 \text{ g/mole})(1000 \text{ mg/g}) = 16.81 \text{ mg/ℓ}$$

5/ Calculate partial pressure of CO_2 gas in equilibrium with mixed liquor using equation 4-127.

$$P_g = \frac{16.81}{1688} = 0.00995 \text{ atm}$$

6/ Compute the moles of CO_2 per ft³ of off-gas.

$$n = \frac{(0.00995)(1 \text{ cm}^3)}{(82.06)(273 + 20)} = 4.14 \times 10^{-7} \text{ mole/cm}^3$$

or

$$n = 4.14 \times 10^{-7} \text{ mole/cm}^3 \times \frac{10^6 \text{ cm}^3}{35.3 \text{ ft}^3} = 0.0117 \text{ mole/ft}^3$$

which is

$$\frac{(0.0117)(44)}{454} = 0.00113 \text{ lb/ft}^3$$

7/ Estimate the ft³ of air required per gallon of wastewater treated.

$$\frac{40{,}000}{(10{,}000{,}000)(0.0750)(0.232)(0.1)} = 2.3 \text{ ft}^3\text{/gal}$$

where

0.0750 = specific weight of air, lb/ft³

0.232 = percent of O_2 in air, fraction

0.1 = transfer efficiency of aeration system, fraction

8/ Calculate weight of CO_2 theoretically produced per gallon of wastewater.

$$\begin{bmatrix}\text{weight of} \\ CO_2\end{bmatrix} = \begin{bmatrix}\text{weight of} \\ \text{dissolved } CO_2\end{bmatrix} + \begin{bmatrix}\text{weight of } CO_2 \\ \text{in off-gas}\end{bmatrix}$$

$$= 1.40 \times 10^{-4} + (2.3)(0.00113)$$

$$= 2.74 \times 10^{-3} \text{ lb/gal}$$

9/ Determine CO_2 quantity in lb/gal based on 1.38 lb CO_2 produced/lb O_2 consumed.

$$\frac{40{,}000 \text{ lb } O_2}{10{,}000{,}000 \text{ gal}} \times 1.38 = 5.52 \times 10^{-3} \text{ lb/gal}$$

10/ Compare the weight of CO_2 (lb/gal) calculated in step 8 to that calculated in step 9.

$$0.00274 \neq 0.00552$$

Since the two quantities are not equal, a different steady-state pH must be assumed in step 2 and the calculation repeated until the two are equal. The assumed pH under such conditions is the steady-state pH that can be expected in the mixed liquor before nitrification.

Having completed calculations such as those presented in the previous example, a decision can then be made as to the requirement for lime addition to increase the alkalinity of the system in order to maintain the pH and inorganic carbon within an acceptable range for nitrification.

Nitrification with Suspended Growth Systems

The discussion presented in this section will apply only to activated sludge processes and completely mixed aerated lagoons. Other types of lagoon systems are not considered because in those systems, nitrification is not consistent (Stone et al., 1975).

NITRIFICATION KINETICS

It has been found that the growth of nitrifying bacteria can be represented by Monod kinetics (Downing and Hopwood, 1964), where energy of maintenance is neglected:

$$\mu = \frac{\mu_{\max} S}{K_s + S} \tag{2-12}$$

where μ = specific growth rate of nitrifying bacteria, time^{-1}
$\mu_{\max}$ = maximum specific growth rate of nitrifying bacteria, time^{-1}
K_s = saturation constant numerically equal to the growth-limiting nutrient concentration at which $\mu = \mu_{\max}/2$, mass volume^{-1}
S = residual growth-limiting nutrient concentration, mass volume^{-1}; for nitrification this is considered to be the energy source

Under steady-state conditions, experimental evidence has shown nitrite accumulation to be insignificant. This suggests that the rate-limiting step for the conversion of ammonium to nitrate is the oxidation of ammonium to nitrite by the genus *Nitrosomonas*. For this situation nitrification can be described by the equation

$$(\mu)_{NS} = (\mu_{\max})_{NS} \frac{[NH_4^+\text{-N}]}{K_N + [NH_4^+\text{-N}]} \tag{4-129}$$

where $(\mu)_{NS}$ = specific growth rate of *Nitrosomonas*, time^{-1}
$(\mu_{\max})_{NS}$ = maximum specific growth rate of *Nitrosomonas*, time^{-1}
$[NH_4^+\text{-N}]$ = ammonium nitrogen concentration surrounding the microorganisms, mass volume^{-1}
K_N = saturation constant, mass volume^{-1}

Earlier it was shown that for a completely mixed activated sludge process operating at steady-state, μ is equal to θ_c^{-1}. This implies that if a nitrifying culture is to be established, the reciprocal of the operating BSRT must be less than the maximum specific growth rate of *Nitrosomonas*. If the reciprocal of the operating BSRT is greater than $(\mu_{\max})_{NS}$, the rate of organism wasting from the system will be greater than the rate of production, so that washout of the nitrifying culture will soon occur.

Equation 4-129 describes specific growth rate in terms of $(\mu_{\max})_{NS}$, K_N, and the growth-limiting nutrient concentration taken to be the energy source for

Nitrosomonas (i.e., the residual concentration of ammonium). However, specific growth rate is also affected by: (1) the residual concentration of the electron acceptor (i.e., molecular oxygen), (2) the operating temperature, and (3) the operating pH.

Influence of Residual Oxygen Concentration. The influence of residual oxygen concentration on the specific growth rate of *Nitrosomonas* is given by the relationship

$$(\mu)_{NS} = (\mu_{\max})_{NS} \frac{[O_2]}{K_{O_2} + [O_2]} \tag{4-130}$$

where $[O_2]$ = dissolved oxygen concentration surrounding the microorganisms, mass volume^{-1}

K_{O_2} = saturation constant for oxygen, mass volume^{-1}. Values for K_{O_2} range from 0.1 to 2.0 mg/ℓ, where values of 0.2 to 0.4 mg/ℓ are those most commonly quoted

Wild et al. (1971) found that there was no inhibition to nitrification at DO levels greater than 1 mg/ℓ. Results of this study are given in Figure 4-52. On the other hand, Nagel and Haworth (1969) found that increasing the DO concentration above 1 mg/ℓ would increase the ammonium oxidation rate, as shown in Figure 4-53. The results of the study by Nagel and Haworth (1969) seem to contradict the well-established concept of critical oxygen concentration, and as such, some plausible explanation for these two different observations is warranted.

Nitrification with the activated sludge process can be accomplished in

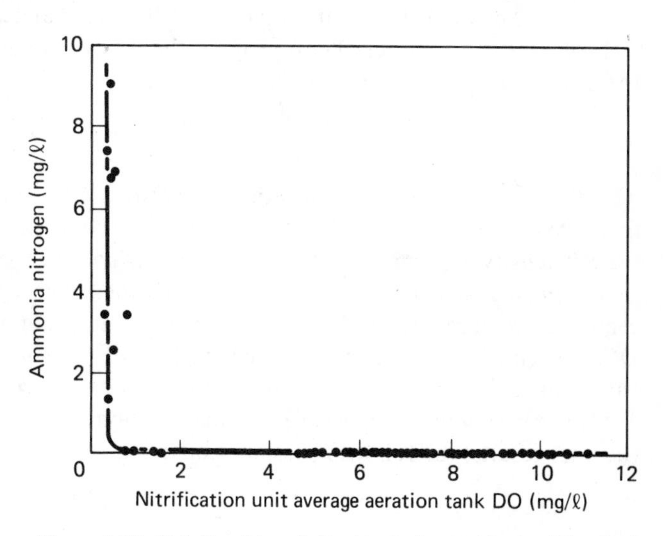

Figure 4-52. Relationship of Residual Ammonia to Dissolved Oxygen. (After Wild et al., 1971.)

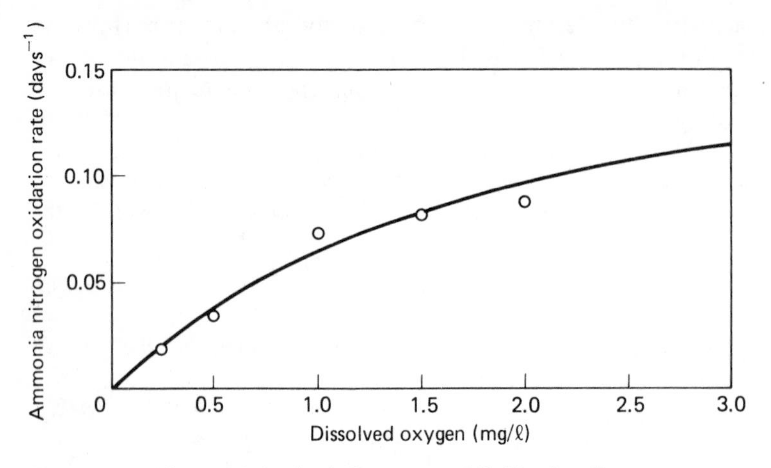

Figure 4-53. Effect of Dissolved Oxygen on Nitrification Rate. (Data of Nagel and Haworth, 1969, as presented by Brown and Caldwell, 1975.)

either of two ways: (1) in a combined carbon oxidation/nitrification process, or (2) in a separate-stage nitrification process. In the combined carbon oxidation/nitrification process, activated sludge floc size varies but is generally considered to fall within the range of 20 to 160 μm (Mueller et al., 1966; Englande and Eckenfelder, 1972). As a possible model, consider that the activated sludge floc is spherical in shape and is composed of three distinct concentric shells: an outermost layer several cells thick where cell growth is occurring, a second shell of viable but nonproliferating cells and an anaerobic central core.

Biryukov and Shtoffer (1973) propose the following equation to describe the distribution of oxygen in such a microcolony:

$$\frac{Dd^2C}{\rho\, dr^2} + \frac{2D\, dC}{\rho r\, dr} = Q \qquad (4\text{-}131)$$

where C is the oxygen concentration (mg O_2/cm^3) at some distance r (cm) from the center of the floc, D is the oxygen diffusivity (cm^2/s) in the floc, ρ is the cell density (mg/cm^3) in the floc, and Q is the specific oxygen consumption (mg O_2/mg cell-s). These workers suggest that the concept of critical O_2 concentration requires that cell respiration proceed at a rate independent of the DO concentration as long as the concentration remains above the critical value. Since the values normally quoted for critical O_2 concentration (Finn, 1954) are very low, it is possible to make, without a considerable change in the final results, the following assumptions:

$$Q = 0 \text{ at } C = 0 \qquad (4\text{-}132)$$

and

$$Q = Q \text{ at } C > 0 \qquad (4\text{-}133)$$

Solving the system of equations 4-131, 4-132, and 4-133 under the following boundary conditions,

$$C = 0 \text{ at } r = r_0 \tag{4-134}$$

$$\frac{dC}{dr} = 0 \text{ at } r = r_0 \tag{4-135}$$

$$C = C_s \text{ at } r = R \tag{4-136}$$

where r_0 is the distance from the center of the floc to the outer limit of the anaerobic zone in cm, R is the distance from the center of the floc to the floc surface in cm, and C_s is the O_2 concentration at the floc surface in mg O_2/cm^3. Biryukov and Shtoffer obtained the following general solution:

$$C = \frac{A}{6}r^2 + \frac{A}{3}\frac{r_0^3}{r} - \frac{A}{2}r_0^2 \text{ at } r_0 < r \leq R \tag{4-137}$$

and

$$C = 0 \text{ at } 0 \leq r \leq r_0 \tag{4-138}$$

where

$$A = \frac{Q\rho}{D} \tag{4-139}$$

For this equation to support the proposed floc model, r_0 must have some finite value when considering floc sizes within the normal range of those found in an activated sludge system.

Taking into account the boundary condition given by equation 4-136, the relationship

$$C_s = \frac{A}{6}R^2 + \frac{A}{3}\frac{r_0^3}{R} - \frac{A}{2}r_0^2 \tag{4-140}$$

can be established.

For activated sludge, Mueller et al. (1968) found wet density values ranging from 1.04 to 1.10 g/cm³ and dry density values from 1.38 to 1.65 g/cm³.

Although to the authors' knowledge no one has ever determined values for O_2 diffusivity in activated sludge floc, Mueller et al. (1968) have found that the value of 1.2×10^{-5} cm²/s gave reasonable results when used to calculate limiting O_2 concentration for larger-size floc.

The specific O_2 consumption will vary, but it can be approximated for a particular set of conditions from O_2 uptake and mixed liquor volatile suspended solids data. The O_2 concentration at the surface of the activated sludge floc will be less than the concentration in the bulk liquid and will be greatly influenced by the degree of turbulence in the aeration tank. Figure 4-54 illustrates the influence of mixing on the O_2 concentration at the floc surface. Note that the double crosshatching reflects the depth of O_2 penetration into the floc for condition *A*, which is a condition of less mixing than *B*.

For the degree of turbulence normally encountered in an activated sludge aeration tank, a conservative estimate for the oxygen concentration at the

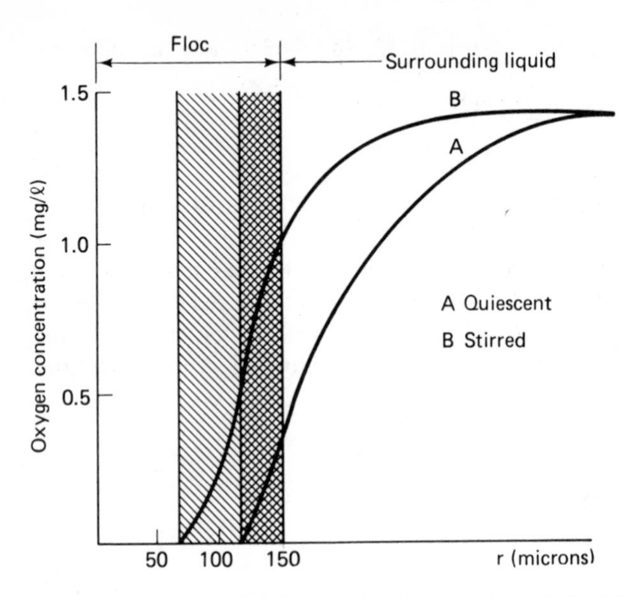

Figure 4-54. Effect of Stirring on O_2 Concentration at the Floc Surface. (After Biryukov and Shtoffer, 1973.)

floc surface would probably be 90% of the O_2 concentration in the liquid bulk. Hence, if one assumes that the average floc is 100 μm in diameter and the DO in the liquid bulk is 2 mg/ℓ, then, using the values for the various coefficients reported in the literature and letting $Q = 8.3 \times 10^{-5}$ mg O_2/mg VSS-s, the radius of the anaerobic zone of the floc would be approximately 86 μm. Hence, the model can be supported mathematically.

In the combined carbon oxidation/nitrification process the nitrifying bacteria compose only a very small percentage of the total biomass (approximately 5%). Under such a circumstance it is easy to visualize that a large fraction of the nitrifying organisms would lie some depth below the surface of the activated sludge floc. Thus, an increase in the DO concentration would increase the depth of penetration as well as the DO concentration within the sludge floc. This would increase the fraction of nitrifiers in the aerobic zone which are exposed to a DO level above critical. Hence, it is logical that increasing the DO concentration in a combined process will increase the rate of nitrification. The Pomona Water Renovation Plant where Nagel and Haworth (1969) performed their study is such a process.

In the separate-stage nitrification process the biomass has a larger nitrifying fraction, which would suggest that increasing the DO level above 1.0 mg/ℓ, an upper limit that is generally accepted for critical oxygen concentration, would not result in an increase in the ammonium oxidation rate. The data presented by Wild et al. (1971) came from such a system.

It can, therefore, be argued that effects of DO concentration on the specific growth rate of *Nitrosomonas* should be considered when combined carbon oxidation/nitrification processes are used but can be neglected in separate-stage nitrification since the process is always designed to maintain DO levels greater than 1 mg/ℓ. Furthermore, DO effects in the combined process can probably be neglected if the aeration system is designed to maintain a minimum DO level of 2.0 mg/ℓ, which should always be the case.

Influence of Operating Temperature. Hultman (1971) proposes that the maximum specific growth rate for *Nitrosomonas*, $(\mu_{max})_{NS}$, and the saturation constant, K_N, are affected by temperature according to the following relationships:

$$(\mu_{max})_{NS} = (\mu_{max})_{NS(20°C)}\,10^{0.033(T-20)} \tag{4-141}$$

where $(\mu_{max})_{NS}$ = maximum specific growth rate of *Nitrosomonas* at operating temperature, $days^{-1}$

$(\mu_{max})_{NS(20°C)}$ = maximum specific growth rate of *Nitrosomonas* at 20°C, day^{-1}

T = operating temperature, °C

and

$$K_N = 10^{0.051(T)-1.158} \tag{4-142}$$

where K_N represents the saturation constant in mg/ℓ as N.

The relationship between maximum specific growth rate and temperature reported by Downing and Hopwood (1964) is given by the following expression:

$$(\mu_{max})_{NS} = (\mu_{max})_{NS(15°C)}e^{0.12(T-15)} \tag{4-143}$$

However, in a later study Knowles et al. (1965) propose that the maximum specific growth rate of *Nitrosomonas* be adjusted for temperature by applying an equation of the form

$$(\mu_{max})_{NS} = (\mu_{max})_{NS(15°C)}e^{0.095(T-15)} \tag{4-144}$$

Stankewich (1972) reports that a typical value for $(\mu_{max})_{NS(15°C)}$ is 0.18 day^{-1} for activated sludge nitrification systems.

In this text equation 4-141 will be used to adjust for temperature effects on $(\mu_{max})_{NS}$. It should be understood, however, that this is simply a matter of personal preference, as no one has yet demonstrated the superiority of one equation over the others for general application.

Influence of Operating pH. The operating pH has been shown to affect the growth rate of nitrifying bacteria. To adjust the maximum specific growth rate of *Nitrosomonas* for pH variations, Hultman (1971) suggested the following equation:

$$(\mu_{\max})_{NS} = \frac{(\mu_{\max})_{NS} \text{ at optimum pH}}{1 + 0.04(10^{(\text{pH})_{\text{opt}} - \text{pH}} - 1)} \tag{4-145}$$

where $(\mu_{\max})_{NS}$ at optimum pH = maximum specific growth rate of *Nitrosomonas* at optimum pH, day^{-1}

$(\text{pH})_{\text{opt}}$ = optimum pH for growth of *Nitrosomonas*; values range from 8.0 to 8.4

pH = operating pH

Since this relationship is based on the premise that any deviation from optimum pH reduces microbial activity according to the mechanism of noncompetitive inhibition, no adjustment in K_N for pH variation is required. In other words, the degree of inhibition because of a pH change is independent of the substrate concentration. The relationship between maximum rate of nitrification and pH at a constant temperature is shown in Figure 4-55.

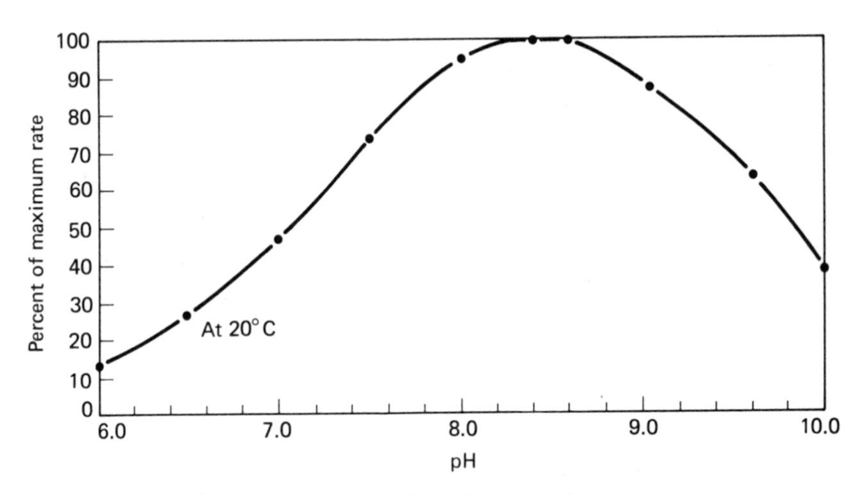

Figure 4-55. Percent of Maximum Rate of Nitrification at Constant Temperature Versus pH. (After Wild et al., 1971.)

During experimental studies Stankewich (1972) and Haug and McCarty (1972) have observed that even though nitrifying bacteria initially prefer a pH near 8, they are capable of acclimating to lower environmental pHs wherein they will reestablish their maximum growth rate. Haug and McCarty (1972) noted that a pH shift from 7.0 to 6.0 required approximately 10 days of acclimation, which must occur within the time interval of one BSRT or else washout will occur. To ensure that an adequate growth of nitrifiers can be maintained in the aeration tank, Bennett (1977) recommends that a means of sludge recycle be provided from the aerobic digester if such a unit is available. Aerobic digester sludge is generally rich in nitrifying bacteria, and as such will serve to reseed the aeration sludge if an upset is experienced which results in a loss of nitrifiers from the system.

Organism Yield. If energy of maintenance is neglected, the relationship between the absolute growth rate of *Nitrosomonas* and the rate of ammonium oxidation can be expressed as

$$\left(\frac{dX}{dt}\right)_{NS} = Y_N\left(\frac{dN}{dt}\right)_{NS} \tag{4-146}$$

where $\left(\frac{dX}{dt}\right)_{NS}$ = absolute growth rate of *Nitrosomonas*, mass volume^{-1} time^{-1}

$\left(\frac{dN}{dt}\right)_{NS}$ = oxidation rate of ammonium, mass volume^{-1} time^{-1}

Y_N = yield coefficient; biomass (*Nitrosomonas*) produced per unit of ammonium oxidized

Dividing both sides of equation 4-146 by the concentration of *Nitrosomonas* gives

$$(\mu)_{NS} = Y_N(q)_{NS} \tag{4-147}$$

where $(q)_{NS}$ = specific ammonium oxidation rate, time^{-1}

Biokinetic Constants for Nitrification. Table 4-11 presents normal values for the various biokinetic constants applicable to the nitrification process. It should be noted that although a value of 0.05 mg VSS/mg [NH_4^+-N] oxidized is given for the yield value, Painter (1970) suggests a range of 0.04 to 0.13 for this value, whereas Stankewich (1972) reports a range of values from 0.04 to 0.29.

TABLE 4-11

BIOKINETIC CONSTANTS FOR NITRIFICATION

Constant	*Value*
$(\mu_{max})_{NS[20°C, (pH)_{opt}]}$	0.3–0.5 day^{-1}
K_N	0.5–2.0 mg/ℓ
Y_N	≈0.05 mg VSS/mg [NH_4^+-N]
$(pH)_{opt}$	8.0–8.4

Source: After Hultman (1973).

Distribution of Nitrifying Bacteria. Neglecting the ammonium utilized in cell synthesis of nitrifying microorganisms, the distribution of nitrifying bacteria in a suspended growth system can be estimated from the relationship

$$(F)_N = \frac{Y_N(E_N)[(TKN)_0 - 0.122Y_{obs}(E_H)(S_0)]}{Y_{obs}(E_H)(S_0) + Y_N(E_N)[(TKN)_0 - 0.122Y_{obs}(E_H)(S_0)]} \tag{4-148}$$

where $(F)_N$ = fraction of total biomass that is composed of nitrifiers
E_N = nitrification efficiency, fraction
$(TKN)_0$ = total Kjeldahl nitrogen concentration in influent, mg/ℓ as N
Y_{obs} = observed yield coefficient related to the operating BSRT and calculated from equation 2-63
E_H = organic removal efficiency, fraction
S_0 = influent BOD_u, mg/ℓ

Through the use of equation 4-148, calculations can be made which show that the fraction of biomass which is composed of nitrifiers is considerably larger in separate-stage nitrification when compared to that in the combined process. It should be understood that equation 4-148 was developed by assuming that biomass production due to the metabolic activity of the heterotrophic microorganisms occurs before nitrification.

NITRIFICATION WITH THE COMPLETELY MIXED ACTIVATED SLUDGE PROCESS

For the completely mixed activated sludge process θ_c^m was defined as the BSRT at which biomass was removed from the system faster than it could be produced. For this condition the limiting nutrient concentration surrounding the microorganisms is considered equal to the nutrient concentration in the influent. This concept can be applied to nitrification. If energy of maintenance is neglected and the limiting nutrient concentration is considered to be the ammonium concentration, then

$$(\mu)_{NS} = \frac{1}{(\theta_c^m)_N} = (\mu_{max})_{NS} \frac{[NH_4^+\text{-N}]_0}{K_N + [NH_4^+\text{-N}]_0} \tag{4-149}$$

where $[NH_4^+\text{-N}]_0$ = influent ammonium concentration, (mg/ℓ as N)

$$(\theta_c^m)_N = \frac{[NH_4^+\text{-N}]_0 + K_N}{[NH_4^+\text{-N}]_0 (\mu_{max})_{NS}} \tag{4-150}$$

Since K_N is usually small compared to $[NH_4^+\text{-N}]_0$, it is possible to approximate equation 4-150 by the relationship

$$(\theta_c^m)_N \simeq \frac{1}{(\mu_{max})_{NS}} \tag{4-151}$$

which implies that if a nitrifying culture is to be established, the system must be operated at a BSRT greater than the minimum value predicted by equation 4-151 when the maximum specific growth rate of *Nitrosomonas* has been corrected for temperature and pH. This effect is illustrated in Figure 4-56. The shape of the response curve in this figure suggests that when the process is operated below $(\theta_c^m)_N$, no nitrification will occur. Thus, nitrification can be viewed as basically an all or nothing effect.

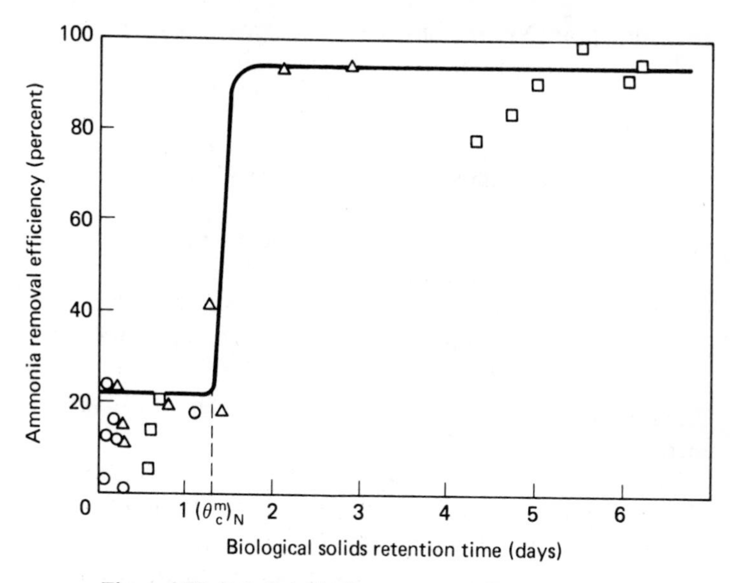

Figure 4-56. Relationship Between Ammonia Removal and Biological Solids Retention Time. (After Wuhrmann, 1968).

Hultman (1973) has presented an expression that describes the efficiency of the nitrification process in a completely mixed reactor operating at steady-state. To develop Hultman's equation, it is assumed that specific growth rate is described by equation 4-129,

$$(\mu)_{NS} = (\mu_{\max})_{NS} \frac{[NH_4^+\text{-N}]_e}{K_N + [NH_4^+\text{-N}]_e} \qquad (4\text{-}129)$$

Recalling that for a completely mixed system operating at steady-state

$$(\mu)_{NS} = \frac{1}{\theta_c}$$

Equation 4-129 can be written as

$$\frac{1}{\theta_c} = (\mu_{\max})_{NS} \frac{[NH_4^+\text{-N}]_e}{K_N + [NH_4^+\text{-N}]_e} \qquad (4\text{-}152)$$

which can be solved for $[NH_4^+\text{-N}]_e$ to give

$$[NH_4^+\text{-N}]_e = \frac{K_N}{\theta_c(\mu_{\max})_{NS} - 1} \qquad (4\text{-}153)$$

Since

$$E_N = 1 - \frac{[NH_4^+\text{-N}]_e}{[NH_4^+\text{-N}]_0} \qquad (4\text{-}154)$$

substituting for $[NH_4^+\text{-N}]_e$ gives

$$E_N = 1 - \frac{K_N}{[NH_4^+\text{-N}]_0[\theta_c(\mu_{max})_{NS} - 1]} \tag{4-155}$$

where θ_c = operating BSRT, days
$[NH_4^+\text{-N}]_0$ = ammonium nitrogen available as an energy source for nitrification, mg/ℓ
K_N = saturation constant adjusted to operating temperature according to equation 4-142, mg/ℓ as N
$(\mu_{max})_{NS}$ = maximum specific growth rate of *Nitrosomonas* adjusted to operating temperature and pH according to equations 4-141 and 4-145, $days^{-1}$

Using this relationship, the degree of nitrification can be determined for any operating BSRT above θ_c^m.

Equation 4-17 can be modified and used to determine the required volume of the nitrification unit when separate-stage treatment is employed. Equation 4-17 can be expressed in the form

$$\left(\frac{dN}{dt}\right)_{NS} = \frac{Q([NH_4^+\text{-N}]_0 - [NH_4^+\text{-N}]_e)}{V_N} \tag{4-156}$$

where Q = flow applied to nitrification stage, MGD
$[NH_4^+\text{-N}]_e$ = effluent ammonium concentration, mg/ℓ
V_N = volume of nitrification stage, MG

Dividing both sides of equation 4-156 by X_N gives

$$\frac{(dN/dt)_{NS}}{X_N} = (q)_{NS} = \frac{Q([NH_4^+\text{-N})_0 - [NH_4^+\text{-N}]_e)}{X_N V_N} \tag{4-157}$$

where X_N = nitrifier biomass concentration in the nitrification unit, mg/ℓ

Rearranging and substituting for $(q)_{NS}$ from equation 4-147 gives the equation

$$V_N = \frac{Y_N Q([NH_4^+\text{-N}]_0 - [NH_4^+\text{-N}]_e)}{(\mu)_{NS} X_N} \tag{4-158}$$

This expression may then be used to calculate the volume of the nitrification stage required for a specified nitrogen removal efficiency.

PROCESS SELECTION

In many cases the selection of either a combined carbon oxidation/nitrification process (see Figure 4-57) or a separate-stage nitrification process (see Figure 4-58) is governed mainly by the preference of the design engineer. However, each process has certain advantages and disadvantages that should be considered before a selection is made. Stall and Sherrard (1974) suggest the use of a combined process because sludge production is minimized in this

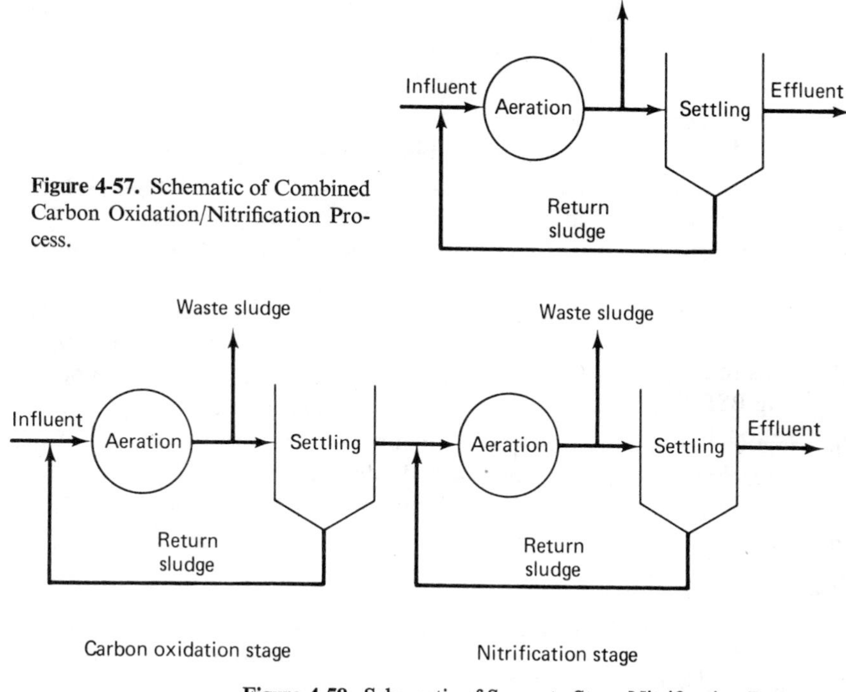

Figure 4-57. Schematic of Combined Carbon Oxidation/Nitrification Process.

Figure 4-58. Schematic of Separate-Stage Nitrification Process.

system. This is true, yet the separate-stage nitrification process may be desirable in many situations because of the following reasons:

1/ This type of system offers the capability of separate control and optimization of each process.

2/ To maximize the nitrogen removal efficiency, a tank geometry conducive to plug flow can be used in second-stage construction without fear of producing a high oxygen demand at the head of the tank (Brown and Caldwell, 1975).

3/ Nitrifying bacteria comprise a much larger fraction of the biomass in the second stage and as a result the process is less temperature-sensitive (Barth, 1968).

4/ Biodegradable organic materials which may be toxic to nitrifying bacteria are removed in the carbon oxidation stage (Boon and Burgess, 1974).

5/ A lower residual DO concentration may be used in either of the two stages than would be required for the combined process to ensure that DO did not affect $(\mu_{max})_{NS}$.

6/ Operating experience seems to indicate that separate-stage nitrification has a greater reliability.

7/ When the overall design is controlled by BOD removal kinetics, the separate-stage nitrification process will require less total aeration tank volume and total oxygen.

For small wastewater flows a combined process is generally preferred. The method of choice for this situation might be an extended aeration process modification followed by sand filtration. Such an arrangement should produce an effluent with a BOD_5 and TKN less than 15 mg/ℓ and 5 mg/ℓ, respectively.

In previous sections biological nitrification has been discussed and the kinetics of nitrification for suspended-growth system presented. Example 4-12 illustrates how these concepts may be applied to the design of both a combined and separate-stage nitrification. This example also suggests that the engineer has two choices when designing an activated sludge system to treat municipal wastewater. These are (1) design to avoid nitrification, or (2) design to achieve nitrification. There is no reason to follow any other approach.

Example 4-12

Compare the total aeration tank volume and oxygen requirement between the separate stage and combined processes if a 95% nitrification efficiency is desired. The following design criteria are applicable:

$$Q = 20 \text{ MGD}$$

$$S_0 = 200 \text{ mg/}\ell \text{ BOD}_u$$

$$TKN_0 = 20 \text{ mg/}\ell \text{ as } N$$

$$S_e = 10 \text{ mg/}\ell$$

$$K = 0.04\ \ell\text{/mg-day at } 20°\text{C}$$

$$K_d = 0.05 \text{ day}^{-1} \text{ at } 20°\text{C}$$

$$Y_T = 0.5$$

critical summer operating temperature = 25°C

critical winter operating temperature = 16°C

operating pH = 7.6 in carbon removal stage

operating pH = 7.0 in nitrification stage

solution

I/ *Separate-Stage Process*

A/ *Carbon Removal Stage*

1/ Compute $(\mu_{max})_{NS}$ for both winter and summer conditions assuming $(\mu_{max})_{NS}$ at 20°C is 0.4 day^{-1} and that $(\text{pH})_{opt}$ is 8.2. These values are arbitrary and their selection is left to the preference of the engineer.

a/ Determine the correction factor for pH.

$$\frac{(\mu_{max})_{NS}}{(\mu_{max})_{NS(pH)_{opt}}} = \frac{1}{1 + 0.04(10^{(pH)_{opt} - pH} - 1)}$$

$$\text{factor} = \frac{1}{1 + 0.04(10^{8.2-7.6} - 1)}$$

$$= 0.892$$

b/ Determine the correction factor for temperature.

$$\frac{(\mu_{max})_{NS}}{(\mu_{max})_{NS(20^\circ C)}} = 10^{0.033(T-20)}$$

Winter:

$$\text{factor} = 10^{0.033(16-20)}$$

$$= 0.737$$

Summer:

$$\text{factor} = 10^{0.033(25-20)}$$

$$= 1.46$$

c/ Estimate the corrected $(\mu_{max})_{NS}$.
Winter:

$$(\mu_{max})_{NS} = (0.4)(0.892)(0.737)$$

$$= 0.262 \text{ day}^{-1}$$

Summer:

$$(\mu_{max})_{NS} = (0.4)(0.892)(1.46)$$

$$= 0.521 \text{ day}^{-1}$$

2/ Determine the operating BSRT below which nitrification will not occur, $(\theta_c^m)_N$, for both winter and summer conditions.

$$(\theta_c^m)_N = \frac{1}{(\mu_{max})_{NS}}$$

Winter:

$$(\theta_c^m)_N = \frac{1}{0.262}$$

$$= 3.8 \text{ days}$$

Summer:

$$(\theta_c^m)_N = \frac{1}{0.521}$$

$$= 1.91 \text{ days}$$

Thus, operating at a BSRT of approximately 4 days during the winter and 2 days during the summer should prevent nitrification in the first stage. The larger the θ_c value used for the first stage, however, the larger the aeration tank volume. Hence, select the smallest θ_c which will provide a sludge with reasonably good flocculating characteristics as well as prevent nitrification. In this case a θ_c of two days will be used in both winter and summer. Although a BSRT of 2 days would normally be too low for effective biomass agglomeration, this does not affect process efficiency as long as the second-stage solids–liquid separation step is functioning properly and solids recycle in the first stage is possible.

3/ Compute the specific substrate utilization rate for both winter and summer conditions.

$$\frac{1}{\theta_c} = Y_T q - K_d$$

or

$$q = \frac{1/\theta_c + K_d}{Y_T}$$

a/ Correct K_d for temperature variations.

Winter:

$$K_d = (K_d)_{20°C}(1.05)^{T-20}$$
$$= (0.05)(1.05)^{16-20}$$
$$= 0.041 \text{ day}^{-1}$$

Summer:

$$K_d = (0.05)(1.05)^{25-20}$$
$$= 0.064 \text{ day}^{-1}$$

b/ Compute the specific substrate utilization rate.

Winter:

$$q = \frac{1/2 + 0.041}{0.5}$$
$$= 1.08 \text{ day}^{-1}$$

Summer:

$$q = \frac{1/2 + 0.064}{0.5}$$
$$= 1.13 \text{ day}^{-1}$$

4/ Estimate the soluble effluent substrate concentration for both winter and summer conditions.

$$S_e = \frac{q}{K}$$

a/ Correct K for temperature variations.

Winter:

$$K = (K)_{20°C}(1.03)^{T-20}$$
$$= (0.04)(1.03)^{16-20}$$
$$= 0.035\ \ell/\text{mg-day}$$

Summer:

$$K = (0.04)(1.03)^{25-20}$$
$$= 0.046\ \ell/\text{mg-day}$$

b/ Calculate the soluble effluent substrate concentration.
Winter:

$$S_e = \frac{1.08}{0.035}$$
$$= 30.8 \text{ mg}/\ell$$

Summer:

$$S_e = \frac{1.13}{0.046}$$
$$= 24.5 \text{ mg}/\ell$$

5/ Determine the organic removal efficiency for both winter and summer conditions.

$$E_0 = \frac{S_0 - S_e}{S_0}(100)$$

Winter:

$$E_0 = \frac{200 - 30.8}{200}(100)$$
$$= 84.6\%$$

Summer:

$$E_0 = \frac{200 - 24.5}{200}(100)$$
$$= 87.7\%$$

6/ Assume some reasonable value for MLVSS concentration (e.g., 1000 to 2000 mg/ℓ) and determine the aeration tank volume requirement for both winter and summer conditions. The larger value will control the design. For this problem a 2000-mg/ℓ MLVSS concentration will be assumed.

$$V_a = \frac{Q(S_0 - S_e)}{Xq}$$

Winter:

$$V_a = \frac{20(200 - 30.8)}{(2000)(1.08)}$$
$$= 1.57 \text{ MG}$$

Summer:

$$V_a = \frac{20(200 - 24.5)}{(2000)(1.13)}$$
$$= 1.55 \text{ MG}$$

Therefore, winter conditions control first-stage volume requirements.

7/ For the required aeration tank volume, determine the MLVSS concentration for the temperature which did not control the design.

$$X = \frac{Q(S_0 - S_e)}{V_{a(\text{design})}q}$$
$$= \frac{20(200 - 24.5)}{(1.57)(1.13)}$$
$$= 1978 \text{ mg}/\ell$$

8/ Calculate the observed yield coefficient for both winter and summer conditions.

$$Y_{obs} = \frac{Y_T}{1 + K_d\theta_c}$$

Winter:

$$Y_{obs} = \frac{0.5}{1 + (0.041)(2)}$$

$$= 0.46$$

Summer:

$$Y_{obs} = \frac{0.5}{1 + (0.064)(2)}$$

$$= 0.44$$

9/ Estimate the biomass production for both winter and summer conditions.

$$\Delta X = Y_{obs}Q(8.34)(S_0 - S_e)$$

Winter:

$$\Delta X = (0.46)(20)(8.34)(200 - 30.8)$$

$$= 12{,}982 \text{ lb/day}$$

Summer:

$$\Delta X = (0.44)(20)(8.34)(200 - 24.5)$$

$$= 12{,}880 \text{ lb/day}$$

10/ For both winter and summer conditions, determine the steady-state sludge wasting rate, assuming that wasting is done from the aeration tank.

$$Q_w = \frac{\Delta X}{(8.34)X}$$

Winter:

$$Q_w = \frac{12{,}982}{(8.34)(2000)}$$

$$= 0.78 \text{ MGD}$$

Summer:

$$Q_w = \frac{12{,}880}{(8.34)(1978)}$$

$$= 0.78 \text{ MGD}$$

11/ Estimate nitrogen removal as a result of heterotrophic cell synthesis for both winter and summer conditions.

$$\text{nitrogen removal} = (0.122)\ \Delta X$$

Winter:

$$\text{nitrogen removal} = (0.122)(12{,}982)$$

$$= 1583 \text{ lb/day}$$

Summer:

$$\text{nitrogen removal} = (0.122)(12{,}880) = 1571 \text{ lb/day}$$

12/ For both winter and summer conditions, compute the nitrogen concentration in the effluent from the first stage, assuming 100% solids capture in the clarifier.

$$\text{effluent nitrogen concentration} = \frac{\begin{bmatrix}\text{nitrogen}\\ \text{loading}\end{bmatrix} - \begin{bmatrix}\text{nitrogen removal}\\ \text{for cell synthesis}\end{bmatrix}}{(8.34)Q}$$

Winter:

$$\text{effluent nitrogen concentration} = \frac{(20)(20)(8.34) - 1583}{(8.34)(20)} = 10.5 \text{ mg}/\ell \text{ as N}$$

Summer:

$$\text{effluent nitrogen concentration} = \frac{(20)(20)(8.34) - 1571}{(8.34)(20)} = 10.5 \text{ mg}/\ell \text{ as N}$$

13/ Compute the oxygen requirement for both winter and summer conditions.

$$\Delta O_2 = 8.34Q(S_0 - S_e) - 1.42\,\Delta X$$

Winter:

$$\Delta O_2 = (8.34)(20)(200 - 30.8) - (1.42)(12{,}982) = 9787 \text{ lb/day}$$

Summer:

$$\Delta O_2 = (8.34)(20)(200 - 24.5) - (1.42)(12{,}880) = 10{,}983 \text{ lb/day}$$

14/ Compare calculated values for winter and summer conditions and determine design values for aeration tank volume, biomass production, and oxygen requirement.

aeration tank volume = 1.57 MG

biomass production = 12,982 lb/day

oxygen requirement = 10,983 lb/day

B. *Nitrification Stage*

15/ Compute $(\mu_{max})_{NS}$ for both winter and summer conditions assuming that $(\mu_{max})_{NS}$ at 20°C is 0.4 day^{-1} and that $(pH)_{opt}$ is 8.2.

a/ Determine the correction factor for pH.

$$\frac{(\mu_{\max})_{NS}}{(\mu_{\max})_{NS(\mathrm{pH})_{\mathrm{opt}}}} = \frac{1}{1 + 0.04(10^{(\mathrm{pH})_{\mathrm{opt}} - \mathrm{pH}} - 1)}$$

$$\text{factor} = \frac{1}{1 + 0.4(10^{8.2-7.0} - 1)}$$

$$= 0.63$$

b/ Correction factors for temperature are assumed to be the same as those for the first stage.

$$\text{winter factor} = 0.737$$

$$\text{summer factor} = 1.46$$

c/ Estimate the corrected $(\mu_{\max})_{NS}$.

Winter:

$$(\mu_{\max})_{NS} = (0.4)(0.63)(0.737)$$

$$= 0.186 \text{ day}^{-1}$$

Summer:

$$(\mu_{\max})_{NS} = (0.4)(0.63)(1.46)$$

$$= 0.368 \text{ day}^{-1}$$

16/ Calculate K_N for both winter and summer conditions.

$$K_N = 10^{0.051(T) - 1.158}$$

Winter:

$$K_N = 10^{0.051(16) - 1.158}$$

$$= 0.45 \text{ mg}/\ell \text{ as N}$$

Summer:

$$K_N = 10^{0.051(25) - 1.158}$$

$$= 1.31 \text{ mg}/\ell \text{ as N}$$

17/ For both winter and summer conditions, compute the required operating BSRT to give the desired nitrification efficiency.

$$E_N = 1 - \frac{K_N}{[\mathrm{NH_4^+\text{-}N}]_0[\theta_c(\mu_{\max})_{NS} - 1]}$$

Winter:

$$0.95 = 1 - \frac{0.45}{(10.5)[\theta_c(0.186) - 1]}$$

$$\theta_c = 9.98 \text{ days}$$

Summer:

$$0.95 = 1 - \frac{1.31}{(10.5)[\theta_c(0.368) - 1]}$$

$$\theta_c = 9.63 \text{ days}$$

A year-round θ_c of 10 days will be selected for design.

18/ Compute the nitrification efficiency for both winter and summer conditions using $\theta_c = 10$ days.

Winter:

$$E_N = 1 - \frac{0.45}{(10.5)[(10)(0.186) - 1]}$$
$$= 95\%$$

Summer:

$$E_N = 1 - \frac{1.31}{(10.3)[(10)(0.368) - 1]}$$
$$= 95.2\%$$

19/ Using the steady-state relationship, $(\mu)_{NS} = 1/\theta_c$, select values for Y_N and the nitrifier concentration, X_N, and then calculate the aeration tank volume for both winter and summer conditions. In this problem it will be assumed that $Y_N = 0.05$ and the design nitrifier concentration will be 100 mg/ℓ.

$$V_N = \frac{Y_N(Q - Q_w)([NH_4^+\text{-N}]_0 - [NH_4^+\text{-N}]_e)}{(\mu_{NS})X_N}$$

Winter:

$$V_N = \frac{(0.05)(20 - 0.78)[(0.95)(10.5)]}{(0.1)(100)}$$
$$= 0.96 \text{ MG}$$

Summer:

$$V_N = \frac{(0.05)(20 - 0.78)[(0.952)(10.5)]}{(0.1)(100)}$$
$$= 0.96 \text{ MG}$$

Note: Experience has shown that the nitrifier fraction seldom exceeds 10 percent of the total biomass. Thus, X_N Should be selected to give $(F)_N$ values less than 0.1.

20/ Heterotrophic considerations in the second stage.

a/ For both winter and summer conditions, determine the specific substrate utilization rate using the operating BSRT calculated in step 17.

$$q = \frac{1/\theta_c + K_d}{Y_T}$$

Winter:

$$q = \frac{\frac{1}{10} + 0.041}{0.5}$$
$$= 0.28 \text{ day}^{-1}$$

Summer:

$$q = \frac{\frac{1}{10} + 0.064}{0.5}$$
$$= 0.33 \text{ day}^{-1}$$

b/ Compute the soluble effluent substrate concentration for both winter and summer conditions.

$$S_e = \frac{q}{K}$$

Winter:

$$S_e = \frac{0.28}{0.035}$$
$$= 8 \text{ mg}/\ell$$

Summer:

$$S_e = \frac{0.33}{0.046}$$
$$= 7 \text{ mg}/\ell$$

c/ Check calculated S_e values with the desired S_e value.

Winter:

$$8 \text{ mg}/\ell < 10 \text{ mg}/\ell \qquad \text{o.k.}$$

Summer:

$$7 \text{ mg}/\ell < 10 \text{ mg}/\ell \qquad \text{o.k.}$$

If the desired S_e value had been exceeded, it would have been necessary to compute the specific substrate utilization rate necessary to achieve the S_e value required. This value of q could then be used to determine the operating θ_c and this θ_c value applied to the efficiency formula in step 17. All calculations from step 17 through step 20 would then be repeated.

d/ Estimate the heterotrophic biomass concentration for both winter and summer conditions.

$$X_H = \frac{(Q - Q_w)(S_0 - S_e)}{qV_N}$$

Winter:

$$X_H = \frac{(20 - 0.78)(30.8 - 8)}{(0.28)(0.96)}$$
$$= 1630 \text{ mg}/\ell$$

Summer:

$$X_H = \frac{(20 - 0.78)(24.5 - 7)}{(0.33)(0.96)}$$
$$= 1062 \text{ mg}/\ell$$

e/ Calculate the observed yield coefficient for both winter and summer conditions.

$$Y_{obs} = \frac{Y_T}{1 + K_d\theta_c}$$

Winter:

$$Y_{obs} = \frac{0.5}{1 + (0.041)(10)}$$
$$= 0.35$$

Summer:

$$Y_{\text{obs}} = \frac{0.5}{1 + (0.064)(10)}$$

$$= 0.30$$

21/ Compute the heterotrophic biomass production for both winter and summer conditions.

$$\Delta X = Y_{\text{obs}}(Q - Q_w)(8.34)(S_0 - S_e)$$

Winter:

$$\Delta X = (0.35)(20 - 0.78)(8.34)(30.8 - 8)$$

$$= 1279 \text{ lb/day}$$

Summer:

$$\Delta X = (0.30)(20 - 0.78)(8.34)(24.5 - 7)$$

$$= 841 \text{ lb/day}$$

22/ For both winter and summer conditions, determine the fraction of biomass that is composed of nitrifiers.

$$(F)_N = \frac{Y_N(E_N)[(TKN)_0 - (0.122)\,Y_{\text{obs}}(E_H)(S_0)]}{Y_{\text{obs}}(E_H)(S_0) + Y_N(E_N)[(TKN)_0 - (0.122)\,Y_{\text{obs}}(E_H)(S_0)]}$$

Winter:

$$(F)_N = \frac{(0.05)(0.95)[(10.5) - (0.122)(0.35)(30.8 - 8)]}{(0.35)(30.8 - 8) + (0.05)(0.95)[(10.5) - (0.122)(0.35)(30.8 - 8)]}$$

$$= 0.054$$

Summer:

$$(F)_N = \frac{(0.05)(0.952)[(10.5) - (0.122)(0.3)(24.5 - 7)]}{(0.3)(24.5 - 7) + (0.05)(0.952)[(10.3) - (0.122)(0.3)(24.5 - 7)]}$$

$$= 0.082$$

23/ Check the nitrifier concentration for winter conditions.

$$X_N = \frac{1630}{1 - 0.054} - 1630$$

$$= 93 \text{ mg}/\ell$$

The 93 mg/ℓ value is very near the design concentration of 100 mg/ℓ and supports the validity of equation 4-148 for computing the nitrifier fraction.

24/ Determine the oxygen requirement for both winter and summer conditions.

$$\Delta O_2 = (8.34)(Q - Q_w)(S_0 - S_e) - 1.42\,\Delta X$$

$$+ 38.1(Q - Q_w)(E_N)[NH_4^+\text{-N}]_0$$

Winter:

$$\Delta O_2 = (8.34)(20 - 0.78)(30.8 - 8) - 1.42(1279) + (38.1)(20 - 0.78)(0.95)(10.5) = 9143 \text{ lb/day}$$

Summer:

$$\Delta O_2 = (8.34)(20 - 0.78)(24.5 - 7) - 1.42(841) + (38.1)(20 - 0.78)(0.952)(10.5) = 8931 \text{ lb/day}$$

25/ Compare calculated values for winter and summer conditions and determine design values for aeration tank volume, biomass production, and oxygen requirement.

aeration tank volume = 0.96 MG

biomass production = 1279 lb/day

oxygen requirement = 9143 lb/day

26/ Determine the total design requirements by combining values tabulated in steps 14 and 25.

total aeration tank volume = 2.53 MG

total biomass production = 14,261 lb/day

total oxygen requirement = 20,126 lb/day

II/ *Combined Process*

27/ Compute $(\mu_{max})_{NS}$ for both winter and summer conditions, assuming that $(\mu_{max})_{NS}$ at 20°C is 0.4 day^{-1} and that $(pH)_{opt}$ is 8.2.

a/ Determine the correction factor for pH using the operation pH for the nitrification stage.

From step 15a, the pH correction factor is 0.63.

b/ Correction factors for temperature are assumed to be the same as those computed in step 1b.

winter factor = 0.737

summer factor = 1.46

c/ Estimate the corrected $(\mu_{max})_{NS}$.

From step 15c:

winter $(\mu_{max})_{NS} = 0.186 \text{ day}^{-1}$

summer $(\mu_{max})_{NS} = 0.368 \text{ day}^{-1}$

28/ Compute K_N for both winter and summer conditions.

From step 16:

winter $K_N = 0.45$ mg/ℓ

summer $K_N = 1.31$ mg/ℓ

29/ Determine the specific substrate utilization rate required to achieve the desired effluent quality for both winter and summer conditions.

$$q = KS_e$$

Winter:

$$q = (10)(0.035)$$
$$= 0.35 \text{ day}^{-1}$$

Summer:

$$q = (10)(0.046)$$
$$= 0.46 \text{ day}^{-1}$$

30/ Estimate the BSRT associated with the q values calculated in step 29 for both winter and summer conditions.

$$\frac{1}{\theta_c} = Y_T q - K_d$$

Winter:

$$\frac{1}{\theta_c} = (0.5)(0.35) - 0.041$$
$$\theta_c = 7.5 \text{ days}$$

Summer:

$$\frac{1}{\theta_c} = (0.5)(0.46) - 0.064$$
$$\theta_c = 6.0 \text{ days}$$

31/ Calculate the observed yield coefficient for both winter and summer conditions.

$$Y_{\text{obs}} = \frac{Y_T}{1 + K_d\theta_c}$$

Winter:

$$Y_{\text{obs}} = \frac{0.5}{1 + (0.041)(7.5)}$$
$$= 0.38$$

Summer:

$$Y_{\text{obs}} = \frac{0.5}{1 + (0.064)(6.0)}$$
$$= 0.36$$

32/ Determine the biomass production for both winter and summer conditions.

$$\Delta X = Y_{\text{obs}} Q(8.34)(S_0 - S_e)$$

Winter:

$$\Delta X = (0.38)(20)(8.34)(200 - 10)$$
$$= 12{,}043 \text{ lb/day}$$

Summer:

$$\Delta X = (0.36)(20)(8.34)(200 - 10)$$
$$= 11{,}409 \text{ lb/day}$$

33/ Estimate nitrogen removal as a result of heterotrophic cell synthesis for both winter and summer conditions.

$$\text{nitrogen removal} = (0.122)\,\Delta X$$

Winter:

$$\text{nitrogen removal} = (0.122)(12{,}043)$$
$$= 1469 \text{ lb/day}$$

Summer:

$$\text{nitrogen removal} = (0.122)(11{,}409)$$
$$= 1392 \text{ lb/day}$$

34/ Approximate the nitrogen available for nitrification for both winter and summer conditions.

$$\text{available nitrogen} = \frac{\begin{bmatrix}\text{nitrogen}\\ \text{loading}\end{bmatrix} - \begin{bmatrix}\text{nitrogen removal}\\ \text{for cell synthesis}\end{bmatrix}}{(8.34)(Q)}$$

Winter:

$$\text{available nitrogen} = \frac{(20)(20)(8.34) - 1469}{(20)(8.34)}$$
$$= 11.1 \text{ mg}/\ell$$

Summer:

$$\text{available nitrogen} = \frac{(20)(20)(8.34) - 1392}{(20)(8.34)}$$
$$= 11.6 \text{ mg}/\ell$$

35/ For both winter and summer conditions, calculate the required operating BSRT to give the desired nitrification efficiency.

$$E_N = 1 - \frac{K_N}{[NH_4^+\text{-N}]_0[\theta_c(\mu_{\max})_{NS} - 1]}$$

Winter:

$$0.95 = 1 - \frac{0.45}{(11.1)[\theta_c(0.186) - 1]}$$
$$\theta_c = 9.7 \text{ days}$$

Summer:

$$0.95 = 1 - \frac{1.31}{(11.6)[\theta_c(0.368) - 1]}$$
$$\theta_c = 8.8 \text{ days}$$

Therefore, an operating θ_c of 10 days will be selected for both winter and summer conditions. This is the same value of θ_c determined in step 17.

36/ For both winter and summer conditions determine the actual specific substrate utilization rate using the operating BSRT found in step 35.

From step 20a:

$$\text{winter } q = 0.28 \text{ day}^{-1}$$
$$\text{summer } q = 0.33 \text{ day}^{-1}$$

37/ Compute the soluble effluent substrate concentration for both winter and summer conditions.

From step 20b:

$$\text{winter } S_e = 8 \text{ mg}/\ell$$

$$\text{summer } S_e = 7 \text{ mg}/\ell$$

38/ Check calculated S_e values with the desired S_e value.

Winter:

$$8 \text{ mg}/\ell < 10 \text{ mg}/\ell \qquad \text{o.k.}$$

Summer:

$$7 \text{ mg}/\ell < 10 \text{ mg}/\ell \qquad \text{o.k.}$$

39/ Assume some reasonable value for MLVSS concentration (e.g., 1000 to 2000 mg/ℓ) and determine the aeration tank volume requirement for both winter and summer conditions. The larger value will control the design. For this problem a 2000-mg/ℓ MLVSS concentration will be assumed.

$$V = \frac{Q(S_0 - S_e)}{Xq}$$

Winter:

$$V = \frac{20(200 - 8)}{(2000)(0.28)}$$

$$= 6.86 \text{ MG}$$

Summer:

$$V = \frac{20(200 - 7)}{(2000)(0.33)}$$

$$= 5.84 \text{ MG}$$

Thus, winter conditions control the volume requirement.

40/ Calculate the actual observed yield coefficient for both winter and summer conditions.

From step 20f:

$$\text{winter } Y_{\text{obs}} = 0.35$$

$$\text{summer } Y_{\text{obs}} = 0.30$$

41/ Compute the fraction of biomass that is composed of nitrifiers for winter conditions.

$$(F)_N = \frac{(0.05)(0.95)(11.1)}{(0.35)(200 - 8) + (0.05)(0.95)(11.1)}$$

$$= 0.008$$

The nitrogen incorporated into heterotrophic synthesis was considered in step 34.

42/ Compute the nitrifier concentration for winter conditions.

$$X_N = (0.008)(2000)$$

$$= 16 \text{ mg}/\ell$$

43/ Calculate the aeration tank volume required for nitrification and compare this value to that computed for carbon removal.

$$V_N = \frac{(0.05)(20)(0.95)(11.1)}{16(0.1)}$$

$$= 6.59 \text{ MG}$$

Therefore, the aeration tank volume computed in step 39 is adequate.

44/ Estimate the actual biomass production for both winter and summer conditions.

$$\Delta X = Y_{obs}Q(8.34)(S_0 - S_e)$$

Winter:

$$\Delta X = (0.35)(20)(8.34)(200 - 8)$$

$$= 11{,}209 \text{ lb/day}$$

Summer:

$$\Delta X = (0.30)(20)(8.34)(200 - 7)$$

$$= 9658 \text{ lb/day}$$

Even though the differences in biomass production will result in a small change in the amount of nitrogen removed for cell synthesis, there is no need to readjust this value as the change is insignificant.

45/ Compute the oxygen requirement for both winter and summer conditions.

$$\Delta O_2 = 8.34Q(S_0 - S_e) - 1.42\,\Delta X + 38.1\,Q(E_N)[NH_4^+\text{-N}]$$

Winter:

$$\Delta O_2 = (8.34)(20)(200 - 8) - (1.42)(11{,}209) + (38.1)(20)(0.951)(11.1)$$

$$= 24{,}151 \text{ lb/day}$$

Summer:

$$\Delta O_2 = (8.34)(20)(200 - 7) - (1.42)(9658) + (38.1)(20)(0.957)(11.6)$$

$$= 26{,}936 \text{ lb/day}$$

46/ Compare calculated values for winter and summer conditions and determine design values for aeration tank volume, biomass production, and oxygen requirement.

aeration tank volume = 6.86 MG

biomass production = 12,043 lb/day

oxygen requirement = 26,936 lb/day

Comparing the design values given in step 26 with the values tabulated in step 46 shows that the overall aeration tank volume for the separate stage process is 4.33 MG less than that required for the combined process, and the total oxygen requirement is

6801 lb/day less than the combined process requirements. The combined process, however, does show a sludge production which is 2218 lb/day less than that for the separate-stage process, and the separate-stage process does require an extra clarifier.

4-7 Biological Denitrification

As noted earlier, a nitrified effluent is far preferable to one containing a large ammonium concentration. However, ammonium and nitrate are interchangeable nitrogenous nutrients for algae, and as a result, highly nitrified effluents can still accelerate algal blooms. Therefore, in some cases high nitrate levels in a plant effluent may also be undesirable. For example, when a treatment plant discharges into a receiving stream with a low available nitrogen concentration and with a flow much larger than the effluent, the presence of nitrates in the effluent generally does not adversely affect stream quality. On the other hand, if the nitrate concentration in the stream is significant, it may be desirable to control the nitrogen content of the effluent. Even more critical is the case where treatment plant effluent is discharged directly into relatively still bodies of water such as lakes or reservoirs.

To control algal growth through nutrient removal, it is necessary to determine the minimum level of nutrient needed to trigger algal blooms. The most widely quoted figures for such minimum levels are 0.30 mg/ℓ for inorganic nitrogen and 0.015 mg/ℓ for soluble orthophosphate (Sawyer, 1947).

Another argument for the control of nitrogen in the aquatic environment is the occurrence of infantile methemoglobinemia, which is a serious problem resulting from high concentrations of nitrates in drinking water. Such a danger led the Public Health Service to establish a 10 mg NO_3^--N/ℓ standard for drinking water in 1962.

If a nitrified effluent is determined to be unacceptable, a means of nitrogen removal must be provided in the treatment train. The four basic processes used for this purpose are (1) ammonia stripping, (2) selective ion exchange, (3) breakpoint chlorination, and (4) biological nitrification/denitrification. The latter is the only biological method available and as such will be the only nitrogen removal method discussed.

Microbiological Aspects of Denitrification

Biological nitrification/denitrification, as the name implies, is a two-step process. The first step to be accomplished is nitrification, which, as previously discussed, is the conversion of ammonia to nitrate through the action of nitrifying bacteria. The second step is nitrate conversion (denitrification), which is carried out by facultative heterotrophic bacteria under anoxic conditions.

Nitrate conversion takes place through both assimilatory and dissimilatory cellular functions. In assimilatory denitrification, nitrate is reduced to ammonia, which then serves as a nitrogen source for cell synthesis. Thus, nitrogen is removed from the liquid stream by incorporating it into cytoplasmic material.

In dissimilatory denitrification, nitrate serves as the electron accpetor in energy metabolism and is converted to various gaseous end products but principally molecular nitrogen, N_2, which is then stripped from the liquid stream. Because the microbial yield under anoxic conditions is considerably lower than under aerobic conditions, a relatively small fraction of the nitrogen is removed through assimilation. Dissimilatory denitrification is, therefore, the primary means by which nitrogen removal is achieved, accounting for 70 to 75% of the total removal.

Denitrification results from the metabolic activity of certain facultative heterotrophic bacteria which utilize organic carbon both as a carbon and energy source. Nitrate is used by these organisms as an electron acceptor in energy metabolism. The principal biochemical pathways involved are the same as those used in aerobic respiration. The major difference here is that nitrate replaces oxygen in the electron transport chain. Thus, denitrification falls into the category of anaerobic respiration rather than aerobic respiration. As oxygen will generally have a lower redox potential than other inorganic electron acceptors, more ATP will be realized from aerobic respiration. This accounts for the fact that microbial yield is less under anoxic conditions.

Since the bacteria responsible for denitrification use organic carbon as both a carbon and energy source, and because in a typical secondary effluent very little organic carbon is available, an external source of organic carbon is usually added prior to the dentrification process. For economic reasons methanol has received the widest application as a supplementary carbon source. However, other organic compounds have also been found satisfactory for use (McCarty, 1969).

Because dissimilatory denitrification is the primary means by which nitrogen removal is achieved biologically, it is worthwhile to consider the use of methanol as an energy source (i.e., electron donor). In this case dissimilatory denitrification can be considered as the two-step process illustrated by equations 4-159 and 4-160 (Metcalf and Eddy, 1972):

$$6NO_3^- + 2CH_3OH \longrightarrow 6NO_2^- + 2CO_2 + 4H_2O \qquad (4\text{-}159)$$

$$6NO_2^- + 3CH_3OH \longrightarrow 3N_2 + 3CO_2 + 3H_2O + 6OH^- \qquad (4\text{-}160)$$

$$\text{overall} \quad 6NO_3^- + 5CH_3OH \longrightarrow 5CO_2 + 3N_2 + 7H_2O + 6OH^- \qquad (4\text{-}161)$$

Equation 4-161 clearly shows that nitrate is the electron acceptor because it gains electrons and is reduced to molecular nitrogen, whereas methanol acts as the electron donor because it loses electrons and is oxidized to carbon dioxide.

Using data obtained from laboratory studies, McCarty (1969) has proposed that the overall nitrate removal reaction (including both assimilation and dissimilation reactions) can be described by the following empirical equation:

$$NO_3^- + 1.08CH_3OH + H^+ \longrightarrow 0.065C_5H_7O_2N + 0.47N_2 + 0.76CO_2 + 2.44H_2O \quad (4\text{-}162)$$

Denitrification in a Suspended-Growth System

Biological denitrification can be accomplished either in a suspended-growth system under anoxic conditions or in a columnar (attached-growth) system under anoxic conditions. In this section only denitrification in a suspended-growth system is considered. A discussion of denitrification in attached-growth systems is presented in Chapter 7.

Depending upon whether nitrification is accomplished by a combined process or a separate-stage process, the number of reactors in the treatment scheme will be 3 or 4, respectively. Schematics of the respective flow regimes are illustrated in Figure 4-59 and 4-60. The aeration tank between the denitrification reactor and the final clarifier serves to strip nitrogen gas from the waste stream to prevent rising sludge in the clarifier, to aerobically remove small amounts of excess methanol and to raise the DO level so that no further denitrification will occur in the final clarifier.

The denitrification reactor is generally covered to minimize oxygen absorption from the atmosphere. However, care must be taken to provide proper venting so that the nitrogen and carbon dioxide released during denitrification can escape.

Mixing is provided in the denitrification tank by mixers similar to those used in the flocculation process at water treatment plants. Power requirements of $\frac{1}{4}$ to $\frac{1}{2}$ hp/1000 ft^3 of tank volume have been found adequate to maintain solids in suspension (Metcalf and Eddy, 1973).

Barnard (1975) states that the rate of denitrification follows zero-order kinetics with respect to nitrate concentration and organic carbon content as long as carbon is not limiting. Thus,

$$\left(\frac{dNO_3}{dt}\right)_{DN} = KX \quad (4\text{-}163)$$

where $\left(\frac{dNO_3}{dt}\right)_{DN}$ = rate of denitrification, mass volume^{-1} time^{-1}

K = denitrification reaction rate constant, time^{-1}

X = biomass concentration, mass volume^{-1}

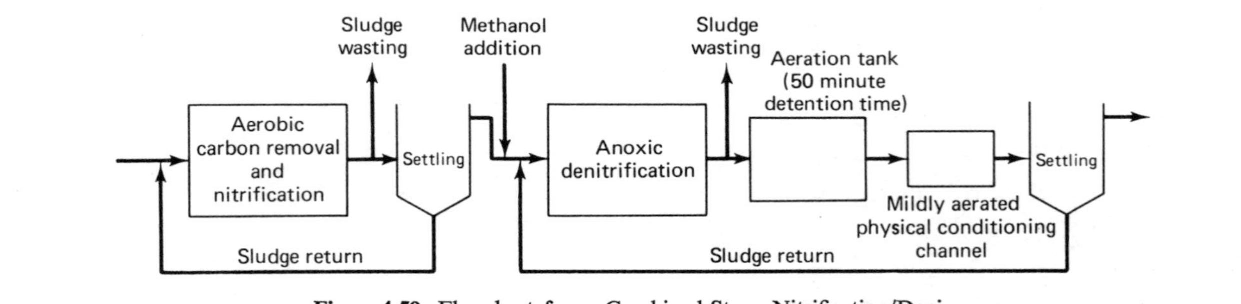

Figure 4-59. Flowsheet for a Combined-Stage Nitrification/Denitrification Process. (After Brown and Caldwell, 1975.)

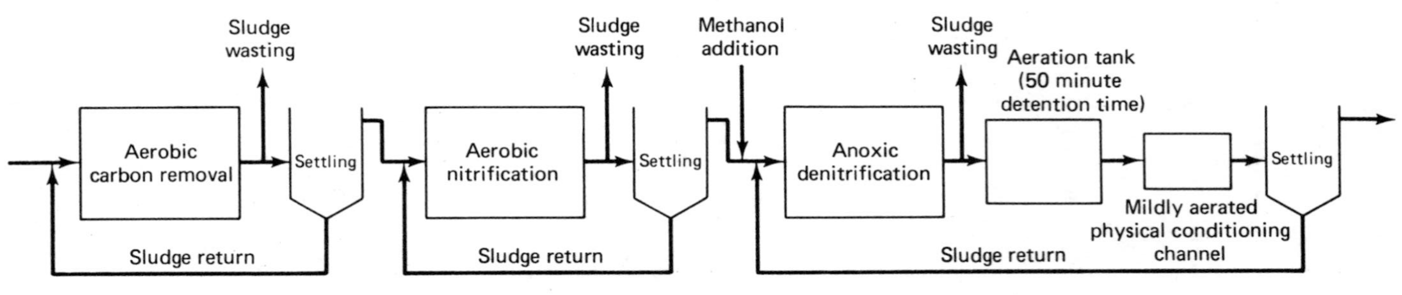

Figure 4-60. Flowsheet for a Separate-Stage Nitrification/Denitrification Process. (After Brown and Caldwell, 1975.)

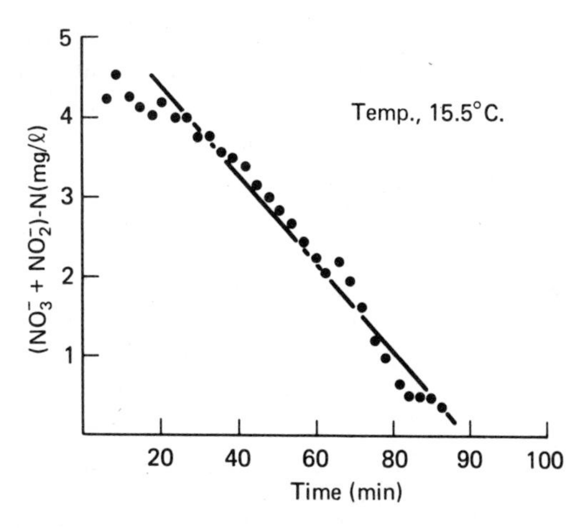

Figure 4-61. Methanol Denitrification of Residual Nitrate and Nitrite Nitrogen. (After Bishop et al., 1976.)

A typical zero-order response for denitrification is presented in Figure 4-61.

For a completely mixed reactor with sludge return, a material balance expression for nitrate entering and leaving the denitrification tank is given by

$$\left(\frac{d\mathrm{NO}_3}{dt}\right) V = QN_0 + RQN_e - \left(\frac{d\mathrm{NO}_3}{dt}\right)_{DN} V - (1+R)QN_e \quad (4\text{-}164)$$

where Q represents the wastewater flow into the denitrification unit; N_0 and N_e, the influent and effluent nitrate nitrogen concentrations, respectively; V, the reactor volume; and R, the recycle ratio. The material-balance expression given by equation 4-164 assumes that denitrification occurs only within the boundaries of the reactor.

Substituting for $(d\mathrm{NO}_3/dt)_{DN}$ from equation 4-163, at steady-state equation 4-164 can be expressed as

$$\frac{Q(N_0 - N_e)}{XV} = K \quad (4\text{-}165)$$

or

$$(q)_{DN} = K \quad (4\text{-}166)$$

where $(q)_{DN}$ represents the specific denitrification rate and has the unit time^{-1}.

Dawson and Murphy (1973) have reported that the specific denitrification rate is related to temperature by the equation

$$(q)_{DN} = 0.07(1.06)^{T-20} \quad (4\text{-}167)$$

where $(q)_{DN}$ = specific rate of denitrification, (mg/ℓ NO_3^--N removed/mg/ℓ MLVSS-h

T = operating temperature, °C

This expression can be considered valid between 5 and 30°C. Dawson and Murphy (1973) indicate that denitrification effectively ceases at 3°C.

To determine the biomass production and methanol requirements for denitrification in a suspended-growth system, McCarty (1969) proposes the following expressions:

$$C_b = 0.53N_0 + 0.32N_1 + 0.19\text{DO} \tag{4-168}$$

where C_b = biomass production, mg/ℓ
N_0 = influent nitrate nitrogen concentration, mg NO_3^--N/ℓ
N_1 = influent nitrite nitrogen concentration, mg NO_2^--N/ℓ
DO = influent dissolved oxygen concentration, mg/ℓ

and

$$C_m = 2.47N_0 + 1.53N_1 + 0.87\text{DO} \tag{4-169}$$

where C_m = required methanol concentration, mg/ℓ

Metcalf and Eddy (1973) suggest that 6.5 to 7.5 is the optimum pH range for denitrifying bacteria. If the operating pH is expected to be outside the optimum range, $(q)_{DN}$ as determined from equation 4-167 must be adjusted accordingly. The appropriate correction factor may be obtained from Figure 4-62 for the pH range 6.1 to 7.9.

Process design is generally based on complete mixing, even though most of the time denitrification basins are constructed to give plug flow. Typical design parameters for a biological nitrification/denitrification process employing separate-stage nitrification are given in Table 4-12. Under process

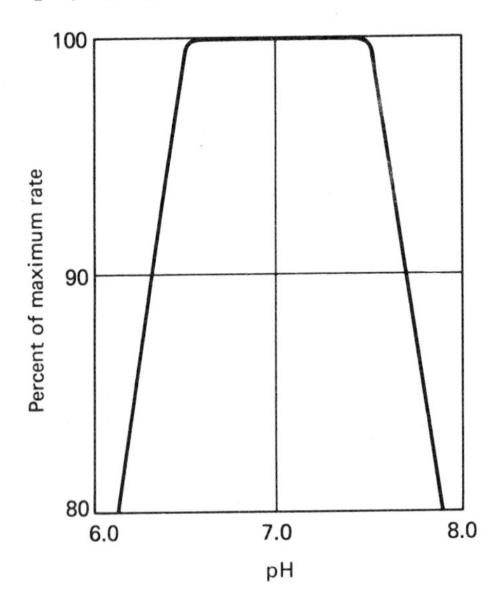

Figure 4-62. Percent of Maximum Rate of Denitrification Versus pH. (After Metcalf and Eddy, 1973.)

TABLE 4-12

TYPICAL DESIGN PARAMETERS FOR A BIOLOGICAL NITRIFICATION/DENITRIFICATION PROCESS EMPLOYING SEPARATE-STAGE NITRIFICATION

Process	θ_c (*days*)	*t* (*h*)	*MLVSS* (*mg/ℓ*)	*pH*	Q_r/Q
Carbon removal	2–5	1–3	1000–2000	6.5–8.0	0.25–1.0
Nitrification	10–20	0.5–3	1000–2000	7.4–8.6	0.5 –1.0
Denitrification	1–5	0.2–2	1000–2000	6.5–7.0	0.5 –1.0

Source: After Metcalf and Eddy (1972).

conditions nitrogen removal efficiency can be expected to range from 80 to 90%.

Example 4-13

Design the denitrification unit for a separate stage nitrification/denitrification process. The average wastewater flow is 10 MGD, and it has been determined that the effluent from the nitrification process will contain 30 mg/ℓ nitrate nitrogen (NO_3^--N), 2 mg/ℓ dissolved oxygen, and no nitrite nitrogen. A 90% nitrogen removal efficiency is desired at the critical winter operating temperature of 10°C and pH 7.0.

solution

1/ Based on the critical winter operating temperature, compute $(q)_{DN}$ from equation 4-167.

$$(q)_{DN} = 0.07(1.06)^{10-20}$$
$$= 0.039\ \text{h}^{-1}$$

2/ Assuming a desired MLVSS concentration of 1000 mg/ℓ, determine the required tank volume by rearranging equation 4-165.

$$V = \frac{10[30 - (1 - 0.9)(30)]}{(1000)(0.039)(24)}$$
$$= 0.3\ \text{MG}$$

3/ Check the detention time against the typical values given in Table 4-12.

$$t = \frac{0.3}{10}$$
$$= 0.03\ \text{day} = 0.72\ \text{h}$$

The detention time is within the range 0.2 to 2.0 h.

4/ Compute the expected biomass production (lb/day) from a modification of equation 4-168.

$$\Delta X = (8.34)(10)[(0.53)(30) + (0.19)(2)]$$
$$= 1358 \text{ lb/day}$$

5/ Determine the required methanol feed (lb/day) from a modification of equation 4-169.

$$M = (8.34)(10)[(2.47)(30) + 0.87(2)]$$
$$= 6325 \text{ lb/day}$$

6/ Calculate the required sludge wastage rate.

$$Q_w = \frac{\Delta X}{(8.34)(X)}$$
$$= \frac{1358}{(8.34)(1000)} = 0.16 \text{ MGD}$$

Neglecting the solids lost in the effluent, equation 4-2 reduces to

$$Q_w = \frac{V}{\theta_c}$$

or

$$\theta_c = \frac{V}{Q_w}$$

Thus,

$$\theta_c = \frac{0.3}{0.16} = 2 \text{ days}$$

4-8
Anaerobic Contact Process

The anaerobic waste treatment process is an effective method for the treatment of many organic wastes. This treatment is mediated by facultative and anaerobic microorganisms, which, in the absence of oxygen, convert organic materials into gaseous end products such as carbon dioxide and methane. Pfeffer et al. (1967) reports that the major advantages of anaerobic treatment over aerobic treatment are (1) less biomass produced per unit of substrate (organic material) utilized, which also means a decrease in the requirement for nitrogen and phosphorus; (2) economic value of the methane gas generated in the treatment process; and (3) higher organic loading potential because the process is not limited by the oxygen transfer capability at high oxygen utilization rates. Disadvantages of the anaerobic process are the elevated temperatures required to maintain microbial activity at a reasonable rate and the incompleteness of organic stabilization at economical treatment times.

Although anaerobic waste treatment is used primarily for stabilization and volume reduction of municipal and industrial sludges, it has been applied successfully in treating the entire flow of medium-strength packinghouse, brewery, distillery, fatty-acid, wood-fiber, and synthetic-milk wastewaters. In such cases an anaerobic contact process was employed. The anaerobic contact process differs from anaerobic sludge digestion in that sludge recycle is used and lower substrate concentrations are applied. A typical flowsheet for the process is presented in Figure 4-63. The degasifier is needed to minimize floating solids in the clarifier.

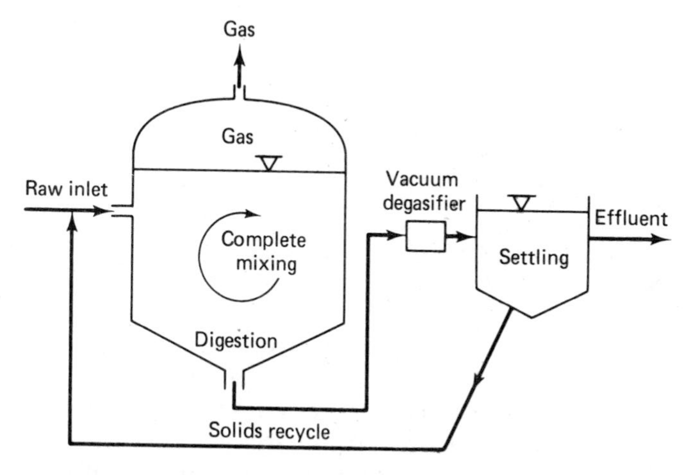

Figure 4-63. Anaerobic Contact Process. (After Eckenfelder, 1966.)

Fundamental Microbiology

The anaerobic treatment of complex wastes, resulting in the production of methane and carbon dioxide, involves two distinct stages. In the first stage, complex waste components, including fats, proteins, and polysaccharides, are first hydrolyzed to their component subunits by a heterogeneous group of facultative and anaerobic bacteria. These bacteria then subject the products of hydrolysis (triglycerides, fatty acids, amino acids, and sugars) to fermentations, β-oxidations, and other metabolic processes leading to the formation of simple organic compounds, mainly short-chain (volatile) acids and alcohols. The first stage is commonly referred to as "acid fermentation." In this stage organic material is simply converted to organic acids, alcohols, and new bacterial cells so that little stabilization of BOD or COD is realized. However, in the second stage the end products of the first stage are converted to gases (mainly methane and carbon dioxide) by several different species of strictly anaerobic bacteria. Thus, it is here that true stabilization of the organic material occurs. This stage is generally referred to as "methane fermentation." The two stages of anaerobic waste treatment are illustrated in Figure 4-64. It should be understood that even though the anaerobic process is presented as being sequential in nature, both stages take place simultaneously and synchronously in an active, well-buffered system.

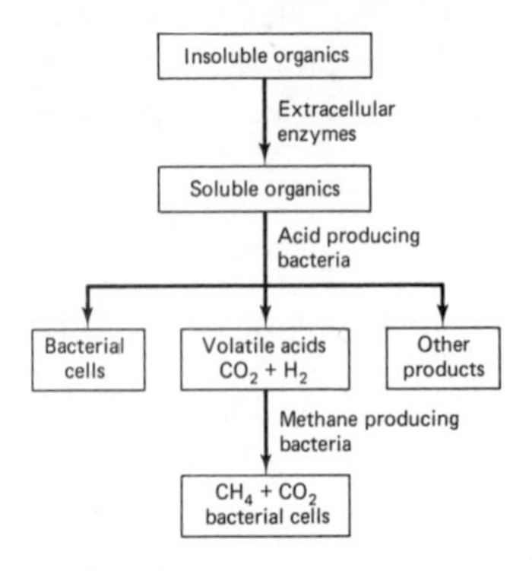

Figure 4-64. Sequential Mechanism of Anaerobic Waste Treatment. (After Andrews, 1975.)

In Chapter 2 the importance of pyruvate as a pivotal compound in metabolism was discussed. It was noted that if an external electron acceptor is present, pyruvate is converted to acetyl CoA and enters the Krebs cycle. However, when no external electron acceptor is present (as is the case for acid fermentation), pyruvate may undergo any of several alternative reactions which serve to regenerate NAD from NADH. Figure 2-7 shows many products which can be derived from pyruvic acid. The primary acids produced during acid fermentation are propionic and acetic (note that Figure 2-7 shows that CO_2 is released during the formation of propionic acid). The significance of these acids as precursors for methane formation is illustrated in Figure 4-65. McCarty (1968) points out that the percentages given in this figure can be expected to vary between different wastes. However, by far the largest percentage of the total methane formed can always be expected from methane fermentation of acetic and propionic acids. Formic acid fermentation and methane fermentation associated with β-oxidation of some long-chain fatty acids probably account for most of the small percentage of methane not derived from acetic or propionic acid.

McCarty (1968) reports that only one group of methane bacteria is necessary for the methane fermentation of acetic acid, whereas propionic acid, which is fermented through acetic acid, requires two different groups of methane bacteria. The methane fermentation reactions for these two acids are:

Acetic acid:

$$CH_3COOH \longrightarrow CH_4 + CO_2 \qquad (4\text{-}170)$$

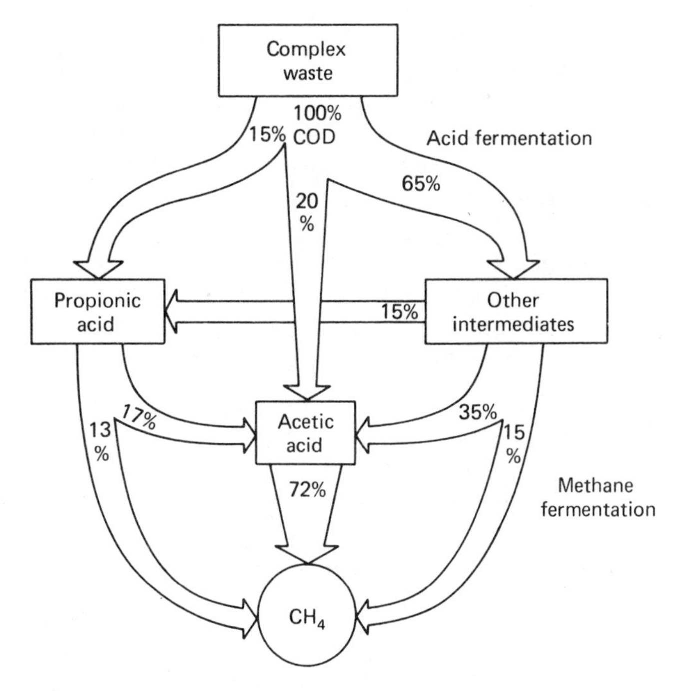

Figure 4-65. Pathways in Methane Fermentation of Complex Wastes. Percentages Represent Conversion of Waste COD by Various Routes. (After McCarty, 1968.)

Propionic acid:

1st step $$CH_3CH_2COOH + 0.5H_2O \longrightarrow CH_3COOH + 0.25CO_2 + 0.75CH_4 \qquad (4\text{-}171)$$

2nd step $$CH_3COOH \longrightarrow CH_4 + CO_2 \qquad (4\text{-}172)$$

overall $$CH_3CH_2COOH + 0.5H_2O \longrightarrow 1.25CO_2 + 1.75CH_4 \qquad (4\text{-}173)$$

The bacteria responsible for acid fermentation are relatively tolerant to changes in pH and temperature and have a much higher rate of growth than the bacteria responsible for methane fermentation. As a result, methane fermentation is generally assumed to be the rate-controlling step in anaerobic waste treatment processes.

Process Kinetics: The Rate-Limiting Step Approach

In any multistep reaction, the rate of progression of the slowest step will control the rate of progression of the overall reaction. McCarty (1966a) and O'Rourke (1968) have presented data which strongly suggest that methane fermentation of short- and long-chain fatty acids is indeed the rate-limiting (slowest) step in anaerobic waste treatment. Furthermore, O'Rourke (1968) found that lipid degradation was feasible (proceeds at a reasonable rate) only above 20°C. Thus, considering that 35°C is normally accepted as the optimum temperature for mesophilic anaerobic waste treatment, Lawrence (1971) proposes that over the temperature range 20° to 35°C, the kinetics of methane fermentation of long- and short-chain fatty acids will adequately describe the overall kinetics of anaerobic treatment. This implies that the kinetic equations that were developed to describe the completely mixed activated sludge process with sludge return are equally applicable to the anaerobic contact process outline in Figure 4-63. These equations are:

$$\frac{1}{\theta_c} = Y_T \frac{kS_e}{K_s + S_e} - K_d \tag{4-9}$$

$$S_e = \frac{(1 + K_d\theta_c)K_s}{\theta_c(Y_T k - K_d) - 1} \tag{4-14}$$

$$X = \frac{\theta_c Y_T Q(S_0 - S_e)}{V_a(1 + K_d\theta_c)} \tag{4-20}$$

$$\frac{1}{\theta_c} = \frac{Q}{V_a}\left(1 + R - R\frac{X_r}{X}\right) \tag{4-32}$$

$$\frac{1}{\theta_c^m} = Y_T \frac{kS_0}{K_s + S_0} - K_d \tag{4-34}$$

Equation 4-34 describes a situation where process failure occurs because the BSRT is reduced to a value where the microorganisms are removed from the system faster than they can reproduce (i.e., washout occurs). This minimum BSRT is reached when $S_e = S_0$. McCarty (1968) notes that K_d is small and can generally be neglected with little error. In this case θ_c^m can be estimated from the expression

$$\theta_c^m = \frac{1}{Y_T k}\frac{K_s + S_0}{S_0} \tag{4-174}$$

Values for θ_c^m determined for anaerobic processes are much larger than those for aerobic processes because the yield per unit of substrate utilized is much lower in anaerobic treatment. This is shown in equation 4-174, where θ_c^m varies inversely with Y_T. For a particular design situation, θ_c values from 2 to 10 times θ_c^m are commonly used to ensure process failure does not occur.

Lawrence and McCarty (1969) state that when $S_0 \gg K_s$ (i.e., when the influent substrate concentration is great enough to be non-growth-limiting), the θ_c^m value is given by the reciprocal of the maximum specific growth rate and designated as $(\theta_c^m)_{\lim}$.

$$(\theta_c^m)_{\lim} = \frac{1}{\mu_{\max}} \approx \frac{1}{Y_T k} \tag{4-175}$$

$(\theta_c^m)_{lim}$ values for methane fermentation of various substrates are presented in Table 4-13.

By assuming that the values of Y_T, K_d, and k are equal for all fatty acid fermentations, O'Rourke (1968) proposes that equation 4-14 can be modified into the form

$$(S_e)_{overall} = \frac{1 + K_d\theta_c}{\theta_c(Y_T k - K_d) - 1}(K_c) \qquad (4\text{-}176)$$

where $K_c = \sum K_s$ for all fatty acids found or produced in the waste to be treated.

TABLE 4-13

MINIMUM VALUES OF θ_c FOR METHANE FERMENTATION OF VARIOUS SUBSTRATES

	$(\theta_c^m)_{lim}$ *(days)*			
Energy substrate	*35°C*	*30°C*	*25°C*	*20°C*
Acetic acid	3.1	4.2	4.2	—
Propionic acid	3.2	—	2.8	—
Butyric acid	2.7	—	—	—
Long-chain fatty acids	4.0	—	5.8	7.2
Hydrogen	0.95[a]	—	—	—
Sewage sludge	4.2[b]	—	7.5[b]	10[b]
Sewage sludge	2.6[c]	—	—	—

[a]37°C.
[b]Computed values of θ_c^m for $S_0 = 18.1$ g/ℓ COD.
[c]Experimentally observed value.
Source: After Lawrence (1971).

To define the effects of temperature on methane fermentation, O'Rourke (1968) gives the following empirical equations:

$$(k)_T = (6.67 \text{ day}^{-1})10^{-0.015(35-T)} \qquad (4\text{-}177)$$

and

$$(K_c)_T = (2224 \text{ mg COD}/\ell)10^{0.046(35-T)} \qquad (4\text{-}178)$$

Equations 4-176, 4-177, and 4-178 are valid for a complex waste with high lipid content over the temperature range 20 to 35°C and, according to Lawrence (1971), are satisfactory for anaerobic treatment design. For a low-lipid-content waste the values of k and K_s for acetate and propionate presented in Table 4-14 provide the basic design information. In this case equation 4-14 and not equation 4-176 should be employed.

The range and average values of Y_T and K_d determined for the methane fermentation of acetic, propionic, and butyric acid are presented in Table 4-15. These data were determined by Lawrence (1971), who observed little

TABLE 4-14

AVERAGE VALUES OF SUBSTRATE UTILIZATION COEFFICIENTS FOR ACETIC, PROPIONIC, AND BUTYRIC ACIDS

	35°C				30°C				25°C			
	k (*mg/mg-day*)		K_s (*mg/ℓ*)		*k* (*mg/mg-day*)		K_s (*mg/ℓ*)		*k* (*mg/mg-day*)		K_s (*mg/ℓ*)	
Volatile acid substrate	*"Removed" as HAc*[a]	*As COD to* CH_4	*As HAc*	*As COD*	*"Removed" as HAc*	*As COD to* CH_4	*As HAc*	*As COD*	*"Removed" as HAc*	*As COD to* CH_4	*As HAc*	*As COD*
Acetic	8.1	8.7	154	165	4.8	5.1	333	356	4.7	5.0	869	930
Propionic	9.6	7.7	32	60	—	—	—	—	9.8	7.8	613	1145
Butyric	15.6	8.1	5	13	—	—	—	—	—	—	—	—

[a]HAc, acetic acid.

Source: After Lawrence (1971).

TABLE 4-15

RANGE AND AVERAGE VALUES OF Y_T AND K_d IN METHANE FERMENTATION OF VOLATILE ACIDS

Parameter	*Range*	*Average*
Y_T (mg/mg)	0.040–0.054	0.044
K_d (day^{-1})	0.010–0.040	0.019

Source: After Lawrence (1971).

variation in the values with temperature. As a result, he suggests that for design purposes Y_T and K_d can be considered as constant and unaffected by temperature variations and that $Y_T = 0.044$ and $K_d = 0.019$ day^{-1} be used for low-lipid wastes, whereas $Y_T = 0.04$ and $K_d = 0.015$ day^{-1} are applicable for high-lipid-content, complex wastes such as municipal sewage sludge.

Gas Production

McCarty (1968) proposes that the quantity of methane released during anaerobic treatment can be estimated from the oxygen equivalent of methane gas.

$$\underset{16}{CH_4} + \underset{64}{2O_2} \longrightarrow CO_2 + 2H_2O \qquad (4\text{-}179)$$

Thus, 1 mole (16 g) of methane (CH_4) is equivalent to 64 g of COD, or $\frac{1}{64}$ mole of CH_4 is equivalent to 1 g of COD. The volume of methane gas produced for each pound of COD or ultimate BOD oxidized can be determined by considering that at standard temperature and pressure (0°C and 1 atm), 1 mole of any gas occupies 22.4 ℓ. Therefore, $\frac{1}{64}$ mole of CH_4 occupies 22.4/64 = 0.35 ℓ, or 0.35 ℓ of CH_4 are formed per gram of BOD_u oxidized. Hence, (0.35 ℓ/g)(454 g/lb) = 159 ℓ/lb. Since 28.32 ℓ are equivalent to 1 ft^3, the volume of methane formed per pound of BOD_u oxidized is 5.63 ft^3. By applying Charles' law, the volume of gas formed at temperatures other than 0°C can be computed.

$$V_2 = \frac{T_2}{T_1} V_1 \qquad (4\text{-}180)$$

where V_2 = gas volume at fermentation temperature, liters
V_1 = gas volume at standard conditions, 22.4 ℓ
T_1 = temperature at standard conditions, 273°K
T_2 = fermentation temperature, °K

Methane production can be estimated from a relationship similar to equation 4-90 for the daily oxygen requirement.

$$G = G_0[\Delta S - 1.42(\Delta X)] \tag{4-181}$$

where G = total methane produced, ft^3/day
G_0 = cubic feet of methane produced per pound of COD or BOD_u oxidized, ft^3/lb
ΔS = COD or BOD_u removed, lb/day
ΔX = biomass produced, lb/day

At steady-state the amount of biomass produced is equal to the amount of biomass wasted each day. Since the organic material channeled into cell synthesis is not oxidized, no methane is realized from this function. The constant 1.42 represents the oxygen equivalents per unit of biomass oxidized. Therefore, the product of 1.42 times ΔX gives the portion of total BOD_u removed each day which does not contribute to methane production. Since the gas produced during anaerobic waste treatment is only about two-thirds methane, the total volume of gas produced per day is given by $G/0.67$.

The total available heat content can be estimated by considering that 1 ft^3 of methane (not total gas volume) has a net heating value of 960 Btu at standard conditions. When it is required that heat be supplied in order to raise the temperature of the incoming wastewater to fermentation temperature, the heat requirement can be calculated from the equation

$$R_H = Q_m W_s (T_2 - T_1) \tag{4-182}$$

where R_H = heat required to raise the incoming wastewater temperature to the fermentation temperature, Btu/day
Q_m = average total mass flow of wastewater, lb/day
W_s = specific heat of water, 1 Btu/lb-°F
T_2 = fermentation temperature, °F
T_1 = incoming wastewater temperature, °F

Environmental Factors

The methane bacteria are quite sensitive to changes in pH. It has been found that the rate of methane fermentation is relatively constant within the pH range of 6.0 to 8.5 but drops very rapidly outside this range (see Figure 4-66).

Sufficient alkalinity is essential for proper pH control in anaerobic treatment because it acts as a buffer to the system. Alkalinity is produced from the breakdown of organic material and at typical fermentation pHs (around 7.0) is present primarily in the form of bicarbonates. An example of alkalinity production from the fermentation of nitrogenous compounds is provided by equation 4-183.

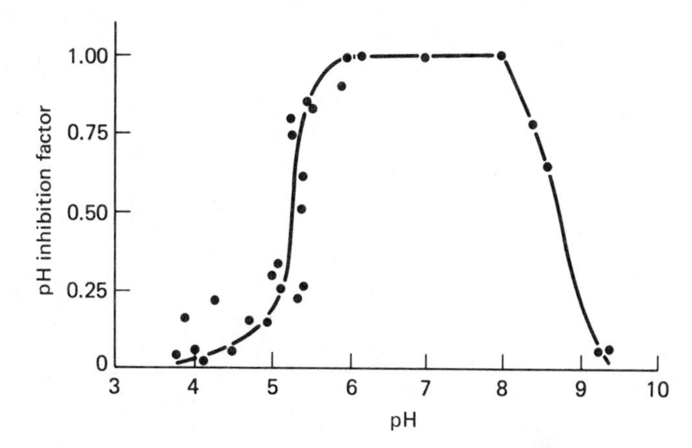

Figure 4-66. Effect of pH on the Rate of Methane Fermentation. (After Clark and Speece, 1971.)

$$\underset{\text{alanine}}{CH_3\text{–}\overset{\overset{NH_2}{|}}{C}H\text{–}COOH} + \underset{\text{glycine}}{2\overset{\overset{NH_2}{|}}{C}H_2\text{–}COOH} + 5H_2O$$

$$\longrightarrow 3CH_4 + CO_2 + 3NH_4^+\!:\!HCO_3^- \qquad (4\text{-}183)$$

A fundamental property of gases is that each component of a gas mixture occupies the entire volume of the mixture. The percentage composition by volume refers to the volumes of the separate gases before mixing. The volume of each component is therefore increased by mixing. The pressure must therefore be multiplied by a fraction smaller than 1 to correct for the volume change. This fraction is the original volume divided by the final volume, which is the same as the original volume percent divided by 100. Thus, the partial pressure of each gas in a mixture is proportional to the percent by volume of that gas in the mixture. Applying this relationship and the relationship between alkalinity present in the system and the CO_2 content of the off-gas discussed in the section on nitrification, expected alkalinity values may be computed.

Example 4-14

Determine the equilibrium alkalinity concentration as $CaCO_3$ if the off-gas from an anaerobic treatment process contains 27% carbon dioxide and the fermentation pH is 7.2. The operating temperature is 20°C at 1 atm pressure.

solution

1/ Applying equation 4-127, determine the dissolved CO_2 concentration.

$$C_{CO_2} = HP_{CO_2}$$
$$= (1688)(0.27)(1\text{ atm}) = 456\text{ mg}/\ell$$

or

$$C_{CO_2} = (0.456 \text{ g}/\ell)/(44 \text{ g/mole}) = 0.01 \text{ mole}/\ell$$

2/ Assuming that the concentration of carbonic acid is approximately the same as the concentration of dissolved CO_2, compute the HCO_3^- concentration from equation 3-28.

$$\text{pH} = \text{p}K_1 - \log\left(\frac{[H_2CO_3^*]}{[HCO_3^-]}\right)$$

$$\log [HCO_3^-] = 7.2 - 6.38 - 2.0$$
$$= -1.18$$

or

$$[HCO_3^-] = 0.066 \text{ mole}/\ell$$
$$= (0.066 \text{ mole}/\ell)(61 \text{ g/mole}) = 4026 \text{ mg}/\ell$$

3/ Convert bicarbonate concentration to mg/ℓ as $CaCO_3$.

$$\text{alkalinity} = 4026\left(\frac{50}{61}\right) = 3300 \text{ mg}/\ell \text{ as } CaCO_3$$

The off-gas from anaerobic treatment is typically between 30 and 40% CO_2 by volume. Therefore, within the operating pH range 6.6 to 7.4, the alkalinity concentration can be expected to vary from approximately 1000 to 5000 mg/ℓ as $CaCO_3$.

Total alkalinity is measured as the amount of acid required to lower the sample pH to the methyl orange end point (pH ≈ 4.3). However, organic acid salts contribute to the alkalinity measured by acid titration. To determine the bicarbonate alkalinity from such a titration, McCarty (1964) recommends the following equation:

$$BA = TA - (0.85)(0.833)TVA \qquad (4\text{-}184)$$

where BA = bicarbonate alkalinity, (mg/ℓ as $CaCO_3$)
TA = total alkalinity measured in titration, mg/ℓ as $CaCO_3$
TVA = total volatile (short-chain) acid concentration, mg/ℓ as acetic acid

The constant 0.833 is a conversion factor, whereas 0.85 is included to account for the fact that approximately 85% of the volatile acid alkalinity is measured at the methyl orange end point.

It has been found that the rate of methane formation is affected by cation concentration (Kugelman and McCarty, 1965). Such an effect is illustrated in Figure 4-67, which shows that over a range of relatively low concentrations, cations have a stimulatory effect on the system. However, an optimum concentration exists which, when exceeded, will result in a decrease in the rate of methane fermentation. The degree to which the reaction is retarded at high cation concentrations depends on the extent to which the optimum concentration is exceeded. The relative effects of some common cations on

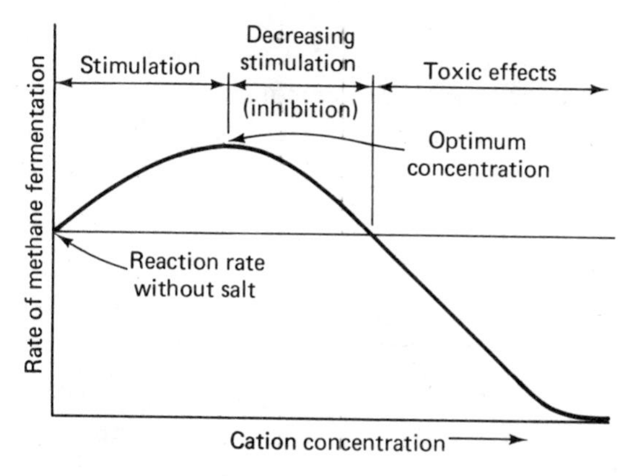

Figure 4-67. Effects of Cations on Methane Fermentation. (After Kugelman and McCarty, 1965.)

the rate of methane fermentation, as summarized by McCarty (1964), are given in Table 4-16.

The effect of ammonia concentration on the rate of methane fermentation is similar to that observed for cation concentration (see Table 4-17). An

TABLE 4-16

EFFECTS OF CATIONS ON METHANE FERMENTATION

Cation	*Stimulatory concentration (mg/ℓ)*	*Moderately inhibitory concentration (mg/ℓ)*	*Strongly inhibitory concentration (mg/ℓ)*
Calcium	100–200	2500–4500	8,000
Magnesium	75–150	1000–1500	3,000
Potassium	200–400	2500–4500	12,000
Sodium	100–200	3500–5500	8,000

Source: After McCarty (1964).

TABLE 4-17

EFFECTS OF AMMONIA ON METHANE FERMENTATION

Observed effect	*Ammonia concentration (mg/ℓ) as N*
Beneficial	50–200
No adverse effects	200–1000
Inhibitory at high values of pH	1500–3000
Toxic	Above 3000

Source: After McCarty (1964).

important difference is that the fermentation pH determines the percentage distribution between ammonia and the ammonium ion. Free ammonia, which is the toxic form, is favored by high pH values.

Specific environmental factors considered to be optimum for methane fermentation are presented in Table 4-18.

TABLE 4-18

ENVIRONMENTAL FACTORS FOR METHANE FERMENTATION

Variable	*Optimum*	*Extreme*
Temperature (°C)	30–35	25–40
pH	6.8–7.4	6.2–7.8
Oxidation–reduction potential MV	−520 to −530	−490 to −550
Volatile acids (mg/ℓ as acetic)	50–500	2000[a]
Alkalinity (mg/ℓ as $CaCO_3$)	2000–3000	1000–5000

[a]The value of 2000 mg/ℓ is generally considered to be the maximum tolerable value, although the maximum volatile acid concentration that will be tolerated by methane bacteria depends on numerous factors, and must be established for a given system.

Source: After Malina (1962).

Process Design Considerations

The value of θ_c selected for design of an anaerobic contact process generally ranges from 2 to 10 times the θ_c^m value. At lower temperatures, θ_c^m increases significantly because of the decrease in the maximum specific rate of substrate utilization, k. As a result, anaerobic treatment below 20°C is usually not feasible. Since most anaerobic processes operate best at approximately 35°C, design engineers should consider heating wastes that have temperatures significantly less than this. A major factor in making such a decision is whether or not a sufficient quantity of methane can be produced to support the heating function without relying on a large amount of supplemental fuel. Lawrence (1971) reports that the COD of the waste to be treated should be 5000 mg/ℓ or greater if a significant waste temperature adjustment is to be required and the use of additional fuel is undesirable.

From a kinetic point of view, for a given θ_c there is theoretically no upper limit to the organic loading rate that may be imposed upon a biological process. However, as was observed for the activated sludge process, certain nonkinetic factors will limit the loading rate. For the anaerobic contact process Lawrence (1971) states that the nonkinetic factors of importance are

(1) nutrient availability, (2) toxicity, (3) pH control, (4) efficiency of solids–liquid separation, and (5) rate of solids recycle. To ensure adequate solids–liquid separation, the MLVSS concentration is normally controlled to a level of 3000 to 4000 mg/ℓ. The recycle rates used in the anaerobic contact process are usually in the range 2:1 to 4:1 (recycle flow rate to raw wastewater flow rate.)

Because the process is very sensitive and difficult to control, all factors must be considered very carefully before an anaerobic contact process is selected for treating a particular waste. Schroeder (1977) states that variations in either the organic or hydraulic loading over a short period of time may cause operational problems. Furthermore, the anaerobic contact process is generally not suitable for treating wastes from seasonal processes (e.g., small cannery operations) because of the long startup time required. Graef and Andrews (1974), however, have developed control strategies for anaerobic systems which can be used to overcome many of the operating difficulties.

The performance of anaerobic contact systems treating industrial wastes is summarized in Table 4-19.

TABLE 4-19

TREATMENT PERFORMANCE FOR THE ANAEROBIC CONTACT PROCESS

			BOD_5		
Waste	*Hydraulic retention time (days)*	*Digestion tempera-ture (°F)*	*Raw waste (mg/ℓ)*	*lb/1000 ft³/day added*	*Percent removed*
Maize starch	3.3	73	6,280	110	88
Whisky distillery	6.2	92	25,000	250	95
Cotton kiering	1.3	86	1,600	74	67
Citrus	1.3	92	4,600	214	87
Brewery	2.3		3,900	127	96
Starch gluten	3.8	95	14,000[a]	100[a]	80[a]
Wine	2.0	92	23,400[a]	730[a]	85[a]
Yeast	2.0	92	11,900[a]	372[a]	65[a]
Molasses	3.8	92	32,800[a]	546[a]	69[a]
Meat packing	1.3	92	2,000	110	95
	0.5	92	1,380	156	91
	0.5	95	1,430	164	95
	0.5	85	1,310	152	94
	0.5	75	1,110	131	91

[a]Volatile suspended solids, rather than BOD_5.

Source: After Eckenfelder (1966).

Example 4-15

Design an anaerobic contact process to treat a wastewater flow of 0.2 MGD from a meat-packing plant. The raw wastewater flow has a temperature of 20°C and a COD of 3000 mg/ℓ. It is desired that the fermentation temperature be maintained at 35°C and the MLVSS concentration at 3500 mg/ℓ. A safety factor of 5 is to be used in computing the design θ_c value.

solution

1/ Meat-packing wastes are similar to municipal wastewater in terms of composition; therefore, this waste will be considered in the category of a high-lipid-content complex waste. The appropriate biokinetic constants are:

$$(k)_{35^\circ} = (6.67 \text{ day}^{-1})10^{-0.015(35-35)}$$

or

$$(k)_{35^\circ} = 6.67 \text{ day}^{-1}$$

$$(K_c)_{35^\circ} = (2224 \text{ mg COD}/\ell)10^{0.046(35-35)}$$

or

$$(K_c)_{35^\circ} = 2224 \text{ mg COD}/\ell$$

$$Y_T = 0.04$$

$$K_d = 0.015 \text{ day}^{-1}$$

2/ Compute θ_c^m from equation 4-174 and, using a safety factor of 5, determine the design value for θ_c.

$$\theta_c^m = \frac{1}{(0.04)(6.67)} \frac{2224 + 3000}{3000}$$

$$= 6.5 \text{ days}$$

Therefore,

$$\theta_c = (5)(6.5) = 32.5 \text{ days}$$

3/ Determine the soluble effluent substrate concentration from equation 4-176.

$$(S_e)_{\text{overall}} = \frac{[1 + (0.015)(32.5)](2224)}{(32.5)[(0.04)(6.67) - (0.015)] - 1}$$

$$= 461 \text{ mg}/\ell$$

4/ Applying equation 4-20, estimate the required reactor volume.

$$V_a = \frac{(32.5)(0.04)(0.2)(3000 - 461)}{(3500)[1 + (0.015)(32.5)]}$$

$$= 0.13 \text{ MG}$$

5/ For an operating BSRT of 32.5 days, compute the observed yield coefficient from equation 2-63.

$$Y_{\text{obs}} = \frac{(0.04)}{1 + (0.015)(32.5)}$$

$$= 0.027$$

6/ Determine the biomass production using equation 4-85.

$$\Delta X = (0.2)(8.34)(0.027)(3000 - 461)$$
$$= 114 \text{ lb/day}$$

7/ Estimate the nutrient requirements from equations 4-100 and 4-101.

$$\text{N} = (0.122)(114) = 14 \text{ lb/day}$$
$$\text{P} = (0.023)(114) = 2.6 \text{ lb/day}$$

8/ Calculate the cubic feet of methane produced per pound of COD removed.

$$V_2 = \frac{308}{273}(22.4)$$
$$= 25.3\ \ell$$

Therefore,

$$\text{methane production} = \frac{(25.3/64)(454)}{28.32}$$
$$= 6.34 \text{ ft}^3\text{/lb COD removed}$$

9/ The total methane production is given by equation 4-181.

$$G = 6.34[(3000 - 461)(8.34)(0.2) - (1.42)(114)]$$
$$= 25{,}823 \text{ ft}^3\text{/day}$$

10/ Compute the total available heat content.

$$\text{available heat content} = 960\frac{\text{Btu}}{\text{ft}^3} \times 25{,}823\frac{\text{ft}^3}{\text{day}} \times \frac{5.63}{6.34}$$
$$= 22{,}013{,}904 \text{ Btu/day}$$

In this case heat content is based on gas production at standard conditions because methane has a heat value of 960 Btu/ft^3 at standard conditions.

11/ Estimate the heat required to raise the incoming wastewater temperature to 35°C by applying equation 4-182.

$$Q_m = (200{,}000 \text{ gal/day})\frac{1 \text{ ft}^3}{7.5 \text{ gal}}(62.4 \text{ lb/ft}^3)$$
$$= 1{,}664{,}000 \text{ lb/day}$$

Therefore,

$$R_H = (1{,}664{,}000 \text{ lb/day})\frac{1 \text{ Btu}}{\text{lb-°F}}(95°\text{F} - 68°\text{F})$$
$$= 44{,}928{,}000 \text{ Btu/day}$$

As methane production will satisfy approximately one-half the heat requirement, a considerable amount of additional fuel will be required to maintain the desired fermentation temperature.

PROBLEMS

4-1 A waste treatability study has been conducted using four laboratory-activated sludge units similar to the unit shown in Figure 4-43. The data obtained from the lab study have been averaged for each of the four units and are as follows:

Unit	ΔX (lb/day)	Volume (gal)	Q (gal/day)	MLVSS (mg/ℓ)	MLSS (mg/ℓ)	S_0 (mg/ℓ)	S_e (mg/ℓ)	OUR (mg/ℓ-h)
1	0.0033	4	24	1000	1300	100	17	22.9
2	0.0066	4	24	1000	1300	150	25	30.2
3	0.0099	4	24	1000	1300	200	33	37.5
4	0.0133	4	24	1000	1300	250	42	44.8

Determine the biokinetic constants K, K_d, Y_T, a, and b for the wastewater used and the microbial culture developed in this study.

4-2 The following criteria are applicable to the design of a completely mixed activated sludge process:

$$S_0 = 250 \text{ mg/ℓ BOD}_u \qquad K_d = 0.08 \text{ day}^{-1}$$
$$S_e = 25 \text{ mg/ℓ BOD}_u \qquad \theta_c = 8 \text{ days}$$
$$[\text{TKN}]_0 = 26 \text{ mg/ℓ as N}$$
$$Y_T = 0.5$$
$$Y_N = 0.05$$

If the solids yield from nitrification is neglected in computing the biomass production, what percent error is incurred if it is assumed that all the TKN is utilized in the nitrification process and the process is 100% efficient?

4-3 A completely mixed activated sludge process is to be designed to treat a municipal wastewater flow of 20 MGD and BOD_u of 150 mg/ℓ. The following design criteria are applicable:

$$\text{MLVSS} = 2000 \text{ mg/ℓ}$$
$$\text{MLVSS} = 0.8 \text{ MLSS}$$
$$S_e = 5 \text{ mg/ℓ BOD}_u$$
$$K = 0.04 \text{ ℓ/mg-day}$$
$$K_d = 0.05 \text{ day}^{-1}$$
$$Y_T = 0.5$$
$$[\text{TKN}]_0 = 20 \text{ mg/ℓ as N}$$
$$\text{influent P} = 10 \text{ mg/ℓ as P}$$

There is no significant variation in operating temperature between summer and winter. Furthermore, it has been found that the wastewater contains a compound that retards nitrification but otherwise has no detrimental effect.

a/ Determine the required aeration tank volume.

b/ Plot each of the following as a function of θ_c for θ_c values of 5, 8, 10, 12, and 15 days.

1/ Effluent substrate concentration.
2/ Steady-state biomass concentration.
3/ Observed yield coefficient.
4/ Sludge production.
5/ Pounds of nitrogen removed by cell synthesis.
6/ Pounds of phosphorus removed by cell synthesis.
7/ Oxygen requirement.

c/ Plot X_r/X versus θ_c for θ_c values of 5, 8, 10, 12, and 15 days using R values of 0.3, 0.4, and 0.5.

4-4 How many pounds of activated sludge must be maintained in an extended aeration plant designed to treat a waste flow of 500,000 gpd with a total BOD_u of 1500 lb over a 24-h period? The required BOD_u reduction is 85%, the mean microbial decay rate of the sludge is 10%/day, and the sludge yield coefficient is 0.55.

4-5 A wastewater was found to have the following characteristics after primary clarification:

Q (*MGD*)	S_0 (*mg/ℓ*)	*TKN* (*mg/ℓ*)	*P* (*mg/ℓ*)	*Nondegradable solids* (*mg/ℓ*)
2.33	98	39	8	102

Design a combined completely mixed activated sludge process to achieve 95% nitrification using the following biokinetic constants:

$$X = 2000 \text{ mg}/\ell$$
$$Y_T = 0.5$$
$$K_d = 0.07 \text{ day}^{-1} \text{ at } 20°\text{C}, \theta = 1.05$$
$$K = 0.03\ \ell/\text{mg-day at } 20°\text{C}, \theta = 1.03$$
$$(\mu_{\max})_{NS[20°\text{C}, (\text{pH})_{\text{opt}}]} = 0.4 \text{ day}^{-1}$$
$$Y_N = 0.05$$
$$(\text{pH})_{\text{opt}} = 8.0$$
$$\text{operating pH} = 7.4$$
$$\text{winter operating temperature} = 40°\text{F}$$
$$\text{summer operating temperature} = 65°\text{F}$$

The soluble effluent BOD_u for this problem should not exceed 10 mg/ℓ. Correct for the nitrogen incorporated into heterotrophic cells by synthesis when considering the nitrogen available for nitrification.

4-6 A completely mixed activated sludge plant has been designed to treat a wastewater flow of 3.0 MGD. The plant has three aeration tanks, each of 0.3 MGD volume. No primary clarifiers are used, but pretreatment for grit removal is provided. Determine the sludge production and oxygen requirement if the following design criteria are applicable:

$$S_0 = 240 \text{ mg}/\ell \text{ BOD}_u$$

$$\text{influent SS} = 200 \text{ mg}/\ell$$

$$30\% \text{ of influent SS is nondegradable}$$

$$\text{operating pH} = 7.1$$

$$\text{influent TKN} = 25 \text{ mg}/\ell \text{ as N}$$

$$\text{design temperature: winter} = 7°\text{C}$$

$$\text{summer} = 18°\text{C}$$

$$\text{nitrification efficiency} = 95\%$$

$$Y_T = 0.5$$

$$K = 0.04 \ \ell/\text{mg-day at } 20°\text{C}, \theta = 1.03$$

$$K_d = 0.1 \text{ day}^{-1} \text{ at } 20°\text{C}, \theta = 1.05$$

$$Y_N = 0.05$$

$$(\mu_{\max})_{NS[20°\text{C}, (\text{pH})_{\text{opt}}]} = 0.4 \text{ day}^{-1}$$

$$(\text{pH})_{\text{opt}} = 8.0$$

Assume that the plant operates as a step-aeration process and that the soluble effluent BOD_u should not exceed 15 mg/ℓ. Neglect the nitrogen incorporated into heterotrophic cells by synthesis when considering the nitrogen available for nitrification.

4-7 Given the constant values, $(\mu_{\max})_{NS} = 0.3 \text{ day}^{-1}$, $K_N = 1.0 \text{ mg}/\ell$, $Y_N = 0.05$ and operating pH = 8.0, what is the minimum operating BSRT that will allow a nitrifying culture to become established in a completely mixed activated sludge process operating at steady-state?

REFERENCES

ADAMS, C. E., JR., AND W. W. ECKENFELDER, JR., *Process Design Techniques for Industrial Waste Treatment*, Enviro Press, Nashville, Tenn., 1974.

AGARDY, F. J., R. D. COLE, AND E. A. PEARSON, "Kinetic and Activity Parameters of Anaerobic Fermentation Systems," SERL Report No. 63-2, Sanitary Engineering Research Laboratory, University of California, Berkeley, Calif., 1963.

ALBERTSON, J. G., J. R. MCWHIRTER, E. K. ROBINSON, AND N. P. VAHDIECK, "Investigation of the Use of High Purity Oxygen Aeration in the Conventional Activated Sludge Process," FWQA Report No. 17050 DN W05/70, Washington, D.C., 1970.

ANDREWS, J. F., "Kinetic Models of Biological Waste Treatment Processes," *Biotechnology and Bioengineering, Symposium No. 2*, John Wiley & Sons, Inc., New York, 1971.

ANDREWS, J. F., "Control Strategies for the Anaerobic Digestion Process," *Water and Sewage Works,* Mar. 1975, p. 62.

ANDREWS, J. F., H. O. BUHR, AND M. K. STENSTROM, "Control Systems for the Reduction of Effluent Variability from the Activated Sludge Process," paper presented at the International Conference on Effluent Variability from Wastewater Treatment Processes and Its Control, New Orleans, La., Dec. 1974.

BALL, J. E., M. J. HUMENICK, AND R. E. SPEECE, "The Kinetics and Settleability of Activated Sludge Developed Under Pure Oxygen Conditions," Technical Report No. EHE-72-18, University of Texas, Austin, Tex., 1972.

BANKS, N., "U.K. Work on the Use of Oxygen in the Treatment of Wastewater Associated with the CCMS Advanced Wastewater Treatment Project," *Water Pollution Control,* **75**, 228 (1976).

BARNARD, J. L., "Nutrient Removal in Biological Systems," *British Journal of Water Pollution Control,* **45**, 143 (1975).

BARTH, E. F., R. C. BRENNER, AND R. F. LEWIS, "Chemical–Biological Control of Nitrogen in Wastewater Effluent," *Journal of the Water Pollution Control Federation,* **40**, 2040, (1968).

BENEDICT, A. H., AND D. A. CARLSON, "Temperature Acclimation in Aerobic Bio-oxidation Systems," *Journal of the Water Pollution Control Federation,* **45**, 10 (1973).

BENEFIELD, L. D., AND C. W. RANDALL, "Design Procedure for a Contact-Stabilization Activated Sludge Process," *Journal of the Water Pollution Control Federation,* **48**, 147 (1976).

BENEFIELD, L. D., AND C. W. RANDALL, "Evaluation of a Comprehensive Kinetic Model for the Activated Sludge Process, *Journal of the Water Pollution Control Federation,* **49**, 1636 (1977).

BENEFIELD, L. D., C. W. RANDALL, AND P. H. KING, "The Stimulation of Filamentous Microorganisms in Activated Sludge by High Oxygen Concentration," *Water, Air, and Soil Pollution,* **5**, 113 (1975).

BENEFIELD, L. D., C. W. RANDALL, AND P. H. KING, "Temperature Considerations in the Design and Control of Completely-Mixed Activated Sludge Plants," paper presented at 2nd Annual National Conference on Environmental Engineering Research, Development and Design, ASCE, University of Florida, Gainesville, Fla., 1975.

BENEFIELD, L. D., C. W. RANDALL, AND P. H. KING, "The Effect of High Purity Oxygen on the Activated Sludge Process," *Journal of the Water Pollution Control Federation,* **49**, 269 (1977).

BENNETT, E. R., personal communication, University of Colorado, Boulder, Colo., 1977.

BIRYUKOV, V. V., AND L. D. SHTOFFER, "Effects of Stirring on Distribution of Nutrients and Metabolities in Bacterial Suspension During Cultivation," *Applied Biochemistry and Microbiology,* July 1973, p. 9.

BISHOP, D. F., J. A. HEIDMAN, AND J. B. STAMBERG, "Single-Stage Nitrification-Denitrification," *Journal of the Water Pollution Control Federation,* **48**, 520 (1976).

BISOGNI, J. J., JR., AND A. W. LAWRENCE, "Relationship Between Biological Solids

Retention Time and Settling Characteristics of Activated Sludge," *Water Research*, **5**, 753 (1971).

BOON, A. G., "Technical Review of the use of Oxygen in the Treatment of Wastewater," *Water Pollution Control*, **75**, 206 (1976).

BOON, A. G., AND D. R. BURGESS, "Treatment of Crude Sewage in Two High-Rate Activated Sludge Plants Operated in Series," *Water Pollution Control*, **74**, 382 (1974).

BROCK, T. D., *Biology of Microorganisms, Third Edition,* Prentice-Hall, Inc., Englewood Cliffs, N.J., 1979.

BROWN AND CALDWELL, Consulting Engineers, *Process Design Manual for Nitrogen Control*, EPA Technology Transfer Series, 1975.

BUCKSTEEG, W., "Determination of Sludge Activity: A Possibility of Controlling Activated Sludge Plants," in *Proceedings of the 3rd International Conference on Water Pollution Research*, Munich, 1966.

BURCHETT, M. E., AND G. TCHOBANOGLOUS, "Facilities for Controlling the Activated Sludge Process by Mean Cell Residence Time," *Journal of the Water Pollution Control Federation*, **46**, 973 (1974).

CLARK, R. H., AND R. E. SPEECE, "The pH Tolerance of Anaerobic Digestion," *Proceedings of the 5th International Conference on Water Pollution Research*, 1971.

DAIGGER, G. T., AND C. P. L. GRADY, JR., "A Model for the Bio-oxidation Process Based on Product Formation Concepts," *Water Research*, **11**, 1049, 1977.

DAWSON, R. N. AND K. L. MURPHY, "Factors Affecting Biological Denitrification of Waste Water," in *Advances in Water Pollution Research*, ed. by S. H. Jenkins, Pergamon Press, London, 1973, p. 671.

DICK, R. I., AND M. T. SUIDAN, "Modeling and Simulation of Clarification and Thickening Processes," in *Mathematical Modeling for Water Pollution Control*, ed. by Thomas Keinath and Martin Wanielista, Ann Arbor Science Publishers, Inc., Ann Arbor, Mich., 1975.

DICK, R. I., AND P. A. VESILIND, "The Sludge Volume Index: What Is It?" *Journal of the Water Pollution Control Federation*, **41**, 7 (1969).

DICK, R. I., AND K. W. YOUNG, "Analysis of Thickening Performance of Final Settling Tanks," paper presented at the 27th Annual Purdue Industrial Waste Conference, Purdue University, West Lafayette, Ind., May 2–4, 1972.

DOWNING, A. L., AND A. P. HOPWOOD, "Some Observations on the Kinetics of Nitrifying Activated Sludge Plants," *Schweizerische Zeitschrift für Hydrologie*, **26**, 271 (1964).

DOWNING, A. L., H. A. PAINTER, AND G. KNOWLES, "Nitrification in the Activated-Sludge Process," *Journal of the Institute Sewage Purification*, **4**, 130 (1964).

ECKENFELDER, W. W., JR., AND D. J. O'CONNOR, *Biological Waste Treatment*, Pergamon Press, New York, 1961.

ECKENFELDER, W. W., JR., *Industrial Water Pollution Control*, McGraw-Hill Book Company, New York, 1966.

ECKENFELDER, W. W., JR., B. L. GOODMAN, AND A. J. ENGLANDE, "Scale-up of Biological Wastewater Treatment Reactors," in *Advances in Biochemical Engineering*,

Vol. 2, ed. by T. K. Ghose, A. Fiechter, and N. Blakebrough, Springer-Verlag, New York, 1972.

ENGLANDE, A. J., AND W. W. ECKENFELDER, JR., "Oxygen Concentrations and Turbulance as Parameters of Activated Sludge Scale-Up," paper presented at Water Resources Symposium No. 6, University of Texas, Austin, Tex., Nov. 1972.

Environmental Engineers' Handbook, Vol. I, ed. by Bela G. Liptak, Chilton Book Company, Radnor, Pa., 1974.

ENVIRONMENTAL PROTECTION AGENCY, *Status of Oxygen Activated Sludge Wastewater Treatment*, EPA-6254-77-033a, Technology Transfer, Environmental Research Information Center, 1977.

FINN, R. K., "Agitation-Aeration in the Laboratory and in Industry," *Bacteriological Review*, **18**, 254, (1954).

GAUDY, A. F., JR., AND E. T. GAUDY, "ΔCOD Gets Nod over BOD Test," *Industrial Water Engineering*, Aug./Sept. 1972, p. 30.

GOODMAN, B. L., AND A. J. ENGLANDE, JR., "A Unified Model of the Activated Sludge Process," *Journal of the Water Pollution Control Federation*, **46**, 312 (1974).

GRADY, C. P. L., JR., AND D. R. WILLIAMS, "Effects of Influent Substrate Concentration on the Kinetics of Natural Microbial Population in Continuous Culture," *Water Research*, **9**, 171 (1975).

GRAEF, S. P., AND J. F. ANDREWS, "Stability and Control of Anaerobic Digestion," *Journal of the Water Pollution Control Federation*, **46**, 666 (1974).

GRAU, P., M. DOHANYOS, AND J. CHUDOBA, "Kinetics of Multicomponent Substrate Removal by Activated Sludge," *Water Research*, **9**, 637 (1975).

Great Lakes–Upper Mississippi River Board of State Sanitary Engineers: *Recommended Standards for Sewage Works* (*Ten-State Standards*), 1971.

GRIEVES, R. B., W. F. MILBURY, AND W. O. PIPES, "A Mixing Model for Activated Sludge," *Journal of the Water Pollution Control Federation*, **36**, 619 (1964).

HAMMER, M. J., *Water and Wastewater Technology*, John Wiley & Sons, New York, 1975.

HARTMANN, L., AND G. LAUBENBERGER, "Toxicity Measurements in Activated Sludge," *Journal of the Sanitary Engineering Division, ASCE*, **94**, SA2, 247 (1968).

HAUG, R. T., AND P. L. MCCARTY, "Nitrification with the Submerged Filter," *Journal of the Water Pollution Control Federation*, **44**, 2086 (1972).

HERBERT, D., "Some Principles of Continuous Culture," in *Recent Progress in Microbiology*, ed. by G. Tunevall, Blackwell Scientific Publishers, Oxford, England, 1958.

HOFFMEIER, L. H., "Upgrading Colorado Municipal Wastewater Treatment Plants to Meet State Standards," master's thesis, University of Colorado, Boulder, Colo., 1974.

HULTMAN, B., "Kinetics of Biological Nitrogen Removal," Inst. Vattenforsorjmings-och Avloppsteknik samt Vattenkemi, KTH, Pub. 71: 5, Stockholm, 1971.

HULTMAN, B., "Biological Nitrogen Reduction Studies as a General Microbiological

Engineering Process," in *Environmental Engineering*, ed. by G. Linder and K. Nyberg, D. Reidel Publishing Co., Dordrecht, Holland, 1973.

JENKINS, D., AND W. GARRISON, "Control of Activated Sludge by Mean Cell Residence Time," *Journal of the Water Pollution Control Federation*, **40**, 1905 (1968).

JEWELL, W. J., AND S. E. MACKENZIE, "Microbial Yield Dependence on Dissolved Oxygen in Suspended and Attached Systems," in *Applications of Commercial Oxygen to Water and Wastewater Systems*, University of Texas, Austin, Tex., 1973, p. 62.

KNOWLES, G., A. L. DOWNING, AND M. J. BARRETT, "Determination of Kinetic Constants for Bacteria in a Mixed Culture, with the Aid of an Electronic Computer," *Journal of General Microbiology*, **38**, 263 (1965).

KUGELMAN, I. J., AND P. L. MCCARTY, "Cation Toxicity and Stimulation in Anaerobic Waste Treatment," *Journal of the Water Pollution Control Federation*, **37**, 97 (1965).

LAWRENCE, A. W., "Application of Process Kinetics to Design of Anaerobic Processes," in *Anaerobic Biological Treatment Processes*, F. G. Pohland, Symposium Chairman, American Chemical Society, Cleveland, Ohio, 1971.

LAWRENCE, A. W., "Modeling and Simulation of Slurry Biological Reactors," in *Mathematical Modeling for Water Pollution Control*, ed. by Thomas Keinath and Martin Wanielista, Ann Arbor Science Publishers, Inc., Ann Arbor, Mich., 1975.

LAWRENCE, A. W., AND P. L. MCCARTY, "Kinetics of Methane Fermentation in Anaerobic Treatment," *Journal of the Water Pollution Control Federation*, **41**, R1 (1969).

LAWRENCE, A. W., AND P. L. MCCARTY, "Unified Basis for Biological Treatment Design and Operation," *Journal of the Sanitary Engineering Division, ASCE*, **96**, SA3, 757 (1970).

LAWRENCE, A. W., AND T. R. MILNES, "Discussion Paper," *Journal of the Sanitary Engineering Division, ASCE*, **97**, 121 (1971).

MALINA, J. F., JR., "Variables Affecting Anaerobic Digestion," *Public Works*, **93**, 9, 113 (Sept. 1962).

MALINA, J. F., JR., "Anaerobic Waste Treatment," in *Manual of Treatment Processes*, Vol. I, ed. by W. W. Eckenfelder, Environmental Science Services Corporation, Stamford, Conn., 1970.

MCCARTY, P. L., "Anaerobic Waste Treatment Fundamentals" (four parts), *Public Works*, Sept. 1964, p. 107; Oct. 1964, p. 123; Nov. 1964, p. 91; Dec. 1964, p. 95.

MCCARTY, P. L., "Kinetics of Waste Assimilation in Anaerobic Treatment," in *Developments in Industrial Microbiology*, Vol. 7, American Institute of Biological Sciences, Washington, D.C., 1966a, p. 144.

MCCARTY, P. L., "Sludge Concentration—Needs, Accomplishments and Future Goals," *Journal of the Water Pollution Control Federation*, **38**, 493 (1966b).

MCCARTY, P. L., "Anaerobic Treatment of Soluble Wastes," in *Advances in Water Quality Improvement*, ed by E. F. Gloyna and W. W. Eckenfelder, University of Texas Press, Austin, Tex., 1968.

MCCARTY, P. L., BECK, L., AND ST. AMANT, P., "Biological Denitrification of Wastewaters by Addition of Organic Materials," Paper presented at the 24th

Annual Industrial Waste Conference, Purdue University, West Lafayette, Ind., 1969.

McCarty, P. L., "Phosphorus and Nitrogen Removal by Biological Systems," in *Proceedings, Wastewater Reclamation and Reuse Workshop*, Lake Tahoe, Calif., June 25–27, 1970.

McKinney, R. E., "Discussion Paper," *Journal of the Environmental Engineering Division, ASCE*, **100**, 789 (1974).

Metcalf and Eddy, Inc., *Wastewater Engineering*, McGraw-Hill Book Company, New York, 1972.

Metcalf and Eddy, Inc., "Design of Nitrification and Denitrification Facilities," EPA Technology Transfer Seminar Publication, 1973.

Milbury, W. F., W. O. Pipes, and R. B. Grieves, "Compartmentalization of Aeration Tanks," *Journal of the Sanitary Engineering Division, ASCE*, **91**, SA3, 45 (1965).

Muck, R. E., and C. P. L. Grady, Jr., "Temperature Effects on Microbial Growth in CSTR's," *Journal of the Environmental Engineering Division, ASCE*, **100**, EE5, 1147 (Oct. 1974).

Mueller, J. A., K. G. Voelkel, and W. C. Boyle, "Nominal Diameter of Floc Related to Oxygen Transfer," *Journal of the Sanitary Engineering Division, ASCE*, SA2, **92**, 4756 (1966).

Mueller, J. A., W. C. Boyle, and E. N. Lightfoot, "Oxygen Diffusion Through Zoogloeal Flocs," *Biotechnology and Bioengineering*, **10**, 331 (1968).

Mynhier, M. D., and C. P. L. Grady, Jr., "Design Graphs for Activated Sludge Process," *Journal of the Environmental Engineering Division, ASCE*, **101**, 829 (1975).

Nagel, C. A., and J. G. Haworth, "Operational Factors Affecting Nitrification in the Activated Sludge Process," paper presented at the 42nd Annual Conference of the Water Pollution Control Federation, Dallas, Tex., 1969.

O'Rourke, J. T., "Kinetics of Anaerobic Treatment at Reduced Temperatures," thesis presented to Stanford University, Stanford, Calif., in partial fulfillment for the degree of Doctor of Philosophy, 1968.

Painter, H. A., "A Review of Literature on Inorganic Nitrogen Metabolism in Microorganism," *Water Research*, **4**, 393 (1970).

Parker, D. S., and M. S. Merrill, "Oxygen and Air Activated Sludge: Another View," *Journal of the Water Pollution Control Federation*, **48**, 2511 (1976).

Patterson, J. W., P. L. Brezonik, and H. D. Putnam, "Sludge Activity Parameters and Their Application to Toxicity Measurements in Activated Sludge," paper presented at 24th Annual Industrial Waste Conference, Purdue University, West Lafayette, Ind., 1969.

Pavoni, J. L., M. W. Tenney, and W. F. Echelberger, Jr., "Bacterial Exocellular Polymers and Biological Flocculation," *Journal of the Water Pollution Control Federation*, **44**, 414 (1972).

Pfeffer, J. T., M. Leiter, and J. R. Worlund, "Population Dynamics in Anaerobic Digestion," *Journal of the Water Pollution Control Federation*, **39**, 1305 (1967).

PIRT, S. J., "The Maintenance Energy of Bacteria in Growing Cultures," *Proceedings of the Royal Society, London*, **B163**, 224 (1965).

RIESING, R. R., "Relationship Between Influent Substrate Concentration, Growth Rate and Effluent Quality in a Chemostat," M.S.E. thesis, Purdue University, West Lafayette, Ind., 1971.

SAWYER, C. N., "Fertilization of Lakes by Agricultural and Urban Drainage," *Journal of the New England Water Works Association*, **61**, 109 (1947).

SAYIGH, B. A., AND J. F. MALINA, JR, "Temperature Effects on the Kinetics and Performance of the Completely Mixed Continuous Flow Activated Sludge Process," paper presented at the 49th Annual Water Pollution Control Federation Conference, Minneapolis, Minn., 1976.

SCHROEDER, E. D., *Water and Wastewater Treatment*, McGraw-Hill Book Company, New York, 1977.

SHERRARD, J. H., AND E. D. SCHROEDER, "Cell Yield and Growth Rate in Activated Sludge," *Journal of the Water Pollution Control Federation*, **45**, 1889 (1973).

SHERRARD, J. H., E. D. SCHROEDER, AND A. W. LAWRENCE, "Mathematical and Operational Relationships for the Completely-Mixed Activated Sludge Process," *Water and Sewage Works*, **R-84** (1974).

SPEECE, R. E., AND M. J. HUMENICK, "Solids Thickening in Oxygen Activated Sludge," *Journal of the Water Pollution Control Federation*, **46**, 43 (1974).

STALL, T. R., AND J. H. SHERRARD, "One Sludge or Two Sludge?" *Water and Wastes Engineering*, **41**, (Apr. 1974).

STALL, T. R., AND J. H. SHERRARD, "Evaluation of Control Parameters for the Activated Sludge Process," *Journal of the Water Pollution Control Federation*, **50**, 450 (1978).

STANKEWICH, M. J. JR., "Biological Nitrification with the High Purity Oxygenation Process," paper presented at the 27th Annual Purdue Industrial Waste Conference, Purdue University, West Lafayette, Ind., 1972.

STONE, R. W., D. S. PARKER, AND J. A. COTTERAL, "Upgrading Lagoon Effluent for Best Practicable Treatment," *Journal of the Water Pollution Control Federation*, **47**, 2019, (1975).

SYKES, R. M., "Theoretical Heterotrophic Yields," *Journal of the Water Pollution Control Federation*, **47**, 591 (1975).

TOPIWALA, H., AND C. G. SINCLAIR, "Temperature Relationship in Continuous Culture," *Biotechnology and Bioengineering*, **13**, 795 (1971).

VESILIND, P. A., *Treatment and Disposal of Wastewater Sludges*, Ann Arbor Science Publishers, Inc., Ann Arbor, Mich., 1974.

WALKER, L. F., "Hydraulically Controlling Solids Retention Time in the Activated Sludge Process," *Journal of the Water Pollution Control Federation*, **43**, 30 (1971).

WEDDLE, C. L. AND D. L. JENKINS, "The Viability and Activity of Activated Sludge," *Water Research*, **5**, 621 (1971).

WILD, H. E., C. N. SAWYER, AND T. C. MCMAHON, "Factors Affecting Nitrification Kinetics," *Journal of the Water Pollution Control Federation*, **43**, 1845 (1971).

WUHRMANN, K., "Grundlager für die Dimensionierung der Belüflung bei Belebtschlammanlagen," *Schweiz. A. Hydrol.*, **26**, 310 (1964).

Aeration

In aerobic metabolism oxygen acts primarily as an electron acceptor for catabolism. As the activated sludge process is designed to be substrate-limiting, metabolism sets the rate of oxygen demand. The function of the aeration system is to transfer oxygen to the liquid at such a rate that oxygen never becomes the limiting factor in process operation (i.e., never limits the rate of organic utilization or other metabolic functions). It is the engineer's responsibility to ensure that aeration systems are adequate. This requires an understanding of the basic principles of gas transfer as well as some familiarity with the different types of aeration devices that are available.

5-1 Fundamentals of Gas Transfer

All solutes tend to diffuse through solutions until the composition is homogeneous throughout. The rate at which solutes diffuse across a uniform cross-sectional area depends on the molecular size and shape and the concentration gradient of that substance. Matter moves spontaneously from a region of high concentration toward a region of lower concentration, and the more the concentration is decreasing, the more the diffusion rate increases. This can be expressed by writing the concentration gradient term as $-\partial C/\partial Y$, where C is the concentration and Y the distance. If $\partial M/\partial t$ represents the rate at which M grams of solute cross the reference plane, *Fick's first law of diffusion* states that

$$\frac{\partial M}{\partial t} = -D_L A \frac{\partial C}{\partial Y} \tag{5-1}$$

where $\dfrac{\partial M}{\partial t}$ = rate of mass transfer, mass time^{-1}

D_L = diffusivity constant, area time^{-1}

A = cross-sectional area across which the solute is diffusing, area

$\dfrac{\partial C}{\partial Y}$ = concentration gradient (i.e., the change in concentration with distance), mass volume^{-1} length^{-1}

The simplest concept of a gas transfer process is the *stationary liquid film theory.* This theory suggests that at the interface between the gas phase and the liquid phase, there exists a stationary liquid film in which gas molecules are concentrated. The gas concentration is not homogeneous throughout the liquid film but rather decreases from the saturation concentration given by Henry's law to some lower concentration at the film/bulk liquid boundary. Figure 5-1 illustrates the stationary liquid film theory. In this figure C_s represents the saturation concentration of the gas in the liquid as predicted by Henry's law, C the concentration of the gas in the bulk of the liquid, Y_f the thickness of the film, and C_L the gas concentration at the film/bulk boundary. Applying Fick's first law of diffusion to this situation gives

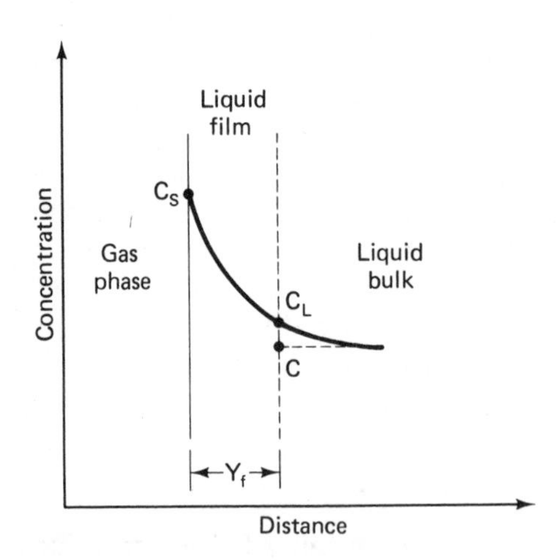

Figure 5-1. Schematic Representation of Gas Transfer Through Stationary Liquid Film.

$$\frac{\partial M}{\partial t} = -D_L A \frac{\partial C}{\partial Y_f} \tag{5-2}$$

In this case A represents the interfacial area of contact between the gas and liquid phases.

Since the liquid film thickness is small (only a few molecules thick), it is possible to approximate the differential quantity $\partial C/\partial Y_f$ with the linear approximation

$$\frac{\partial C}{\partial Y_f} \simeq \frac{C_s - C}{Y_f} \tag{5-3}$$

The system, as described by the linear approximation, is shown in Figure 5-2. For the linear approximation, equation 5-2 reduces from a partial differential equation in time and space to an ordinary differential equation in time.

$$\frac{dM}{dt} = -D_L A \frac{C_s - C}{Y_f} \tag{5-4}$$

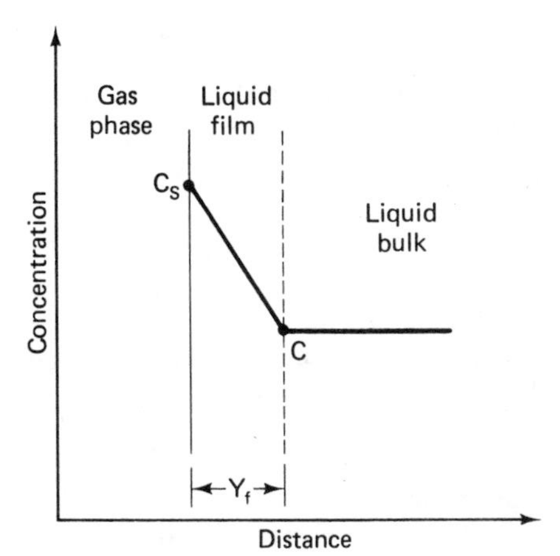

Figure 5-2. Linear Approximation of Gas Transfer Through Stationary Liquid Film.

Dividing both sides of equation 5-4 by V (the volume of the liquid phase), the equation becomes

$$\frac{1}{V}\frac{dM}{dt} = -D_L \frac{A}{V}\frac{C_s - C}{Y_f} \tag{5-5}$$

Since the term $[1/V(dM/dt)]$ has the units of mass volume^{-1} time^{-1} or concentration per unit time, this term can be expressed by the differential expression dC/dt:

$$\frac{1}{V}\frac{dM}{dt} = \frac{dC}{dt} \tag{5-6}$$

Substituting for the $[1/V(dM/dt)]$ term from equation 5-6 gives

$$\frac{dC}{dt} = -D_L \frac{A}{V}\frac{C_s - C}{Y_f} \tag{5-7}$$

Because the value of the film thickness is normally unknown, it is usually combined with D_L to define a new constant term.

$$K_L = \frac{D_L}{Y_f} \tag{5-8}$$

where K_L represents the gas transfer coefficient and has the units of length time^{-1}. K_L can be incorporated into equation 5-7 to give an expression which has the form

$$\frac{dC}{dt} = -K_L\frac{A}{V}(C_s - C) \tag{5-9}$$

Equation 5-9 represents the change in concentration to be expected as molecules diffuse from a region of high concentration to a region of low concentration so that the concentration is decreasing with time. When the gas concentration increases with time during the aeration process, the negative sign is dropped and equation 5-9 reduces to

$$\frac{dC}{dt} = K_L\frac{A}{V}(C_s - C) \tag{5-10}$$

In most cases the interfacial area of contact, A, is difficult to determine. To circumvent this problem, a second constant, K_La, is introduced. This constant has a value equal to the product of K_L and A/V:

$$K_La = K_L\frac{A}{V} \tag{5-11}$$

K_La is defined as the overall gas transfer coefficient and has the units of time^{-1}. Integrating this constant into equation 5-10 gives

$$\frac{dC}{dt} = K_La(C_s - C) \tag{5-12}$$

The overall gas transfer coefficient can be considered as an overall conductance (i.e., as the inverse of an overall resistance). Hence, when the resistance to gas transfer is large, K_La will be small, and vice versa.

Equation 5-12 can be rearranged into the form

$$\frac{dC}{C_s - C} = K_La\,dt \tag{5-13}$$

Integrating equation 5-13 gives

$$-\ell n\,(C_s - C) = K_Lat + \text{constant of integration} \tag{5-14}$$

If $C = C_0$ at $t = 0$, the constant of integration has the value $-\ell n\,(C_s - C_0)$; then

$$-\ell n\,(C_s - C) = K_Lat - \ell n\,(C_s - C_0) \tag{5-15}$$

or

$$\ell n\left(\frac{C_s - C_0}{C_s - C}\right) = K_Lat \tag{5-16}$$

or

$$\log\left(\frac{C_s - C_0}{C_s - C}\right) = \frac{K_La}{2.3}t \tag{5-17}$$

This implies that a semilog plot of $(C_s - C_0)/(C_s - C)$ versus t will give a linear trace with a slope equal to $K_La/2.3$. Some typical C_s values for pure water are given in Table 5-1.

TABLE 5-1

SATURATION CONCENTRATION OF DISSOLVED OXYGEN EXPOSED TO AN ATMOSPHERE CONTAINING 21% OXYGEN (CHLORIDE CONCENTRATION = 0 AND 1 ATM. PRESSURE)

Temperature (°C)	C_s (mg/ℓ)
0	14.62
2	13.84
4	13.13
6	12.48
8	11.87
10	11.33
12	10.83
14	10.37
16	9.95
18	9.54
20	9.17
22	8.83
24	8.53
26	8.22
28	7.92
30	7.63

5-2 Factors Affecting Oxygen Transfer

Equation 5-7 shows that the rate of change in the dissolved oxygen concentration during aeration is directly proportional to the interfacial area of contact, A, and the oxygen deficit, $(C_s - C)$, and that it is inversely proportional to the thickness of the liquid film, Y_f. Hence, any factor that affects these parameters will affect the rate of oxygen transfer. Such factors are (1) oxygen saturation, (2) temperature, (3) wastewater characteristics, and (4) degree of turbulence. Because of their importance, each of these factors will be discussed in some detail.

Oxygen Saturation

The saturation concentration of oxygen in water depends upon salinity, temperature, and the partial pressure of the oxygen in contact with the water.

Eckenfelder and O'Connor (1961) suggest that the saturation concentration may be obtained from the following equation:

$$(C_s)_{760} = \frac{475 - 2.65S}{33.5 + T} \tag{5-18}$$

where $(C_s)_{760}$ = saturation value of oxygen at a total atmospheric pressure of 760 mm Hg, mg/ℓ
S = dissolved solids concentration in the water, g/ℓ
T = temperature, °C

Many workers correct for the presence of dissolved salts by introducing a β factor, defined as

$$\beta = \frac{\text{saturation concentration in wastewater}}{\text{saturation concentration in tap water}} \tag{5-19}$$

The value of oxygen saturation given by equation 5-18 may be corrected for prevailing pressure by applying the expression

$$C_s = (C_s)_{760} \frac{P - \bar{p}}{760 - \bar{p}} \tag{5-20}$$

where P represents the prevailing barometric pressure in mm Hg and $\bar{p}$ represents the saturated water vapor pressure at the temperature of the water (see Table 5-2 for $\bar{p}$ values). It is important to understand that the oxygen saturation value given by equation 5-20 is the value at the water surface. This value is used for surface aerator design but must be corrected for diffused air and submerged turbine systems because of the increase in partial pressure due to liquid submergence. An equation that can be used to adjust for this increase in partial pressure will be presented in a later section.

TABLE 5-2

VAPOR PRESSURE OF WATER IN CONTACT WITH AIR

Temperature (°C)	*Vapor pressure (mm Hg)*
0	4.5
5	6.5
10	9.2
15	12.8
20	17.5
25	23.8
30	31.8

Farkas (1966) found that the dissolved oxygen saturation values in heat-sterilized sludge and tap water coincide within a range of 0.1 mg/ℓ. Thus, there is little need to correct the saturation oxygen concentration for MLSS.

Temperature

Temperature affects the overall oxygen transfer coefficient according to the following expression (Eckenfelder, 1966):

$$K_La_{(T)} = (K_La)_{20°C}(1.020)^{T-20} \tag{5-21}$$

where T represents the water temperature in degrees Celsuis. O'Connor and Dobbins (1956) have proposed another approach wherein K_La is corrected for both temperature and viscosity effects:

$$\frac{(K_La)_1}{(K_La)_2} = \left(\frac{T_1\mu_2}{T_2\mu_1}\right)^{1/2} \tag{5-22}$$

where T represents the temperature in °K and μ the absolute viscosity.

It is interesting to note that the effects of temperature on C_s and K_La are of approximately the same magnitude but in an opposite direction, so that these effects tend to cancel each other. Because of this, temperature variations can usually be neglected in K_La and C_s corrections as long as the value of each parameter is based on the same temperature (Marais, 1975).

Wastewater Characteristics

Under process conditions, the value of K_La is usually less for wastewater than for tap water. This occurs because of the presence of soluble organic compounds, particularly surface-active materials. Stukenberg et al. (1977) report that mixed liquor suspended solids have little effect on oxygen transfer and can generally be neglected in K_La corrections.

Surface-active agents such as short-chain fatty acids and alcohols are soluble in both water and oil solvents. The hydrocarbon part of the molecule is responsible for its solubility in oil, while the polar carboxyl or hydroxyl group has sufficient affinity to water to drag a short-length nonpolar hydrocarbon chain into aqueous solution. These molecules will concentrate at an air/water interface, where they are able to locate their hydrophilic group in the aqueous phase, which allows the hydrophobic hydrocarbon chain to extend into the vapor phase (see Figure 5-3). This situation is energetically more favorable but creates a concentration of molecules or "film" that retards molecular diffusion. Hence, resistance to oxygen transfer is increased and, consequently the value of K_La is decreased.

To compensate for the effects of surface-active agents on oxygen transfer, an α factor is introduced, where

$$\alpha = \frac{K_La \text{ of wastewater}}{K_La \text{ of tap water}} \tag{5-23}$$

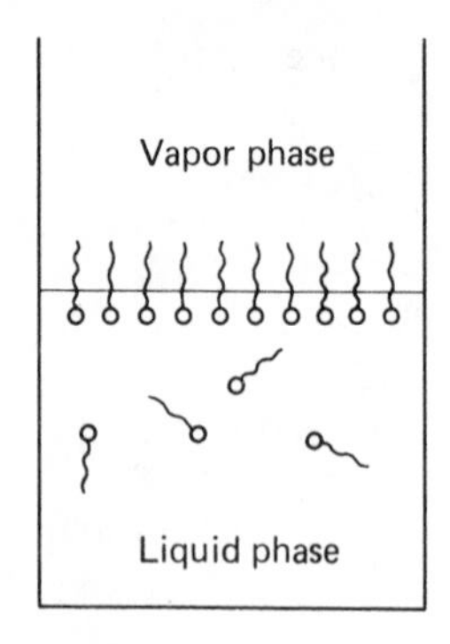

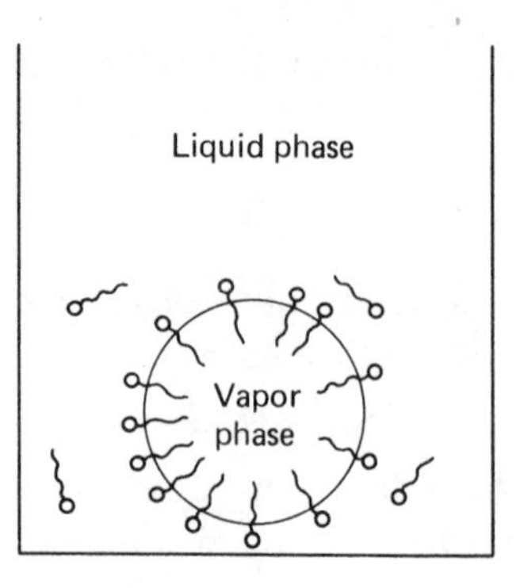

Figure 5-3. Adsorption of Surface-Active Agents at Air/Water Interface.

Turbulence

Eckenfelder and Ford (1970) report that the degree of turbulence in the aeration tank will influence the value of α as follows:

1/ Under near-quiescent conditions (a lower degree of turbulence), fluid motion has little effect on α because the resistance to diffusion in the bulk of the liquid is greater than the film resistance.

2/ Increasing fluid agitation to a moderate degree decreases the resistance to diffusion in the liquid bulk so that film resistance will control the diffusion rate. At this point α is depressed to a minimum value.

3/ A further increase in fluid agitation will produce a high degree of turbulence and break up the film. Under such conditions α will approach unity.

The effects of turbulence on α, as developed by Mancy and Okun (1960), are illustrated by Fig. 5-4.

If surface-active materials are not present in the liquid under aeration, agitation will tend to reduce the film thickness. According to equation 5-7, any reduction in film thickness will increase the rate of oxygen transfer. Hence, the denominator term in equation 5-23 will always increase with an increase in turbulence.

The tendency for surface-active molecules to pack into an interface favors an expansion of the interface; this must therefore be balanced against the tendency for the interface to contract under normal surface tension forces. The overall effect is a decrease in the surface tension which makes entrained air bubbles easier to shear into smaller sizes. Thus, under highly turbulent conditions the A/V ratio may increase for wastewater (containing surface-active material), which may result in values greater than 1.0 for α.

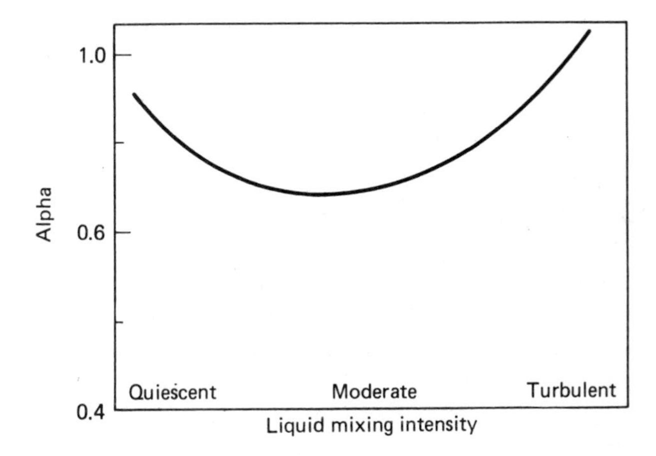

Figure 5-4. Effects of Liquid Mixing Intensity on α. (After Mancy and Okun 1960.)

Oxygen Transfer Rates

The oxygen transfer rate of a particular aeration device quoted by a manufacturer applies only for standard conditions and a specific tank geometry. Standard conditions mean that the aerator was tested with tap water at zero dissolved oxygen concentration, 20°C and 760 mm Hg atmospheric pressure. Thus, a manufacturer's figure for the rate of oxygen transfer must be modified for process conditions. This can be done by incorporating the factors that affect the oxygen transfer rate into equation 5-12 to give

$$\left(\frac{dC}{dt}\right)_{\text{actual}} = \alpha(K_L a)_{20^\circ\text{C}}\left(\frac{P-\bar{p}}{760-\bar{p}}\beta C_s - C\right) \tag{5-24}$$

This expression neglects temperature effects on $K_L a$ and C_s and assumes both parameters are evaluated at 20°C.

Under standard conditions where $C = 0$, equation 5-12 reduces to

$$\left(\frac{dC}{dt}\right)_{\text{standard}} = (K_L a)_{20^\circ\text{C}} C_s \tag{5-25}$$

To determine the design oxygen transfer rate, equations 5-24 and 5-25 can be combined in the following manner:

$$\frac{(dC/dt)_{\text{actual}}}{(dC/dt)_{\text{standard}}} = \frac{\alpha\left(\frac{P-\bar{p}}{760-\bar{p}}\beta C_s - C\right)}{C_s} \tag{5-26}$$

or

$$\left(\frac{dC}{dt}\right)_{\text{actual}} = \left(\frac{dC}{dt}\right)_{\text{standard}} \frac{\alpha\left(\frac{P-\bar{p}}{760-\bar{p}}\beta C_s - C\right)}{C_s} \tag{5-27}$$

If the oxygen transfer rate is expressed on a per unit horsepower basis, equation 5-27 implies that more power will be required under process conditions than under standard conditions.

5-3

Determination of K_La and α Values

Either the steady-state or non-steady-state test is used to determine aeration equipment characteristics under process conditions. The basic formulation used in the steady-state test is developed by modifying equation 5-12 such that

$$\left(\frac{dC}{dt}\right)_{\text{overall}} = K_La(C_s - C) - R \tag{5-28}$$

where R represents the oxygen utilization rate of the biomass and has the units mass volume^{-1} time^{-1}. The C_s term reflected in equation 5-28 is specific for process conditions. At steady-state conditions the rate of oxygen transfer by the aeration system is equal to the rate of oxygen utilization by the biomass, implying that $(dC/dt)_{\text{overall}}$ is equal to zero. Thus, equation 5-28 can be solved for K_La to give

$$K_La = \frac{R}{C_s - C} \tag{5-29}$$

where K_La = overall oxygen transfer rate in wastewater at process conditions, time^{-1}. This K_La value includes the effects of surface active material, TDS, temperature, and partial pressure

Generally, the oxygen utilization rate of an activated sludge culture is quite rapid and difficult to measure. To circumvent this problem, it is recommended that the biomass be deprived of an external source of substrate for 1 to 2 h before testing. At the end of this time the oxygen utilization should be due only to endogenous respiration and therefore be slow and easy to measure. Samples should be taken at several points in the tank for the DO analysis. Stukenburg et al. (1977) recommend the sampling points indicated in Figure 5-5. In all aeration systems one sampling point should be under the aerator in order to measure the DO concentration in the liquid just prior to aeration.

In the non-steady-state test, final effluent or supernatant from settled mixed liquor is generally used. The laboratory procedure for this test is as follows:

1/ Adjust the liquid temperature to that expected in the field.

2/ Deoxygenate the liquid in the test basin using sodium sulfide with a cobalt chloride catalyst. The cobalt chloride dose should be no

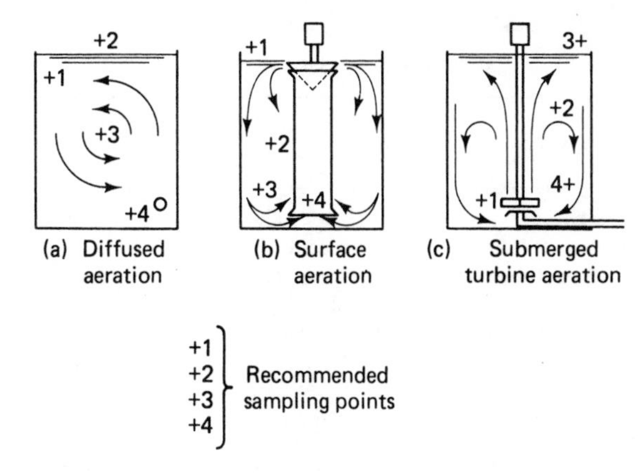

Figure 5-5. Recommended Sampling Points for Steady-State Analysis. (After Stukenberg et al., 1977.)

greater than 0.05 mg/ℓ. The reaction between sodium sulfite and oxygen is as follows:

$$Na_2SO_3 + 0.5\,O_2 \longrightarrow Na_2SO_4 \qquad (5\text{-}30)$$

Theoretically, 7.9 mg/ℓ of sodium sulfite is required for each mg/ℓ of oxygen present. However, it is common practice to add 1.5 to 2.0 times this amount to ensure complete deoxygenation.

3/ Oxygenate the liquid using the same type of aeration device to be used under process conditions.

4/ Tabulate the dissolved oxygen concentration at various time intervals and sampling points until oxygen saturation is reached.

5/ A plot of $\log[(C_s - C_0)/(C_s - C)]$ versus time will give a linear trace of slope $K_La/2.3$ (see Figure 5-6).

6/ Repeat the same procedure using tap water as the test liquid.

Using equations 5-19 and 5-23, values for α and β may be determined. Some typical values for these parameters are presented in Tables 5-3 and 5-4. The

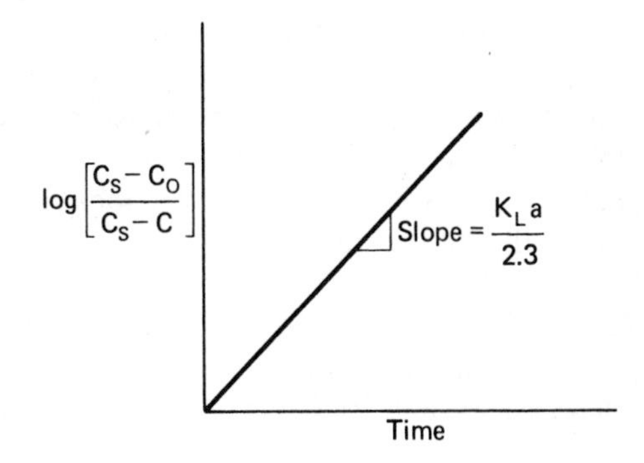

Figure 5-6. Determination of the Overall Oxygen Transfer Coefficient, K_La.

TABLE 5-3

TYPICAL α VALUES FOR SOME SPECIFIC WASTEWATERS

Unit	*Waste*	*Alpha* (α)	*Conditions*
Sparjers[1]	Activated sludge effluent	1.320	10–25 scfm/unit, 15 ft depth, 25 ft width
Plate tubes[1]	Activated sludge effluent	0.860	6–14 scfm/unit, 15 ft depth, 25 ft width
INKA system[2]	Kraft Black Liquor		
	20 ppm	0.875	6 ft depth, 6.8 ft width, 2.6 ft submergence
	100 ppm	0.750	6 ft depth, 6.8 ft width, 2.6 ft submergence
	200 ppm	0.625	6 ft depth, 6.8 ft width, 2.6 ft submergence
Small bubble diffuser[2]	Kraft Black Liquor		
	20 ppm	0.880	100 ℓ volume, 3 ft depth, 10 × 10 cm bubble air diffuser
	100 ppm	0.813	100 ℓ volume, 3 ft depth, 10 × 10 cm bubble air diffuser
	200 ppm	0.662	100 ℓ volume, 3 ft depth, 10 × 10 cm bubble air diffuser
Lightning lab stirrer[3]	Nylon manufacturing wastewater		10 ℓ volume, surface mixing
		0.560	212 rpm
		0.715	220 rpm
Aeration cone[4]	Tap water + 5 ppm anionic detergent	1.330	12,500 ft^3 volume, 6 ft diameter, 36 rpm

[1]Barnhart (1965) [2]Bewtra (1964) [3]Randall (1971) [4]Eckenfelder (1968)

values reported in Table 5-4 were determined by Stukenberg et al. (1977) at a completely mixed activated sludge plant in Texas. Lister and Boon (1973) have found α factors of approximately 0.3 and 0.8 at the inlet and outlet ends of a diffused air, plug-flow activated sludge process treating municipal wastewater. These investigators also found that α factors approaching 0.8 might be expected for completely mixed systems. Such results imply that the overall oxygen transfer coefficient will increase as the organic material is removed from the wastewater and suggest that the average rate of oxygen transfer will be greater in completely mixed systems.

As noted earlier (Figure 5-4), the value of α depends to some extent on the degree of turbulence in the aeration tank. Since it is impossible to accurately simulate process conditions in the laboratory, any value for α determined in this manner will at best be only a reasonable approximation.

TABLE 5-4

SURFACE AERATOR MIXED LIQUOR TEST DATA

Oxygen uptake rate (mg/ℓ-h)	*Temperature (°C)*	α	βC_s *(mg/ℓ)*
40	19.8	0.89	7.9
41	19.8	0.86	7.9
36	19.8	0.85	7.9
40	18.7	0.78	8.2
43	19.0	0.90	8.2
48	19.4	0.89	8.1
56	19.0	0.93	8.1
50	19.5	0.93	8.0
64	20.5	0.90	7.9
59	20.6	0.94	7.9
52	19.3	0.84	8.0
52	20.0	0.99	7.9

Source: After Stukenberg et al. (1977).

5-4 Design of Aeration Systems

The costliest item in the activated sludge process is the aeration system. Because of this, its design is critical if the treatment facility is to be cost-effective.

Manufacturers of aeration devices will usually quote a figure for the oxygen transfer rate in terms of the mass of oxygen that the aerator can introduce into water per unit of time per unit of power input. This figure quoted by the manufacturer will be valid only under standard conditions and the specified tank geometry. It will, therefore, be necessary to adjust the manufacturers' figures to those which more realistically describe actual process conditions.

Current methods used to transfer oxygen in aerobic biological wastewater treatment processes include (1) compressed air diffusion, (2) submerged turbine aeration, (3) low-speed surface aeration, and (4) motor-speed surface aeration. Each of these methods are illustrated in Figure 5-7.

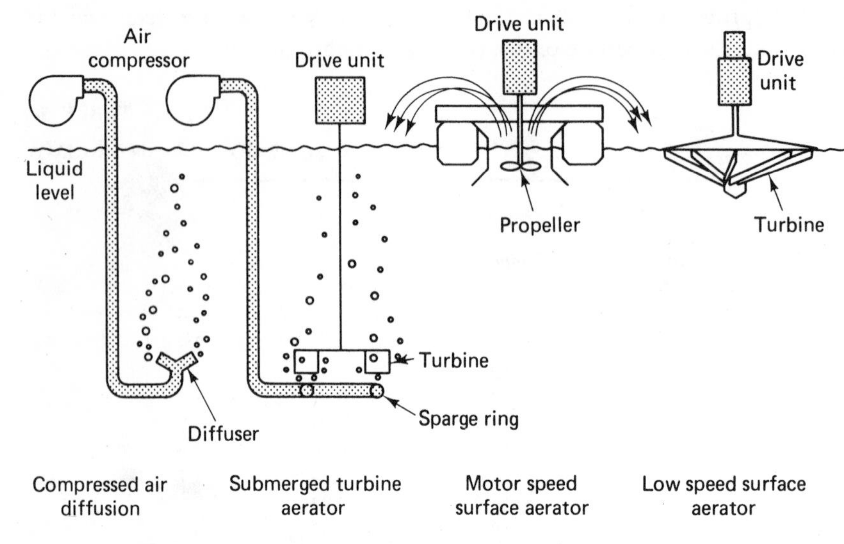

Figure 5-7. Methods of Artificial Aeration. (After Lipták, 1974.)

As a guide to system selection, Stukenberg et al. (1977) suggest that diffused aeration not be used when the oxygen utilization rate exceeds 40 mg/ℓ-hr. Low-speed surface aerators are acceptable as long as the oxygen utilization rate is less than 80 mg/ℓ-hr. However, when the oxygen utilization rate exceeds 80 mg/ℓ-hr, submerged turbine aeration should be the method of choice. Furthermore, in areas where freezing temperatures are experienced for long periods during winter months, either diffused or submerged turbine aeration is preferred over surface aeration. (Surface aeration is an efficient heat dissipation process and may significantly lower the temperature of the liquid in the process.)

Diffused Aeration

Diffused air systems operate by blowing compressed air through diffusers. Compressed air is provided by compressors operating at a pressure sufficient to overcome the head created by frictional losses in the air piping system and the static head of liquid above the diffuser. The diffusers are positioned near the bottom and along the side wall of the aeration tank to effect oxygen transfer and mixing (see Figure 5-8), or are evenly spaced across the bottom of the aeration tank. They may be attached to either fixed or retractable mountings. Such systems will generally provide adequate mixing with air flows of 20 to 30 standard cubic feet per minute (scfm) per 1000 ft^3 of tank volume. Other commonly used figures for air flow in these systems are 1 ft^3/gal wastewater treated or 1000 ft^3/lb BOD_5 applied to the aeration tank.

Figure 5-8. Diffused air activated sludge process. (Courtesy of FMC Corporation.)

There are basically two types of diffusers. One type produces small bubbles by passing compressed air through a porous medium prior to its discharge into the liquid. The porous medium used is either a material composed of silicon dioxide or aluminum oxide grains held by a ceramic binder or plastic-wrapped (e.g., Saran-wrapped) or plastic-cloth tubes. Diffusers of this type are termed *fine-bubble diffusers* and produce bubbles ranging in size from 2.0 to 2.5 mm. Even though the rate of oxygen transfer is greater with a smaller bubble size (increased interfacial area of contact), a large headloss is experienced through a fine-bubble diffuser, which increases the power requirement for these devices. They also clog fairly easily, and this increases the headloss even more.

The second type of diffuser, termed a *coarse-bubble diffuser*, produces bubbles up to 25 mm in size. Such diffusers may be of the nozzle, orifice, valve, or hydraulic shear type. Since the interfacial area of contact is less with a larger bubble size, the rate of oxygen transfer for this type of diffuser is less than that for fine-bubble diffusers. However, the coarse-bubble diffuser has the advantage of requiring less maintenance and less power. Examples of coarse-bubble diffusion units are:

1/ Sparjer: a large-orifice diffusion unit where air is emitted at high velocity from four short tube orifices at 90° centers.

2/ INKA system: a series of pipes that form a lateral grating. The pipes

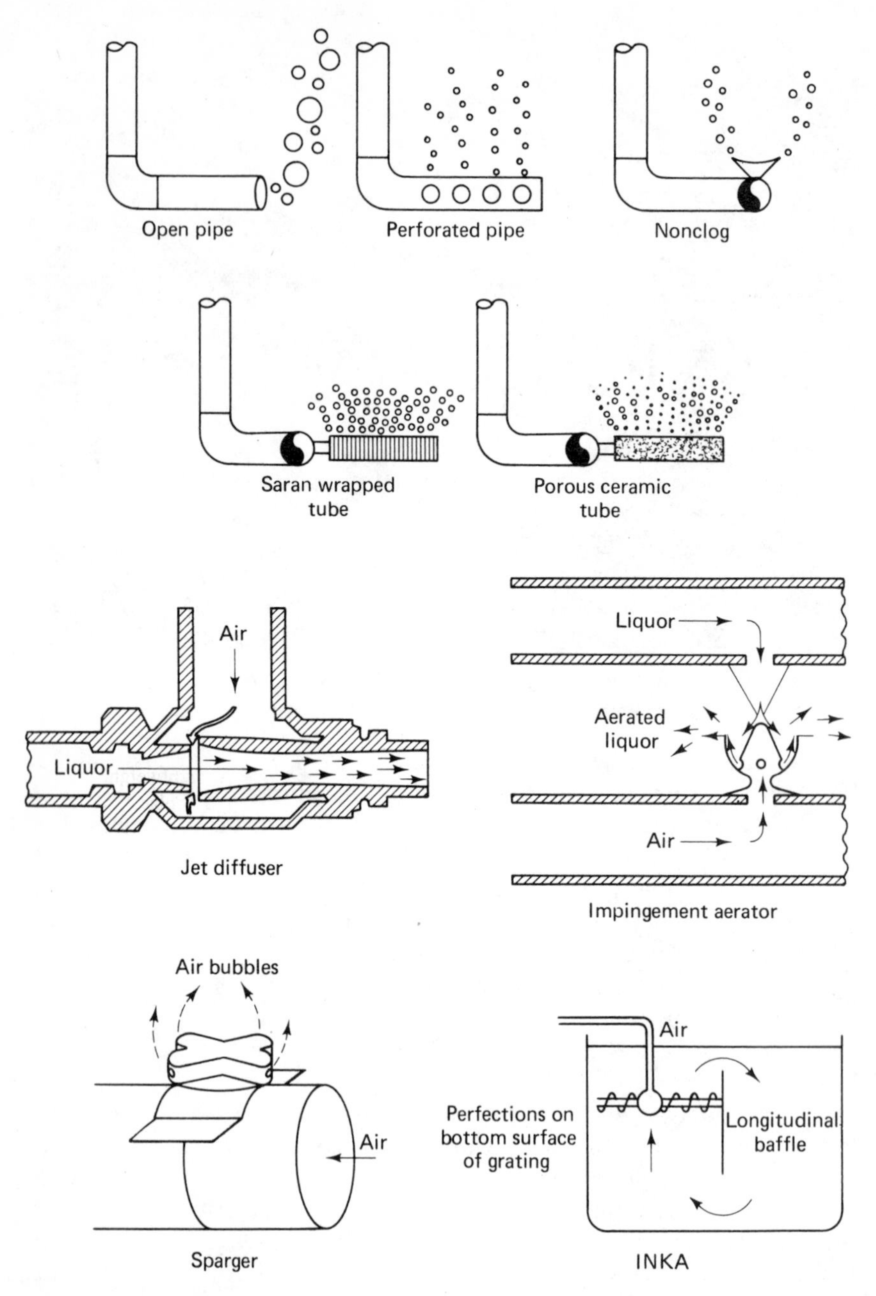

Figure 5-9. Schematic Representation of Different Air Diffusion Devices. (After Eckenfelder and O'Connor, 1961; Shell and Cassady, 1973.)

are perforated on the underside and air is delivered through the perforations at high velocities. The pipe grid is generally mounted between 2 and 6 ft below the liquid surface. Hence, the air can be delivered at a fairly low pressure (Eckenfelder and O'Connor, 1961).

Both fine-bubble and coarse-bubble diffusion units are illustrated in Figures 5-9 to 5-15.

Figure 5-10. Deflectofuser diffuser: A lightweight coarse bubble diffuser manufactured from Cycolac, which is extremely resistant to corrosion. (Courtesy of FMC Corporation.)

The performance of a particular type of diffused air system is affected by diffuser spacing, tank width, tank depth, and rate of air flow. Diffuser spacing depends on the particular type of device used and is governed by mixing considerations and the need to prevent bubble coalescence. Typical minimum and maximum spacings are 6 in. and 24 in., respectively (Eckenfelder and O'Connor, 1961). The relationship among oxygen transfer, air flow, and diffuser spacing is illustrated in Figure 5-16 for Saran-wrapped diffusion units in a 15 ft deep by 24 ft wide aeration tank.

If adequate mixing is to be maintained, the width/depth ratio should not exceed 2. Tank width also affects oxygen transfer. This is illustrated in Figure 5-17, which shows the effect of air flow on oxygen transfer for 8- and 24-ft-wide aeration tanks using both Saran tube and sparjer aeration devices.

The rate of oxygen transfer can be increased by increasing the contact time between the air bubbles and the water. This can be achieved by locating the diffuser at a greater tank depth. However, this increases the static head above the diffuser and requires that the compressor operate at a greater pressure to overcome the additional head. Thus, a greater power expenditure

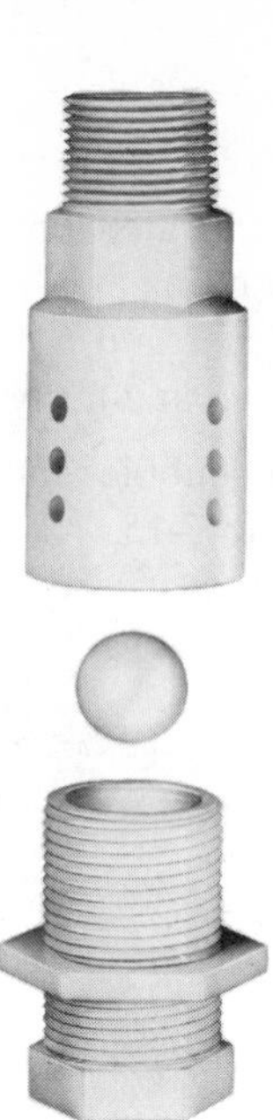

Figure 5-11. Adjustaire Diffuser: A coarse bubble diffuser molded from Delrin for optimum corrosion resistance. (Courtesy of FMC Corporation.)

Figure 5-12. Flexofuser diffuser: A fine bubble diffuser consisting of a tube body, fine sheath, and holding clamp. (Courtesy of FMC Corporation.)

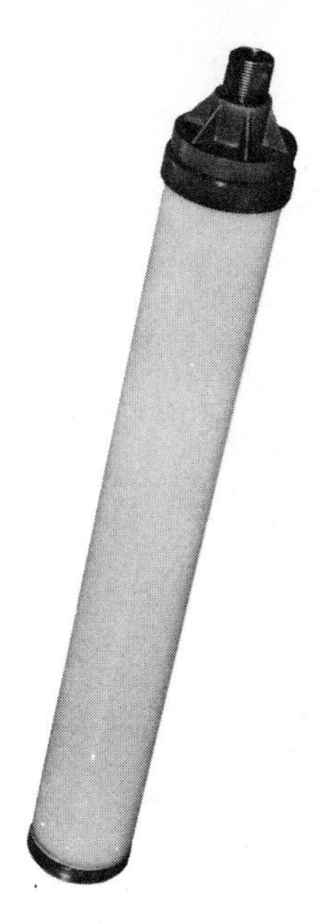

Figure 5-13. Pearlcomb diffuser: Manufactured from an acrylonitrile coplolymer in a cylindrical shape, this diffuser will provide a high oxygen transfer efficiency. (Courtesy of FMC Corporation.)

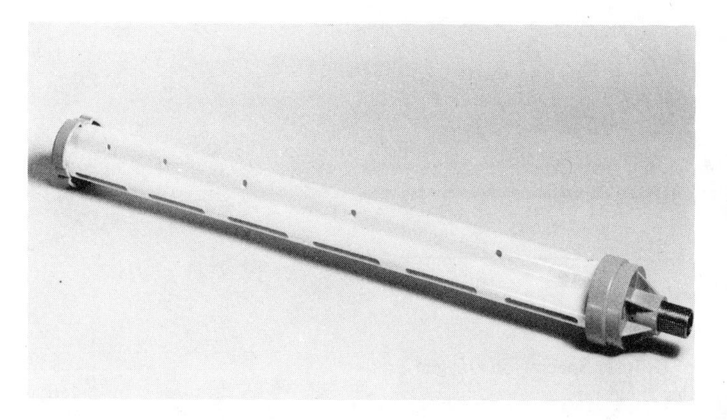

Figure 5-14. Convertofuser diffuser: A wide band coarse bubble diffuser developed to obtain the added oxygen transfer efficiency available with a wide-band air diffuser system. (Courtesy of FMC Corporation.)

is realized with increased diffuser depth. Alternatively, if the diffuser is located at a shallow depth, for the same power expenditure more air can be discharged, resulting in a more violent mixing action. Shell and Cassady (1973) suggest that a diffuser depth between 8 and 16 ft usually gives the optimum balance between mixing and oxygen transfer rate. The relationship between

Figure 5-15. Discfuser diffuser: A coarse bubble diffuser which operates so that in the event of interruption in the air supply the hydrostatic head over the diffuser will close the disc against the body, thereby preventing excessive amounts of the aerated liquor from entering the air diffusion piping system. (Courtesy FMC Corporation.)

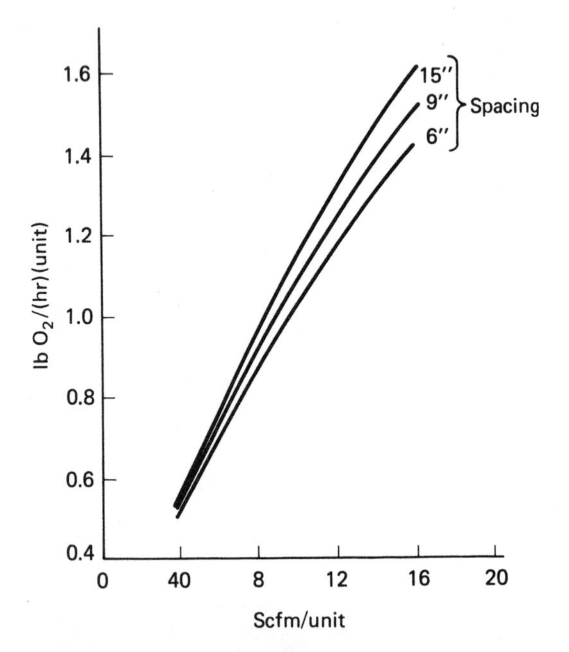

Figure 5-16. Effect of Diffuser Spacing on Oxygen Transfer. (After Eckenfelder, 1966.)

oxygen transfer rate and diffuser depth is given in Figure 5-18 for both Saran tube and sparjer aeration devices.

When coarse-bubble diffusers are used, an increase in air flow rate creates additional turbulence, which may shear the larger bubbles into smaller ones, thereby increasing the interfacial area of contact and resulting in an increase in the rate of oxygen transfer. Such an effect is illustrated in Figures 5-19 and 5-20 for a sparjer and INKA aeration system, respectively. It should be understood, however, that an increase in air flow rate does not necessarily mean an increase in the rate of oxygen transfer. High air flow rates along one

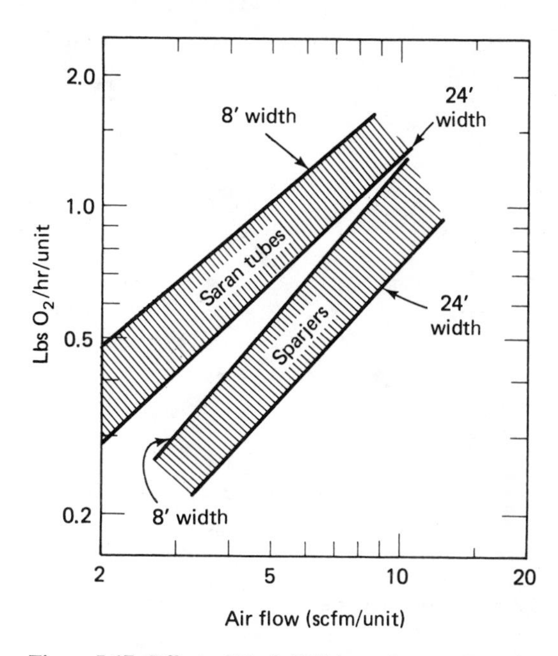

Figure 5-17. Effect of Tank Width on Oxygen Transfer in Diffused Air Systems. (After Bewtra, 1964.)

side of an aeration tank will reduce the percent of oxygen absorbed and may result in a decrease in the rate of oxygen transfer. Busch (1968) has presented an excellent discussion on the effects of increasing air flow rates in diffused air systems. He points out that the fraction of oxygen applied which is removed by the biomass is set by the rate of oxygen utilization and, when measured under maximum loading conditions at an activated sludge plant

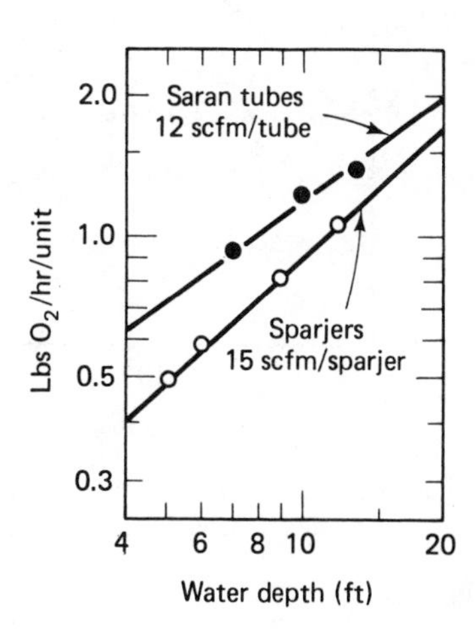

Figure 5-18. Effect of Liquid Depth on Oxygen Transfer in Diffused Air Systems. (After Bewtra, 1964.)

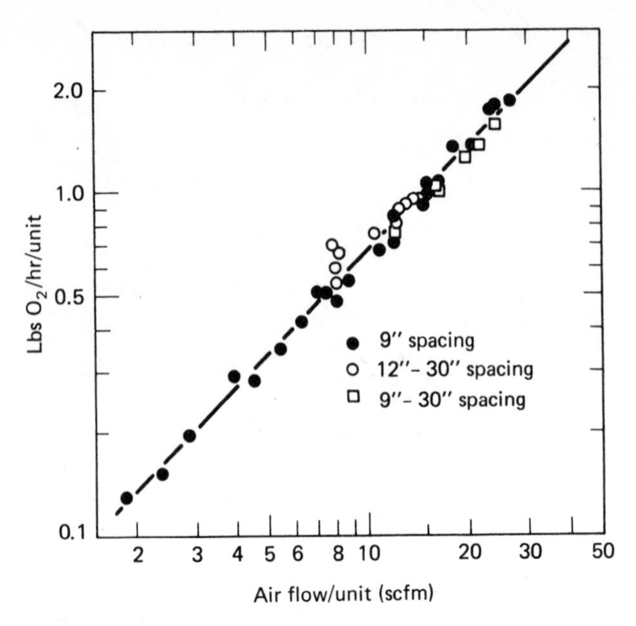

Figure 5-19. Effect of Air Flow on Oxygen Transfer from Sparjers in Water. (After Bewtra, 1964 and Barnhardt, 1965.)

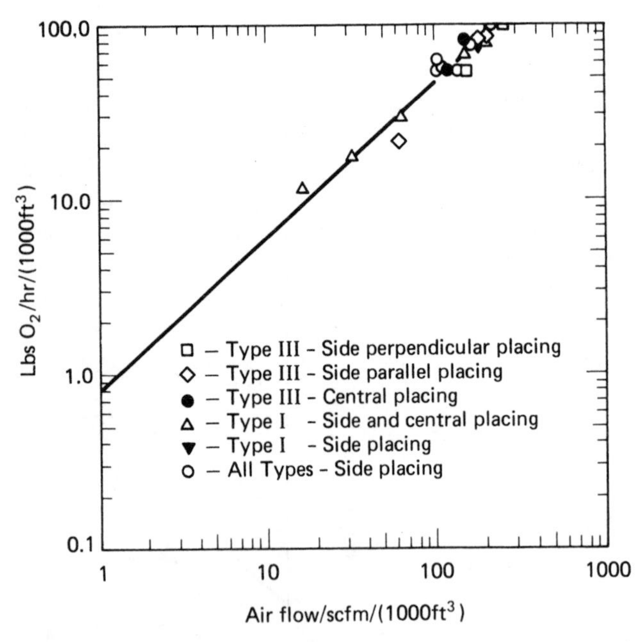

Figure 5-20. Effect of Air Flow on Oxygen Transfer from the INKA Aeration System. (After Ganczarczyk, 1964.)

treating municipal wastewater, will generally be between 5 and 8%. There is a definite relationship between the rate of oxygen consumption, air flow rate, and percentage oxygen absorbed. Morgan and Bewtra (1963) presented the data in Figure 5-21 and noted that the rapid oxygen-consuming sulfite reaction removed only 22% of the oxygen from air passing through water when the air flow rate was low, and this value dropped to 18% when the flow rate was increased. When an oxygen-consuming reaction is not present, the system is even more sensitive to air flow rate. Figure 5-21 suggests that as long as the oxygen demand is satisfied, the rate of air flow should be great enough to provide the required degree of mixing but no greater.

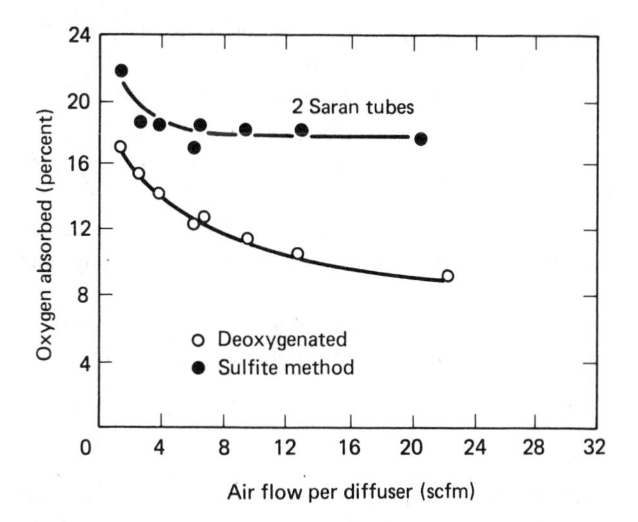

Figure 5-21. Comparison of Deoxygenated Water Method with Sulfite Method in Measuring Efficiency of Diffused Aeration with Saran Tubes. (After Morgan and Bewtra, 1963.)

It is common to express the rate of oxygen transfer in terms of pounds of oxygen transferred per hour per diffuser unit for diffused air systems. Equation 5-27 can be used to adjust the oxygen transfer rate under standard conditions to that value expected under process conditions. In this case the value of C_s given in the numerator of equation 5-27 must also reflect process conditions. Diffused aeration systems use oxygen under air pressure and studies have shown that the oxygen saturation concentration which should be used to evaluate these systems is very nearly the one-third depth value. To determine the oxygen saturation concentration for such a system, Stukenberg et al. (1977) recommend the use of the following equation:

$$C_{sa} = C_s\left(\frac{\text{Pb}}{59.84} + \frac{\text{O}_t}{42}\right) \tag{5-31}$$

where C_{sa} = oxygen saturation at the one-third depth, mg/ℓ
C_s = oxygen saturation at the liquid surface, mg/ℓ
Pb = air pressure at the point of release from diffuser, in. Hg
O_t = percentage of oxygen in the air leaving the liquid surface, %; usually it is assumed that between 6 and 10% of the oxygen is absorbed and that air initially contains 21% oxygen

Values for C_s and Pb can be determined from the relationships

$$C_s = \frac{P - \bar{p}}{760 - \bar{p}} \frac{475 - 2.65S}{33.5 + T_s} \qquad (5\text{-}32)$$

where P = prevailing barometric pressure, mm Hg
$\bar{p}$ = saturated water vapor pressure at standard conditions, mm Hg
T_s = wastewater temperature at standard conditions (i.e., at 20°C)
S = dissolved solids in the wastewater, g/ℓ

and

$$\text{Pb} = \left[\frac{H}{2.3} + \frac{P}{760}(14.7)\right](2.036) \qquad (5\text{-}33)$$

where H represents the liquid depth in feet at the point of bubble release. The constant 2.036 is the conversion from pounds per square inch to inches of mercury.

Once a value for C_{sa} has been determined, the oxygen transfer rate under process conditions is computed from the following modified form of equation 5-27:

$$(\text{T.R.})_{\text{actual}} = (\text{T.R.})_{\text{standard}}\alpha\frac{C_{sa} - C}{9.2} \qquad (5\text{-}34)$$

In this equation oxygen transfer rate, (T.R.), is expressed as pounds of oxygen transferred per hour per diffuser unit. C represents the desired minimum dissolved oxygen concentration to be maintained in the aeration tank and is usually taken as 2.0 mg/ℓ. The value of 9.2 in the denominator is the oxygen saturation concentration at standard conditions. No term for temperature adjustment is included because the C_{sa} value is computed from an expression that considers the temperature at 20°C, which is the same temperature at which K_La is determined.

The theoretical horsepower requirement for a diffused aeration system can be computed from a formula that describes adiabatic compression (Humenick, 1977):

$$\text{Thp} = 0.00436Q_1P_1\left(\frac{k}{k-1}\right)\left[\left(\frac{P_2}{P_1}\right)^{(k-1)/k} - 1\right] \qquad (5\text{-}35)$$

where Thp = theoretical compressor horsepower requirement
k = ratio of specific heats at constant pressure to constant volume; for adiabatic compression of diatomic molecules, k has a value near 1.395

P_1 = absolute inlet pressure, psia
P_2 = absolute outlet pressure, psia; this term can be estimated from the following expression when frictional losses are neglected:

$$P_2 = \frac{\gamma_w H}{144} + P_1 \tag{5-36}$$

γ_w = specific weight of water, lb/ft³. Typical values of γ_w are presented in Table 5-5.
H = liquid depth to point of bubble release, ft
Q_1 = air flow at the intake, cfm

TABLE 5-5

SPECIFIC WEIGHT OF WATER

Temperature (°C)	γ_w (*lb/ft*³)
0.0	62.42
4.4	62.43
10.0	62.41
15.5	62.37
21.1	62.30
26.6	62.22
32.2	62.11

Other useful expressions are

$$\frac{P_2}{P_1} = \left(\frac{T_2}{T_1}\right)^{k/(k-1)} \tag{5-37}$$

where T_1 = air temperature at the intake, degrees Rankin (i.e., 460 + °F)
T_2 = air temperature at the discharge point, °R

$$Q_2 = \frac{nRT_2}{P_2} \tag{5-38}$$

where Q_2 = rate of air flow at the discharge point, cfm
R = gas constant, which has a value of 10.73
n = moles of oxygen transferred, lb moles/min; this term can be estimated from the following expression where the value for oxygen absorbed is the same as that used to compute O_t:

$$n = \frac{\text{oxygen required (lb/day)}}{(32\text{ lb/lb-mole})(1440\text{ min/day})(0.21)(\text{fraction } O_2 \text{ absorbed})} \tag{5-39}$$

and

$$Q_1 = \frac{P_2 Q_2 T_1}{P_1 T_2} \tag{5-40}$$

Air flows (in cfm) delivered by a diffuser system can be converted to standard cubic feet per minute by applying the following formula:

$$Q_s = \frac{P_2}{P_0}\frac{T_0}{T_2}Q_2 \qquad (5\text{-}41)$$

where Q_s = air flow at discharge point, scfm
P_0 = standard pressure = 14.7 psi
T_0 = standard temperature = 32°F = 492°R

The brake horsepower (required horsepower input to compressor) can be obtained from the expression

$$\text{Bhp} = \frac{\text{Thp}}{e} \qquad (5\text{-}42)$$

where e = compressor efficiency, fraction; for centrifugal compressors and air flows greater than 15,000 cfm, values between 0.7 and 0.8 are used; for rotary positive displacement compressors and air flows less than 15,000 cfm, values between 0.67 and 0.74 are generally applied

When used in a biological process, the oxygen transfer rate of a diffused aeration system is usually between 1.5 and 2.5 lb oxygen transferred/blower hp/h.

Example 5-1

Design calculations for a completely mixed activated sludge process show that 32,000 lb/day of oxygen will be required. A diffused air system will be employed for aeration. The diffusers will be located 15 ft below the liquid surface, and it is estimated that frictional losses in the piping system are equivalent to 5 ft of water. System design will be based on an ambient air temperature of 75°F and prevailing pressure of 0.95 atm. The manufacturer states that the diffuser units to be used will transfer 1.5 lb O_2/h per diffuser unit when operating at an air flow of 10 scfm under standard conditions at a liquid depth of 15 ft. Studies on process effluent indicate that α has a value near 0.8 and the dissolved solids concentration of the liquid is 600 mg/ℓ. If the compressor efficiency is 0.7, determine the brake horsepower and the number of diffusers required.

solution

1/ Compute the one-third depth oxygen saturation concentration assuming that 8% of the oxygen is absorbed as the air bubbles pass through the aeration tank.

a/ Compute the oxygen saturation concentration at the liquid surface from equation 5-32.

$$C_s = \frac{722 - 17.5}{760 - 17.5}\,\frac{475 - (2.65)(0.6)}{33.5 + 20}$$

$$= 8.4 \text{ mg}/\ell$$

b/ Calculate the air pressure at the point of bubble release using equation 5-33.

$$\text{Pb} = \left[\frac{20}{2.3} + \left(\frac{722}{760}\right)14.7\right](2.036)$$

$$= 46.0 \text{ in. Hg}$$

or

Pb = 41.6 in. Hg when $H = 15$ is used; this value of H should be used in computing C_{sa} because this is the actual head at the point of bubble release

c/ Determine O_t by assuming that 8% of the oxygen passing through the aeration tank is absorbed.

$$O_t = (21\%)(1 - 0.08) = 19.3\%$$

d/ Applying equation 5-31, estimate the oxygen concentration at one-third depth.

$$C_{sa} = (8.4)\left(\frac{41.6}{59.84} + \frac{19.3}{42}\right)$$

$$= 9.7 \text{ mg}/\ell$$

2/ Adjust the oxygen transfer rate furnished by the manufacturer for standard conditions to process conditions. Assume that a minimum dissolved oxygen concentration of 2.0 mg/ℓ is required and that this level can be maintained under maximum loading conditions if the aeration system design is based on a DO residual of 3.5 mg/ℓ. Equation 5-34 is applicable in this calculation.

$$(\text{T.R.})_{\text{actual}} = (1.5)(0.8)\frac{9.7 - 3.5}{9.2}$$

$$= 0.81 \text{ lb } O_2/\text{h/diffuser unit}$$

3/ Estimate the number of diffusers required.

$$\text{number of diffusers} = \frac{32{,}000 \text{ lb/day}}{(0.81)(24 \text{ h/day})} = 1646$$

4/ Calculate the required air flow at the point of discharge.

a/ Estimate the required air flow under standard conditions.

$$Q_s = (10 \text{ scfm/unit})(1646 \text{ units}) = 16{,}460 \text{ scfm}$$

b/ Determine the air temperature at the discharge point using equation 5-37.

$$\frac{22.6}{13.9} = \left(\frac{T_2}{535}\right)^{1.395/0.395}$$

$$T_2 = 613°\text{R}$$

c/ Convert the standard rate of air flow at the discharge point to actual flow by applying equation 5-41.

$$16{,}460 = \left(\frac{22.6}{13.9}\right)\left(\frac{492}{613}\right)Q_2$$

$$Q_2 = 12{,}700 \text{ cfm}$$

5/ Applying equation 5-40, calculate the rate of air flow at the compressor intake.

$$Q_1 = \frac{(22.6)(12{,}700)(535)}{(13.9)(613)}$$

$$= 17{,}956 \text{ cfm}$$

6/ Compute the compressor brake horsepower.

a/ From equation 5-35, determine the theoretical horsepower requirement.

$$\text{Thp} = (0.00436)(17{,}956)(13.9)\frac{1.395}{1.395 - 1}\left[\left(\frac{22.6}{13.9}\right)^{0.28} - 1\right]$$

$$= 560$$

b/ Calculate the brake horsepower from equation 5-42.

$$\text{Bhp} = \frac{560}{0.7}$$

$$= 800$$

Submerged Turbine Aeration

In submerged turbine aeration, compressed air is discharged from spargers located beneath impellers which shear the bubbles into smaller ones which are dispersed throughout the aeration tank. Such an aeration system is presented in Figure 5-22.

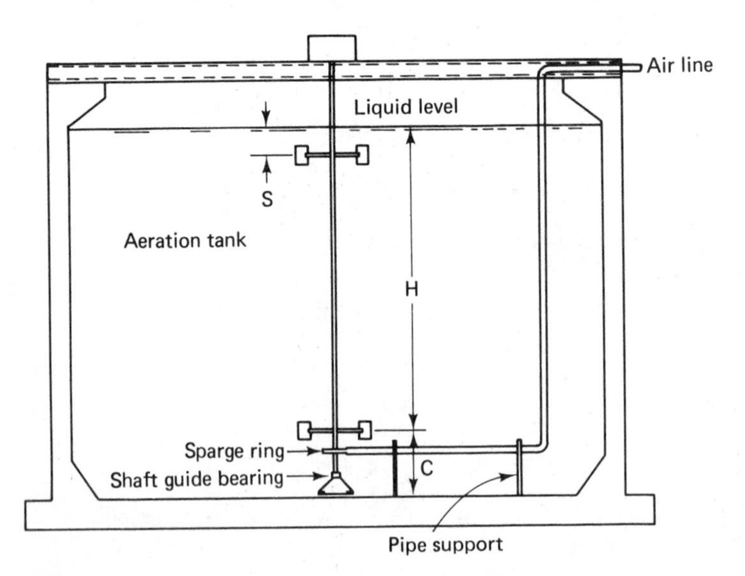

Figure 5-22. Typical Submerged Turbine Aeration System. (After Eckenfelder, 1966.)

Since the degree of mixing provided by this type of aeration unit is controlled by the power input to the turbines and is independent of the compressed air flow, there are no limitations to tank geometry, such as the width/depth ratio imposed upon a diffused aeration system. The turbine diameter to equivalent tank diameter generally varies from 0.1 to 0.2 for this type of system.

Eckenfelder and Ford (1968) have reported that the optimum oxygen transfer rate occurs when the ratio of turbine horsepower to compressor horsepower is approximately 1. Such data are presented in Figure 5-23. In this figure P_d is given by the following ratio:

$$P_d = \frac{(\text{hp})_T}{(\text{hp})_C} \tag{5-43}$$

where $(\text{hp})_T$ = turbine horsepower
$(\text{hp})_C$ = compressor horsepower

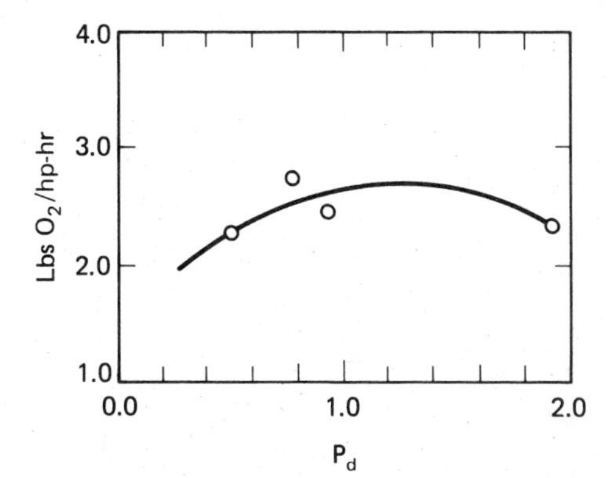

Figure 5-23. Effect of Turbine to Blower Horsepower Ratio on the Oxygen Transfer Rate of a Submerged Turbine Aeration System. (After Quirk, 1962.)

The oxygen transfer rate of a single impeller submerged turbine aeration device can be expected to range from 1.5 to 2.0 lb oxygen transferred/hp-h, whereas dual impeller turbines will provide from 2.5 to 3.0 lb O_2/hp-h.

Submerged turbines are especially suited for application where a large fluctuation is expected in the rate of oxygen utilization. For both diffuser and submerged turbine systems the oxygen transfer rate is varied by adjusting the compressor horsepower. For the turbine system this means a change in the P_d ratio. The engineer should, therefore, ensure that the anticipated operating range be within the maximum range of oxygen transfer rate as related to P_d (see Figure 5-23). Submerged turbine systems are also applicable where icing may prevent the use of surface aerators.

Example 5-2

A dual impeller submerged turbine aeration system is to be used in a completely mixed activated sludge process. The manufacturer states that the oxygen transfer rate for the units to be used is 2.5 lb O_2/hp-h under standard conditions. If the oxygen requirement is 15,000 lb/day, determine the total horsepower requirement. Assume that compressed air is to be released 20 ft below the liquid surface and that frictional losses can be neglected. System design is to be based on an ambient air temperature of 75°F, a prevailing pressure of 0.95 atm, an α value of 0.8, a dissolved solids concentration in the liquid of 600 mg/ℓ, and a compressor efficiency of 0.7.

solution

1/ Compute the one-third-depth oxygen saturation concentration assuming that 8% of the oxygen is absorbed as the air bubbles pass through the aeration tank.

Following steps 1a, 1b, 1c and 1d as outlined in Example 5-1, C_{sa} is found to be 10.3 mg/ℓ.

2/ Adjust the standard condition oxygen transfer rate to process conditions by assuming a residual DO of 2.0 mg/ℓ.

$$(\text{T.R.})_{\text{actual}} = (2.5)(0.8)\frac{10.3 - 2.0}{9.2}$$

$$= 1.8 \text{ lb } O_2/\text{hp-h}$$

3/ Compute the total horsepower requirement for the aeration system.

$$\text{total hp} = \frac{15{,}000 \text{ lb/day}}{(1.8 \text{ lb } O_2/\text{hp-h})(24 \text{ hr/day})}$$

$$= 347$$

For a proper design the horsepower split between the compressor and the turbine should be 1:1. This implies that the total horsepower requirement is 348, where 174 is compressor horsepower and 174 is turbine horsepower. The figure of 174 for the turbine is the operating horsepower. Since operational experience has shown that the operating horsepower is approximately 70% of the ungassed horsepower, the ungassed horsepower will be 174/0.7 = 249. Similarly, the required compressor horsepower is 174/0.7 = 249.

Surface Aerators

Earlier it was stated that there are basically two types of surface aerators: low speed and high or motor speed. The motor-speed aerator has no gear reducer between the motor and the impeller and is therefore cheaper. However, in many cases the mixing capability and oxygen transfer rate of these units are less than those of the low-speed units. Furthermore, as a result of the close proximity of the impeller and housing (see Figures 5-24 and 5-25), clogging is quite often a problem when the wastewater contains large amounts of suspended material. Because of this, the discussion in this section will relate mainly to low-speed aeration units such as those shown in Figures 5-26, 5-27, and 5-28.

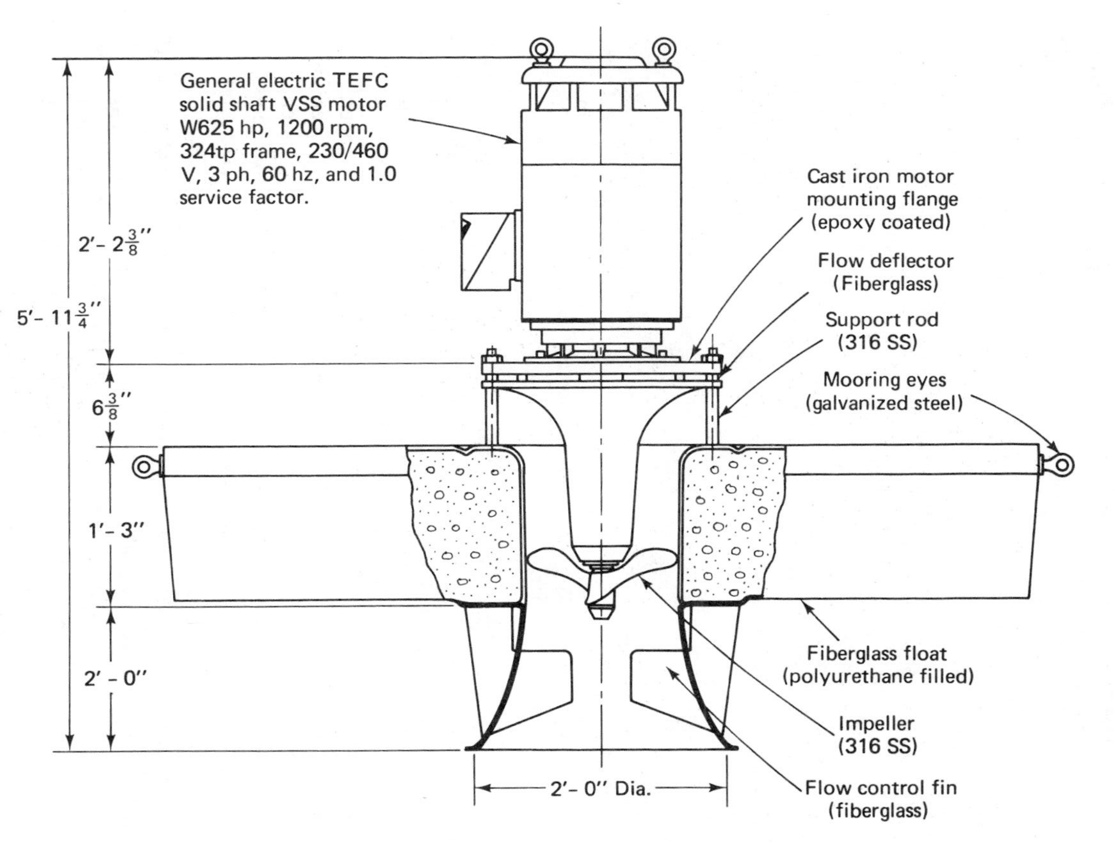

Figure 5-24. High-Speed Aerator. (Courtesy of CLOW Corporation.)

Figure 5-25. CLOW motor speed floating aerator. (Courtesy CLOW Corporation.)

The major difference between aeration accomplished by low-speed surface aerators and aeration by diffused air systems is that surface aeration occurs through point source oxygenation (Busch, 1968). This means that only liquid pumped through the aerator will be oxygenated. Eckenfelder and Ford (1967) have found that this is not the case for motor speed units and determined that up to 60% of the transfer was due to the surface impact of the pumped water for these systems.

Since low-speed surface aerators oxygenate the liquid by pumping, the volume of liquid under aeration is an important consideration. This relationship is usually expressed as horsepower per thousand gallons of volume. To illustrate this point, performance data for several types of surface aerators in water are presented in Figures 5-29 and 5-30. The average ranges of horsepower per thousand gallons (hp/1000 gal) are shown in Figure 5-31 for the activated sludge process, the extended aeration process, and for aerated lagoons.

The oxygen transfer rate of a surface aerator is expressed as pounds of oxygen transferred per horsepower per hour and is usually reported at standard conditions. To correct for process conditions, the following formula is used:

$$(\text{T.R.})_{\text{actual}} = (\text{T.R.})_{\text{standard}}\alpha \frac{C_s - C}{9.2} \tag{5-44}$$

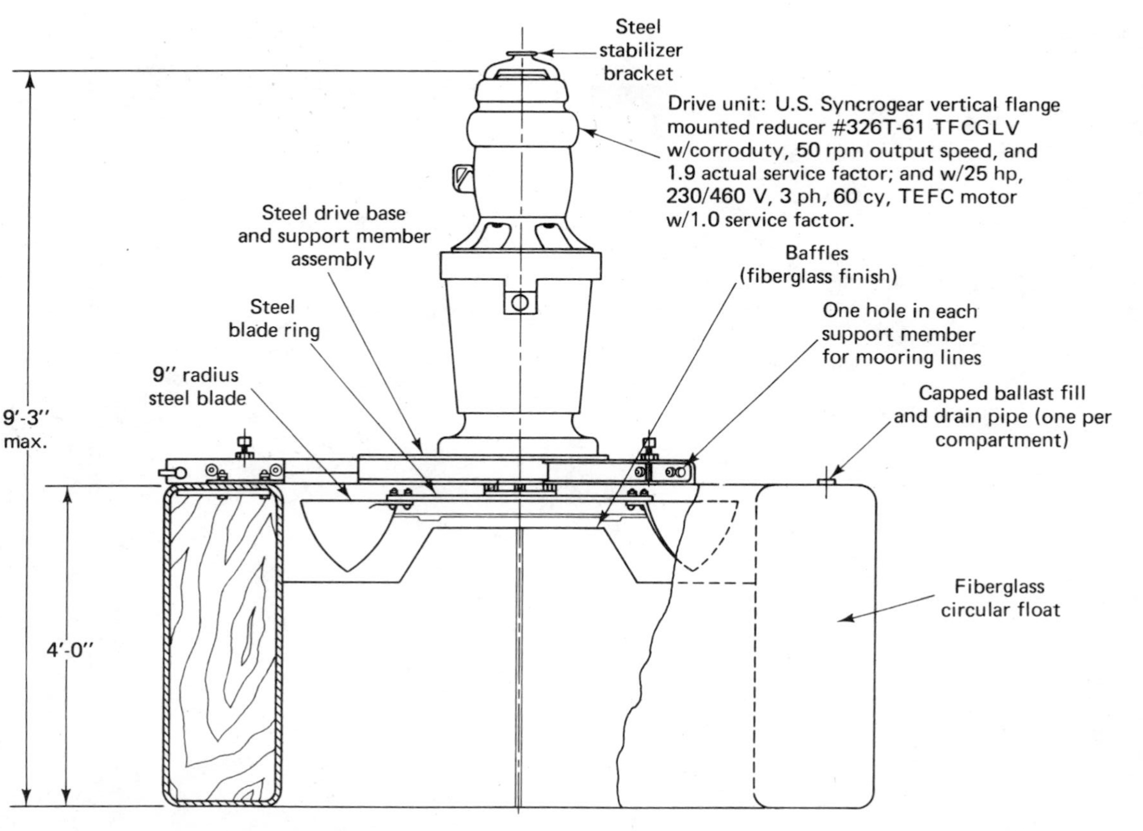

Figure 5-26. Sigma-Pac Floating Aerator. (Courtesy of CLOW Corporation.)

Figure 5-27. CLOW low speed floating aerator. (Courtesy CLOW Corporation.)

Figure 5-28. CLOW low speed floating aerator. (Courtesy CLOW Corporation.)

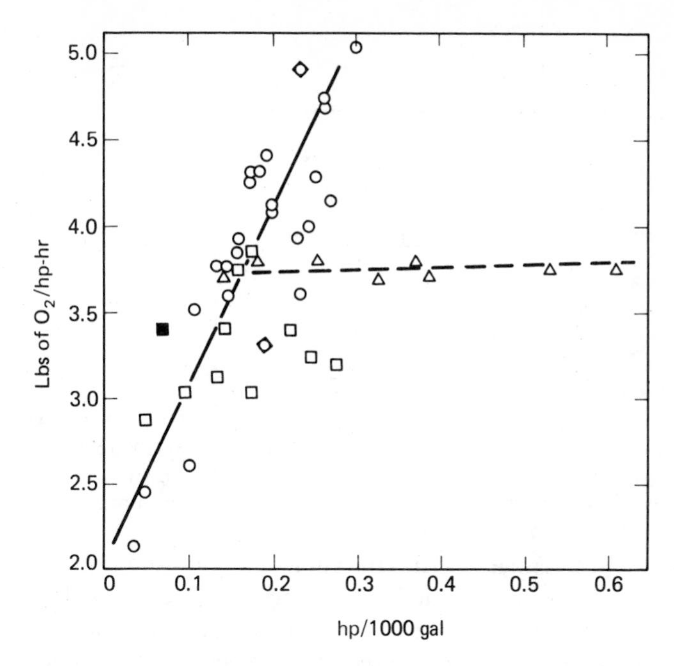

Figure 5-29. Performance Data for Several Types of Surface Aerators. (After Eckenfelder and Ford, 1967.)

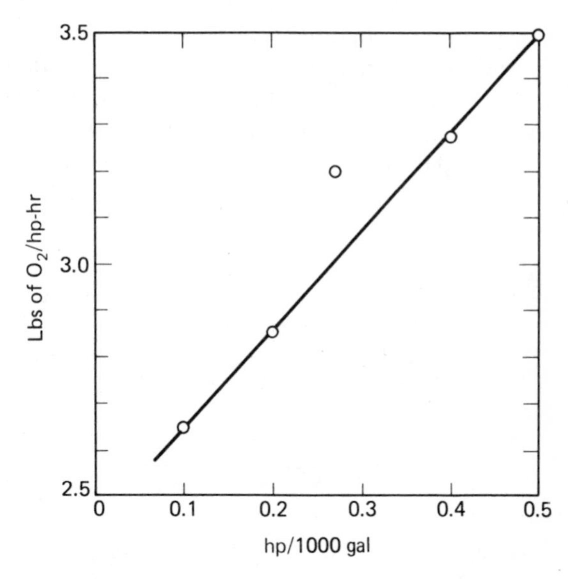

Figure 5-30. Surface Aeration Performance Data. (After Eckenfelder and Ford, 1967.)

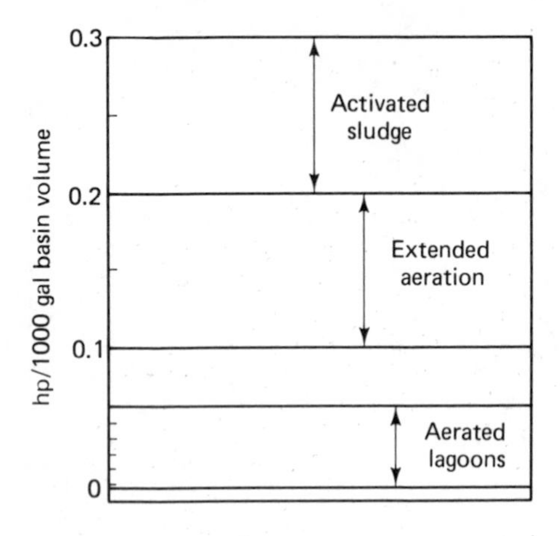

Figure 5-31. Average Ranges of hp/1000 gal for Different Types of Aeration Systems. (After Eckenfelder and Ford, 1967.)

where C_s is the oxygen saturation concentration at the surface of the liquid under process conditions and is computed from equation 5-32.

When adjusting the oxygen transfer rate for process conditions, it is also important to consider speed and submergence. Most surface aeration units have a speed and submergence that gives an optimum oxygen transfer rate. Such a relationship between speed and oxygenation capacity is illustrated in Figure 5-32. Most manufacturers report oxygen transfer rate on the basis of these optimum conditions.

The oxygen transfer rate of low-speed aerators range from 2.5 to 3.5 lb O_2/hp-h. During periods of fluctuation in the rate of oxygen utilization, the transfer rate can be varied by adjusting either the degree of submergence or the rotational speed. The latter requires the use of variable-speed motors.

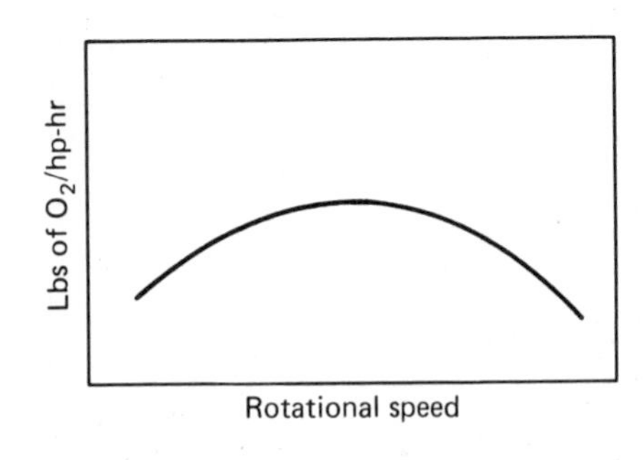

Figure 5-32. Effect of Rotational Speed on the Oxygen Transfer Rate of a Surface Aerator. (After Eckenfelder and Ford, 1967.)

Example 5-3

A surface aeration system is to be employed to satisfy an oxygen demand of 12,000 lb/day in an activated sludge plant. If a manufacturer states that a particular type of aerator will provide 3.0 lb O_2/hp-h at standard conditions, determine the total horsepower requirement for this system. System design is to be based on an ambient air temperature of 75°F, a prevailing pressure of 0.95 atm, an α value of 0.8, and a dissolved solids concentration in the liquid of 600 mg/ℓ. The total aeration tank volume is 1.2 million gal.

solution

1/ Compute the oxygen saturation concentration at the liquid surface. From step 1a in Example 5-1, a value of 8.4 mg/ℓ is determined.

2/ Adjust the standard condition oxygen transfer rate to process conditions by assuming a residual DO of 2.0 mg/ℓ.

$$(\text{T.R.})_{\text{actual}} = (3.0)(0.8)\frac{8.4 - 2.0}{9.2}$$

$$= 1.67 \text{ lb } O_2/\text{hp-h}$$

3/ Determine the total horsepower requirement.

$$\text{hp} = \frac{12{,}000 \text{ lb/day}}{(1.67 \text{ lb } O_2/\text{hp-h})(24 \text{ h/day})}$$

$$= 300$$

4/ Compute the power level (i.e., the horsepower input per 1000 gal of aeration tank volume) and check Figure 5-31 to ensure that the calculated value is a reasonable one for activated sludge plants.

$$\text{PL} = \frac{300}{1200} = 0.25$$

According to Figure 5-31, this is a typical value for activated sludge.

Mixing Considerations

Mixing is also an important consideration in the design of aeration systems. Good mixing is required to maintain the biomass in suspension as well as to distribute oxygen throughout the liquid volume. However, the degree of mixing required for oxygen dispersion is considerably less than that required for complete mixing. For a specific power input this means that the zone of oxygen dispersion will be larger than the zone of complete mixing. Such performance data are given in Table 5-6.

Since all aerators are low-head pumps, mixing can be related to pumping capacity. Busch (1968) reports that the pumping capacity of a submerged turbine aerator is between 1 and 10 cfs/hp, whereas the pumping capacity for diffused aeration and low-speed surface aeration is 4.7 and 4.5 cfs/hp, respectively. These values can be used to compute the theoretical turnover time for a given volume. Shell and Cassady (1973) suggest that a turnover time of 7.5 min or less is generally sufficient for complete mixing if the system is well baffled.

TABLE 5-6

TFNI AERATOR PERFORMANCE DATA
(AQUA-AEROBIC SYSTEMS, INC.)

Hp	N_C[a] ($lb\ O_2/hp\text{-}h$)	Z_{CM}[b] (ft)	D (ft)	Z_{OD}[c] (ft)	Q[d] (gpm)
20	3.2	72	10	230	8,320
25	3.4	80	10	255	9,830
30	3.5	88	10	280	12,570
40	3.8	102	10	325	14,000
50	3.5	105	12	330	18,560
60	3.5	115	12	350	20,560
75	3.0	130	12	380	22,550
100	3.1	150	12	440	41,000
125	3.3	165	12	490	47,500
150	3.2	185	12	530	57,000

[a]Transfer rate at standard conditions.
[b]Zone of complete mix.
[c]Zone of complete oxygen dispersion.
[d]Pumping rate through unit.

Example 5-4

For the surface aeration system described in Example 5-3, the manufacturer has furnished a pumping capacity figure of 4.5 cfs/hp. For this system, determine the theoretical turnover time. Is the resulting value within the range required for complete mixing?

solution

1/ Determine the theoretical turnover time.

$$\text{turnover time} = \frac{1{,}200{,}000\ \text{gal}}{4.5\,\dfrac{\text{ft}^3}{\text{sec-hp}} \times 0.25\,\dfrac{\text{hp}}{1000\ \text{gal}} \times 1200\ (1000\ \text{gal}) \times 60\,\dfrac{\text{sec}}{\text{min}} \times 7.5\,\dfrac{\text{gal}}{\text{ft}^3}}$$

$$= 2\ \text{min}$$

The turnover time is less than 7.5 min; therefore, complete mixing should occur.

PROBLEMS

5-1 In a research laboratory controlled at 15°C, testing of an aerator gave the following data:

Time (min)	0	6	12	24	30	40
DO (mg/ℓ)	1.5	3.5	5.1	7.2	7.9	8.5

What is K_La as determined by the research staff? Assume zero chloride concentration and standard conditions.

5-2 A completely mixed activated sludge plant is to be constructed completely enclosed so that an atmosphere of pure oxygen can be maintained over the mixed liquor. A surface aeration system is to be used for oxygenation. Before the plant was enclosed, the system was tested. During the test period it was determined that if a residual dissolved oxygen concentration of 2.0 mg/ℓ was maintained, the oxygen transfer rate was 2.0 lb O_2/hp-h for a prevailing atmospheric pressure of 760 mm Hg and dissolved solids concentration in the liquor of 750 mg/ℓ. If a prevailing pressure of 790 mm Hg is maintained, what will be the horsepower requirement if the residual DO level under pure oxygen conditions is to be increased to 6.0 mg/ℓ and the total oxygen demand is 15,000 lb/day? The operating temperature for both situations is 15°C.

5-3 A steady-state test was used to determine the overall oxygen transfer coefficient for a completely mixed activated sludge process. Two hours before the test was performed, flow was diverted from the aeration tank so that the oxygen utilization rate of the biomass would be due mainly to endogenous respiration. This rate was measured at 6 mg/ℓ-h. During the test period the average DO level in the aeration tank was 3.0 mg/ℓ. Compute K_La if the oxygen saturation concentration in the mixed liquor at the time of the test was 8.0 mg/ℓ.

5-4 It is possible to use a non-steady-state method to determine K_La by testing in the mixed liquor. This can be accomplished by turning down the aerators and allowing the DO level to drop to a low value. Aeration is then resumed and the change in DO concentration with time is recorded. To use these data, equation 5-28 is rearranged into the form

$$\frac{dC}{dt} = (K_LaC_s - R) - K_LaC$$

A plot is then made of oxygen concentration versus time which will give a curve of slope dC/dt. The value of dC/dt at any value of C can be determined by computing the slope of the tangent to the curve at that point. Using these data a plot of dC/dt versus C gives a linear trace with a slope equal to $-K_La$. If the saturation oxygen concentration is 8.4 mg/ℓ, determine K_La for the aeration system tested if the following data are applicable:

DO (mg/ℓ)	Time (min)
4.5	0.0
5.3	0.5
6.0	1.0
6.6	1.5
7.0	2.0

REFERENCES

BARNHART, E. B., Unpublished Master's Thesis, Sanitary Engineering, Manhattan College (1965).

BEWTRA, J. K., and W. R. NICHOLAS, "Oxygenation from Diffused Air in Aeration Tanks," *Journal of the Water Pollution Control Federation*, **63**, 1195 (Oct., 1964).

BUSCH, A. W., *Aerobic Biological Treatment of Wastewaters*, Olygodynamics Press, Houston, Tex., 1968.

ECKENFELDER, W. W., Jr., *Industrial Water Pollution Control*, McGraw-Hill Book Company, New York, 1966.

ECKENFELDER, W. W., Jr., and D. L. FORD, "Engineering Aspects of Surface Aeration Design," in *Proceedings, 22nd Industrial Waste Conference*, Purdue University, West Lafayette, Ind. 1967.

ECKENFELDER, W. W., Jr., and D. L. FORD, "New Concepts in Oxygen Transfer and Aeration," in *Advances in Water Quality Improvement*, ed. by E. F. Gloyna and W. W. Eckenfelder, University of Texas Press, Austin, Tex., 1968.

ECKENFELDER, W. W., Jr., and D. L. FORD, *Water Pollution Control: Experimental Procedures for Process Design*, Pemberton Press, Jenkins Publishing Company, Austin, Texas, 1970.

ECKENFELDER, W. W., Jr., and D. J. O'CONNOR, *Biological Waste Treatment*, Pergamon Press, New York, 1961.

FARKAS, P., "Methods for Measuring Aerobic Decomposition Activity of Activated Sludge in an Open System," In *Advances in Water Pollution Research*, **19**, II-309 (1966).

GANCZARCZYK, JERZY, "Some Features of Low Pressure Aeration," *Second International Conference on Water Pollution Research*, Tokyo, Japan (1964).

HUMENICK, M. J., Jr., *Water and Wastewater Treatment*, Marcel Dekker, Inc., New York, 1977.

LIPTÁK, B. G., *Environmental Engineers' Handbook, Volume 1*, Chilton Book Company, Radnor, Pa., 1974.

LISTER, A. R., and A. G., BOON, "Aeration in Deep Tanks: An Evaluation of a Fine-Bubble Diffused Air System," *British Journal of Water Pollution Control*, **72**, 590 (1973).

MANCY, K. H., and D. A. OKUN, "The Effects of Surface Active Agents on the Rate of Oxygen Transfer," in *Advances in Biological Waste Treatment*, ed. by W. W. Eckenfelder and J. McCabe, Pergamon Press, New York, 1960.

MARAIS, G. V. R., "Aeration Devices: Basic Theory," *British Journal of Water Pollution Control*, **74**, 172 (1975).

MORGAN, P. F. and J. K. BEWTRA, "Diffused Air Oxygen Transfer Efficiencies," in *Advances in Biological Waste Treatment*, ed. by W. W. Eckenfelder and J. McCabe, Pergamon Press, New York, 1963.

O'CONNOR, D. J., and W., DOBBINS, "The Mechanics of Reaeration in Natural Streams," *Journal of the Sanitary Engineering Division, ASCE*, **82**, SA6, (1956).

QUIRK, T. P., "Optimization of Gas-Liquid Contacting Systems," Unpublished Report, Quirk, Lawler and Matusky, Engineers, New York (1962).

RANDALL, C. W., and P. H. KING, Unpublished Consulting Report, Virginia Polytechnic Institute and State University, Blacksburg, Va. (1971).

SHELL, G., and T, CASSADY, "Selecting Mechanical Aerators," *Industrial Water Engineering*, July/Aug. 1973. 21

STUKENBERG, J. R., V. N., WAHBEH, and R. E., MCKINNEY, "Experiences in Evaluating and Specifying Aeration Equipment," *Journal of the Water Pollution Control Federation*, **49**, 66 (1977).

Treatment Ponds and Aerated Lagoons

Treatment ponds have been used to treat wastewater for many years, particularly as wastewater treatment systems for small communities. Since their beginning many terms have been used to describe the different types of systems employed in wastewater treatment. For example, in recent years *oxidation pond* has been widely used as a collective term for all types of ponds. Originally, an oxidation pond was a pond that received partially treated wastewater, whereas a pond that received raw wastewater was known as a *sewage lagoon*. *Waste stabilization pond* has been used as an all-inclusive term which refers to a pond or lagoon that is used to treat organic waste by biological and physical processes. These processes would commonly be referred to as self-purification if they took place in a stream. To avoid confusion, the classification to be employed in this discussion will be as follows (Caldwell et al., 1973):

1/ *Aerobic ponds:* shallow ponds, less than 3 ft in depth, where dissolved oxygen is maintained throughout the entire depth mainly by the action of photosynthesis.

2/ *Facultative ponds:* ponds 3 to 8 ft deep, which have an anaerobic lower zone, a facultative middle zone, and an aerobic upper zone maintained by photosynthesis and surface reaeration.

3/ *Anaerobic ponds:* deep ponds receiving high organic loadings such that anaerobic conditions prevail throughout the entire pond depth.

4/ *Maturation or tertiary ponds:* ponds used for polishing effluents from other biological processes. Dissolved oxygen is furnished through photosynthesis and surface reaeration. This type of pond is also known as a *polishing pond.*

5/ *Aerated lagoons:* ponds oxygenated through the action of surface or diffused air aeration.

6-1 Aerobic Ponds

The *aerobic pond* is a shallow pond in which light penetrates to the bottom, thereby maintaining active algal photosynthesis throughout the entire system. During the daylight hours large amounts of oxygen are supplied by the photosynthesis process, and wind mixing of the shallow water mass generally provides a high degree of surface reaeration during the hours of darkness. Stabilization of the organic material entering an aerobic pond is accomplished mainly through the action of aerobic bacteria.

In an aerobic pond bacteria and algae exist in a mutually beneficial or symbiotic relationship. Through the process of photosynthesis (Chapter 2), algae synthesize organic material from carbon dioxide, inorganic nutrients, and water using light energy, and then either form cytoplasm or excrete the organic compounds for subsequent heterotrophic utilization. In this process water is oxidized, resulting in the release of electrons, protons, and molecular oxygen (see Figure 2-13). Heterotrophic bacteria remove the organic material present in the wastewater and utilize this material in both energy and synthesis functions. That portion of the organic material channeled into the energy function is oxidized to carbon dioxide, water, and other inorganic forms. In these catabolic reactions of energy metabolism, oxygen produced during photosynthesis acts as an electron acceptor. The CO_2 and inorganic nutrients released as a result of bacterial metabolism are then utilized by the algae during their growth process. Hence, under normal light conditions the metabolic action of these two microbial groups complement each other (see Figure 6-1).

Diurnal Variations in Aerobic Ponds

Since photosynthesis requires solar radiation, oxygen evolved from this process only appears during daylight hours. At night the algae compete with the bacteria for both dissolved oxygen and organic compounds, which

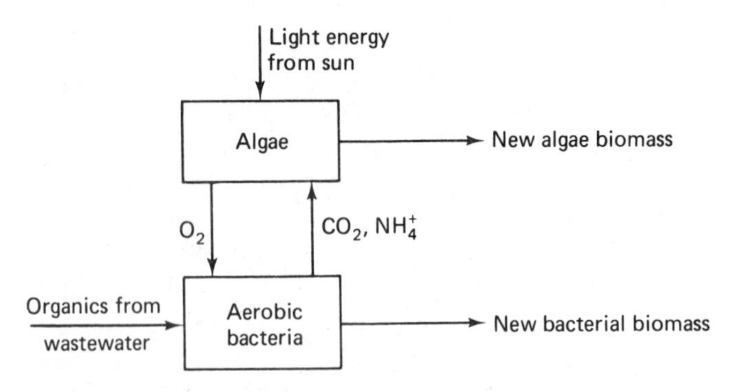

Figure 6-1. Algae/Bacteria Symbiotic Relationship in an Aerobic Pond.

depletes the oxygen budget. This results in a diurnal variation in dissolved oxygen concentration from values that may far exceed saturation during the day to much lower values or depletion at night.

Algae obtain carbon for photosynthesis from bacterial respiration and the carbonic acid system. Most of the algae indigenous to treatment ponds require inorganic carbon in the form of carbon dioxide. Thus, during the day when photosynthesis is occurring and CO_2 is being utilized in the synthesis reaction, the pH will increase. During this period significant amounts of ammonia may be released because equilibrium will shift to the gaseous state. At night, when CO_2 is produced from both bacterial and algal respiration, the pH will decrease. Therefore, as is the case with dissolved oxygen, a diurnal variation in pH also occurs. The diurnal variation of these parameters is important, as extreme values in either direction may be detrimental to microbial activity.

The heterotrophic respiration of algae is an interesting phenomenon that significantly complicates the modeling of the process. During photosynthesis the algae act as "organic factories" that spew large quantities of organics into the surrounding medium. When photosynthesis ceases, the algae metabolize the excreted or other organic compounds and form cellular mass while evolving CO_2. Then, with the return of light, large quantities of CO_2 are available for photosynthesis. Abeliovich and Weisman (1977) have shown that when *Scenedesmus obliquus* was grown under high-rate oxidation pond conditions, with glucose present as an organic carbon source, approximately 15% of the algal biomass carbon was derived from heterotrophic glucose metabolism. It was concluded, however, that algae plays but a minor role in wastewater BOD reduction in such a system.

Design Relationships

Because of the complex nature of the treatment system, no truly rational design procedure has been developed for aerobic ponds, but several empirical procedures have been proposed. Of the different empirical methods, the one developed by Oswald and Gotaas (1957) seems to have the most rational basis. Assuming that the contents of the pond are completely mixed and that no settling occurs, these workers developed an empirical method relating the efficiency of solar energy utilization to the surface area of the pond. It will be this method, as summarized by Rich (1963), which will form the basis of the design procedure for aerobic ponds presented in this section.

From the relationships proposed by Oswald and Gotaas (1957), Rich (1963) reports that an energy balance between algae mass produced and energy utilized may be expressed as

$$hW_a = ES_R A \tag{6-1}$$

where h = unit heat of combustion of algae cells, cal/g
W_a = algal biomass produced, g/day
E = efficiency of energy conversion, fraction
S_R = solar radiation incident on the pond surface, cal/cm^2-day, or langleys
A = surface area of pond, cm^2

Oxygen production and algal growth are related by the expression

$$W_{O_2} = pW_a \tag{6-2}$$

where W_{O_2} = oxygen production as a result of photosynthesis, g/day
p = oxygenation factor representing the oxygen produced per day per unit of biomass synthesized

An expression for the surface area of the pond can be obtained by substituting from equation 6-2 for the W_a term in equation 6-1 and solving for A.

$$A = \frac{hW_{O_2}}{pES_R} \tag{6-3}$$

The unit heat of combustion of algal cells is a variable term since it depends upon cell composition, which is affected by numerous environmental factors. Oswald and Gotaas (1957) relate the heat content of algal cells to an R-value:

$$h = \left(\frac{R}{7.89} + 0.4\right)1000 \tag{6-4}$$

R represents the degree of reduction of the cellular material formed as a result of the synthesis function and may be estimated from the expression

$$R = (100)\frac{(\%\ \text{carbon})(2.66) + (\%\ \text{hydrogen})(7.94) - (\%\ \text{oxygen})}{398.9} \tag{6-5}$$

Jewell and McCarty (1968) and Foree and McCarty (1968) have reported mean, minimum, and maximum values for the major elements found in algae cells. These values, expressed in percentage of ash-free dry weight are: carbon, 53, 42.9, 70.2; hydrogen, 8, 6.0, 10.5; oxygen, 31, 17.8, 34.0; nitrogen 8, 0.6, 16.0; and phosphorus, 2.0, 0.16, 5.0. As noted earlier, growth conditions in the system will affect each of these values.

The oxygenation factor depends on the composition of the algal biomass (organic material) synthesized during the photosynthetic process. As an example, Oswald and Gotaas (1957) assume that for a particular culture, the algae have the following cellular composition on an ash-free dry weight basis: 59.3% carbon, 5.24% hydrogen, 26.3% oxygen, and 9.1% nitrogen. The cell formula is determined by first dividing each percentage by the corresponding atomic weight. Therefore,

$$C = \frac{59.3}{12} = 4.94$$

$$H = \frac{5.24}{1} = 5.24$$

$$O = \frac{26.3}{16} = 1.64$$

$$N = \frac{9.1}{14} = 0.65$$

To avoid elemental fractions less than unity, each fraction is multiplied by 1.54 to increase the nitrogen fractional weight to 1.0.

$$C = (4.94)(1.54) = 7.6$$

$$H = (5.24)(1.54) = 8.1$$

$$O = (1.64)(1.54) = 2.5$$

$$N = (0.65)(1.54) = 1.0$$

The corresponding cellular structure is given by $C_{7.6}H_{8.1}O_{2.5}N$. Assuming that water, ammonia, and carbon dioxide are the sources of oxygen, nitrogen, and carbon, respectively, the overall photosynthetic process may be represented by the general expression

$$aCO_2 + (0.5b - 1.5d)H_2O + dNH_3 \longrightarrow C_aH_bO_cN_d + (a + 0.25b - 0.75d - 0.5c)O_2 \qquad (6\text{-}6)$$

or for this example

$$7.6CO_2 + 2.5H_2O + NH_3 \longrightarrow \underset{153.3}{C_{7.6}H_{8.1}O_{2.5}N} + \underset{(7.6 \times 32)}{7.6O_2} \qquad (6\text{-}7)$$

From equation 6-7, the oxygen released per unit of algal mass formed is $(243.2/153.3) = 1.58$. Because algal cells are generally composed of 85% volatile material and 15% fixed material, the oxygen yield per gram of TSS (ash included) is $(1.58)(0.85) = 1.34$. For most environmental conditions conducive to photosynthesis, values for p have been observed to range from 1.25 to 1.75.

Factors that affect the amount of solar radiation incident on a horizontal surface are (1) season of the year, (2) elevation, (3) geographical location, and (4) meteorological conditions. As noted in Chapter 2, the relatively small portion of the electromagnetic spectrum that can be seen is called the visible region, and it is this region of the spectrum that supplies the light energy for photosynthesis. Energy values related to the northern hemisphere for this region of the electromagnetic spectrum are presented in Table 6-1. These are ideal values which must be adjusted for elevation and cloudiness by applying the following correction factors:

Correction for cloudiness:

$$(S_R)_c = (S_R)_{\text{min}} + r[(S_R)_{\text{max}} - (S_R)_{\text{min}}] \tag{6-8}$$

Correction for elevation up to 10,000 ft:

$$(S_R)_{\text{design}} = (S_R)_c(1 + 0.001e) \tag{6-9}$$

In these equations e represents the elevation above sea level in feet and r the fraction of time the weather is clear, that is, (uncloudy daylight hours)/(total daylight hours). For a specified location, the total possible daylight hours for a particular month may be obtained from Figure 6-2. Weather Bureau records may be used to predict the degree of cloud cover to be expected for specific months of the year.

When light intensity is less than a certain saturation value (400 to 600 ft-candles for algae), the rate of photosynthesis is directly proportional to this intensity. Beyond the saturation point the rate remains constant until the light intensity reaches an inhibitory level (somewhere between 1000 and 4000 ft-candles). At this point the rate of photosynthesis decreases with increasing intensity. Such an effect is illustrated in Figure 6-3, where high, inhibitory intensities are realized at the surface and intensities much less than saturation are experienced at moderate depths below the surface. Oswald and Gotaas (1957) propose that the fraction of energy in visible light over and above that energy corresponding to the saturation point utilized in algal photosynthesis is given by

$$f = \frac{I_s}{I_0}\left[\ell\text{n}\left(\frac{I_0}{I_s}\right) + 1\right] \tag{6-10}$$

where f = fraction of available light utilized
I_0 = light intensity at the pond surface
I_s = saturation intensity

A graphical solution to equation 6-10 is provided in Figure 6-4. It must be understood, however, that direct application of equation 6-10 is limited because environmental factors other than light also limit growth (e.g., nutrient deficiency). Rich (1963) reports that the actual energy conversion efficiency, E, for aerobic ponds ranges between 0.02 and 0.09 and that 0.04 is an average value.

For the aerobic pond to function effectively as a treatment process,

TABLE 6-1

SOLAR RADIATION ON A HORIZONTAL SURFACE AT SEA LEVEL IN LANGLEYS PER DAY

NORTH LATITUDE		MONTH																							
		Jan		Feb		Mar		Apr		May		Jun		Jul		Aug		Sep		Oct		Nov		Dec	
Degree	Range	vis[c]	tot[d]	vis	tot	vis	tot	vis	tot	vis	tot	vis	tot	vis	tot	vis	tot	vis	tot	vis	tot	vis	tot	vis	tot
0	max[e]	255	685	266	700	271	708	266	690	249	645	236	626	238	630	252	666	269	690	265	694	256	683	253	667
	min[f]	210	580	219	583	206	536	188	462	182	480	103	274	137	368	167	432	207	533	203	530	202	543	195	527
2	max	250	670	263	693	271	706	267	697	253	655	241	642	244	646	255	673	269	693	262	688	251	666	249	646
	min	206	560	213	560	204	534	188	464	184	484	108	288	141	375	169	442	206	531	200	523	198	526	189	505
4	max	244	650	259	688	270	704	268	701	258	665	247	656	250	657	258	678	269	695	260	680	246	650	244	628
	min	200	540	206	543	202	532	187	466	187	492	113	300	146	385	171	448	204	529	196	513	194	510	183	480
6	max	238	630	254	675	268	702	270	705	262	675	252	668	255	669	261	683	269	697	256	670	240	634	238	610
	min	193	520	199	530	200	530	186	467	189	500	118	310	150	395	172	452	202	524	191	500	188	494	176	460
8	max	230	610	249	665	267	700	270	709	266	685	258	678	260	680	263	688	267	695	252	660	234	616	231	590
	min	187	495	192	510	196	523	185	467	191	506	124	320	154	405	174	456	200	518	186	486	182	478	169	440
10	max	223	595	244	655	264	694	271	711	270	694	262	688	265	690	266	693	266	693	248	650	228	600	225	570
	min	179	475	184	490	193	513	183	464	192	512	129	330	158	414	176	460	196	510	181	474	176	462	162	420
12	max	216	572	239	645	262	690	271	710	273	702	267	700	269	700	267	697	264	691	244	640	221	585	217	550
	min	172	455	176	470	189	500	181	462	193	518	133	343	161	421	176	464	193	502	176	462	169	446	154	400
14	max	208	555	233	630	258	680	271	709	276	710	272	710	273	708	269	700	262	688	240	627	214	567	209	536
	min	163	430	167	450	184	487	179	460	194	524	137	354	164	429	177	467	189	496	170	449	162	430	146	380
16	max	200	530	226	610	255	670	272	707	279	718	276	720	277	715	270	703	259	684	234	615	206	554	200	520
	min	154	400	159	430	180	473	177	456	194	528	141	363	167	435	177	469	185	489	164	434	154	410	138	360
18	max	192	515	220	590	250	664	272	705	282	723	280	728	280	723	272	705	256	680	229	605	198	538	192	500
	min	144	380	150	410	174	459	174	452	194	530	145	375	170	442	177	471	180	479	157	418	146	390	129	340
20	max	183	500	213	575	246	652	271	703	284	730	284	738	282	729	272	706	252	674	224	596	190	520	182	480
	min	134	360	140	390	168	440	170	447	194	532	148	383	172	450	177	472	176	467	150	400	138	370	120	320
22	max	174	480	206	560	241	644	270	701	286	734	286	747	285	736	273	707	248	668	218	582	183	500	172	460
	min	123	335	132	370	162	426	167	440	193	530	152	392	173	454	176	472	170	455	143	380	128	350	110	300
24	max	166	460	200	545	236	625	268	697	288	738	290	753	287	742	273	708	244	659	212	568	175	480	161	440
	min	111	310	123	340	156	410	164	433	191	525	155	403	176	459	174	471	165	443	136	360	119	326	101	280
26	max	156	440	192	530	230	615	266	690	288	741	292	760	288	749	273	706	240	652	205	552	166	460	149	420
	min	99	280	114	310	149	390	160	425	189	518	158	409	177	463	172	469	160	429	128	332	109	300	90	260
28	max	146	420	184	510	224	603	264	683	289	743	294	764	288	755	272	704	236	635	199	537	157	440	138	400
	min	87	250	106	290	142	373	156	415	187	506	161	418	178	467	169	466	154	415	120	310	99	278	80	236

Source: After Oswald and Gotaas (1957).

TABLE 6-1 (Continued)

NORTH LATITUDE		MONTH																							
Degree	Range	Jan		Feb		Mar		Apr		May		Jun		Jul		Aug		Sep		Oct		Nov		Dec	
		vis[c]	tot[d]	vis	tot	vis	tot	vis	tot	vis	tot	vis	tot	vis	tot	vis	tot	vis	tot	vis	tot	vis	tot	vis	tot
30	max	136	400	176	490	218	587	261	575	290	744	296	768	289	759	271	702	231	625	192	524	148	420	126	380
	min	76	220	96	260	134	362	151	405	184	490	163	425	178	469	166	462	147	399	113	290	90	256	70	210
32	max	126	380	169	470	212	570	258	663	290	744	296	772	289	761	269	700	226	615	185	510	138	400	114	360
	min	63	180	87	240	126	340	146	395	181	475	166	431	178	472	163	458	140	385	104	270	80	224	60	184
34	max	114	360	160	450	204	553	254	657	290	743	297	775	289	763	267	696	221	602	178	490	128	380	101	338
	min	53	155	78	215	118	320	141	385	176	462	168	439	178	472	159	448	134	368	96	250	70	202	47	158
36	max	103	335	150	430	196	538	250	650	288	741	298	776	289	765	264	690	215	590	170	470	118	360	88	314
	min	44	133	70	200	111	300	136	375	172	444	170	443	177	470	155	438	127	350	88	230	60	180	39	134
38	max	90	310	140	415	189	520	246	640	287	738	298	778	288	766	262	684	210	576	162	450	106	336	77	290
	min	36	120	62	180	103	280	131	365	166	428	171	448	175	464	152	429	120	330	80	216	50	158	30	111
40	max	80	280	130	390	181	500	241	630	286	732	298	778	288	765	258	680	203	562	152	430	95	313	66	270
	min	30	105	53	160	95	270	125	355	162	415	173	450	172	455	147	416	112	310	72	202	42	134	24	94
42	max	68	255	119	370	172	485	236	618	283	728	298	777	287	761	254	670	196	547	144	410	84	289	56	244
	min	24	90	45	140	88	250	120	344	157	405	174	451	167	442	143	403	105	290	65	187	34	112	19	78
44	max	55	228	106	340	165	470	230	607	280	722	298	777	285	755	250	660	189	530	132	390	72	263	47	218
	min	20	80	37	130	80	230	114	325	153	395	175	453	164	430	139	389	98	270	58	173	28	98	15	62
46	max	45	200	94	315	156	450	224	598	278	716	298	776	284	749	245	650	181	512	122	370	61	238	39	194
	min	16	74	30	110	72	210	108	315	150	385	175	455	161	420	134	374	90	250	52	158	23	86	11	48
48	max	35	180	82	290	149	430	218	582	274	710	297	776	282	740	241	640	174	496	111	350	50	210	32	170
	min	12	64	25	99	64	190	102	307	146	378	176	458	158	410	129	358	81	230	45	144	18	70	9	37
50	max	28	164	70	265	141	410	210	568	271	703	297	776	280	733	236	625	166	480	100	329	40	183	26	144
	min	10	54	19	80	58	173	97	300	144	371	176	458	155	403	125	342	73	210	40	130	15	60	7	30
52	max	22	140	60	240	134	390	202	555	267	695	296	776	278	725	232	615	158	460	87	307	32	160	21	121
	min	8	45	14	62	51	158	92	295	141	366	176	460	153	398	120	326	65	190	34	120	12	53	4	27
54	max	16	120	50	215	126	370	194	542	263	687	296	776	276	720	224	602	150	440	76	285	25	140	16	99
	min	6	40	11	45	46	145	88	289	139	360	176	460	150	394	116	312	58	170	29	106	9	43	3	23
56	max	12	102	43	200	120	350	188	528	258	680	295	775	273	714	218	587	141	420	64	261	20	120	12	78
	min	4	35	8	35	41	132	85	283	136	352	175	460	148	390	110	297	51	150	24	95	7	36	2	18
58	max	9	80	37	170	113	330	182	516	254	670	294	774	270	710	212	575	134	402						
	min	3	28	6	28	37	118	82	277	134	346	175	460	146	385	106	285	44	132						
60	max	7	64	32	150	107	310	176	500	249	660	294	773	268	708	205	556	126	386						
	min	2	20	4	20	33	105	79	270	132	340	174	460	144	380	100	270	38	116						

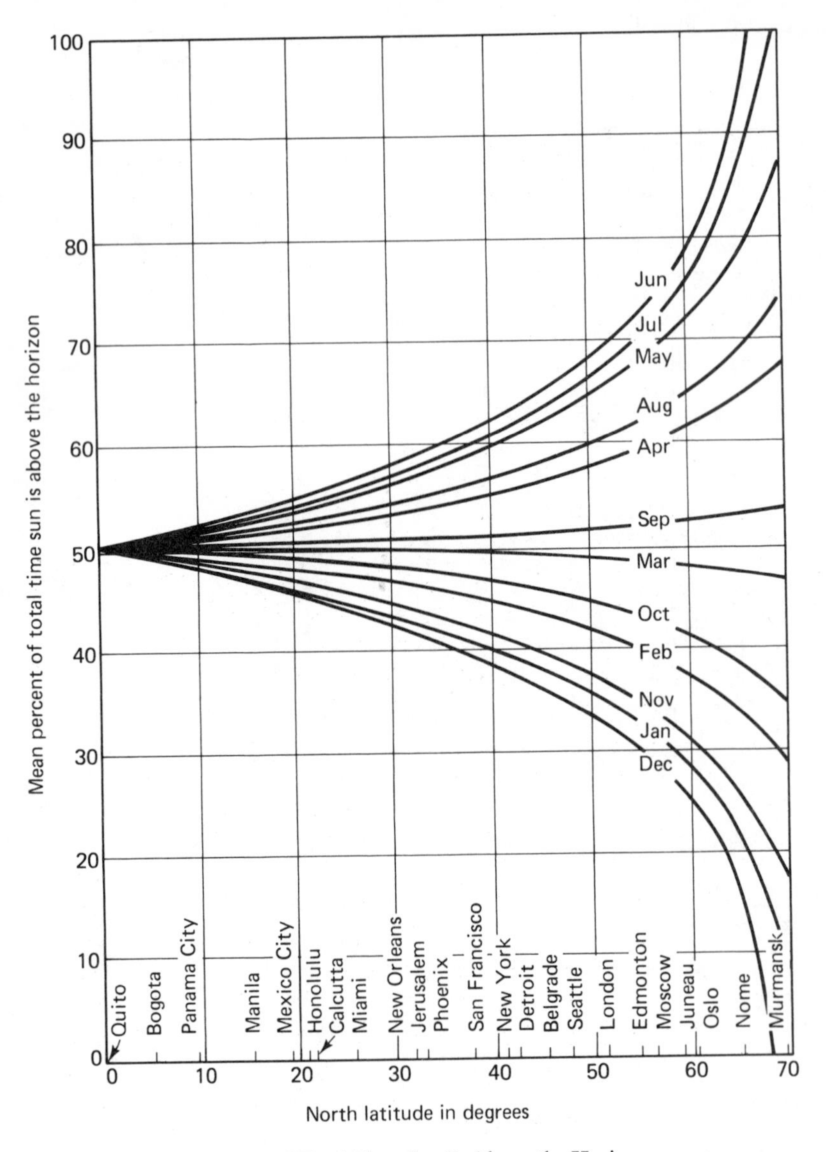

Figure 6-2. Mean Percent of Total Time Sun Is Above the Horizon at a Specified Latitude. (After Oswald and Gotaas, 1957.)

ignoring algal heterotrophic metabolism, oxygen must be provided at the same rate at which the bacteria utilize it in energy metabolism. Therefore, W_{O_2} can be approximated as the ultimate BOD removed per unit time.

Applying empirical formulations proposed by Oswald (1963), Roesler and Preul (1970) developed the following expression for BOD removal in

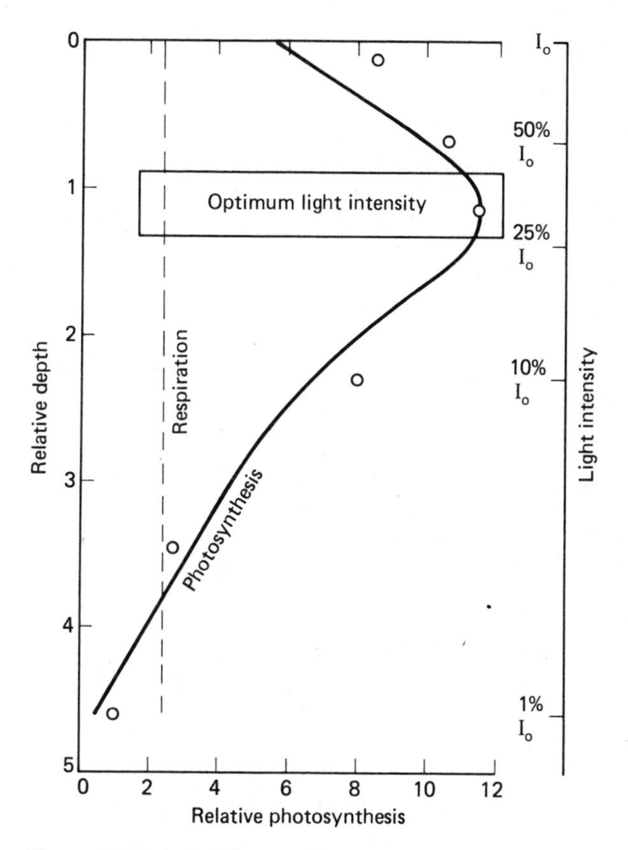

Figure 6-3. Relative Photosynthesis from Surface to Depth of Penetration of 1% Surface Light. (After Bartsch, 1961.)

an aerobic pond:

$$(\mathrm{BOD}_u)_r = \frac{32.808\ \ell\mathrm{n}\ (I_0/24)}{d} \tag{6-11}$$

where $(\mathrm{BOD}_u)_r$ = ultimate BOD removed, mg/ℓ

d = aerobic depth or depth of aerobic pond, ft

I_0 = light intensity at pond surface, ft-C

To apply equation 6-11 in design, it is necessary that a value of I_0 be obtained. Oswald and Gotaas (1957) suggest the following procedure for making this determination:

1/ Select the appropriate values of $(S_R)_{\max}$ and $(S_R)_{\min}$, based on total solar radiation, from Table 6-1.

2/ Correct the total solar radiation value for cloudiness and elevation by applying equations 6-8 and 6-9.

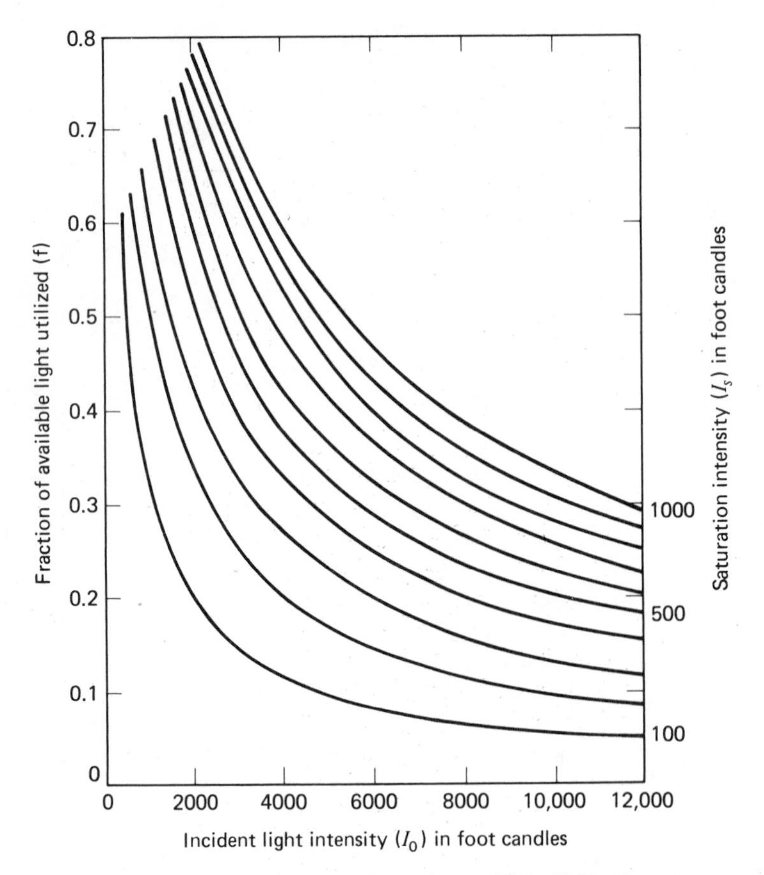

Figure 6-4. Influence of Saturation Intensity on Light Utilization by Algae. (After Oswald and Gotaas, 1957.)

3/ Multiply the corrected solar radiation value by 10.

4/ Multiply the value obtained in step 3 by the fraction of the time the sun is above the horizon as given by Figure 6-2. The resulting value is I_0 expressed as ft-candles.

Design Considerations

Typical design criteria for aerobic ponds are presented in Table 6-2. The shallow depth (between 0.5 and 1.5 ft) is required as these ponds are designed to maximize the production of algae, which requires that light penetrate the entire pond depth. However, Nusbaum (1957) reports several serious operational problems which may result because of a shallow pond depth.

TABLE 6-2

DESIGN CRITERIA FOR AEROBIC PONDS

Parameter	*Value*
Depth (ft)	0.5–1.5
Retention time (days)	2–6
BOD_u loading (lb/acre-day)	100–200
BOD_u removal (%)	80–95
Algae concentration (mg/ℓ)	100–200
Recirculation ratio	0.2–2.0
Effluent suspended solids concentration (mg/ℓ)	150–350

These are:

1/ Nuisance vegetation: depths less than 3 ft do not prevent the emergence of aquatic vegetation, which creates an excellent breeding ground for mosquitoes.

2/ Inhibitory temperature: in certain areas of the country, pond temperatures may become high enough during the summer months to inhibit the growth of certain algae.

3/ Oxygen retention: greater depths permit the retention of larger amounts of oxygen during times of supersaturation.

4/ Shock loads: greater depths provide a larger volume, which permits a more rapid dispersion of the incoming wastewater.

To achieve the best performance, the pond contents should be mixed periodically. Thermal stratification will occur in the absence of mixing. In this situation high temperatures exist near the surface and cooler temperatures near the pond bottom. The density of water decreases with an increase in temperature. Thus, at the pond surface the liquid density is reduced with a rise in temperature. Because of this, the nonmotile algae will settle to some depth below the surface. Furthermore, motile algae will move away from the high temperature at the pond surface and form a dense layer at some greater depth. This dense algae layer will prevent light from penetrating the entire depth of the pond, thereby reducing the number of algae in the photic zone (zone of light penetration). As a result, oxygen production and hence organic utilization will be reduced. Wind is probably the most important factor that influences mixing within a pond. Eckenfelder (1970) suggests that an unobstructed contact length (fetch) of about 650 ft is required to mix a pond 3 ft in depth.

If primary treatment is not provided, large quantities of solids will settle to the bottom of the pond and form a sludge layer. In deeper ponds methane fermentation generally prevents the sludge layer from becoming excessive. However, sludge buildup is a problem in aerobic ponds because methanogenic organisms which are strict anaerobes are normally unable to establish themselves to any significant extent. Even with primary treatment, sludge buildup will occur, but in this case the rate will be somewhat slower.

Oswald and Gotaas (1957) indicate that the recirculation of pond effluent is important in the operation of aerobic ponds because it permits seeding of the influent wastewater with algae cells. The high DO concentration in the recycle stream also serves to increase the DO concentration of the raw wastewater after the two streams are mixed. Nusbaum (1957) suggests a minimum recirculation ratio of 0.5 when primary effluent is to be treated.

Metcalf and Eddy (1972) recommend that individual aerobic ponds be less than 10 acres in size to minimize short-circuiting caused by wind action. For design situations where surface areas greater than 10 acres are required, parallel systems should be used where each pond in the treatment scheme has a surface area less than 10 acres.

Typically, aerobic ponds have length/width ratios of 2 to 3:1 (Mara, 1976). Pond embankments should be constructed with maximum and minimum slopes of 3:1 and 6:1, respectively (see Figure 6-5, where $n = 3$ for maximum slope and $n = 6$ for minimum slope). It is generally considered that surface areas calculated for design are actually middepth areas (i.e., the surface area that would exist if the pond had vertical sides).

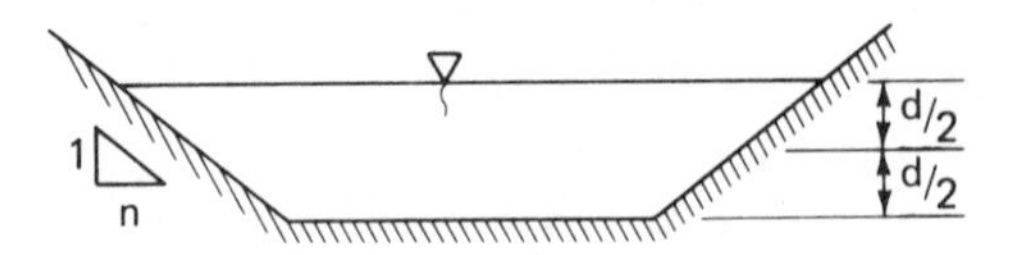

Figure 6-5. Cross Section of Typical Aerobic Pond. (After Mara, 1976.)

Climatic conditions are a major consideration in pond design. In areas where freezing conditions persist for long periods during the winter months, pond volumes should be sufficient to store the total flow during this period. In such areas, ponds function primarily during the summer months and are generally biologically dormant during the winter. According to Oswald (1972), the use of ponds that are designed for year-round continuous flow should be restricted to areas where the visible solar radiation is greater than 100 cal/cm^2/day 90% of the time and freezing conditions never persist for any significant length of time. Such conditons are satisfied for approximately 40% of North America. Figure 6-6 is presented as an aid to locating the specific areas of the United States where these conditions are met.

As an aerobic pond is designed to maximize algal production, the effluent from such a pond contains a high concentration of algae cells. Effluent limitations required for secondary treatment are 30 mg/ℓ or less for both

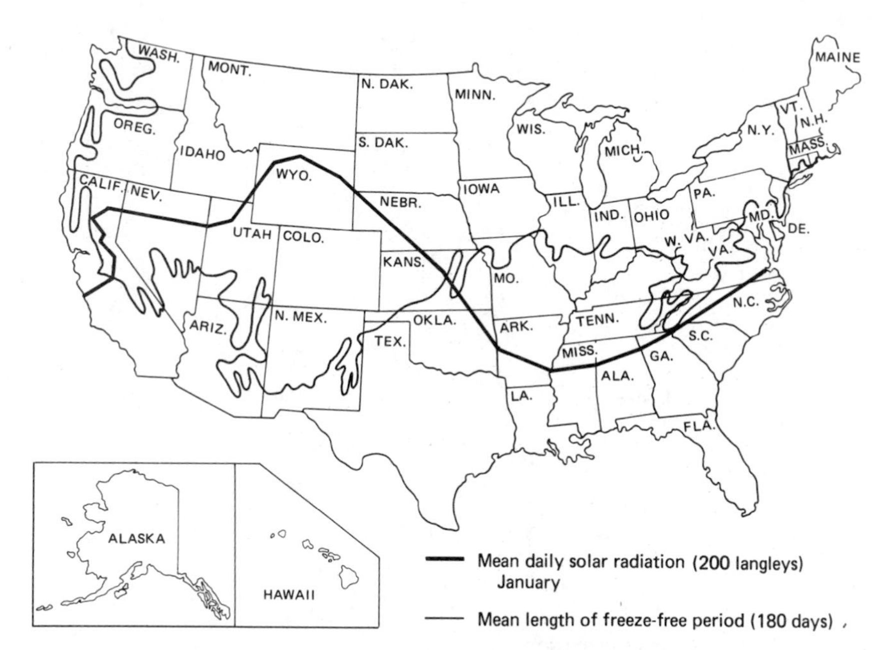

Figure 6-6. Climatic Conditions for Various Parts of the United States. (After Dildane and Franzmathes, 1970.)

BOD_5 and suspended solids on a 30-day average (or at least 85% removal, whichever is more stringent) and 45 mg/ℓ or less for both BOD_5 and suspended solids on a 7-day average. Most ponds cannot comply with the suspended solids limitation because of algae. As a consequence, supplemental treatment of the pond effluent for algae removal will be required. Such treatment is costly and generally requires operational skills beyond that available in small communities.

Example 6-1

After primary treatment the wastewater flow of 1.0 MGD from a small community (elevation 50 ft) near New Orleans, Louisiana, is expected to have an ultimate BOD of 200 mg/ℓ. If the critical design month is December when cloud cover exists 50% of the time and the temperature averages 45°F, what size aerobic pond is required for 90% BOD removal?

solution

1/ Compute the required oxygen production, W_{O_2}, in grams per day.

$$W_{O_2} = (0.9)(200 \text{ mg}/\ell)(8.34\ \ell\text{-lb/mg-MG})(1 \text{ MGD})(454 \text{ g/lb})$$
$$= 681{,}545 \text{ g/day}$$

2/ Assuming a mean composition of 53% carbon, 8% hydrogen, 31% oxygen, and 8% nitrogen, determine the molecular formula for the algae cells.

$$C = \frac{53}{12} = 4.42$$

$$H = \frac{8}{1} = 8$$

$$O = \frac{31}{16} = 1.94$$

$$N = \frac{8}{14} = 0.57$$

To avoid elemental fractions less than unity, multiply each term by 1/0.57 = 1.75.

$$C = (4.42)(1.75) = 7.7$$

$$H = (8)(1.75) = 14$$

$$O = (1.94)(1.75) = 3.4$$

$$N = (0.57)(1.75) = 1.0$$

The molecular formula is, therefore, $C_{7.7}H_{14}O_{3.4}N$.

3/ Compute the oxygenation factor, p.

$$p = \frac{(a + 0.25b - 0.75d - 0.5c)O_2}{C_aH_bO_cN_d}$$

$$= \frac{(8.15)(32)}{178.4}$$

$$= 1.46$$

4/ Determine the R-value by applying equation 6-5.

$$R = 100\left[\frac{(53)(2.66) + (8)(7.94) - 31}{398.9}\right]$$

$$= 43.5$$

5/ Calculate the unit heat of combustion, h, through the use of equation 6-4.

$$h = \left(\frac{43.5}{7.89} + 0.4\right)1000$$

$$= 5900 \text{ cal/g}$$

6/ Select $(S_R)_{max}$ and $(S_R)_{min}$ from Table 6-1 and correct for cloudiness and elevation by applying equations 6-8 and 6-9, respectively. Figure 6-2 shows that New Orleans is located at 30° north latitude. For the month of December at 30° north latitude, Table 6-1 gives the maximum and minimum visible solar radiation values as 126 and 70 langleys, respectively.

Correct for cloudiness: For the month of December, $r = 0.5$.

Therefore,

$$(S_R)_c = 70 + 0.5(126 - 70)$$
$$= 98$$

Correct for elevation:

$$(S_R)_{\text{design}} = 98[1 + 0.001(50)]$$
$$= 103$$

7/ Assume an energy utilization efficiency, E, of 0.04 and compute the required pond area from equation 6-3.

$$A = \frac{(5900)(681{,}545)}{(1.46)(0.04)(103)}$$
$$= 668{,}492{,}000 \text{ cm}^2$$

or

$$A = \frac{668{,}492{,}000}{929} = 720{,}000 \text{ ft}^2$$

or

$$A = \frac{720{,}000}{43{,}560} = 16.5 \text{ acres}$$

8/ Compute the light intensity incident on the pond surface.

For the month of December at 30° north latitude, Table 6-1 gives the maximum and minimum total solar radiation values as 380 and 210 langleys, respectively.

Correct for cloudiness:

$$(S_R)_c = 210 + 0.5(380 - 210)$$
$$= 295$$

Correct for elevation:

$$(S_R)_{\text{total}} = 295[1 + (0.001)(50)]$$
$$= 310$$

Therefore, the light intensity, I_0, is given by

$$I_0 = (310)(10)(0.43)$$
$$= 1333 \text{ ft-candles}$$

9. Estimate the required pond depth by rearranging equation 6-11.

$$d = \frac{32.808[\ell\text{n}\,(I_0/24)]}{(\text{BOD}_u)_r}$$
$$= \frac{(32.808)[\ell\text{n}\,(1333/24)]}{(0.9)(200)}$$
$$= 0.73 \text{ ft}$$

Pond depth is inversely related to the strength of the wastewater to be treated. Strong wastewaters requiring dense algal growths must be

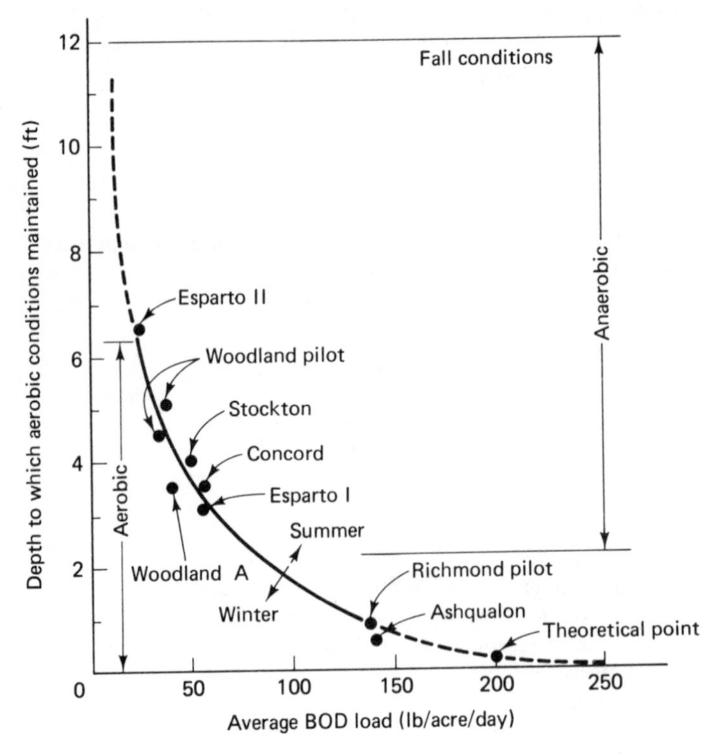

Figure 6-7. Relationship Between Depth of Aerobic Zone and BOD_u Loading. (After Oswald, 1968.)

treated in relatively shallow ponds, whereas weak wastewaters may be treated at greater depths.

The relationship between aerobic depth and BOD_u loading (lb/acre-day) presented in Figure 6-7 was developed by Oswald (1968) for fall conditions. This curve can be used to check computed depths to insure aerobic conditions will prevail. The curve shows that the aerobic depth for winter will be less than that for the fall, while the summer depth will be greater.

6-2
Facultative Ponds

Of the five general classes of lagoons and ponds listed, *facultative ponds* are by far the most common type selected as wastewater treatment systems for small communities. Approximately 25% of the municipal wastewater treatment plants in this country are ponds and about 90% of these ponds are located in communities of 5000 people or less. Facultative ponds are popular for such treatment situations because (1) long retention times provide the ability to handle large fluctuations in wastewater flow and strength with no significant effect on effluent quality, and (2) capital cost and operating and maintenance cost are less than that of other biological systems which would provide equivalent treatment.

A schematic representation of a facultative pond operation is given in Figure 6-8. The raw wastewater enters at one end of the pond. Suspended solids contained in the wastewater settle to the pond bottom, where an anaerobic layer develops. Microorganisms occupying this region do not require molecular oxygen as an electron acceptor in energy metabolism but rather use some other chemical species. Both acid fermentation and methane fermentation occur in the bottom sludge deposits.

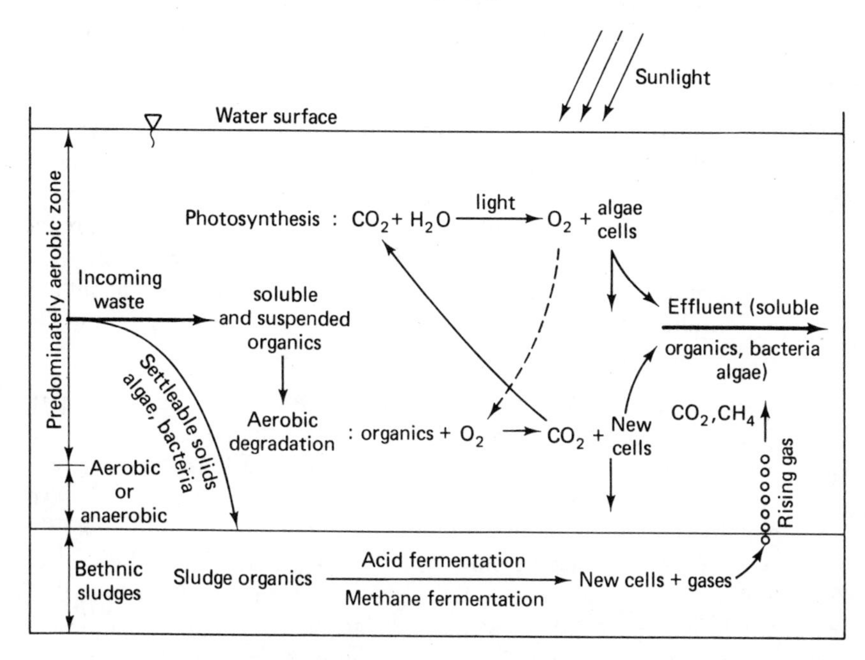

Figure 6-8. Schematic of Facultative Pond Showing the Basic Biological Reactions. (After Hendricks and Pote, 1974.)

The facultative zone exists just above the anaerobic zone. This means that molecular oxygen will not be available in the region at all times. Generally, the zone is aerobic during the daylight hours and anaerobic during the hours of darkness.

Above the facultative zone there exists an aerobic zone which has molecular oxygen present at all times. The oxygen is supplied from two sources.

A limited amount is supplied from diffusion across the air/liquid interface (i.e., the pond surface). However, the majority is supplied through the action of photosynthesis.

Design Relationships

For the design of a facultative pond Gloyna (1972) suggests the following formulation:

$$V = CQS_0\theta^{35-T}ff' \tag{6-12}$$

where V = pond volume, acre-ft

Q = wastewater flow, gal/day

S_0 = influent BOD, mg/ℓ; BOD_5 should be used for weak or pretreated wastewater and ultimate BOD should be used for strong or untreated wastewater.

θ = temperature coefficient; for this parameter Gloyna recommends a value of 1.085

T = design temperature, °C

f = algal toxicity factor to account for the reduction in synthesis of chlorophyll pigment by green algae in the presence of certain industrial organic compounds; Table 6-3 gives the concentration of certain compounds that will affect a 50% chlorophyll reduction; in the absence of such compounds, f will equal unity

f' = sulfide correction factor, which is equal to 1.0 when the influent sulfate concentration is less than 500 mg/ℓ

C = conversion coefficient

In locations where temperature fluctuations are minimal (such as in the tropics), a design depth of 3.5 ft is selected with a numerical value of the coefficient C to be used in design of 5.37×10^{-8}. In locations where temperature fluctuations are large, the design depth selected is 6 ft, while C is given a value of 10.7×10^{-8}.

Equation 6-12 is applicable only for 80 to 90% BOD removal efficiency. The equation contains no correction for the required pond volume for BOD removal efficiencies outside this range. Solar energy for photosynthesis is assumed to be always available above the saturation level. However, this assumption may not be valid in the months of November to February in locations above about 40° north latitude, where ice and snow cover reduce the percent of light transmission.

To understand fully the limitations of equation 6-12, it is necessary to consider the basis upon which it was developed. In 1958 Hermann and Gloyna reported that the optimum temperature required (based on a minimum retention time requirement) for 80 to 90% BOD_5 removal from domestic wastewater with a BOD_5 of 200 mg/ℓ was 35°C. Furthermore, these

TABLE 6-3

CONCENTRATION OF TOXICANT AFFECTING A 50% CHLOROPHYLL REDUCTION

Organic chemical	*Toxic concentration (mg/ℓ)*	*Organic chemical*	*Toxic concentration (mg/ℓ)*
Methanoic acid	220	1-Hexanol	1,275
Ethanoic acid	350	2-Hexanol	3,100
Propanoic acid	250	Heptanol	525
Butanoic acid	340	Octanol	250
2-Methyl propanoic acid	345	Propenoic acid	120
Pentanoic acid	280	Butenoic acid	280
3-Methyl butanoic acid	400	2,3-Dihydroxy butanedioic acid	480
Hexanoic acid	320	Hydroxyethanoic acid	2,700
Heptanoic acid	180	Sulfanilic acid	970
Octanoic acid	220	Methoxyethanoic acid	580
Ethanedioic acid	290	2-Oxopropanoic acid	880
Propanedioic acid	460	Ethanoic anhydride	360
Butanedioic acid	2,200	Propanal	3,450
Pentanedioic acid	1,200	1-Butanal	2,500
Hexanedioic acid	900	2-Methyl propanal	3,450
Heptanedioic acid	700	1-Heptanal	240
Methanol	31,000	1,2-Ethanediol	180,000
Ethanol	27,200	1,2-Propanediol	92,000
1-Propanol	11,200	Phenol	1,060
2-Propanol	17,400	Cresol	800
1-Butanol	8,500	Malthane	160
2-Butanol	8,900	Ortho (pesticide)	320
3-Butanol	24,200	DDT in xylene	120

Source: After Thirumurthi, 1966.

workers proposed that for the same removal efficiency, the required retention time, t, at any temperature, T, was related to the 35°C retention time by the relationship

$$(t)_{\text{design}} = (t)_{35^\circ\text{C}}\theta^{35-T} \tag{6-13}$$

Since $t = V/Q$, equation 6-13 can be expressed as

$$V = (t)_{35^\circ\text{C}}\theta^{35-T}Q \tag{6-14}$$

If the wastewater to be treated has a BOD_5 other than 200 mg/ℓ, a correction factor must be incorporated into equation 6-14.

$$V = (t)_{33^\circ\text{C}}\theta^{35-T}Q\frac{(\text{BOD}_5)_0}{200} \tag{6-15}$$

where $(BOD_5)_0$ represents the BOD_5 of the wastewater entering the pond. The conversion factor, C, given in equation 6-12 is the term $[(t)_{35^\circ\text{C}}/200]$

given in equation 6-15. Mara (1975a) reports a range of $(t)_{35°C}$ values from 3.5 to 7.5 days and suggests that the large variation in these values is the major limitation to the use of equation 6-12. In this same work Mara (1975a) also presents other arguments against the use of this equation in pond design.

Marais and Shaw (1961) assume that complete mixing occurs in facultative ponds and that substrate utilization follows first-order reaction kinetics such that

$$S_e = \frac{S_0}{1 + Kt} \tag{6-16}$$

where S_e = soluble effluent BOD_5, mg/ℓ
S_0 = total influent BOD_5, mg/ℓ
t = hydraulics retention time, days; Marais and Shaw (1961) suggest that this value never be less than 7 days for a pond system.
K = BOD_5 removal reaction-rate constant, day^{-1}

Thirumurthi (1969) states that facultative ponds are nearer in flow pattern to plug flow than to completely mixed flow and he recommends use of the Wehner–Wilhelm equation for arbitrary flow (equation 1-53). To facilitate the use of equation 1-53, Thirumurthi developed Figure 6-9, wherein the term Kt is plotted against S/S_0 for dispersion factors varying from zero for an ideal plug-flow reactor to infinity for a completely mixed reactor. Values for facultative ponds may range from 0.1 to 2.0, but seldom exceed 1.0. Using the figure, for 90% BOD removal and a dispersion factor of 0.25, the value of Kt is approximately 3.4. Remembering that $t = V/Q$, then the volume V is equal to $3.4(Q/K)$.

The difficulty of using the arbitrary-flow approach is that it requires knowledge of both the dispersion factor and K. If complete mixing is assumed, as recommended by Marais and Shaw (1961), a factor of safety is introduced because, for a given BOD removal efficiency, the retention time required under arbitrary flow is less than that under completely mixed conditions (see Chapter 1).

The major difficulty associated with the use of equation 6-16 is assigning a value to the reaction rate constant, K. Marais (1970) notes that the appropriate value of K to be used in the equation depends upon such factors as wastewater characteristics, sludge fermentation feedback, pond temperature, the period the pond has been in operation, and the algae concentration. He proposes that K is more a multivalue function than a single-value function and should be selected for a critical condition that may develop sometime during the pond's operational life as a result of the interaction of the many factors which affect K. Correlating monthly air temperatures with the pond liquid and sludge temperatures, Marais (1970) found that the monthly mean-maximum air temperature related the most successfully to K. Figure 6-10

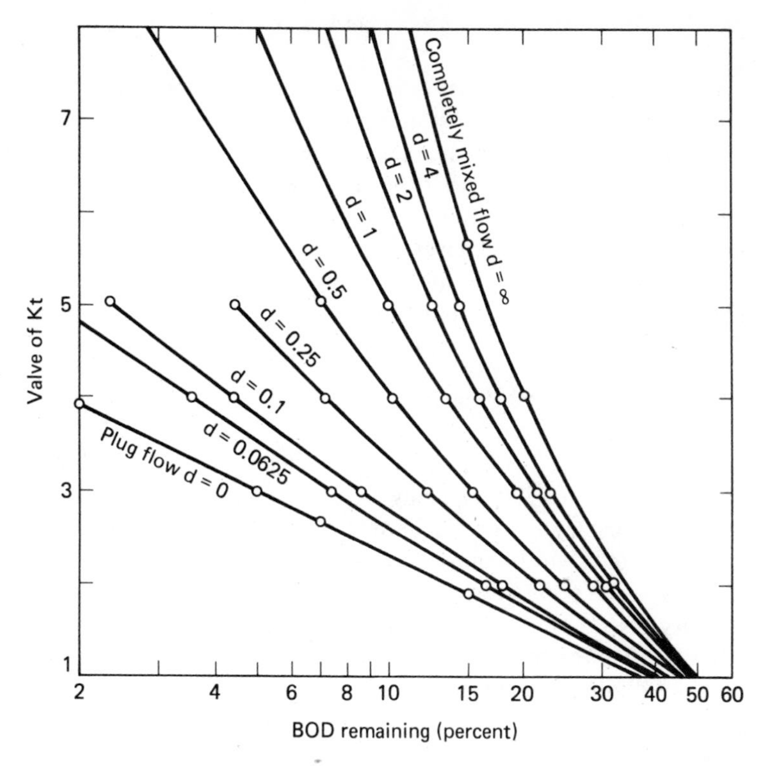

Figure 6-9. Design Formula Chart for Facultative Ponds. (After Thirumurthi, 1969.)

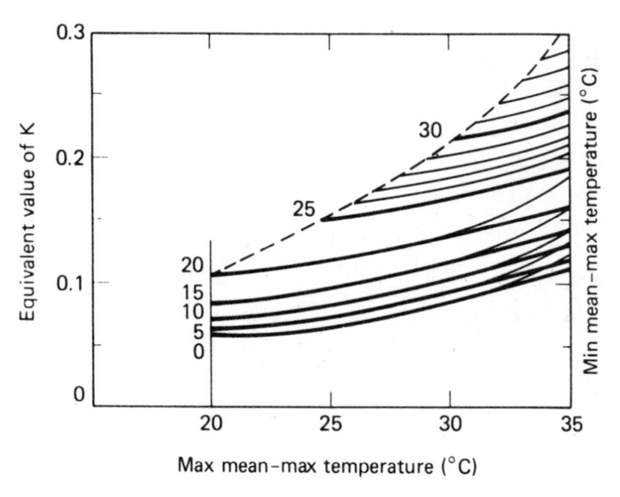

Figure 6-10. BOD Removal Reaction-Rate-Constant Values for the Design of Facultative Ponds. (After Marais, 1970.)

was constructed by Marais (1970) and is given in terms of the yearly maximum and minimum of the monthly mean-maximum air temperatures. These temperature values are available from local Weather Bureau reports.

Mara (1975b) suggests that the "equivalent" K values proposed by Marais (1970) are too conservative in that a design based on such values would give a pond that would never turn anaerobic. Mara feels that 1 or 2 h of anaerobiosis at night will not adversely affect pond operation and therefore recommends that K be obtained from the following expression:

$$K = 0.30(1.05)^{T-20} \tag{6-17}$$

where T is the operating temperature in °C. This equation is valid for temperatures above approximately 15°C.

In Chapter 1 it was shown that plug flow could be approached by employing a series of completely mixed reactors. Since, for a set volume, the removal efficiency will be greater in a plug-flow system, it follows that a series of small ponds will be more efficient than a single large pond.

For a series pond design a major requirement is to determine the smallest primary (first pond in the series) which will not turn anaerobic. Marais and Shaw (1961) propose that the maximum pond BOD_5, $(S_e)_{max}$, consistent with aerobic conditions is given by

$$(S_e)_{max} = \frac{750}{0.6d + 8} \tag{6-18}$$

where d represents the pond depth in feet. Marais (1970) later suggests that the constant 750 be reduced to 700. Mara (1975b) reports that $(S_e)_{max}$ should be within the range of 60 to 70 mg/ℓ, whereas Marais and Shaw (1961) feel that 55 mg/ℓ is a more realistic value.

McGarry and Pescod (1970) have developed an expression for the maximum BOD_5 loading which can be applied to a facultative pond without the pond becoming completely anaerobic, which has the form

$$\lambda = 9.85(1.054)^{T} \tag{6-19}$$

where λ = maximum BOD_5 loading, lb BOD_5/acre-day
T = operating temperature, °F

Mara (1975b) points out that ponds are not generally designed to operate at the failure point and recommends that the coefficient 9.85 be reduced by one-third to provide a factor of safety. Hence, equation 6-19 reduces to

$$\lambda = 6.57(1.054)^{T} \tag{6-20}$$

Marais (1974) states that maximum efficiency in a series of ponds is achieved when the retention time in each pond is the same. Mara (1976) proved this statement by considering a series of two ponds. His proof is presented as follows.

Considering equation 1-46, an expression for the effluent substrate concentration from the final pond can be written as

$$S_e = \frac{S_0}{(1 + Kt_1)(1 + Kt_2)}$$

where t_1 is the retention time in the first pond and t_2 the retention time in the second pond. This equation shows that S_e will be at a minimum when the product of $(1 + Kt_1)$ times $(1 + Kt_2)$ is at a maximum. Thus, let

$$Z = (1 + Kt_1)(1 + Kt_2)$$

and

$$T = t_1 + t_2$$

or

$$t_2 = T - t_1$$

Substituting for t_2 gives

$$Z = [1 + Kt_1][1 + K(T - t_1)]$$

or

$$Z = 1 + Kt_1 + KT - Kt_1 + K^2Tt_1 - K^2t_1^2$$

which reduces to

$$Z = 1 + KT + K^2Tt_1 - K^2t_1^2$$

Therefore,

$$\frac{dZ}{dt} = K^2T - 2K^2t_1$$

For a maximum, $dZ/dt_1 = 0$. Hence,

$$t_1 = \frac{T}{2}$$

or $t_1 = t_2$ for maximum Z and minimum S_e. Taking the second differential (which results in a negative value) shows that Z is indeed a maximum.

Design Considerations

Criteria applicable to the design of facultative ponds are presented in Table 6-4. These criteria fall within the range of those determined by Canter and Englande (1970), who conducted a survey of the design criteria for facultative ponds throughout the United States. In their survey the states were divided into three groups, as shown in Figure 6-11. The distinctions among the three groups were made on the basis of winter temperature: the ponds in group I have extended periods of ice cover during winter, the ponds in group II have only a short period of ice cover, and the ponds in group III have no appreciable ice cover during the winter months. These workers

TABLE 6-4

DESIGN CRITERIA FOR FACULTATIVE PONDS

Parameter	*Value*
Depth (ft)	3–8
Retention time (days)	7–50
BOD_5 loading (lb/acre-day)	20–50
BOD_5 removal (%)	70–95
Algae concentration (mg/ℓ)	10–100
Effluent suspended solids concentration (mg/ℓ)	100–350
Recirculation ratio	0.2–2.0

summarize the recommended design criteria as follows:

1/ Surface loading: 16.7 to 80 lb BOD_5/day/acre, with the lower values applicable to northern states and the higher values to southern states (these values can generally be converted to ultimate BOD by dividing by 0.7).

2/ Retention time: 20 to 180 + days, with the lower values applicable to southern states and the higher values to northern states (many northern states recommend that ponds have the ability to retain the entire winter flow).

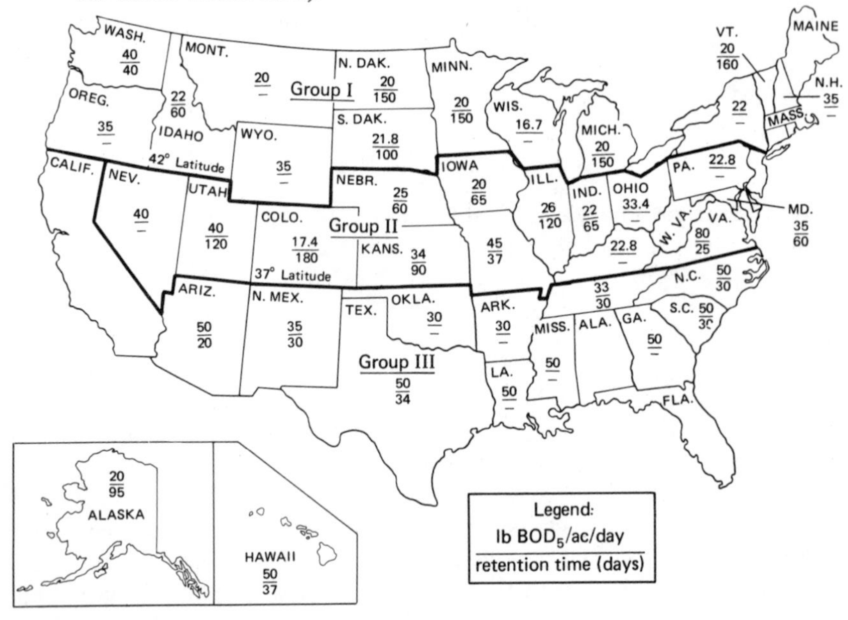

Figure 6-11. Grouping of States on Basis of Winter Temperature. (After Canter and Englande, 1970.)

3/ Liquid depth: 2 to 6 ft, with 1.5 to 3 ft of freeboard.
4/ Levees: 6 to 8 ft minimum top width with interior and exterior slope ratios of 2:1 maximum to 6:1 minimum.
5/ Shape: circular, square, or rectangular shapes are permissible.
6/ Number of ponds: for small installations, single ponds are acceptable; however, multiple-cell units that can operate either in series or parallel are desirable.
7/ Pond bottom: it is generally recommended that the bottom be level and impervious.
8/ Inlet: for circular or square shape ponds, the inlet should be near the center, whereas the inlet for rectangular ponds should be at the one-third point.
9/ Outlet: location should minimize short circuiting, and the structure should have the capability of variable depth draw-off.

It should be noted that ponds located in the southern states usually are designed for higher surface loadings and lower retention times than those located in the northern states.

Although circular, square, or rectangular ponds are acceptable according to most of the states' design criteria, Shindala and Murphy (1969) found that rectangular ponds result in better mixing than do circular or irregularly shaped ones. A length/width ratio of 3:1 is typically used.

Because of the effect of temperature on microbial activity, careful consideration must be given to this parameter during design. To ensure that the pond functions properly throughout the entire year, design must be based on the most severe conditions. Thus, the mean temperature of the coldest month is commonly used.

Solids entering the pond in the raw wastewater will settle to the pond bottom, forming a sludge layer in this region. In this sludge layer anaerobic fermentation will occur, resulting in the formation of methane gas and low-molecular-weight soluble organics which diffuse back into the bulk of the liquid. As a result, the amount of sludge accumulation will be small. Middlebrooks et al. (1965) measured the sludge depth in several ponds and found it to vary from 4 to 7 in. after 82 months of operation. Anaerobic fermentation can, however, create a serious operational problem. When the sludge temperature reaches approximately 22°C, the intensity of anaerobic fermentation has increased to the point that the large quantities of gas produced will cause mats of sludge to rise to the surface. Odors will result unless these mats are immediately dispersed. This can be accomplished with jets of water from a hose or by the agitation created by an outboard motor.

The presence of a small number of green and purple sulfur bacteria in a pond is desirable because these organisms oxidize sulfides (malodorous substances) which are produced during anaerobic respiration where sulfate acts as an external electron acceptor. The green and purple sulfur bacteria are photosynthetic organisms. In contrast to plants, photosynthetic bacteria never use water as the electron donor for photosynthesis, and as a consequence no oxygen is produced. Bacterial photosynthesis is an anaerobic process in which molecular hydrogen, reduced sulfur compounds, or organic compounds are used as external electron donors. The metabolism of the green sulfur bacteria and purple sulfur bacteria involves the use of H_2S as an energy source (electron donor) for the synthesis of cytoplasmic material from CO_2. The oxidation of the sulfur atom is considered to be a two-stage process (Stanier et al., 1963):

$$CO_2 + 2H_2S \longrightarrow (CH_2O) + H_2O + 2S \tag{6-21}$$

and

$$3CO_2 + 2S + 5H_2O \longrightarrow 3(CH_2O) + 4H^+ + 2SO_4^{2-} \tag{6-22}$$

The designation (CH_2O) is used to indicate carbohydrate or microbial cells. Although a small number of green and purple sulfur bacteria are beneficial, the treatment of wastewater having a high sulfide content may result in periodic blooms of these organisms. Such blooms are undesirable because during these periods the oxygen production of the algae will be drastically reduced, resulting in a decrease in the organic removal efficiency.

Flow Patterns

For most situations multiple-pond arrangements are generally preferred. Figure 6-11 illustrates many of the flow patterns employed in multiple-pond operations. McGauhey (1968) suggests that the parallel arrangement provides the maximum distribution of the organic load, whereas the series arrangement produces a higher quality effluent. Figure 6-12 also shows that recirculation can be employed, giving an additional degree of flexibility to pond operation. Recirculation returns active algae cells, which aid in the oxygenation process, and the return flow also serves to reduce the influent BOD concentration. It is common for the rate of recirculation to be four to eight times the average daily flow (Uhte, 1975).

Example 6-2

Design a series facultative pond system to achieve 90% BOD_5 removal from a wastewater flow of 1.5 MGD having an influent BOD_5 of 150 mg/ℓ. The mean temperature of the coldest month is 15°C.

solution

1/ Compute the BOD reaction rate constant, K, from equation 6-17.

$$K = 0.30(1.05)^{15-20}$$
$$= 0.23 \text{ days}^{-1}$$

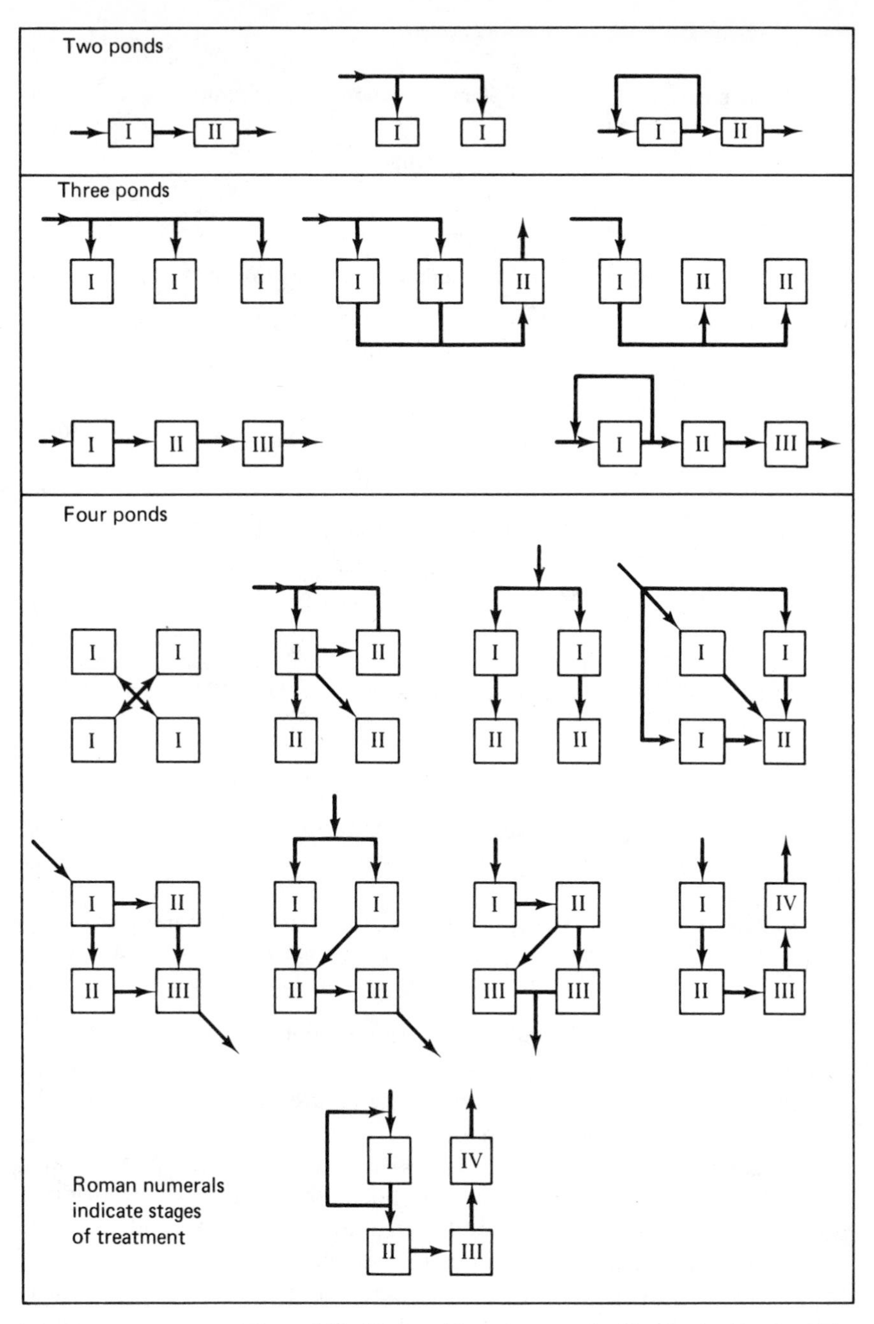

Figure 6-12. Various Flow Patterns for Facultative Ponds. (After McGauhey, 1968.)

2/ Assuming that $(S_e)_{max} = 60$ mg/ℓ, determine the required pond depth using equation 6-18.

$$d = \frac{(700/60) - 8}{0.6}$$

$$= 6 \text{ ft}$$

3/ Rearranging equation 6-16, compute the required retention time in the primary pond.

$$t = \frac{(S_0/S_e) - 1}{K}$$

$$= \frac{(150/60) - 1}{0.23}$$

$$= 6.5 \text{ days}$$

4/ Calculate the required surface area for the primary pond.

$$A = \frac{(6.5)(1{,}500{,}000)}{(7.5)(6)(43{,}560)}$$

$$= 5 \text{ acres}$$

5/ Determine the number of ponds required for 90% BOD_5 removal by applying equation 1-46.

$$(0.1)(150) = \frac{150}{[1 + (0.23)(6.5)]^n}$$

$$n \log (2.15) = \log (10)$$

$$n = 3$$

Use three ponds in series, each with a depth of 6 ft and a surface area of 5.0 acres. This should be the optimum system, since the maximum efficiency is achieved when the retention time in each pond is the same.

A word of caution: In working Example 6-2 it was assumed that the substrate utilization reaction-rate constant, K, was the same in all ponds. Whether or not this is true depends on the nature of the wastewater to be treated. If the wastewater is composed of a heterogenous mixture of easily degradable organic materials, such as those found in domestic wastes, organic removal will occur in the first pond. A portion of this material will be oxidized for energy and the remainder synthesized to new cellular material. The remaining ponds in the series will contribute mainly to the stabilization of the biomass. In this case, the K value for each of the ponds will differ considerably. On the other hand, if the wastewater is composed of a homogeneous mixture of complex organic material that is difficult to degrade, the difference in the K value between the ponds will probably be insignificant. Such considerations are also applicable to series operation when aerated lagoons are employed.

Performance

The secondary treatment standards for municipal installations state that for flows of 1 MGD or more, effluent BOD_5 and suspended solids shall not exceed an arithmetic mean value of 30 mg/ℓ for effluent samples collected in a period of 30 consecutive days nor shall they exceed 15% of the arithmetic mean of the BOD_5 suspended solids values for influent samples collected over the same time period (i.e., 85% removal will be required). Such requirements have raised concern as to the ability of the facultative pond to achieve the necessary effluent quality.

Standards for wastewater flows of less than 1 MGD have been relaxed to permit effluent suspended solids to exceed 30 mg/ℓ if the effluent BOD_5 is maintained at 30 mg/ℓ and the variance does not cause specific water quality standards to be violated. Since the BOD of algal cells is exerted slowly, the possibility of high suspended solids and low BOD_5 in the same effluent exists, and facultative ponds without algal removal are still a viable alternative for small communities.

Barsom (1973), in a state-of-the-art report on lagoon technology, has presented the average median effluent value for BOD and suspended solids from facultative ponds located in various regions of the United States. His data are shown in Figure 6-13 and 6-14. Within the various regions shown, the

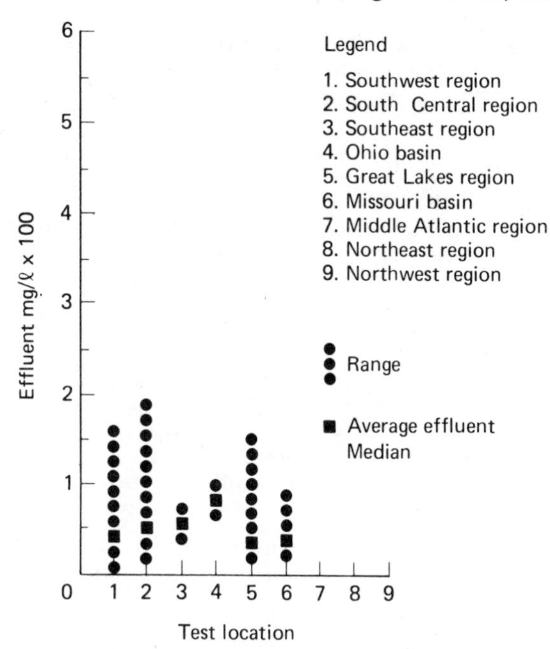

Figure 6-13. Regional Average Median Effluent Values and Ranges of Values for Suspended Solids in Facultative Ponds. (After Barsom, 1973.)

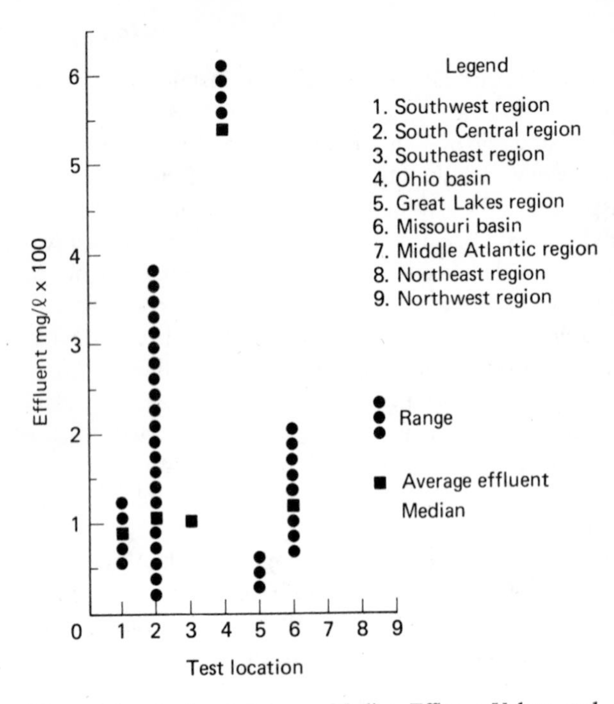

Figure 6-14. Regional Average Median Effluent Values and Ranges of Values for BOD in Facultative Ponds. (After Barsom, 1973.)

average medium effluent BOD and suspended solids concentrations ranged from 25 to 75 mg/ℓ and 40 to 540 mg/ℓ, respectively.

Eckenfelder (1970) notes that the effluent BOD from a facultative pond can be expected to vary during the year, especially in the northern states. Figure 6-15 illustrates a typical seasonal variation resulting from the winter accumulation of deposited sludge and a pond turnover.

However, because of their economical design and simplicity of operation, the facultative pond should always be considered as a treatment alternative for small communities. McKinney (1975) points out that with proper design, a facultative pond should consistently produce a satisfactory effluent. In most cases such a design will require that provisions be made for removing algae from the effluent.

Several methods are available for algal removal from pond effluents. Dryden and Stern (1968) have shown that chemical coagulation followed by sedimentation/filtration will effectively reduce the algae content of pond effluent. However, because personnel are not generally available in smaller communities to maintain such an operation, other methods of algae removal are generally more satisfactory. McKinney (1971) proposes for small ponds

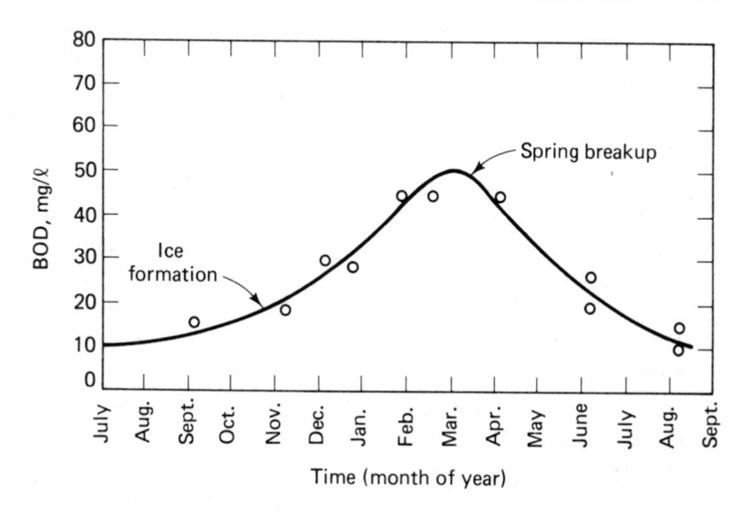

Figure 6-15. Typical Seasonal Variation of Facultative Pond Effluent BOD. (After Fischer et al., 1968.)

that algae removal can be obtained with a series arrangement, where the algae are allowed to settle from the suspending liquid in the final pond. Preliminary studies at the University of Kansas using submerged rock filters (O'Brien, 1975) and at Utah State University using intermittent sand filters (Reynolds et al., 1975) indicate their usefulness in removing algae from pond effluents. For a more detailed discussion of the techniques available for algae removal, see Gloyna, 1976; Middlebrooks et al., 1974, Middlebrooks, 1975; and Parker and Uhte, 1975.

6-3 Anaerobic Ponds

The magnitude of the organic loading and the availability of dissolved oxygen determine whether the biological activity in a treatment pond will occur under aerobic or anaerobic conditions. A pond may be maintained in an *anaerobic* condition by applying a BOD load that exceeds oxygen production from photosynthesis. Photosynthesis can be reduced by decreasing the surface area and increasing the depth. Anaerobic ponds become turbid due to the presence of reduced metal sulfides. This restricts light penetration to the point that algae growth becomes negligible.

Biological activity in an anaerobic environment was discussed in Chapter 4. In that chapter it was noted that anaerobic treatment of complex wastes involves two distinct stages. In the first stage (known as acid fermentation) complex organic materials are broken down to mainly short-chain acids and alcohols. In the second stage (known as methane fermentation) these materials are converted to gases, primarily methane and carbon dioxide. Oswald (1968), after studying anaerobic ponds in California, contends that that proper design must result in environmental conditions favorable to methane fermentation.

Anaerobic ponds are used primarily as a pretreatment process and are particularly suited for the treatment of high-temperature high-strength wastewaters. However, they have been used successfully to treat municipal wastewaters as well. When used as a pretreatment unit, percent reduction in waste strength is more important than effluent quality, since additional treatment will be employed before the waste is discharged. Anaerobic pretreatment offers several advantages: (1) the size of subsequent facultative and aerobic ponds are substantially reduced, (2) floating sludge mats associated with summer operation of facultative ponds are virtually eliminated, and (3) the accumulation of large sludge banks in subsequent treatment ponds is eliminated.

Presently, there are no easily applied relationships that can be used for the design of anaerobic ponds. Design is usually based upon criteria which have been developed by following the successes and failures of earlier ponds, treating similar wastewaters under the same type of climatic conditions.

The operating conditions and design criteria reported for anaerobic ponds vary widely. Recommended depths for ponds treating municipal wastewater range between 3 and 12 ft, while depths up to 20 ft have been used for the treatment of industrial wastes (Malina and Rios, 1976). Depths of 8 to 12 ft offer the advantages of (1) protection of the methanogenic bacteria against oxygen intrusion, (2) better land utilization, and (3) more volume for sludge storage. Oswald (1968) notes that average lapse rates of 1°C/ft are normal during summer months for ponds heated only by sunlight. Since methane fermentation is strongly influenced by temperature, excessive pond depths resulting in cold pond bottoms are undesirable.

Methane fermentation is a biochemical process and, as such, is influenced by temperature changes. Below the temperature for thermal inactivation of methanogenic bacteria, an increase in temperature will result in an increase in the rate of methane fermentation. Data presented in Table 6-5 and Figure 6-16 serve to illustrate this point. Since temperature control is not practical in anaerobic ponds, temperature variations can be expected with season changes. In this regard, nearly all such ponds have been observed to function better during the summer. For example, McIntosh and McGeorge (1964) observed that at a winter temperature of 60°F, BOD reduction was 58% but that it increased to 92% during the summer, when the pond temperature rose to 90°F.

TABLE 6-5

ESTIMATED EFFECT OF TEMPERATURE ON ANAEROBIC TREATMENT

Temperature (°C)	*Rate of methane fermentation relative to the rate at 35°C*	*Retention time required for treatment equivalent to that at 35°C*
5	0.1	10
15	0.4	2.5
25	0.8	1.2
35	1.0	1.0

Source: After McCarty (1966).

Anaerobic ponds have been used successfully in northern climates to treat high-temperature wastes such as that from a slaughterhouse operation. To retain the heat of these waters, floating covers have been used. McIntosh and McGeorge (1964) report the use of a plastic foam raft 3 in. thick to cover 30,000 ft^2 of pond surface. However, artifical covers are not required for meat-packing wastes, as the grease and paunch manure contained in the wastes produce a natural scum cover. Such covers also provide the additional benefit of odor control.

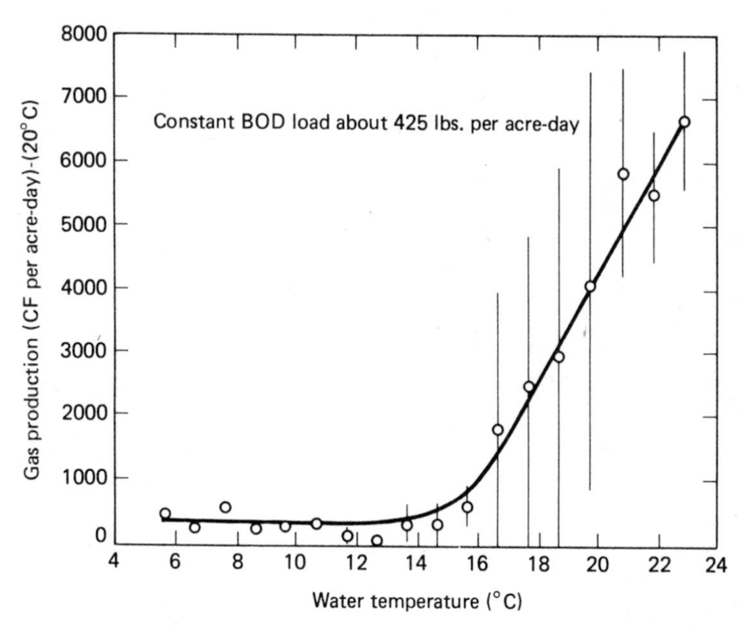

Figure 6-16. Relationship Between Temperature and Gas Production in Ponds. (After Oswald, 1964.)

Depending on geographical location, a wide range of organic loadings have been applied to anaerobic ponds treating municipal wastewaters. In California, for 70% BOD_5 removal, Oswald (1968) recommends 400 lb BOD_5/acre-day under summer conditions and 100 lb BOD_5/acre-day under winter conditions. For summer pond operation in South Africa, Van Eck and Simpson (1966) observed 81 and 62% BOD_5 removal for surface area loadings of 2590 and 1692 lb BOD_5/acre-day, respectively. Operational data reported by Parker and Skerry (1968) for three ponds (3 to 3.5 ft deep) treating municipal wastewater in Australia are presented in Table 6-6. These data show loadings ranging from 438 lb BOD_5/acre-day to 2800 lb BOD_5/acre-day. Anaerobic ponds have also been used to treat a variety of industrial wastes. A summary of operational data for such treatment is presented in Table 6-7. Note that in this table volumetric BOD_5 loading rather than surface BOD_5 loading is given. Volumetric loading is actually a more appropriate design parameter, as the degree of methane fermentation is dependent on the retention time and not on the surface area available for insolation or oxygen transfer.

TABLE 6-6

SUMMARY OF LOADING AND PERFORMANCE DATA FOR ANAEROBIC PONDS TREATING MUNICIPAL WASTEWATER IN AUSTRALIA

	Pond location					
	Werribee 145W		*Wangaratta*			*Kerang*
Item	*Summer*	*Winter*	*Winter*	*Summer*	*Winter*	*Spring*
BOD_5 (mg/ℓ)						
Influent	448	407	358	270	222	190
Effluent	157	291	118	137	149	101
% Removed	65	28	67	49	33	47
Influent SS (mg/ℓ)	436	436	394	—	—	323
Effluent sulfide (mg/ℓ)	14	8	1.0	—	0.3	140
Dissolved O_2 (mg/ℓ)	Nil	Nil	Nil	3.8	Nil	Nil
pH	6.7	6.1	7.0	7.0	6.7	7.1
Water temperature (°C)	20	15	15	24	12	21
Sludge depth (in)						
Near inlet	17	24	10	10	14	30
Near outlet	14	27	7	9	14	30
BOD_5-lb/day/acre						
Loading	482	438	910	710	584	2800
Removed	313	125	630	350	192	1280

Source: After McKinney (1971).

TABLE 6-7

OPERATING DATA FOR ANAEROBIC PONDS TREATING INDUSTRIAL WASTEWATERS

Type of waste	*BOD_5 concentration (mg/ℓ)*		*BOD_5 removal (%)*	*BOD_5 loading (lb/1000 ft^3-day)*	*Temper- ature (°C)*
	Influ- ent	*Efflu- ent*			
Meat	1,096	159	85.5		
	940	458	58.2	16.1	25
	1,880	350	87	31.4	22–27
	—	670	64.9	12.7	—
	1,703	122	92.8	9.0	24
	3,000	300	90	—	25–26
	2,070	158	92.4	—	—
Fruit	3,380	445	86.8	630[a]	
Tomato	728	163	77.6	628[a]	
Citrus	939	241	74.4	662[a]	
Tomato	982	599	39.8	33.9	14–24
Peas	1,444	—	37	23.2	—
Corn	2,164	—	47.5	15.9	—
Chemical and fermentation	10,000	2000	60–80	—	5, 10, 15
Rendering	1,870	—	88	228[a]	19–36
Corn	<4,000	—	58	—	21
products		—	92	—	38
Milking	1,030	160	85	9	29
parlor	1,030	830	20	9	2

[a]lb BOD_5/acre-day.
Source: After McKinney (1971).

A range of retention times between 1.2 and 160 days have been reported for anaerobic ponds treating municipal wastewater. However, Gloyna (1965) states that retention times greater than 5 days are not necessary because at longer retention times, the performance of an anaerobic pond would approach that of a facultative pond. Malina and Rios (1976) suggest that a liquid retention time of 2 days during the summer and 5 days during the winter should be sufficient for anaerobic ponds treating municipal wastewater. Ponds so designed should be effective in removing 70 to 80% of the BOD_5.

The major problem that arises from the use of anaerobic ponds is odor. Odor control can be accomplished by reducing the organic loading so that an aerobic layer can become established at the pond surface or, as recommended by Oswald (1968), recirculate aerobic pond effluent back to the anaerobic pond. Discharge of the recycle near the surface will aid in establishing an aerobic surface layer. To avoid oxygen intrusion into the anaerobic zone, Oswald (1968) recommends recycling when the aerobic pond effluent is warmer than the liquid in the lower reaches of the anaerobic pond. Oxygen intrusion can also be avoided by providing special digestion chambers in the bottom of the anaerobic pond so that deep sludge layers will accumulate. Examples of the special digestion chambers recommended by Oswald (1968) are provided in Figure 6-17.

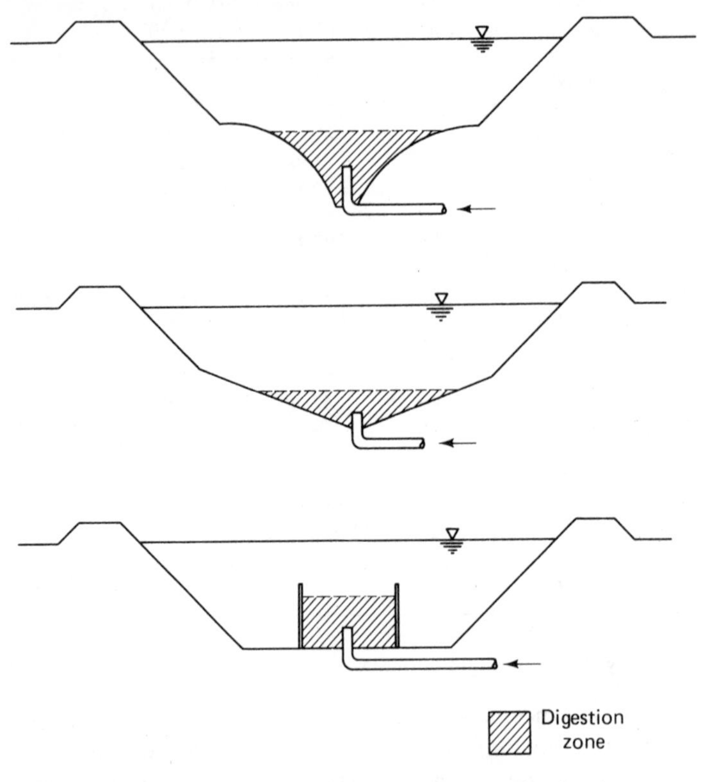

Figure 6-17. Some Methods of Creating a Digestive Chamber in the Bottom of an Anaerobic Pond. (After Oswald, 1968.)

6-4
Polishing Ponds

Ramani (1976) defines *polishing ponds* as ponds that receive wastewater effluent treated to a secondary level and notes that their major function is to achieve considerable improvement in the effluent biochemical, bacteriological, and eutrophic characteristics.

There is no rational procedure for the design of polishing ponds. Mara

(1976) states that BOD_5 removal in these ponds is minimal. To achieve an effluent BOD_5 less than 25 mg/ℓ when the influent BOD_5 is between 50 and 75 mg/ℓ (effluent from the secondary treatment unit), two ponds arranged in series, each with a 7-day retention time, are required. When designing polishing ponds, Mara (1976) recommends that the depth selected be equal to the depth of the associated facultative pond.

The relationship between retention time and fecal bacteria reduction in a treatment pond (aerobic, anaerobic, facultative, or polishing) can be described by *Chick's law* (Mara, 1976):

$$\frac{N_e}{N_0} = \frac{1}{1 + K_b t} \qquad (6\text{-}23)$$

where N_e = number of fecal coliforms, FC/100 ml of effluent
N_0 = number of FC/100 ml of influent
K_b = reaction-rate constant, day^{-1}
t = hydraulic retention time, days

For a series-pond system, equation 1-42 implies that the bacterial die-away can be expressed as

$$\frac{N_e}{N_0} = \frac{1}{(1 + K_b t_1)(1 + K_b t_2) \ldots (1 + K_b t_n)} \qquad (6\text{-}24)$$

where n represents the number of ponds in the series.

Within the temperature range 5 to 21°C, Marais (1974) proposes that K_b is given by

$$K_b = 2.6(1.19)^{T-20} \qquad (6\text{-}25)$$

where T is the pond temperature in °C. Mara (1976) suggests that for municipal wastewaters, a reasonable value for N_0 is 4×10^7 FC/100 ml.

To achieve maximum effluent quality when a pond system is used for wastewater treatment, both Oswald et al. (1970) and Gloyna and Aguirre (1970) recommend series-operated pond units. Typically, a minimum of four ponds should be used: an anaerobic pond, followed by a facultative pond, which in turn is followed by a minimum of two polishing ponds. Gloyna (1965) suggests that the ratio between the surface area of aerobic and anaerobic ponds range between 10:1 and 5:1. He further states that ponds with lower ratios will be sensitive to short-term BOD variations.

6-5 Aerated Lagoons

Pond systems that are oxygenated through the use of mechanical or diffused aeration units are termed *aerated lagoons*. Because of turbidity, turbulence, and other factors, algae growth generally ceases or is greatly reduced when artificial aeration is employed.

There are two basic types of aerated lagoons: (1) *aerobic lagoons*, which are designed with power levels great enough to maintain all the solids in the lagoon in suspension and also to provide dissolved oxygen throughout the liquid volume (see Figure 6-18), and (2) *facultative lagoons*, which are designed with power levels only great enough to maintain dissolved oxygen throughout the liquid volume. In this case the bulk of the solids are not maintained in suspension but settle to the bottom of the lagoon, where they are decomposed anaerobically (see Figure 6-19).

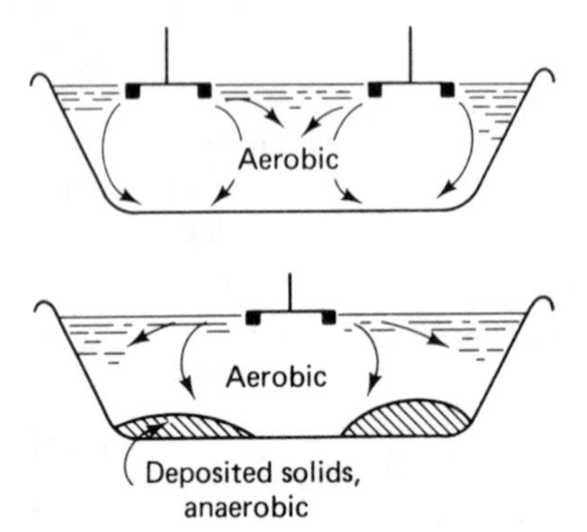

Figure 6-18. Schematic of Aerobic Lagoon. (After Eckenfelder, 1970.)

Figure 6-19. Schematic of Facultative Lagoon. (After Eckenfelder, 1970.)

Aerobic lagoons are usually designed to operate at high F:M ratios or short θ_c's (high-rate systems). These systems achieve little organic stabilization but rather convert soluble organic material into cellular organic material. On the other hand, facultative lagoons are designed for longer θ_c's (low-rate systems) and organic stabilization.

Barnhart (1972) lists the following advantages for the use of aerated lagoons: (1) ease of operation and maintenance, (2) wastewater equalization, and (3) a high capacity for heat dissipation when required. Disadvantages given by Barnhart are (1) large land area requirement, (2) difficult to modify process, (3) high effluent suspended solids concentration, and (4) sensitivity of process efficiency to variations in ambient air temperature.

6-6 Aerobic Lagoons

Because all the solids are maintained in suspension, the retention time in the *aerobic lagoon* required for soluble BOD removal will be less than that required in the facultative lagoon (Kormanik, 1972). However, the power requirement for mixing in the aerobic lagoon will be much greater than that in the facultative lagoon. Furthermore, because all the solids are retained in suspension, the effluent from an aerobic lagoon will have a much higher solids concentration than the effluent from a facultative one. This requires that a solids–liquid separation step follow the aerobic process if a high-quality effluent is to be achieved. The aerobic lagoon is actually a no-recycle-activated sludge system.

Mixing and Aeration

In an aerobic lagoon aeration performs the two functions of oxygen transfer and mixing. In many of the activated sludge process modifications, oxygen transfer controls the aerator design. However, Eckenfelder (1970) points out that for lagoons and extended aeration systems, the mixing requirements control the power level in most situations. McKinney et al. (1971) state that oxygen requirements will control if the retention time in the lagoon is less than 24 h but that mixing requirements will control when the retention time is greater than 24 h.

At present, no rational method exists that will allow the engineer to determine the power level required for total suspension of the solids in the system. Because of this, it is necessary to rely on empirical methods based on experimental and operational data. As a guide to design, the Clow Corporation recommends that for lagoon depths within the range 8 to 18 ft, and suspended solids concentrations within the range 1000 to 5000 mg/ℓ, the power level be within the range 60 to 120 hp/MG to achieve complete mixing. A restriction to using this criteria is that the length of the lagoon not exceed 1.25 times the width. It is believed that such conditions will provide bottom velocities greater than 0.5 ft/s, the minimum necessary for complete mixing. If surface aeration is employed, experimental data have shown that to maintain solids in suspension, lagoon depths should be limited to 12 to 15 ft when no special provisions are made for deep-mixing capabilities. If draft tubes or some special equipment such as the baffle ring assembly (see Fig. 6-20) are used, the depth can be extended to 17 to 20 ft.

Malina et al. (1972) indicates that a minimum power level of 30 hp/MG is required for complete mixing of aerated lagoons. Data supporting this requirement are presented in Figure 6-21. Deviation from a completely mixed regime is given in terms of the number of completely mixed tanks. (Recall that a plug-flow regime can be approximated by a number of completely mixed reactors in series.)

Since aerators oxygenate the liquid by pumping, the relationship between volume of liquid under aeration and aeration horsepower will have an effect on the rate of oxygen transfer. The average range of power levels for different aeration systems are presented in Figure 5-31. The power-level range given in this figure for aerated lagoon systems is nearly equal to or slightly less than the minimum levels shown to exert a significant effect on oxygen transfer

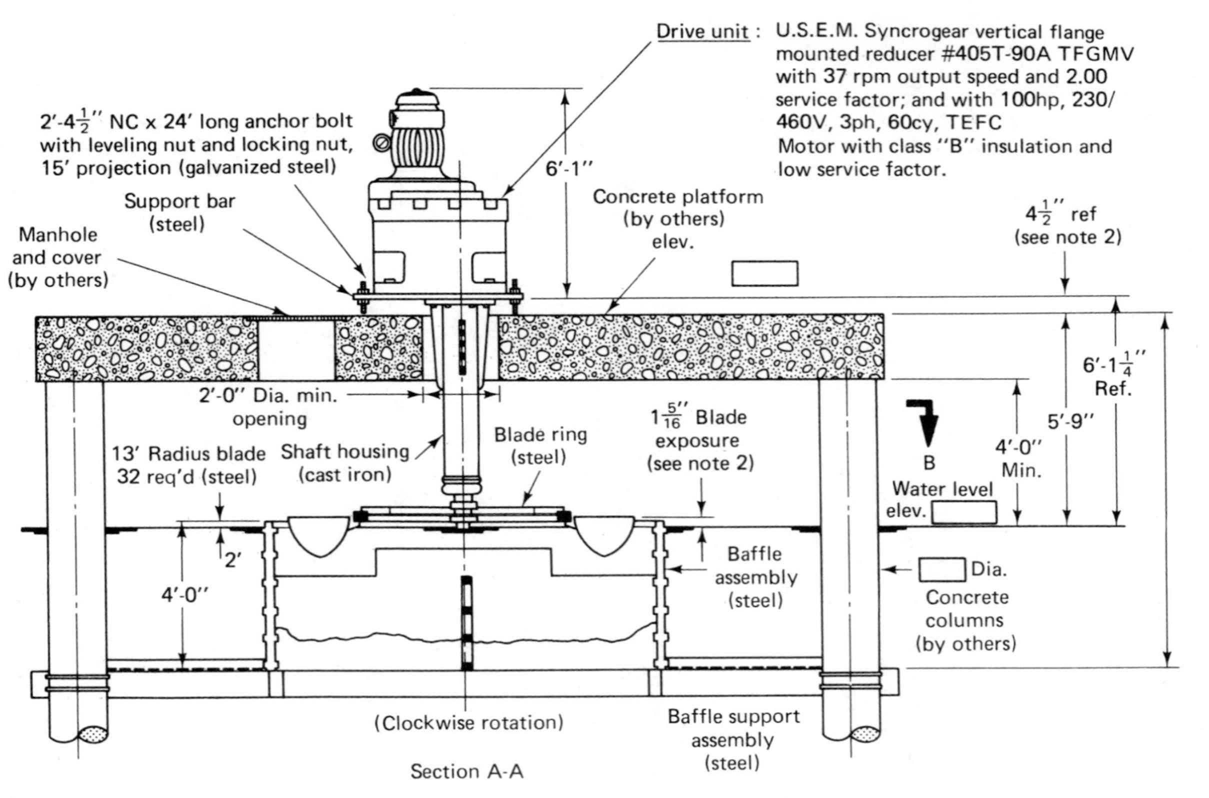

Figure 6-20. Sigma II S Aerator. (Courtesy of CLOW Corporation.)

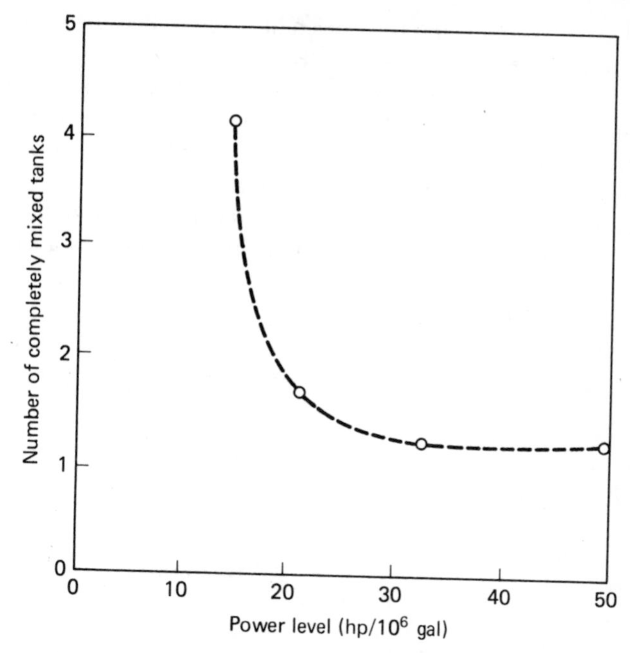

Figure 6-21. Power Level and Degree of Mixing. (After Malina et al., 1972.)

rates (see Figures 5-29 and 5-30). As a result, the effect of power level on oxygen transfer rate can usually be neglected in the design of aeration systems for lagoons.

Mixing is required to maintain solids in suspension as well as to distribute oxygen throughout the liquid volume. The performance data presented in Table 6-8 show that for a specific energy input, the diameter of influence for complete mixing (all solids maintained in suspension) is much less than the

TABLE 6-8

TYPICAL LOW-SPEED SURFACE AERATOR PERFORMANCE DATA

Horsepower (hp)	*Depth (ft)*	*Zone of complete mixing (ft)*	*Zone of oxygen dispersion (ft)*
3.0	6	50	150
5.0	6	70	210
10.0	8	90	260
20.0	10	115	330
25.0	10	130	375

diameter of influence for oxygen dispersion. Because of this, mixing normally controls the aeration system design.

Based on the total horsepower requirement, the number of individual aeration units should be selected, and these units should be located in the lagoon so that there is an overlap of complete mixing zones. In making such a determination, the effect of lagoon depth on the diameter of influence for complete mixing must also be considered. Figure 6-22 illustrates this effect for a specific surface aeration unit. Performance data supplied by manufacturers will specify the depth and horsepower associated with a particular diameter of influence (see Table 6-8).

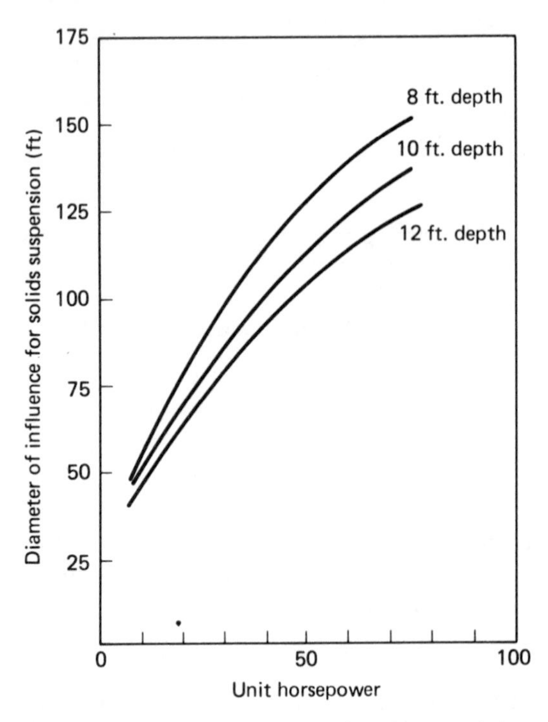

Figure 6-22. Surface Aerator Mixing Characteristics. (After Ford, 1972.)

Sawyer (1968) cautions against the use of aeration units larger than 25 hp by suggesting that such units have a tendency to create too much localized mixing and fail to achieve uniform mixing over the entire zone of influence. McKinney et al. (1971) states that the pumpage per unit horsepower decreases as the horsepower increases. This implies that several small units will be more economical than one large unit. Such an arrangement also provides a certain degree of operational flexibility in the event of equipment failure.

Design Relationships

Because the aerobic lagoon is nothing more than a completely mixed reactor without cellular recycle, design equations for this process can be developed by applying material balances and by using the fundamental kinetic relationships presented in Chapter 2.

A typical flow scheme for an aerobic lagoon is shown in Figure 6-23. In this figure Q represents the rate of flow of wastewater to the lagoon; S_0,

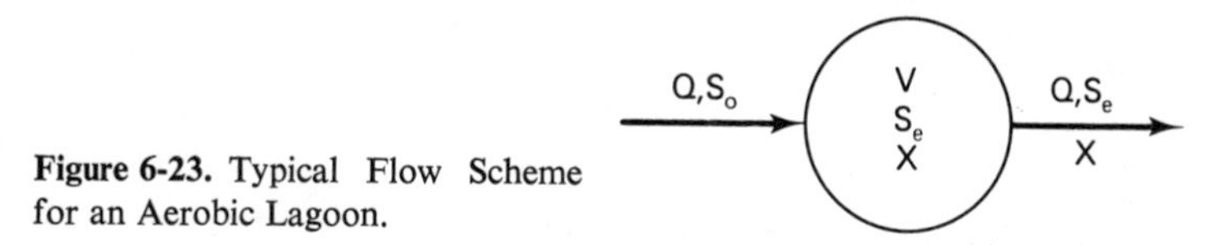

Figure 6-23. Typical Flow Scheme for an Aerobic Lagoon.

the substrate concentration in the raw wastewater; V, the lagoon volume; S_e, the soluble effluent substrate concentration; and X, the biomass concentration in the lagoon. Material-balance expressions developed in this section will be derived from Figure 6-23 and based on the assumptions presented in Chapter 4 for developing the kinetic model for the completely mixed activated sludge process modification.

For an aerobic lagoon process equation 4-1 can be expressed as

$$\theta_c = \frac{XV}{XQ} \tag{6-26}$$

or

$$\theta_c = \frac{V}{Q} \tag{6-27}$$

Since the nominal hydraulic retention time is equal to V/Q, equation 6-27 can be written as

$$\theta_c = t \tag{6-28}$$

where t = hydraulic retention time, time

Hence, biological solids retention time is shown to be equal to hydraulic retention time for completely mixed systems where recycle is not employed.

A material balance for biomass around the entire treatment system (shown in Figure 6-23) gives

$$\begin{bmatrix} \text{net rate of} \\ \text{change in amount} \\ \text{of biomass} \\ \text{within system} \end{bmatrix} = \begin{bmatrix} \text{rate at} \\ \text{which} \\ \text{biomass appears} \\ \text{in system} \end{bmatrix} - \begin{bmatrix} \text{rate at} \\ \text{which} \\ \text{biomass leaves} \\ \text{system} \end{bmatrix} \tag{6-29}$$

which can be expressed mathematically as

$$\left(\frac{dX}{dt}\right)V = \left(\frac{dX}{dt}\right)_g V - QX \tag{6-30}$$

Substituting from equation 2-51 for $(dX/dt)g$ and from equation 6-27 for Q, equation 6-30 becomes

$$\left(\frac{dX}{dt}\right)V = \left[Y_T\left(\frac{dS}{dt}\right)_u - K_dX\right]V - \frac{VX}{\theta_c} \tag{6-31}$$

At steady-state,

$$\begin{bmatrix}\text{rate at which} \\ \text{biomass appears} \\ \text{in system}\end{bmatrix} = \begin{bmatrix}\text{rate at which} \\ \text{biomass leaves} \\ \text{system}\end{bmatrix}$$

which implies that

$$\frac{dX}{dt} = 0$$

Thus, at steady-state, equation 6-31 can be written as

$$0 = \left[Y_T\left(\frac{dS}{dt}\right)_u - K_dX\right]V - \frac{VX}{\theta_c} \tag{6-32}$$

or

$$\frac{1}{\theta_c} = Y_T\frac{(dS/dt)_u}{X} - K_d \tag{6-33}$$

Equation 6-33 is identical to equation 4-8, indicating that the same relationship holds for a completely mixed system both with and without cellular recycle.

A material-balance expression for substrate entering and leaving the lagoon gives

$$\begin{bmatrix}\text{net rate of} \\ \text{change in amount} \\ \text{of substrate} \\ \text{within lagoon}\end{bmatrix} = \begin{bmatrix}\text{rate at which} \\ \text{substrate enters} \\ \text{lagoon}\end{bmatrix} - \begin{bmatrix}\text{rate at which} \\ \text{substrate disappears} \\ \text{from lagoon}\end{bmatrix} \tag{6-34}$$

which can be expressed as

$$\left(\frac{dS}{dt}\right)V = QS_0 - \left(\frac{dS}{dt}\right)_u V - QS_e \tag{6-35}$$

At steady-state,

$$\begin{bmatrix}\text{rate at which} \\ \text{substrate enters} \\ \text{lagoon}\end{bmatrix} = \begin{bmatrix}\text{rate at which} \\ \text{substrate disappears} \\ \text{from lagoon}\end{bmatrix}$$

which implies that

$$\frac{dS}{dt} = 0$$

and equation 6-35 reduces to

$$\left(\frac{dS}{dt}\right)_u = \frac{Q(S_0 - S_e)}{V} \tag{6-36}$$

If equation 2-56 is selected to describe the rate of substrate utilization, equation 6-36 can be written as

$$\frac{Q(S_0 - S_e)}{XV} = KS_e \tag{6-37}$$

Again, note that equation 6-37 and equation 4-25 are identical. This implies that equation 4-23, an expression for soluble effluent substrate concentration, is equally applicable to aerobic lagoon calculations.

Total oxygen requirement for aerobic lagoon systems can be computed from any of the following relationships:

$$\Delta O_2 = \left[\frac{\text{lb BOD}_u\text{ removed}}{\text{day}}\right] - 1.42\,\Delta X + \text{NOD} \tag{6-38}$$

$$\Delta O_2 = 8.34Q[(1 - 1.42Y_T)(S_0 - S_e)] + 8.34(1.42)K_d X V_a + \text{NOD} \tag{6-39}$$

$$\Delta O_2 = a\frac{\text{lb BOD}_u\text{ removed}}{\text{day}} + bXV_a + \text{NOD} \tag{6-40}$$

$$\Delta O_2 = a_1\,\Delta X + b_1 X V_a + \text{NOD} \tag{6-41}$$

In these equations ΔO_2 represents the oxygen requirement in pounds per day, while ΔX is the biomass produced in pounds per day. For this particular treatment system (operating at steady-state) ΔX is equal to the total biomass lost in the effluent each day:

$$\Delta X = 8.34QX \tag{6-42}$$

If either equation 6-40 or 6-41 is used to compute the oxygen requirement, then values for the coefficients a, b, a_1, and b_1 must be known. For domestic wastewaters Balasha and Sperber (1975) report that values for the coefficient, a, range between 0.36 and 0.63 on a BOD_5 basis, while b values vary between 0.13 and 0.28 day^{-1}. Few values for a_1 and b_1 have been reported, so these coefficients must be determined from laboratory studies.

The total effluent BOD_5 from an aerobic lagoon will be the sum of the soluble BOD_5 and the oxygen required for endogenous destruction of the biomass contained in the effluent. To calculate the total effluent BOD_5, equation 4-124 is applied:

$$(\text{BOD}_5)_{\text{eff}} = S_e + 5[1.42K_d\chi_a(\text{TSS})_{\text{eff}}] \tag{4-124}$$

For aerobic lagoons removing a high percentage of soluble BOD_5, Eckenfelder (1970) recommends that equation 4-124 be reduced to

$$(\text{BOD}_5)_{\text{eff}} = S_e + 0.3(\text{TSS})_{\text{eff}} \tag{6-43}$$

When effluent VSS are measured, Balasha and Sperber (1975) propose that total effluent BOD_5 be computed from

$$(\text{BOD}_5)_{\text{eff}} = S_e + 0.54(\text{VSS})_{\text{eff}} \tag{6-44}$$

Temperature Effects

Temperature is a very important consideration in aerobic lagoon design because of the effect it has on treatment kinetics. Barnhart (1972) proposes that lagoon temperatures can be estimated by employing energy-budget relationships. Using this approach, it is assumed that temperature change is mediated by four surface mechanisms, which are heat loss through evaporation, convection, and radiation and heat gain through solar radiation. Such a relationship can be expressed mathematically as

$$H = H_e + H_c + H_r - H_s \tag{6-45}$$

where H = net heat loss
H_e = heat loss by evaporation
H_c = heat loss by convection
H_r = heat loss by radiation
H_s = heat gain by solar radiation

According to Velz (1970), the heat loss due to evaporation can be determined from the relationship

$$H_e = 0.00722 H_v C_1 (1 + 0.1 W)(V_w - V_{pa}) \tag{6-46}$$

where H_e = heat loss due to evaporation (Btu/h-ft^2 of water surface)
W = mean wind velocity, mph
V_w = water vapor pressure corresponding to lagoon temperature, in. Hg
V_{pa} = mean absolute water vapor pressure prevailing in the overlying atmosphere, in. Hg
C_1 = constant that is characteristic of the lagoon; ranges in value from 10 to 15 (for deep lakes and reservoirs the lower value is used, whereas a value of 15 is more appropriate for shallow ponds)
H_v = latent heat of vaporization, Btu/lb

Tables 6-9 and 6-10 give the value for H_v and V_{pa}.

Convection heat loss from a water surface is given by the expression

$$H_c = \left(0.8 + C_2 \frac{W}{2}\right)(T_w - T_a) \tag{6-47}$$

where H_c = heat loss from convection, Btu/h-ft^2
C_2 = constant, ranging in value from 0.16 to 0.32; for a quiescent body of water $C_2 = 0.24$, whereas for flow streams of moderate velocity $C_2 = 0.32$
T_w = lagoon temperature, °F
T_a = ambient air temperature, °F
W = mean wind velocity, mph

TABLE 6-9

LATENT HEAT OF VAPORIZATION, H_v

Temperature (°F)	H_v *(Btu/lb)*	*Temperature (°F)*	H_v *(Btu/lb)*
32	1075.8	100	1037.2
35	1074.1	105	1034.3
40	1071.3	110	1031.6
45	1068.4	115	1028.7
50	1065.6	120	1025.8
55	1062.7	125	1022.9
60	1059.9	130	1020.0
65	1057.1	135	1017.0
70	1054.3	140	1014.1
75	1051.5	145	1011.2
80	1048.6	150	1008.2
85	1045.8	155	1005.2
90	1042.9	160	1002.3
95	1040.1	165	999.3

Source: After Velz (1970).

The net heat loss by radiation can be estimated from the relationship

$$H_r = 1.0(T_w - T_a) \qquad (6\text{-}48)$$

where H_r = heat loss from radiation, Btu/h-ft²

Heat gain from solar radiation cannot be calculated from meteorologic factors, and therefore must be estimated. Generally, the solar radiation measured at the local Weather Bureau is acceptable. In fact, the heat gain due to solar radiation is small, and Barnhart (1972) suggests that for aerated lagoons, this term can usually be neglected.

Neglecting the heat gain due to solar radiation and combining the various heat losses into a single expression for net heat loss, the resulting expression is

$$\text{heat loss} = [0.00722H_vC_1(1 + 0.1W)(V_w - V_{pa}) + \left(0.8 + C_2\frac{W}{2}\right)(T_w - T_a) + (T_w - T_a)]\bar{A} \qquad (6\text{-}49)$$

where $\bar{A}$ = apparent lagoon surface area; the apparent surface area will vary with power level; Figure 6-24 presents scale-up factors for various power levels; the product of the scale-up factor and the lagoon surface area gives the apparent surface area

TABLE 6-10

SATURATED WATER VAPOR PRESSURE, V_{pa} (IN. Hg)

Air temperature (°F)	*Vapor pressure (in. Hg)*	*Air temperature (°F)*	*Vapor pressure (in. Hg)*	*Air temperature (°F)*	*Vapor pressure (in. Hg)*	*Air temperature (°F)*	*Vapor pressure (in. Hg)*
30	0.164	60	0.517	90	1.408	120	3.425
31	0.172	61	0.536	91	1.453	121	3.522
32	0.180	62	0.555	92	1.499	122	3.621
33	0.187	63	0.575	93	1.546	123	3.723
34	0.195	64	0.595	94	1.595	124	3.827
35	0.203	65	0.616	95	1.645	125	3.933
36	0.211	66	0.638	96	1.696	126	4.042
37	0.219	67	0.661	97	1.749	127	4.154
38	0.228	68	0.684	98	1.803	128	4.268
39	0.237	69	0.707	99	1.859	129	4.385
40	0.247	70	0.732	100	1.916	130	4.504
41	0.256	71	0.757	101	1.975	131	4.627
42	0.266	72	0.783	102	2.035	132	4.752

43	0.277	73	0.810	103	2.087	133	4.880
44	0.287	74	0.838	104	2.160	134	5.011
45	0.298	75	0.866	105	2.225	135	5.145
46	0.310	76	0.896	106	2.292	136	5.282
47	0.322	77	0.926	107	2.360	137	5.422
48	0.334	78	0.957	108	2.431	138	5.565
49	0.347	79	0.989	109	2.503	139	5.712
50	0.360	80	1.022	110	2.576		
51	0.373	81	1.056	111	2.652		
52	0.387	82	1.091	112	2.730		
53	0.402	83	1.127	113	2.810		
54	0.417	84	1.163	114	2.891		
55	0.432	85	1.201	115	2.975		
56	0.448	86	1.241	116	3.061		
57	0.465	87	1.281	117	3.148		
58	0.482	88	1.322	118	3.239		
59	0.499	89	1.364	119	3.331		

Source: After Velz (1970).

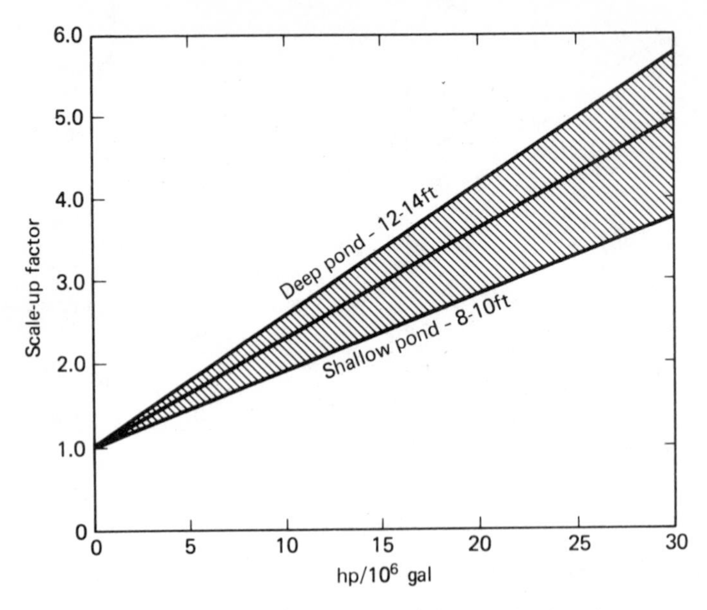

Figure 6-24. Influence of Power Input on Apparent Surface Area. (After Barnhart, 1972.)

Heat input to the lagoon is given by the relationship

$$\text{heat input} = (T_i - T_w)\frac{1\ \text{Btu}}{\text{lb-°F}}(Q) \tag{6-50}$$

where T_i = influent wastewater temperature, °F

$\frac{1\ \text{Btu}}{\text{lb-°F}}$ = specific heat of water

Q = average total mass flow, lb/h

At steady-state,

$$\text{heat input} = \text{heat loss}$$

Therefore,

$$(T_i - T_w)\frac{1\ \text{Btu}}{\text{lb-°F}}(Q) = \Big[0.00722H_vC_1(1 + 0.1W)(V_w - V_{pa}) + \left(0.8 + C_2\frac{W}{2}\right)(T_w - T_a) + (T_w - T_a)\Big]\bar{A} \tag{6-51}$$

In equation 6-51, T_w represents the equilibrium lagoon temperature. However, a value for T_w cannot be obtained directly from this equation since a value for V_w is also involved. In this case a trial-and-error procedure is required where a value of T_w is assumed and the value of the left side of the equation is compared to the right side. When the two sides are equal, the correct value of T_w has been assumed.

Equation 6-51 is cumbersome to use. To simplify equilibrium temperature calculations, Mancini and Barnhart (1968) suggest that the heat transfer coefficients (the surface-area correction factor required because of aeration, wind, and humidity effects) be lumped into a proportionality factor, f. Incorporating this factor into equation 6-51 gives

$$(T_i - T_w)Q = (T_w - T_a)fA \tag{6-52}$$

or

$$T_w = \frac{AfT_a + QT_i}{Af + Q} \tag{6-53}$$

The units of Q in equation 6-53 are MGD and the units of A are ft^2.

Probability plots of f are presented in Figures 6-25 and 6-26 for the eastern United States and the midwestern United States, respectively. When no other data are available for a particular treatment situation, a value of 12×10^{-6} for f is generally used.

Once the lagoon operating temperature has been established, the substrate utilization reaction-rate constant, K, and the microbial decay coefficient,

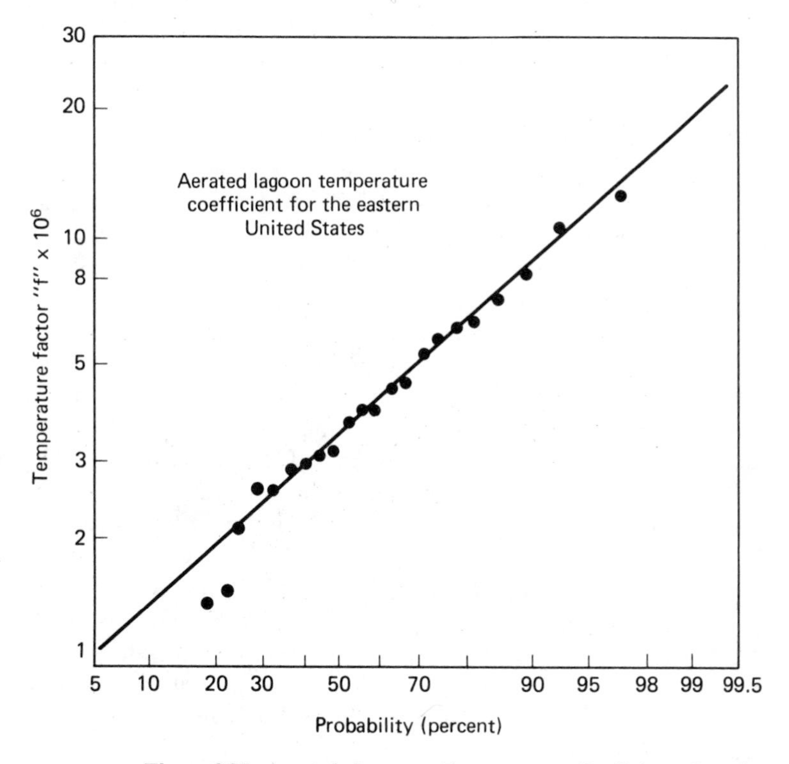

Figure 6-25. Aerated Lagoon Temperature Coefficient for the Eastern United States. (After Mancini and Barnhart, 1968.)

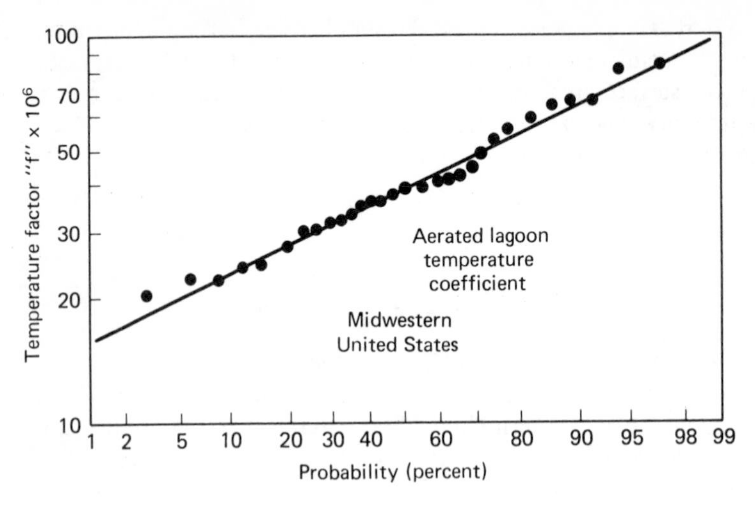

Figure 6-26. Aerated Lagoon Temperature Coefficient for the Midwestern United States. (After Mancini and Barnhart, 1968.)

K_d, may be corrected for temperature by applying equations 4-102 and 4-103, respectively.

Design Considerations

In areas where there are severe winter conditions, deep lagoons are more effective because of heat retention. Under such conditions, diffused aeration using coarse bubble air diffusers are usually preferable to surface aeration, and two cells in series will produce a higher-quality effluent than a single cell (McKinney et al., 1971).

The effluent from an aerobic lagoon is quite high in BOD_5 and suspended solids. It has been shown that the primary cause of the residual BOD_5 is unflocculated microbial solids. Thus, if the effluent is to meet secondary treatment requirements, suspended solids removal will be required. To accomplish this, some designers have included baffle sections which allow the solids to settle prior to discharge, while other designers have employed clarifiers or facultative ponds to achieve solids–liquid separation. If a clarifier is used, operational problems generally develop because at the θ_c where most of these systems operate, the microbial cells have a tendency to remain dispersed and therefore separate poorly. If a facultative pond is placed directly after a single cell aerobic lagoon, the biomass will settle out, but the inorganic nutrients will stimulate a high algal growth. Operational data from such ponds show little change in total effluent BOD_5, only a change in form from bacterial cells to algae cells.

McKinney et al. (1971) recommends that a three-cell series of aerobic lagoons be employed to achieve maximum stabilization of the bacterial solids before discharge to a sedimentation pond. The sedimentation pond following this system is usually 8 to 10 ft deep to provide adequate sludge storage and has a much smaller surface area than would be required for a facultative pond following a single-cell aerobic lagoon. The retention time for a sedimentation pond of this type is usually between 1 and 5 days.

According to McKinney et al. (1971), the optimum balance between power requirements for oxygen transfer and for mixing occurs at a hydraulic retention time of approximately 24 h for an aerobic lagoon treating domestic wastewater. He therefore recommends this value for design. To maximize process efficiency, the retention time in each cell of a series system would also be designed for a 24-h retention time.

During facilities planning when numerous treatment alternatives are under consideration, it is convenient to base the design on criteria previously used to design similar systems. This allows general comparisons to be made between process alternatives without the cost of laboratory experimentation. As an aid to the design engineer, typical design criteria for an aerobic lagoon are given in Table 6-11.

TABLE 6-11

DESIGN CRITERIA FOR AEROBIC LAGOONS

Parameter	*Value*	*Reference*
Loading range ($lb\ BOD_5/day/lb\ MLSS$)	≈ 2.0	Eckenfelder, 1967
Retention time (days)	1–10	McKinney et al., 1971
Lagoon depth (ft)	8–16	Eckenfelder, 1972
Effluent suspended solids (mg/ℓ)	260–300	Metcalf and Eddy, 1972
BOD_5 removal (%)	80–95	Eckenfelder, 1970

The design of an aerobic lagoon depends on the values selected for the biokinetic constants K, Y_T, and K_d. The values for these coefficients are a characteristic of the nature of the wastewater and the microbial culture developed and are independent of the type of treatment. As a result, the biokinetic coefficients presented in Table 4-9 are applicable to the design of aerobic lagoons.

Example 6-3

Design a three-cell aerobic lagoon system based on the following design criteria:

1/ Wastewater flow = 10 MGD.

2/ Wastewater strength = 300 mg/ℓ BOD_u.

3/ Wastewater TKN = 20 mg/ℓ as N.
4/ Wastewater phosphorus = 2 mg/ℓ as P.
5/ Average winter wastewater temperature = 55°F.
6/ Average summer wastewater temperature = 70°F.
7/ Critical winter ambient air temperature = 40°F.
8/ Critical summer ambient air temperature = 80°F.
9/ Substrate utilization constant, $K = 0.3$ ℓ/mg-day at 20°C.
10/ Theoretical yield coefficient, $Y_T = 0.5$.
11/ Microbial decay coefficient, $K_d = 0.1$ day^{-1} at 20°C.
12/ Assume that biokinetic constant values are based on BOD_u.
13/ Operating pH = 7.6.

solution

1/ Using a 24-h retention time and 10 ft depth for each cell, compute the associated surface area.

$$V = Qt$$
$$= (10)(1) = 10 \text{ MG}$$

or

$$V = \frac{10{,}000{,}000}{7.5} = 1{,}333{,}333 \text{ ft}^3$$

The associated surface area is, therefore,

$$A = \frac{1{,}333{,}333}{10} = 133{,}333 \text{ ft}^2$$

or

$$A = \frac{133{,}333}{43{,}560} = 3.1 \text{ acres per lagoon}$$

2/ Assuming an f value of 12×10^{-6}, estimate both the winter and summer operating temperature by applying equation 6-53.

Winter:

$$T_w = \frac{(133{,}333)(12 \times 10^{-6})(40) + (10)(55)}{(133{,}333)(12 \times 10^{-6}) + 10}$$
$$= 53°\text{F or } 11.6°\text{C}$$

Summer:

$$T_w = \frac{(133{,}333)(12 \times 10^{-6})(80) + (10)(70)}{(133{,}333)(12 \times 10^{-6}) + 10}$$
$$= 70°\text{F or } 21.6°\text{C}$$

3/ Correct K and K_d for temperature variations using equations 4-102 and 4-103. The θ value for aerobic lagoons will typically vary between 1.05 and 1.09. This process is somewhat more sensitive to temperature variations than is the activated sludge process because the solids concentration is much lower.

Winter:

$$K = (0.3)(1.07)^{11.6-20}$$
$$= 0.17\ \ell/\text{mg-day}$$
$$K_d = (0.1)(1.05)^{11.6-20}$$
$$= 0.067\ \text{day}^{-1}$$

Summer:

$$K = (0.3)(1.07)^{21.6-20}$$
$$= 0.33\ \ell/\text{mg-day}$$
$$K_d = (0.1)(1.05)^{21.6-20}$$
$$= 0.11\ \text{day}^{-1}$$

4/ Compute the soluble effluent BOD_u from the first cell by applying equation 4-23.

Winter:

$$S_1 = \frac{1 + (0.067)(1)}{(0.5)(0.17)(1)}$$
$$= 12.5\ \text{mg}/\ell$$

Summer:

$$S_1 = \frac{1 + (0.11)(1)}{(0.5)(0.33)(1)}$$
$$= 6.7\ \text{mg}/\ell$$

5/ Determine the biomass concentration (measured as VSS) in the first cell from equation 6-37.

Winter:

$$X_1 = \frac{300 - 12.5}{(12.5)(0.17)(1)}$$
$$= 135\ \text{mg}/\ell$$

Summer:

$$X_1 = \frac{300 - 6.7}{(6.7)(0.33)(1)}$$
$$= 132\ \text{mg}/\ell$$

6/ Determine the biomass concentration in the second cell by assuming that all the soluble BOD_u entering the cell is removed. A steady-state material-balance equation for biomass entering and leaving the second cell can be expressed as

$$0 = QX_1 + Y_T S_1 Q - K_d X_2 V - QX_2$$

or

$$X_2 = \frac{X_1 + Y_T S_1}{(K_d/t) + 1}$$

Winter:

$$X_2 = \frac{135 + (0.5)(12.5)}{(0.067)/1) + 1}$$
$$= 132\ \text{mg}/\ell$$

Summer:

$$X_2 = \frac{132 + (0.5)(6.7)}{(0.11/1) + 1}$$
$$= 122 \text{ mg}/\ell$$

7/ Compute the biomass concentration in the third cell, assuming that the soluble BOD_u entering this cell is zero. A steady-state material-balance equation for biomass entering and leaving the third cell can be expressed as

$$0 = QX_2 - K_d X_3 V - QX_3$$

or

$$X_3 = \frac{X_2}{(K_d/t) + 1}$$

Winter:

$$X_3 = \frac{132}{(0.067/1) + 1}$$
$$= 124 \text{ mg}/\ell$$

Summer:

$$X_3 = \frac{122}{(0.11/1) + 1}$$
$$= 110 \text{ mg}/\ell$$

8/ Assuming that the soluble effluent BOD_5 from the third cell is negligible, estimate the total effluent BOD_5 before solids–liquid separation by applying equation 6-44.

Winter:

$$(BOD_5)_{eff} = (0.54)(124)$$
$$= 67 \text{ mg}/\ell$$

Summer:

$$(BOD_5)_{eff} = (0.54)(110)$$
$$= 59 \text{ mg}/\ell$$

These calculations are based on the assumption that no algal growth occurs in any of the three cells.

9/ Compute the minimum θ_c required for nitrification. This value will establish whether or not NOD must be included in the total oxygen requirement. Assume that $(\mu_{max})_{NS}$ at 20°C is 0.4 day^{-1} and that $(pH)_{opt}$ is 8.2.

a/ Determine the correction factor for pH.

$$\frac{(\mu_{max})_{NS}}{(\mu_{max})_{NS[(pH)_{opt}]}} = \frac{1}{1 + 0.04[10^{(pH)_{opt} - pH} - 1]}$$

$$\text{factor} = \frac{1}{1 + 0.04(10^{8.2-7.6} - 1)}$$
$$= 0.892$$

b/ Determine the correction factor for temperature.

$$\frac{(\mu_{max})_{NS}}{(\mu_{max})_{NS(20°C)}} = 10^{0.033(T-20)}$$

Winter:

$$\text{factor} = 10^{0.033(11.6-20)} = 0.53$$

Summer:

$$\text{factor} = 10^{0.033(21.6-20)} = 1.13$$

c/ Estimate the corrected $(\mu_{max})_{NS}$.

Winter:

$$(\mu_{max})_{NS} = (0.4)(0.892)(0.53) = 0.19\ \text{day}^{-1}$$

Summer:

$$(\mu_{max})_{NS} = (0.4)(0.892)(1.13) = 0.4\ \text{day}^{-1}$$

d/ Calculate the BSRT below which nitrification will not occur.

$$(\theta_c^m)_N = \frac{1}{(\mu_{max})_{NS}}$$

Winter:

$$(\theta_c^m)_N = \frac{1}{0.19} = 5.3\ \text{days}$$

Summer:

$$(\theta_c^m)_N = \frac{1}{0.4} = 2.5\ \text{days}$$

Nitrification can be expected to occur in the third cell during the summer months.

10/ Check the nutrient requirement to ensure that sufficient amounts of nitrogen and phosphorus are available. Since biological growth occurs primarily in the first cell, conditions in the first cell will control the nutrient requirement.

Nutrient requirement:

Winter:

$$\Delta X = (8.34)(10)(135) = 11{,}259\ \text{lb/day}$$

From equation 4-100:

$$\text{nitrogen requirement} = (0.122)(11{,}259) = 1373\ \text{lb/day}$$

From equation 4-101:

$$\text{phosphorus requirement} = (0.023)(11{,}259) = 259\ \text{lb/day}$$

Summer:

$$\Delta X = (8.34)(10)(132) = 11{,}009\ \text{lb/day}$$

From equation 4-100:

$$\text{nitrogen requirement} = (0.122)(11{,}009) = 1343 \text{ lb/day}$$

From equation 4-101:

$$\text{phosphorus requirement} = (0.023)(11{,}009) = 253 \text{ lb/day}$$

Nutrients available:

$$\text{nitrogen} = (10)(8.34)(20) = 1668 \text{ lb/day}$$

$$\text{phosphorus} = (10)(8.34)(2) = 167 \text{ lb/day}$$

These calculations indicate that 86 lb/day of phosphorus must be added to the system. This is usually added in the form of phosphoric acid.

11/ Compute the nitrogen available for nitrification during the summer.

$$N = \frac{1668 - 1343}{(8.34)(10)} = 3.9 \text{ mg}/\ell$$

12/ Calculate the oxygen requirement for each cell of the lagoon from equation 6-39.

For the first cell, neglecting NOD:

Winter:

$$\Delta O_2 = (8.34)(10)\{[1 - (1.42)(0.5)](300 - 12.5)\} + (8.34)(1.42)(0.067)(135)(10) = 8024 \text{ lb/day}$$

Summer:

$$\Delta O_2 = (8.34)(10)\{[1 - (1.42)(0.5)](300 - 6.7)\} + (8.34)(1.42)(0.11)(132)(10) = 8813 \text{ lb/day}$$

For the second cell, neglecting NOD:

Winter:

$$\Delta O_2 = (8.34)(10)\{[1 - (1.42)(0.5)](12.5)\} + (8.34)(1.42)(0.067)(132)(10) = 1349 \text{ lb/day}$$

Summer:

$$\Delta O_2 = (8.34)(10)\{[1 - (1.42)(0.5)](6.7)\} + (8.34)(1.42)(0.11)(122)(10) = 1751 \text{ lb/day}$$

For the third cell, assuming 100% nitrification in summer:
Winter:

$$\Delta O_2 = (8.34)(1.42)(0.067)(124)(10)$$
$$= 984 \text{ lb/day}$$

Summer:

$$\Delta O_2 = (8.34)(1.42)(0.11)(110)(10) + (38.1)(10)(3.9)$$
$$= 2919 \text{ lb/day}$$

13/ Following the procedure outlined in Example 5-3, assume that it has been found that a particular type of surface aerator will deliver 1.8 lb O_2/hp-h under process conditions. Determine the total horsepower requirement for oxygen transfer.
For the first cell:
Winter:

$$\text{hp} = \frac{8024}{(1.8)(24)}$$
$$= 186$$

Summer:

$$\text{hp} = \frac{8813}{(1.8)(24)}$$
$$= 204$$

For the second cell:
Winter:

$$\text{hp} = \frac{1349}{(1.8)(24)}$$
$$= 31$$

Summer:

$$\text{hp} = \frac{1715}{(1.8)(24)}$$
$$= 41$$

For the third cell:
Winter:

$$\text{hp} = \frac{984}{(1.8)(24)}$$
$$= 23$$

Summer:

$$\text{hp} = \frac{2{,}919}{(1.8)(24)}$$
$$= 68$$

14/ Malina et al. (1972) found that power levels near 30 hp/MG are required for complete mixing. Assuming that this value is realistic, compute the required horsepower per cell if complete mixing is to be achieved.

(Note that design criteria recommended by the Clow Corporation suggest a minimum rate of 60 hp/MG.)

$$\text{mixing hp} = (30)(10) = 300$$

Therefore, mixing controls the horsepower requirement in all three cells.

15/ Assuming that a length/width ratio of 3 :1 is to be used, determine whether or not twelve 25-hp aerators will provide adequate mixing in each cell if each aerator has a zone of influence for complete mixing of 120 ft for a 10-ft depth. (*Note:* It is recommended that the length/width ratio not exceed 1.25 in actual practice.)

$$(3W)(W) = 133{,}333 \text{ ft}^2$$

$$W = 211 \text{ ft}$$

Therefore,

$$L = 3(211) = 633 \text{ ft}$$

The aerator layout given in Figure 6-27 suggests that dead spaces will exist and slightly larger units probably should be used.

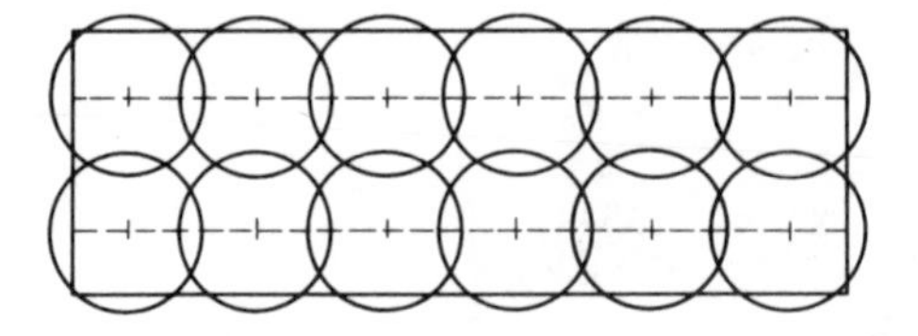

Figure 6-27. Aerator Layout for Example 6-3.

In practice, a temperature drop will be noted across a series lagoon system when surface aeration is employed. If the temperature change is more than three or four degrees, adjustment of the K values may be necessary.

6-7 Facultative Lagoons

The use of facultative lagoons is much more common than the use of aerobic lagoons, mainly because a good effluent can be produced with a low power input. As a consequence of the low power input, both BOD removal and solids separation occurs within the lagoon. The major disadvantage to this process is the long retention time required for wastewater treatment (Kormanik, 1972).

As discussed previously, in the facultative lagoon power is required only to create turbulence levels that are sufficient to disperse dissolved oxygen throughout the liquid phase. The bulk of the solids are not maintained in suspension but settle to the bottom of the lagoon, where they are decomposed anaerobically. The relationship between the suspended solids concentration in the lagoon and the power input is presented in Figure 6-28.

Aerator design for facultative lagoons is based only on oxygen requirement, with no consideration given to mixing.

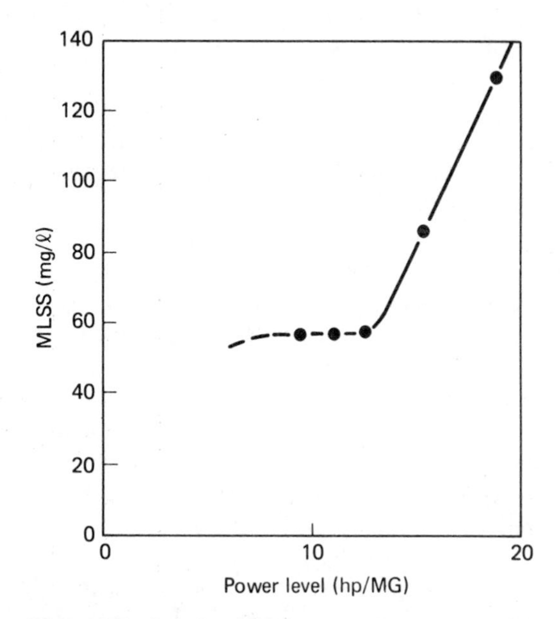

Figure 6-28. Relationship Between Suspended Solids Concentration and Power Input. (After Malina et al., 1972.)

Design Considerations

It may take several years for a facultative lagoon to reach steady-state with respect to solids. Until this condition is reached, substrate utilization rates will continually change because of the sedimentation of the bulk of solids in the influent and a portion of the biomass. The sedimentation occurs in the relatively quiescent portions of the lagoon away from the high-turbulence areas near aerators. A net increase in settled solids will occur even though a portion of the settled solids will be hydrolyzed during anaerobic fermentation. The net increase in settled solids results in a decrease in the liquid volume in the basin, thus causing the zone of high turbulence surrounding the aerators to increase with respect to the total liquid volume in the lagoon. Therefore, more solid material will be maintained in suspension, where it will be decomposed aerobically (Bartsch and Randall, 1971). When the rate of sedimentation equals the rate of resuspension, the lagoon will have reached steady-state conditions.

There are basically three obstacles which prevent the development of a rational procedure for the design of facultative lagoons. These are (1) the relatively long period required to attain steady-state conditions, (2) feedback of soluble BOD to the aerobic layer as a result of anaerobic fermentation, and (3) difficulty in estimating the biomass concentration under aeration. Because of these problems, any design procedure will lack a firm theoretical base. However, it should be noted that the mechanism of BOD removal within facultative lagoons and facultative ponds is basically the same. The primary difference between these two treatment systems is the means by which oxygen is supplied. In the case of a facultative lagoon, oxygen is provided through the action of mechanical aeration, whereas photosynthesis is the major source of oxygen for a facultative pond.

Because of the basic similarity between the two systems, it is felt that equation 6-16 [the relationship proposed by Marais and Shaw (1961) for the design of facultative ponds] is applicable to the design of facultative lagoons. The reaction-rate constant, K, can be estimated from equation 6-17 or Figure 6-10. Such an approach to design should be a conservative one.

McKinney et al. (1971) report that the hydraulic retention time for most facultative lagoons will fall within the range 7 to 20 days. They further state, however, that experience favors the use of 4 to 8 days' retention time, which normally results in 70 to 80% BOD removal.

In a facultative lagoon the oxygen requirement is increased by the feedback of soluble BOD from anaerobic fermentation. As a result, Eckenfelder et al. (1972) recommend that oxygen requirements be computed from the relationship

$$\Delta O_2 = (8.34)FQ(S_0 - S_e) \tag{6-54}$$

where ΔO_2 = oxygen requirement, lb/day
S_0 = influent ultimate BOD, mg/ℓ
S_e = effluent soluble ultimate BOD, mg/ℓ
F = soluble BOD feedback factor, which is given a value of 0.9 during winter operations and a value of 1.4 during summer operations

According to McKinney et al. (1971), oxygen requirements may be computed on the basis of 1.5 lb oxygen/lb BOD_u applied to the system.

Typical design criteria are presented in Table 6-12. Such criteria should aid the engineer engaged in the design of a facultative lagoon treatment system, especially since such a design has very little rational basis.

TABLE 6-12

DESIGN CRITERIA FOR FACULTATIVE LAGOONS

Parameter	*Value*	*Reference*
Loading range (lb BOD_5/acre/day)	30–100	Metcalf and Eddy, 1972
Retention time (days)	7–20 (may be longer in northern climates)	Metcalf and Eddy, 1972
Lagoon depth (ft)	8–16	Eckenfelder et al., 1972
Effluent suspended solids (mg/ℓ)	110–340	Metcalf and Eddy, 1972

PROBLEMS

6-1 A series pond system is employed for wastewater treatment. The system is composed of a facultative pond followed by two polishing ponds, each with a retention time of 10 days. Compute the percent reduction in fecal bacteria during the summer months when the operating temperature in each pond is expected to be 60°F. Assume that $N_0 = 4 \times 10^7$ FC/100 ml.

6-2 Design a series-pond system to treat a meatpacking wastewater flow of 0.5 MGD with a strength of 1000 mg/ℓ BOD_5. The system should consist of an anaerobic pond followed by a facultative pond, which in turn is followed by two polishing ponds. The ponds will be located in an area where the average temperature for the coldest winter month is expected to be 45°F. Present a flow pattern of the final design, indicating if recirculation is to be used. If so, justify its use and the particular recirculation pattern selected.

6-3 Design a single-cell aerobic lagoon system based on the following design criteria:

a/ Wastewater flow = 1 MGD.
b/ Wastewater strength = 400 mg/ℓ BOD_u.
c/ Wastewater TKN = 20 mg/ℓ as N.
d/ Wastewater phosphorus = 10 mg/ℓ as P.
e/ Average winter wastewater temperature = 50°F.
f/ Average summer wastewater temperature = 75°F.
g/ Critical winter ambient air temperature = 40°F.
h/ Critical summer ambient air temperature = 85°F.
i/ Substrate utilization-rate constant, $K = 0.3$ ℓ/mg-day at 20°C., $\theta = 1.07$.
j/ Theoretical yield coefficient, $Y_T = 0.5$.
k/ Microbial decay coefficient, $K_d = 0.1\ \text{day}^{-1}$ at 20°C., $\theta = 1.05$.
l/ $(\mu_{max})_{NS}$ at 20°C and $(\text{pH})_{opt} = 0.4\ \text{days}^{-1}$.

m/ $(pH)_{opt} = 8.2$.
n/ Operating pH = 7.6.
o/ Lagoon depth = 8 ft.
p/ 60 hp/MG required for complete mixing.
q/ Hydraulic retention time = 1 day.
r/ $BOD_5 = 0.7\ BOD_u$.

Vary the lagoon depth from 8 to 16 ft and compute the soluble effluent substrate concentration associated with each depth. Plot S_e versus lagoon depth and explain the observed relationship between these two parameters. Biokinetic constants are based on ultimate BOD. Also, assume that the actual oxygen transfer rate is 1.0 lb O_2/hp/h and use the data in Table 6-8 for aerator design and layout.

6-4 An aerobic pond is to be used to treat a wastewater flow of 0.5 MGD from a community located at 25°N latitude. What size pond is required if the wastewater has a BOD_u of 150 mg/ℓ and 90% removal is required? The critical design month is January when cloud cover exists 40% of the time and the temperature averages 60°F. Assume a mean algal cell composition of 52% carbon, 9% hydrogen, 32% oxygen, and 7% nitrogen. The elevation above sea level is 100 ft and $BOD_5 = 0.7BOD_u$. Use a value of 0.04 for the energy utilization efficiency.

REFERENCES

ABELIOVICH, A., and D. WEISMAN, "Role of Hetrotrophic Nutrition in Growth of the Alga *Scenedesmus obliquus* in High Rate Oxidation Ponds," *Applied and Environmental Microbiology*, **35**, 32 (1978).

BALASHA, E., and H. SPERBER, "Treatment of Domestic Wastes in an Aerated Lagoon and Polishing Pond," *Water Research*, **9**, 43 (1975)

BARNHART, E. L., "Aerated Lagoons," in *Process Design in Water Quality Engineering*, ed. by E. L. Thackston and W. W. Eckenfelder, Jenkins Publishing Company, New York, 1972.

BARSOM, G., "Lagoon Performance and the State of Lagoon Technology," EPA Environmental Protection Technology Series, EPA-R-2-73-144, 1973.

BARTSCH, A. F., "Algae as a Source of Oxygen in Waste Treatment," *Journal of the Water Pollution Control Federation*, **33**, 239 (1961).

BARTSCH, E. H., and C. W. RANDALL, "Aerated Lagoons—A Report on the State of the Art," *Journal of the Water Pollution Control Federation*, **43**, 699 (1971).

CALDWELL, D. H., D. S. PARKER, and W. R. UHTE, "Upgrading Lagoons," EPA Technology Transfer Seminar Publication, 1973.

CANTER, L. W., and A. J. ENGLANDE, JR., "States' Design Criteria for Waste Stabilization Ponds," *Journal of the Water Pollution Control Federation*, **42**, 1840 (1970).

DILDANE, E. D., and J. R. FRANZMATHES, "Current Design Criteria for Oxidation Ponds," in *Proceedings of the Second International Symposium on Waste Treatment Lagoons*, Kansas City, Mo., 1970.

Dryden, F. D., and G. Stern, "Renovated Waste Water Creates Recreational Lake," *Environmental Science and Technology*, **2**, 268 (1968).

Eckenfelder, W. W., Jr., "Comparative Biological Waste Treatment Design," *Journal of the Sanitary Engineering Division, ASCE*, **93**, SA6, 157 (1967).

Eckenfelder, W. W., Jr., *Water Quality Engineering*. Cahners Books, Boston, 1970.

Eckenfelder, W. W., Jr., and D. J. O'Connor, *Biological Waste Treatment*, Pergamon Press, New York, 1961.

Eckenfelder, W. W., Jr., C. D. Magee, and C. E. Adams, Jr., "A Rational Design Procedure for Aerated Lagoons Treating Municipal and Industrial Wastewaters," paper presented at the 6th International Water Pollution Research Conference, 1972.

Fischer, C. P., W. R. Drynan, and G. L. VanFleet, "Waste Stabilization Pond Practices in Canada," in *Advances in Water Quality Improvement*, Vol. I, University of Texas, Austin, Tex., 1968.

Ford, D. L., "Aeration," in *Process Design in Water Quality Engineering*, ed. by E. L. Thackston and W. W. Eckenfelder, Jenkins Publishing Company, New York, 1972.

Foree, E. G., and P. L. McCarty, "The Decomposition of Algae in Anaerobic Waters," Technical Report No. 95, Department of Civil Engineering, Stanford University, Stanford, Calif., 1968.

Gloyna, E. F., "Waste Stablization Pond Concepts and Experiences," *Wastes Disposal Unit Paper*, World Health Organization, Geneva, 1965.

Gloyna, E. F., "Basis for Waste Stabilization Pond Designs," in *Advances in Water Quality Improvement*, ed. by E. F. Gloyna and W. W. Eckenfelder, University of Texas Press, Austin, Tex., 1968.

Gloyna, E. F., "Waste Stabilization Pond Design," in *Process Design in Water Quality Engineering*, ed. by E. L. Thackston and W. W. Eckenfelder, Jenkins Publishing Company, New York, 1972.

Gloyna, E. F., "Facultative Waste Stabilization Pond Design," in *Ponds as a Wastewater Treatment Alternative*, ed. by E. F. Gloyna, J. F. Malina, Jr., and E. M. Davis, The Center for Research in Water Resources, University of Texas, Austin, Tex., 1976.

Gloyna, E. F., and J. Aguirre, "New Experimental Pond Data," in *Proceedings of the Second International Symposium on Waste Treatment Lagoons*, Kansas City, Mo., 1970.

Hendricks, D. W., and W. D. Pote, "Thermodynamic Analysis of a Primary Oxidation Pond," *Journal of the Water Pollution Control Federation*, **46**, 333 (1974).

Hermann, E. R., and E. F. Gloyna, "Waste Stabilization Ponds," *Sewage and Industrial Wastes*, **30**, 511 (1958).

Jewell, W. J., and P. L. McCarty, "Aerobic Decomposition of Algae and Nutrient Regeneration," Technical Report No. 91, Department of Civil Engineering, Stanford University, Stanford, Calif., 1968.

Kormanik, R. A., "Design of Two-Stage Aerated Lagoons," *Journal of the Water Pollution Control Federation*, **44**, 451 (1972).

LEWIS, R. F., "Review of EPA Research and Development Lagoon Upgrading Program for Fiscal Years 1973, 1974, and 1975," in *Symposium Proceedings*, Upgrading Wastewater Stabilization Ponds to Meet New Discharge Standards, PB 240 402, 1975.

MALINA, J. F., JR., and R. A. RIOS, "Anaerobic Ponds," in *Ponds as a Wastewater Treatment Alternative*, ed. by E. F. Gloyna, J. F. Malina, Jr., and E. M. Davis, The Center for Research in Water Resources, University of Texas, Austin, Tex., 1976.

MALINA, J. F., JR., R. KAYSER, W. W. ECKENFELDER, JR., E. F. GLOYNA, and W. R. DRYNAN, *Design Guides for Biological Wastewater Treatment Processes*, Center for Research in Water Resources Report, CRWR-76, University of Texas, Austin, Tex., 1972.

MANCINI, J. L., and E. L. BARNHART, "Industrial Waste Treatment in Aerated Lagoons," in *Advances in Water Quality Improvement*, ed. by E. F. Gloyna and W. W. Eckenfelder, Jr., University of Texas Press, Austin, Tex., 1968.

MARA, D. D., "Discussion," *Water Research*, **9**, 595 (1975a).

MARA, D. D., "Author's Reply," *Water Research*, **9**, 596 (1975b).

MARA, D. D., *Sewage Treatment in Hot Climates*, John Wiley & Sons, Inc., New York, 1976.

MARAIS, G. V. R., "Dynamic Behavior of Oxidation Ponds," in *Proceedings of the Second International Symposium on Waste Treatment Lagoons*, Kansas City, Mo., 1970.

MARAIS, G. V. R., "Faecal Bacterial Kinetics in Stabilization Ponds," *Journal of the Environmental Engineering Division, ASCE*, **100**, 119 (1974).

MARAIS, G. V. R., and V. A. SHAW, "A Rational Theory for the Design of Sewage Stabilization Ponds in Central and South Africa," *Transactions, South Africa Institute of Civil Engineers*, **3**, 205, (1961).

MCCARTY, P. L., "Kinetics of Waste Assimilation in Anaerobic Treatment," in *Developments in Industrial Microbiology*, Vol. 7, American Institute of Biological Sciences, Washington, D.C., 1966, p. 144.

MCGARRY, M. G., and M. B. PESCOD, "Stabilization Pond Design Criteria for Tropical Asia," in *Proceedings of the Second International Symposium on Waste Treatment Lagoons*, Kansas City, Mo., 1970.

MCGAUHEY, P. H., *Engineering Management of Water Quality*, McGraw-Hill Book Company, New York, 1968.

MCINTOSH, G. H., and G. G. MCGEORGE, "Year Round Lagoon Operation," *Food Processing*, (Jan. 1964).

MCKINNEY, R. E., "State of the Art of Lagoon Wastewater Treatment," in *Symposium Proceedings*, Upgrading Wastewater Stabilization Ponds to Meet New Discharge Standards, PB 240 402, 1975.

MCKINNEY, R. E., J. N. DORNBUSH, and J. W. VENNES, "Waste Treatment Lagoons —State of the Art," Missouri Basin Engineering Health Council, EPA WPCRS, 17090EHX, 1971.

METCALF and EDDY, INC., *Wastewater Engineering*, McGraw-Hill Book Company, New York, 1972.

MIDDLEBROOKS, E. J., A. J. PANAGIOTOU, and H. K. WILLIFORD, "Sludge Accumulation in Municipal Sewage Lagoons," *Water and Sewage Works*, **63** (Feb. 1965).

MIDDLEBROOKS, E. J., D. B. PORCELLA, R. A. GEARHEART, G. R. MARSHALL, J. H. REYNOLDS, and W. J. GRENNEY, "Techniques for Algae Removal from Wastewater Stabilization Ponds," *Journal of the Water Pollution Control Federation*, **46**, 2676 (1974).

MIDDLEBROOKS, E. J., D. B. PORCELLA, R. A. GEARHEART, G. R. MARSHALL, J. H. REYNOLDS, and W. J. GRENNEY, "Authors' Response," *Journal of the Water Pollution Control Federation*, **47**, 2333 (1975).

NUSBAUM, I., Discussion of Photosynthesis in Sewage Treatment, *Transactions, ASCE*, **122**, 98 (1957).

O'BRIEN, W. J., "Polishing Lagoon Effluents with Submerged Rock Filters," in *Symposium Proceedings*, Upgrading Wastewater Stabilization Ponds to Meet New Discharge Standards, PB 240 402, 1975.

OSWALD, W. J., "Advances in Stabilization Pond Design," in *Advances in Biological Waste Treatment*, ed. by W. W. Eckenfelder and J. McCabe, Pergamon Press, New York, 1963.

OSWALD, W. J., "Fundamental Factors in Stabilization Pond Design", in *Proceedings, Third Conference on Biological Waste Treatment*, Manhattan College, New York (1960).

OSWALD, W. J., "Advances in Anaerobic Pond System Design," in *Advances in Water Quality Improvement*, ed. by E. F. Gloyna and W. W. Eckenfelder, University of Texas Press, Austin, Tex., 1968.

OSWALD, W. J., "Complete Waste Treatment in Ponds," in *Proceedings of the 6th International Water Pollution Research Conference*, Pergamon Press, London, 1972.

OSWALD, W. J., and H. B. GOTAAS, "Photosynthesis in Sewage Treatment," *Transactions, ASCE*, **122**, 73 (1957).

OSWALD, W. J., A. MERON, and M. D. ZALAT, "Designing Waste Ponds to Meet Water Quality Criteria," in *Proceedings of the Second International Symposium on Waste Treatment Lagoons*, Kansas City, Mo., 1970.

PARKER, C. D., and G. P. SKERRY, "Function of Solids on Anaerobic Lagoon Treatment of Wastewater," *Journal of the Water Pollution Control Federation*, **40**, 192, (1968).

PARKER, D. S. and W. R. UHTE, "Discussion," *Journal of the Water Pollution Control Federation*, **47**, 2330 (1975).

RAMANI, R., "Design Criteria for Polishing Ponds," in *Ponds as a Wastewater Treatment Alternative*, ed. by E. F. Gloyna, J. F. Malina, Jr., and E. M. Davis, The Center for Research in Water Resources, University of Texas, Austin, Tex., 1976.

REYNOLDS, J. H., S. E. HARRIS, D. HILL, D. S. FILIP, and E. J., MIDDLEBROOKS, "Intermittent Sand Filtration to Upgrade Lagoon Effluents—Preliminary Report," *Symposium Proceedings*, Upgrading Wastewater Stabilization Ponds to Meet New Discharge Standards, PB 240 402, 1975.

RICH, L. G., *Unit Processes of Sanitary Engineering*, John Wiley & Sons, Inc., New York, 1963.

ROESLER, J. F., and H. C. PRUEL, "Mathematical Simulation of Waste Stabilization Ponds," in *Proceedings of the Second International Symposium on Waste Treatment Lagoons*, Kansas City, Mo., 1970.

SAWYER, C. N., "New Concepts in Aerated Lagoon Design and Operation," in *Advances in Water Quality Improvement*, ed. by E. F. Gloyna and W. W. Eckenfelder, Jr., University of Texas Press, Austin, Tex., 1968.

SHINDALA, A., and W. C. MURPHY, "Influence of Shape on Mixing and Load of Sewage Lagoons," *Water and Sewage Works*, 391 (Oct. 1969).

STANIER, R. Y., M. DOUDOROFF, and E. A. ADELBERG, *The Microbial World*, Prentice-Hall, Inc., Englewood Cliffs, N.J., 1963.

THIRUMURTHI, D., "Relative Toxicity of Organics to *Chlorella Pyrenoidosa*," Doctoral Dissertation, University of Texas, Austin, Tex., 1966.

THIRUMURTHI, D., "Design Principles of Waste Stabilization Ponds," *Journal of the Sanitary Engineering Division, ASCE*, **95**, 311 (1969).

UHTE, W. R., "Construction Procedures and Review of Plans and Grant Applications," in *Symposium Proceedings*, Upgrading Wastewater Stabilization Ponds to Meet New Discharge Standards, PB 240 402, 1975.

VAN ECK, H., and D. E. SIMPSON, "The Anaerobic Pond System," *Journal and Proceedings of the Institute of Sewage Purification*, Part 3 (1966).

VELZ, C. J., *Applied Stream Sanitation*, Wiley–Interscience, New York, 1970.

Attached-Growth Biological Treatment Processes

Suspended-growth biological reactors are not the only type employed in the treatment of wastewaters. Attached-growth systems are also used. Such reactors require the presence of some type of medium to support the biological growth. Reactors of this type which are of importance include (1) trickling filters, (2) rotating biological contactors, (3) anaerobic filters, (4) submerged filters, (5) biological fluidized beds, and (6) activated biofilters (Williamson and McCarty, 1976a).

7-1 Trickling Filters

The name *trickling filter* is somewhat misleading, as the primary mechanism for organic removal is not by the filtering action of fine pores but rather by diffusion and microbial assimilation. The mechanism of substrate removal can be elucidated by considering an elemental volume of microbial film and liquid, as shown in Figure 7-1. Generally, it is assumed that the flow regime is in the laminar range for the hydraulic loadings normally encountered during

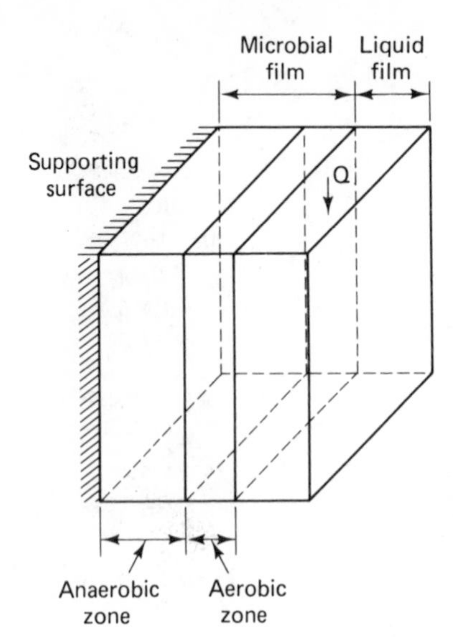

Figure 7-1. Elemental Volume of Microbial Film and Liquid. (After Jank and Drynan, 1973.)

plant operation. It is also assumed that the thickness of the aerobic zone is limited by the depth of penetration of oxygen into the microbial film, which depends upon the coefficient of diffusivity of oxygen in the film, the concentration of oxygen at the solid–liquid interface, and the overall oxygen utilization rate of microorganisms present in the microbial film (Jank and Drynan, 1973). For a specific flow rate and wastewater strength, the thickness of the aerobic zone will attain some unique value. An increase in the strength of the waste will reduce the thickness of this zone, while an increase in flow rate will increase the thickness of the aerobic zone (Jank and Drynan, 1973).

The depth that substrate penetrates into the microbial film is dependent upon (1) the wastewater flow rate, (2) the strength of the wastewater, (3) the coefficient of diffusivity of the substrate molecule in the film, and (4) the rate of substrate utilization by the biomass (Jank and Drynan, 1973). In general, the depth of penetration will increase with both an increase in wastewater strength and flow rate.

As substrate is utilized by the microorganisms, the thickness of the microbial film will increase. After some period of time, the film will reach a thickness such that the substrate is utilized before penetrating the entire depth of the microbial film. The organisms that exist in this starvation zone must utilize their own cytoplasmic material to maintain life-support functions (i.e., they are in the endogenous growth phase). In this phase they lose the ability to cling to the support material and are therefore washed out of the filter. This is known as *sloughing*.

Filter Media

Trickling filtration consists of uniformly distributing wastewater over a bed of support media by a flow distributor (see Figure 7-2). The wastewater forms a thin layer as it flows over the biological growth attached to the surface of the media. Two properties of the filter media are of major importance. These are (1) the specific surface area of the media (the greater the surface area, the greater the amount of biomass per unit volume), and (2) the percent void space (the greater the void space, the higher the hydraulic loading can be without restricting oxygen transfer).

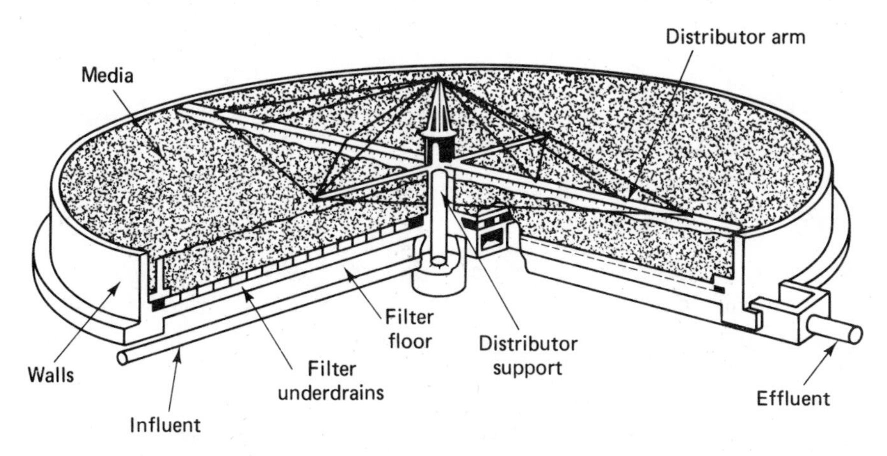

Figure 7-2. Typical Trickling Filter Cross Section. (After Lipták, 1974.)

Stone media and synthetic media are two types of media which are commonly used in trickling filters. Stone-media filters are usually limited to a depth of 3 to 10 ft because of low void space and the requirement for structural support. On the other hand, synthetic media offer the advantages of being lightweight, having larger specific surface areas, and producing a filter with a greater void space. Because of these advantages, trickling filters up to 40 ft deep have been constructed when synthetic media was used as the growth support. Table 7-1 compares the physical properties of certain filter media.

Types of Filters

Trickling filters have been historically classified as high rate or standard rate, based on the organic and hydraulic loading applied to the unit. *High-rate filters* are generally loaded at 10 to 40 million gal/acre/day (MGAD) hydraulically with an organic loading of 23 to 115 lb BOD_5 applied to the filter per day per 1000 ft^3 of filter volume. On the other hand, *standard-*

TABLE 7-1

PHYSICAL PROPERTIES OF VARIOUS TYPES OF TRICKLING FILTER MEDIA

Types of media	*Nominal size (in.)*	*Units per ft^3*	*Unit weight (lb/ft^3)*	*Specific surface area (ft^2/ft^3)*	*Void space (%)*
Granite	1–3	—	90	19	46
	4	—	—	13	60
Blast furnace slag	2–3	51	68	20	49
Aeroblock (vitrified tile)	6 × 11 × 12	2	70	20–22	53
Raschig rings (ceramic)	$1\frac{1}{2} \times 1\frac{1}{2}$	340	40.8	35	68.2
Dowpac 10	$21 \times 37\frac{1}{2}$	2	3.6–3.8	25	94
Dowpac 20	$21\frac{1}{2} \times 38\frac{1}{2}$	2	6	25	94

Source: After Lipták (1974).

rate filters are operated at a surface loading of 1 to 4 million gal/acre/day and an organic load of 7 to 23 lb BOD_5 applied per day per 1000 ft^3 of filter volume (Zajic, 1971). Operational characteristics for both types of filters are presented in Table 7-2.

Recirculation is employed with high-rate filters, whereas standard-rate filters do not use recirculation unless minimal flow (10%) is needed to keep the media wet. It is probable that the benefits of recirculation are restricted to the treatment of organically complex substrates. Figure 7-3 illustrates

TABLE 7-2

COMPARISON OF STANDARD-RATE AND HIGH-RATE TRICKLING FILTERS

Factor	*Standard-rate filter*	*High-rate filter*
Surface loading (MGAD)	1–4	10–40
Organic loading (lb BOD_5-1000 ft^3-day)	7–23	23–115
Depth (ft)	6–10	3–8
Recirculation	None	1:1–4:1
Rock volume	5–10 times	1
Power requirements	None	10–50 hp/MG
Filter flies	Many	Few, larvae washed out
Sloughing	Intermittent	Continuous
Operation	Simple	Some skill

Source: After Metcalf and Eddy (1972).

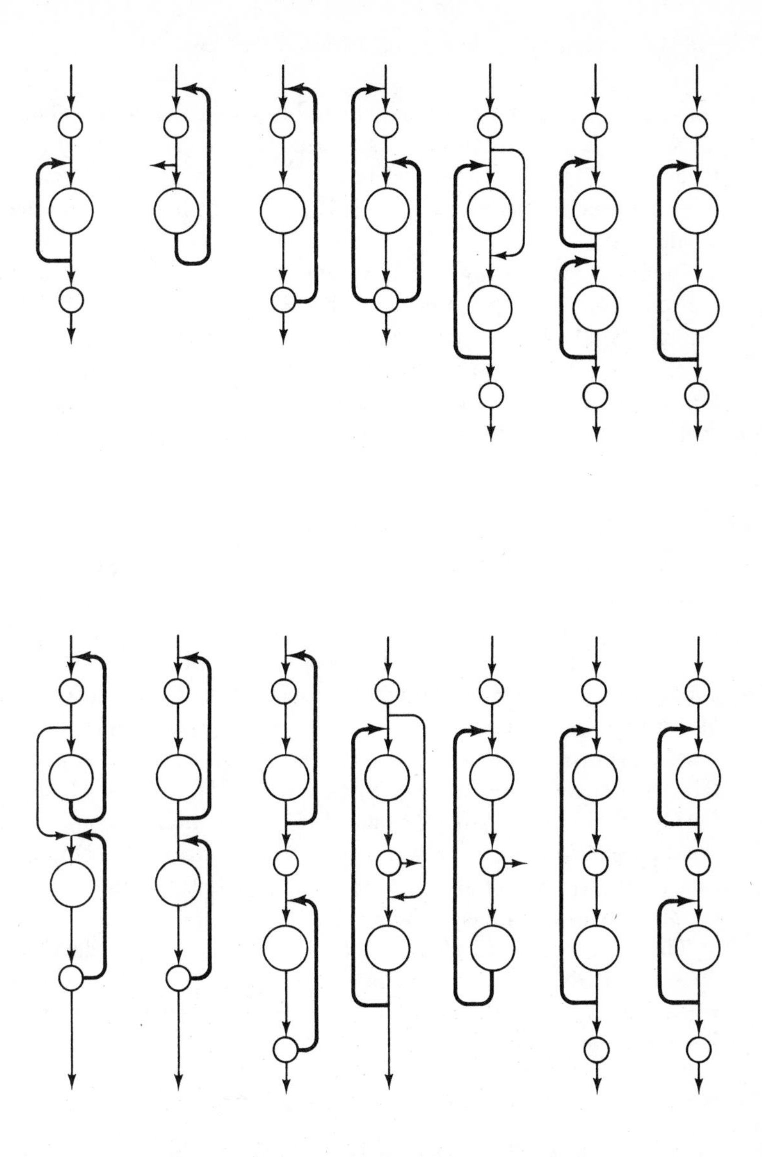

Recirculation design arrangements available are varied.
In the diagram above, larger circles represent filters; small
circles are clarifiers; and heavy lines, recirculation paths.
BOD reductions are somewhat dependent on layout employed.

Figure 7-3. Recirculation Patterns for High-Rate Filters. (After Benzie, 1970.)

recirculation flow patterns that have been used in design. This figure also shows single-stage treatment as well as two-stage treatment (i.e., treatment when two filters are connected in series). Two-stage treatment is used when high efficiencies are required. This is equivalent to increasing the depth of a single filter.

Figure 7-3 also shows cases where direct recirculation is employed (i.e., recirculation back to the head of the filter, as well as recirculation through the clarifiers). When recirculation through the clarifier is used, an increase in clarifier size will result because of the increase in flow. However, Culp (1963) found there was no benefit gained from recycling through the clarifier; therefore, direct recirculation is recommended as the method of choice.

Popular Design Equations

Because of the complex interrelationships between process variables, no good mechanistic model that will accurately predict filter performance has been developed. Thus, when possible, the results of pilot-plant studies should be used for design purposes. However, when pilot-plant studies cannot be conducted, there are design equations which can be used to estimate plant performance. These equations include those given by (1) the National Research Council (Subcommittee, 1946), (2) Great Lakes–Upper Mississippi River Board of State Sanitary Engineers (1971), (3) Velz (1948), (4) Rankin (1955), (5) Fairall (1956), (6) Stack (1957), (7) Schulze (1960), (8) Eckenfelder (1961), (9) Galler and Gotaas (1964), and (10) Kornegay (1975). Of these equations, only the National Research Council, Galler–Gotaas, and Eckenfelder equations will be discussed here.

The National Research Council (NRC) equation was developed from the operational data of stone-media trickling filters treating wastewater from various military installations. For first- or single-stage treatment, the equation has the form

$$E_1 = \frac{1}{1 + 0.0561(W/VF)^{1/2}} \tag{7-1}$$

where E_1 = BOD_5 removal efficiency through filter and secondary clarifier, fraction

W = BOD_5 applied to filter, lb/day; this does not include BOD_5 of the recycled waste

V = volume of filter, 1000 ft^3

F = recirculation factor for a particular stage

The recirculation factor is defined by the expression

$$F = \frac{1 + R}{[1 + (1 - \delta)R]^2} \tag{7-2}$$

where R = recirculation ratio (i.e., the rate of recycle divided by the rate of inflow to the filter)

δ = weighting factor included to compensate for the assumption that the rate of substrate utilization decreased with each pass through the filter (a volume of 0.9 is generally assigned to δ)

In developing an expression to predict the efficiency of second-stage filters, it was observed that the rate of substrate utilization from the effluent of first-stage treatment was less than the rate of substrate utilization observed in the first filter. It was concluded that there was a "decrease in treatability" of the waste applied to the second stage. To compenstate for this, a retardation factor, $[1/(1 - E_1)]^2$, was included in the equation for second-stage efficiency.

$$E_2 = \frac{1}{1 + 0.0561\left[\dfrac{W_2}{V_2F(1 - E_1)^2}\right]^{1/2}} \tag{7-3}$$

or

$$E_2 = \frac{1}{1 + \dfrac{0.0561}{1 - E_1}\left(\dfrac{W_2}{V_2F}\right)^{1/2}} \tag{7-4}$$

where E_2 = BOD_5 removal efficiency through the second stage filter and clarifier, fraction

W_2 = BOD_5 applied to second-stage filter, lb/day

V_2 = volume of second-stage filter, 1000 ft^3

Equation 7-4 was developed on the assumption that an intermediate clarifier will exist between the two filters.

If the NRC equation is used in filter design, only three parameters may be varied in order to calculate the desired efficiency: (1) volume of media, (2) number of stages, and (3) recirculation ratio. Equation 7-1, which describes single-stage treatment, can be rearranged into the form (Baker and Graves, 1968)

$$V_1 = 0.0263QS_0\frac{(1 + 0.1R)^2}{1 + R}\left(\frac{E_1}{1 - E_1}\right)^2 \tag{7-5}$$

where V_1 = filter volume, 1000 ft^3

Q = influent flow rate, MGD

S_0 = influent substrate concentration, mg/ℓ

Equation 7-5 shows that the volume required to maintain a constant treatment efficiency varies directly with wastewater flow and strength if the recycle ratio is held constant. These same results can be observed if equation 7-4, describing second-stage treatment, is rearranged into the form

$$V_2 = 0.0263QS_1\frac{(1 + 0.1R)^2}{1 + R}\left[\frac{E_2}{(1 - E_1)(1 - E_2)}\right]^2 \tag{7-6}$$

where S_1 is the effluent BOD_5 concentration from the first stage.

Increasing the number of stages from one to two will increase the efficiency of the process. A similar response was noted in Chapter 1 for completely mixed reactors operating in series. For this type of system the maximum efficiency is achieved when the retention time in each reactor is the same. Applying these basic fundamentals to a series of trickling filters means that maximum efficiency is achieved when filter volumes and recirculation ratios are equal. This is illustrated in Figure 7-4 through the application of the NRC equation.

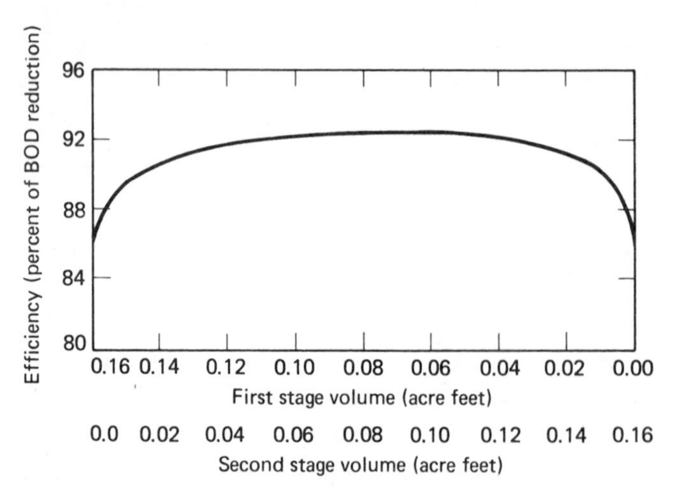

Figure 7-4. Optimization Curve for Two-Stage Filter with a Fixed Total Volume. (After Archer and Robinson, 1970.)

The NRC equation was developed on the premise that recirculation was beneficial. An examination of either equation 7-5 or 7-6 reveals that the equations predict a decrease in the total filter volume required to maintain a constant efficiency if the recycle ratio is increased, up to a value of 5. Above this value no significant decrease in volume is obtained.

Example 7-1

Design a two-stage trickling filter system to treat a wastewater flow of 2 MGD with a strength of 400 mg/ℓ BOD_5. Each filter is to be 8 ft deep and operated at a recirculation ratio of 4: 1. A 90% BOD_5 removal is desired.

solution

For the most effective design, the volume of the two filters should be approximately the same. To determine this volume, it will be necessary to follow a trial-and-error procedure by assuming the efficiency of the first stage and then calculating the efficiency of the second stage and the associated volumes. When the two volumes are approximately the same, the correct design volumes have been reached.

1/ Develop a relationship between E_1 and E_2 by writing an efficiency balance for a "unit" of BOD_5.

$$\underbrace{(1)E_1}_{\text{BOD}_5\text{ removal in first stage}} + \underbrace{(1)(1 - E_1)E_2}_{\text{BOD}_5\text{ removal in second stage}} = \underbrace{(1)E_{\text{total}}}_{\text{total BOD}_5\text{ removal}}$$

or

$$E_1 + (1 - E_1)E_2 = E_{\text{total}}$$

Substituting 0.9 for E_{total} and solving for E_2 gives

$$E_2 = \frac{0.9 - E_1}{1 - E_1}$$

The first trial proceeds as follows:

2/ Assuming that $E_1 = 0.80$, compute the volume of the first-stage filter from equation 7-5.

$$V_1 = (0.0263)(2)(400)\,\frac{[1 + (0.1)(4)]^2}{1 + 4}\left(\frac{0.80}{1 - 0.80}\right)^2$$

$$= 131.95 \text{ thousand ft}^3$$

3/ Calculate the value of E_2 and then use this value to compute V_2 from equation 7-6.

$$E_2 = \frac{0.9 - 0.8}{1 - 0.8}$$

$$= 0.5$$

Therefore,

$$V_2 = (400)(1 - 0.8)(2)(0.0263)\,\frac{[1 + (0.1)(4)]^2}{1 + 4}\left[\frac{0.5}{(1 - 0.8)(1 - 05)}\right]^2$$

$$= 41.23 \text{ thousand ft}^3$$

The second trial proceeds as follows:

4/ Assuming that $E_1 = 0.75$, compute the volume of the first-stage filter from equation 7-5.

$$V_1 = (400)(2)(0.0263)\,\frac{[1 + (0.1)(4)]^2}{1 + 4}\left(\frac{0.75}{1 - 0.75}\right)^2$$

$$= 74.22 \text{ thousand ft}^3$$

5/ Calculate the value of E_2 and then use this value to compute V_2 from equation 7-6.

$$E_2 = \frac{0.9 - 0.75}{1 - 0.75}$$

$$= 0.6$$

Therefore,

$$V_2 = (400)(1 - 0.75)(2)(0.0263)\,\frac{[1 + (0.1)(4)]^2}{1 + 4}\left[\frac{0.6}{(1 - 0.75)(1 - 0.6)}\right]^2$$

$$= 74.22 \text{ thousand ft}^3$$

These calculations show that $E_1 = 0.75$ and $E_2 = 0.6$ are the respective efficiencies for the first and second stages. A volume of 74,300 ft^3 will be selected for design.

6/ Calculate the required diameter of each filter.

$$\text{surface area} = \frac{\text{volume}}{\text{depth}}$$

$$A = \frac{74{,}300}{8}$$

$$= 9288 \text{ ft}^2$$

Therefore,

$$\text{diameter} = \left(\frac{4A}{\pi}\right)^{1/2}$$

$$= \left[\frac{(4)(9288)}{\pi}\right]^{1/2}$$

$$= 109 \text{ ft}$$

Applying multiple regression analysis to operational data from stone-media trickling filter plants, Galler and Gotaas (1964) developed the following equation for a single-stage filter (Weston, 1974):

$$S_e = \frac{K^0(QS_0 + Q_R S_e)^{1.19}}{(Q + Q_R)^{0.78}(1 + D)^{0.67}(r)^{0.25}} \tag{7-7}$$

where S_e = BOD_5 in effluent from filter, mg/ℓ
Q = wastewater flow, MGD
Q_R = recycle flow, MGD
S_0 = inflow BOD_5, mg/ℓ
D = filter depth, ft
r = filter radius, ft

$$K^0 = \frac{0.464(43{,}560/\pi)}{(Q)^{0.28}(T)^{0.15}} \tag{7-8}$$

T = wastewater temperature, °C

Rearranging and solving for volume, equation 7-8 can be written as (Baker and Graves, 1968)

$$V_1 = 0.1355D\left[\frac{(Q)^{0.13}(S_0)^{0.19}[1 + R(1 - E_1)]^{1.19}}{(T)^{0.15}(1 + D)^{0.67}(1 - E_1)(1 + R)^{0.78}}\right]^8 \tag{7-9}$$

where V_1 = volume of first-stage filter, 1000 ft^3

According to Baker and Graves (1968), applying a retardation factor of $1/(1 - E_1)^4$ to the Galler–Gotaas formula will give an expression for second-stage treatment similar to the NRC equation. This equation is

$$V_2 = 0.1355D\left[\frac{(Q)^{0.13}(S_0)^{0.19}[1 + R(1 - E_2)]^{1.19}}{(T)^{0.15}(1 + D)^{0.67}(1 - E_2)(1 + R)^{0.78}(1 - E_1)^{0.5}}\right]^8 \tag{7-10}$$

Equations 7-9 and 7-10 show volume to be a function of wastewater flow and strength as well as filter depth, recirculation, and wastewater temperature. It should be noted that within the range of normal organic and surface loadings, these equations predict an increase in efficiency with an increase in recycle ratio up to a value of approximately 5. Above this value little benefit is gained.

For the two-stage system the optimal volume can be estimated by a ratio of 1:2 when the Galler–Gotaas formula is used as compared to a ratio of 1:1 for the NRC equation. Still, less total volume is required in an optimum two-stage design than for a single-stage design.

Eckenfelder's equations are the only design equations which account for variations in media characteristics. Furthermore, these equations are formulated so that various terms can be evaluated with the use of treatability studies, an approach that is the most desirable method for trickling filter design. Hence, a detailed development of these particular equations is warranted.

In the development of his equations Eckenfelder assumes that the trickling filter can be represented as a plug-flow type of reactor and that substrate utilization follows first-order kinetics such that

$$\frac{S_e}{S_0} = e^{-kXt} \tag{7-11}$$

where S_e = soluble BOD_5 of filter effluent, mg/ℓ
S_0 = inflow BOD_5, mg/ℓ
k = substrate utilization reaction-rate constant, $time^{-1}$
X = active microbial mass, mg/ℓ
t = time of contact between wastewater and microorganisms, time

For a trickling filter, Eckenfelder also proposes that the time of contact can be represented as

$$t = \frac{CD}{Q^n} \tag{7-12}$$

where C, n = constants characteristic of the filter media (see Table 7-3)
D = filter depth, ft
Q = surface loading, gpm/ft^2

Furthermore, Eckenfelder and Barnhart (1963) found that

$$C = C'A_v^m \tag{7-13}$$

where A_v represents the specific surface of the media (see Table 7-3) in ft^2/ft^3 and C' and m are constants which for spheres, rock, and polygrid plastic media without microbial growth have a value of 0.7 and 0.75, respectively.

TABLE 7-3

HYDRAULIC CHARACTERISTICS OF SELECTED FILTER MEDIA

Media	A_v (ft^2/ft^3)	*Exponent* n	*Coefficient* C
Polygrid	30	0.65	9.5
Glass spheres, 0.5 in. (1.3 cm) diameter	85	0.82	22.5
Glass spheres, 0.75 in. (1.9 cm) diameter	60.3	0.80	15.8
Glass spheres, 1.0 in. (2.5 cm) diameter	41.6	0.75	12.0
Porcelain spheres, 3.0 in. (7.6 cm) diameter	12.6	0.53	5.1
Rock, 2.5–4.0 in. (6.3–10.2 cm)	—	0.408	4.15
Dowpac	25	0.50	4.84
Asbestos	25	0.50	5.10
Mead-Cor	30	0.70	5.6
Asbestos	50	0.75	7.2
Asbestos	85	0.80	8.0

Source: After Eckenfelder and Barnhart (1963).

By assuming that the active microbial mass is proportional to the specific surface of the media ($X \propto A_v$) and by substituting for t from equation 7-12 and for C from equation 7-13, equation 7-11 becomes

$$\frac{S_e}{S_0} = e^{-kC'A_v^{1+m}D/Q^n} \tag{7-14}$$

If the specific surface is assumed to remain constant and the media is assumed to have a uniform microbial film cover throughout the filter depth, equation 7-14 can be expressed as

$$\frac{S_e}{S_0} = e^{-K_0'D/Q^n} \tag{7-15}$$

where K_0' = treatability factor which has the units min^{-1} when Q is expressed as gpm/ft^2

Equation 7-15 is valid for a single-stage filter without recirculation. When recirculation is employed, equation 7-15 is modified to the form

$$\frac{S_e}{S_a} = \frac{e^{-K_0'D/Q^n}}{(1 + R) - Re^{-K_0'D/Q^n}} \tag{7-16}$$

where S_a is equal to the BOD_5 of the raw wastewater/recycle mixture (i.e., the BOD_5 of the watewater flow actually applied to the filter). This value is given by the expression

$$S_a = \frac{S_0 + RS_e}{1 + R} \tag{7-17}$$

Although neither equation 7-15 nor equation 7-16 contains a temperature term such as that found in the Galler–Gotaas equation, Eckenfelder proposes that temperature can be accounted for by adjusting the treatability factor, K'_0:

$$K'_0 = K'_{0(20°C)}(1.035)^{T-20} \tag{7-18}$$

where K'_0 = treatability factor determined at 20°C
T = operating temperature, °C

As previously indicated, treatability studies are desirable when trickling filters are to be designed. Using the Eckenfelder equations for design, it is necessary to determine the constants K'_0 and n employing a laboratory-scale trickling filter filled with the media to be used in the prototype. The laboratory procedure required to evaluate these constants is as follows:

1/ Using a lab-scale trickling filter filled with the media of choice, develop a microbial film with the wastewater to be treated. Figure 7-5 illustrates a typical lab-scale trickling filter.

2/ Operate the filter over a range of surface loadings and determine the soluble BOD_5 remaining at various depths in the filter.

3/ To analyze the data, rearrange equation 7-15 into the form

$$2.3 \log \left(\frac{S_e}{S_0}\right) = -K'_0 D Q^{-n} \tag{7-19}$$

Plot % BOD_5 remaining versus sampling depth on semilog paper. A typical plot is shown in Figure 7-6. In this figure the terms *A*, *B*, and *C* represent different surface loading rates in gpm/ft², and $A < B < C$.

4/ Once the slope of each line in Figure 7-6 has been determined, a log-log plot of 2.3 × slope versus surface loading rate is made. Figure 7-7 illustrates such a plot. This plot is based on the relationship

$$\text{slope} = \frac{-K'_0 Q^{-n}}{2.3} \tag{7-20}$$

or

$$2.3 \times \text{slope} = -K'_0 Q^{-n} \tag{7-21}$$

Taking the log of both sides of this expression gives

$$\begin{aligned} \log(2.3 \times \text{slope}) &= -[\log K'_0 + (-n)\log Q] \\ &= -\log K'_0 + n \log Q \end{aligned} \tag{7-22}$$

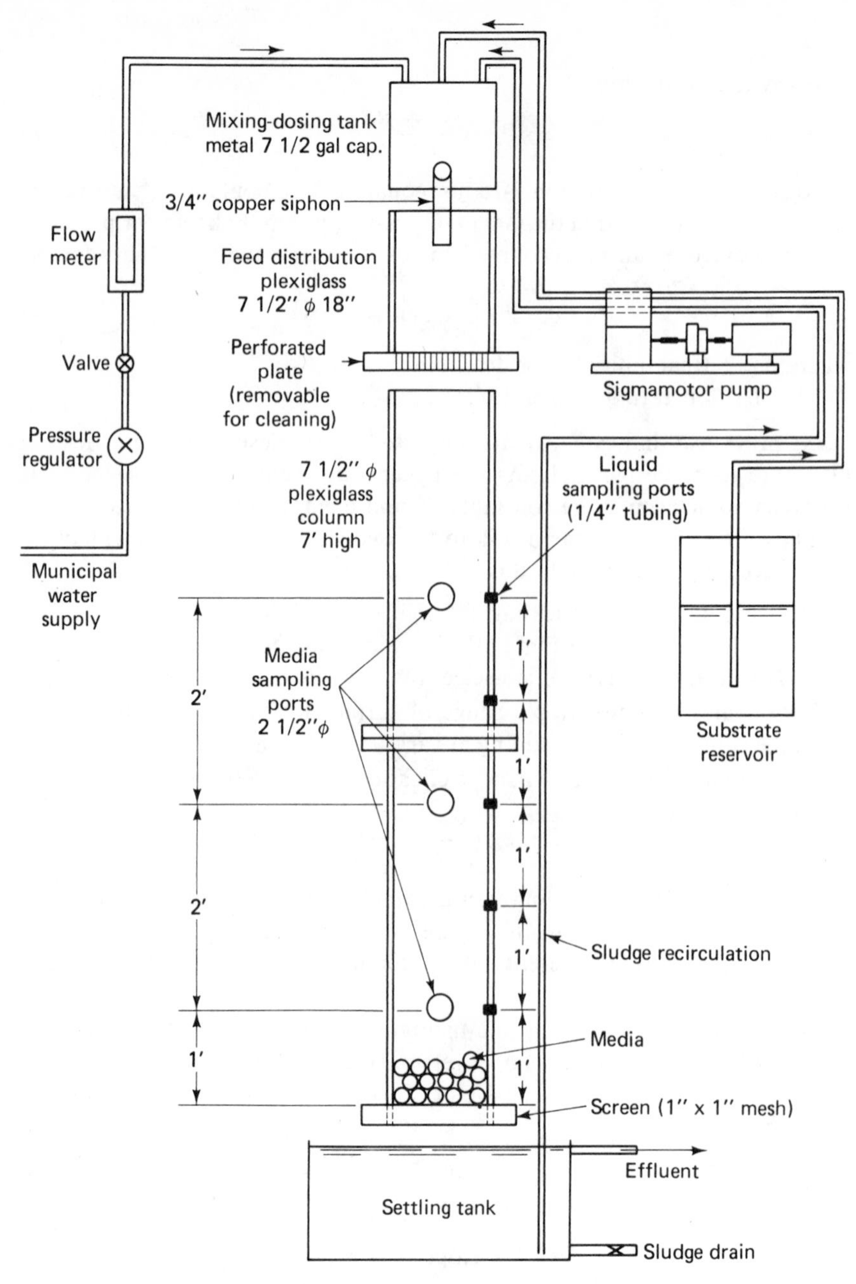

Figure 7-5. Typical Lab-Scale Trickling Filter. (After Cardenas, 1966.)

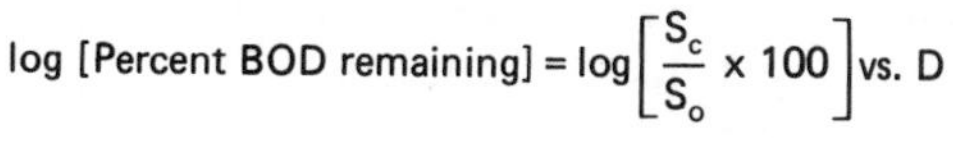

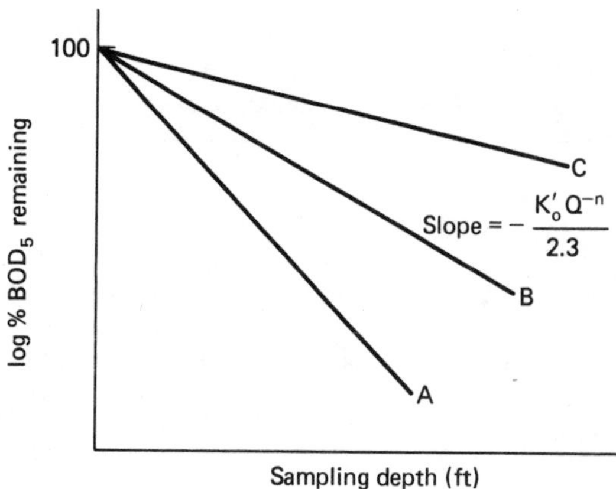

Figure 7-6. Typical Percent BOD Remaining Versus Sampling Depth Plot. (After Eckenfelder, 1972.)

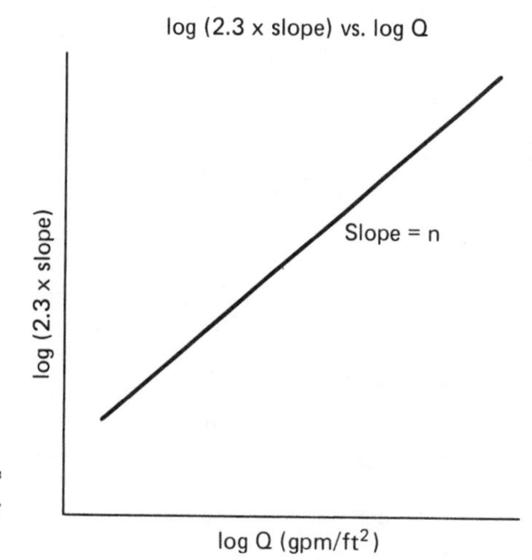

Figure 7-7. Typical Plot of K'_0Q^{-n} Versus Surface Loading. (After Eckenfelder, 1972.)

This is the equation of a straight line of slope n. Therefore, the slope of the linear trace given by the plot of log (2.3 $\times$ slope) versus log Q represents the value of the constant n.

5/ A final plot of % BOD_5 remaining versus D/Q^n is made on semilog paper, and the slope of the line obtained gives the value of $-K'_0/2.3$. Such a plot is given in Figure 7-8.

Typical values of n and K'_0 for different filter media and wastewaters are presented in Tables 7-4 and 7-5.

TABLE 7-4

BOD_5 Treatability Factors of Settled Domestic Wastewater in Trickling Filters with Various Media

Media	*Depth (ft)*	*Range of influent BOD_5 concentration (mg/ℓ)*	*Surface loading (gpm/ft²)*	*n*	*Treatability factor, K'_0, at 20°C (min⁻¹)*
1½ in. Flexirings	8	65–90	0.196–0.42	0.39	0.09
2½ in. clinker	6	220–320	0.015–0.019	0.84	0.021
1½–2½ in. slag	6	112–196	0.08–0.19	1.00	0.014
2½ in. slag	6	220–320	0.015–0.019	0.75	0.029
2½–4 in. rock	12	200	0.48–1.47	0.49	0.036
1–3 in. granite	6	186–226	0.031–0.248	0.4	0.059
¾ in. Raschig rings	6	186–226	0.031–0.248	0.7	0.031
1 in. Raschig rings	6	186–226	0.031–0.248	0.63	0.031
1½ in. Raschig rings	6	186–226	0.031–0.248	0.306	0.078
2¼ in. Raschig rings	6	186–226	0.031–0.248	0.274	0.08
Straight block	6	186–226	0.031–0.248	0.345	0.048
Surfpac	21.6	200	0.49–3.9	0.5	0.05
	12.0	200	0.97–3.9	0.45	0.05
	21.5	—	—	0.50	0.045
	21.5	—	—	0.50	0.088

Source: After Lipták (1974).

TABLE 7-5

Treatability Factors for Various Wastewaters at 20°C

Type of wastewater	*Type of filter media*	*Specific surface (ft²/ft³)*	*K'_0 (min⁻¹)*	*n factor*
Domestic	Surfpac	28.0	0.079	0.5
Fruit canning	Surfpac	28.0	0.0177	0.5
Boxboard	Surfpac	28.0	0.0197	0.5
Steel coke plant	Surfpac	28.0	0.0211	0.5
Textile	Surfpac	28.0	0.0156	0.5
	Surfpac	28.0	0.0394	0.5
	Surfpac	28.0	0.0268	0.5
Pharmaceutical	Surfpac	28.0	0.0292	0.5
Slaughterhouse	Surfpac	28.0	0.0246	0.5

Source: After Eckenfelder (1970).

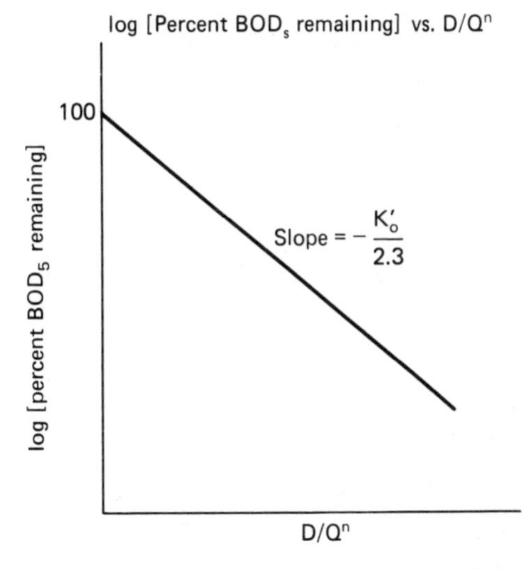

Figure 7-8. Typical Percent BOD Remaining Versus D/Q^n Plot. (After Eckenfelder, 1972.)

Example 7-2

It is required that a wastewater flow of 2 MGD with a strength of 300 mg/ℓ BOD_5 be treated to remove 90% of the BOD_5. To achieve this, a 25-ft-deep filter will be constructed and plastic media will be used. A treatability study has shown that $n = 0.5$ and $K'_0 = 0.05\ \text{min}^{-1}$ at 20°C. If the critical wastewater temperature is expected to be 10°C, what size filter should be required? Recirculation will not be used.

solution

1/ Correct the treatability factor, K'_0, for temperature by applying equation 7-18.

$$K'_0 = 0.05(1.035)^{10-20}$$
$$= 0.0354\ \text{min}^{-1}$$

2/ Compute the acceptable hydraulic loading from equation 7-15.

$$0.1 = \frac{1}{e^{(0.035)(25)/Q^{0.5}}}$$
$$Q = 0.144\ \text{gpm/ft}^2$$

3/ Convert flow in MGD to gpm.

$$2\ \text{MGD} = \frac{2{,}000{,}000}{1440} = 1389\ \text{gpm}$$

4/ Calculate the required filter surface area.

$$A = \frac{1389}{0.144}$$
$$= 9646\ \text{ft}^2$$

5/ Determine the required filter diameter.

$$\text{diameter} = \left(\frac{4A}{\pi}\right)^{1/2}$$

$$= \left[\frac{4(9646)}{\pi}\right]^{1/2}$$

$$= 98.2 \text{ ft}$$

When using a rock filter to treat domestic wastewater, Eckenfelder (1961) recommends the following expression:

$$\frac{S_e}{S_0} = \frac{1}{1 + \dfrac{2.5D^{0.67}}{(Q/A)^{0.5}}} \tag{7-23}$$

where D = filter depth, ft
Q = flow, MGD
A = filter surface area, acres

Baker and Graves (1968) developed the following relationship by solving equation 7-23 for volume:

$$V_1 = 7.0\frac{Q}{D^{0.33}}\left[\frac{E_1/(1 - E_1)}{1 + R}\right]^2 \tag{7-24}$$

where V_1 is the volume of a single-stage filter or the first-stage volume of a two-stage system expressed as thousand cubic feet.

Baker and Graves (1968) also modified the Eckenfelder formula to make it applicable to second-stage treatment by introducing the retardation factor $1/(1 - E_1)^2$. The resulting expression for volume has the form

$$V_2 = 7.0\frac{Q}{D^{0.33}}\left[\frac{E_2}{(1 - E_1)(1 - E_2)(1 + R)}\right]^2 \tag{7-25}$$

Filter volumes computed from equations 7-24 and 7-25 vary directly with flow rate and are a function of depth as $1/D^{0.33}$. The latter dependency causes the required volume to decrease with increasing depth. According to the equations, recirculation will also result in a decrease in the volume. A practical limit is reached at 5:1. In the equations influent BOD_5 is not included as a variable.

For the Eckenfelder formulas, the maximum efficiency for a two-stage system is obtained at filter volume ratios of 1:1. Although the engineer may reduce the required filter volume by increasing the depth and recirculation rate, Baker and Graver (1968) note that increasing either of these parameters generally results in an increase in pumping costs.

There is a considerable difference in the trickling filter volume requirement predicted by the NRC, Galler–Gotaas, and Eckenfelder formulas, and the question of which formula most accurately predicts field performance is as yet unanswered. There are, however, certain factors that an engineer should consider when selecting a particular formula for use in design. These are

(1) the Eckenfelder and Galler–Gotaas formulas consider filter depth as an important design parameter; (2) the Galler–Gotaas formula includes a temperature term; (3) the efficiency values used in developing the NRC equation were all based on settling following filtration, whereas both settled and unsettled effluent values were used in the development of the Galler–Gotaas and Eckenfelder equations; and (4) the Eckenfelder equation is the only one that considers variations in filter media characteristics as an important design variable.

The Trickling Filter as a Treatment Alternative

In the past the trickling filter has been commonly used to provide secondary wastewater treatment, especially in smaller communities with populations less than 10,000. The reason for this popularity was the ease of operation of such units and the lower cost when compared to the activated sludge process. However, with the advent of the 1972 Water Pollution Control Act Amendments, which brought a requirement for more stringent effluent standards, the question of whether or not a trickling filter can be considered as an acceptable secondary treatment process alternative has been raised.

Such a question is not unwarranted when design is based on one of the predictive models, such as the NRC equation, and the filter is to be located in an area with large seasonal temperature variations. It should be recognized, however, that such questions are generally based on the performance of standard-rate trickling filters, which were "accidentally" designed rather than optimally designed. Furthermore, predictive models consider treatment efficiency based on total BOD removal between the inflow to the filter and the effluent of the final clarifier, with no differentiation between soluble and suspended BOD in the plant effluent. Such a model does not allow the engineer to design for substrate assimilation and clarification separately, a practice that would allow emphasis to be shifted to effluent quality if this is an area of concern.

It has been found that trickling filters operate more effectively in a warm climate. Benzie et al. (1963) found a 21% decrease in filter efficiency when comparing the filter efficiency under winter conditions (25 to 31°F) to its efficiency during the summer (67 to 73°F).

Although secondary treatment with activated sludge generally provides a higher level of treatment and certainly provides greater operational flexibility than does the trickling filter (since recirculation is the only means of operational control), perhaps trickling filters should not be discounted as a treatment alternative. If design is based on treatability studies that determine substrate assimilation rates with due consideration given to temperature variation, greater depths are used, and the capability provided for chemical addition to filter effluent when clarification efficiency decreases, then there are cases when a trickling filter may provide the most cost-effective design—especially if, as stated by Smith (1974), "the term *best practicable treatment* is used to represent the ideal situation where all aspects of each pollution control problem are properly weighted and taken into account to arrive at the ideal solution which maximizes the common good and minimizes the common cost."

7-2
Rotating Biological Contactors

In this process a series of closely spaced disks (10 to 12 ft in diameter) are mounted on a common shaft. This unit is installed in a concrete tank so that the surface of the wastewater passing through the tank almost reaches the shaft. This means that about 40% of the total surface area of the disks are always submerged. The shaft continually rotates at 1 to 2 rpm, and a layer of biological growth 2 to 4 mm thick is soon established on the wetted surface of each disk. The biological growth that becomes attached to the disks assimilates the organic materials in the wastewater. Aeration is provided by the rotating action, which exposes the disks to the air after contacting them with the wastewater. Excess biomass is sheared off in the tank, where the rotating action of the disks maintain the solids in suspension. Eventually, the flow of the wastewater carries these solids out of the system and into a clarifier, where they are separated. By arranging several sets of disks in series, it is possible to achieve a high degree of organic removal and nitrification (see Figures 7-9 and 7-10).

One of the rotating biological contactor systems available commercially is the BIO-SURF process. The fundamental module of the BIO-SURF

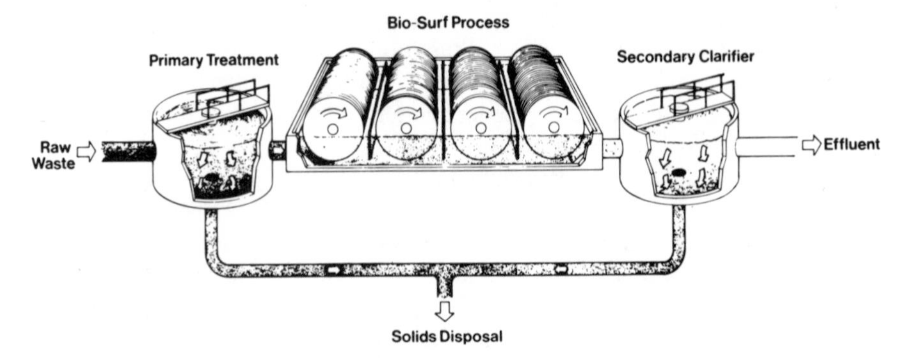

Figure 7-9. Typical BIO-SURF process. (Courtesy of Autotrol Corporation.)

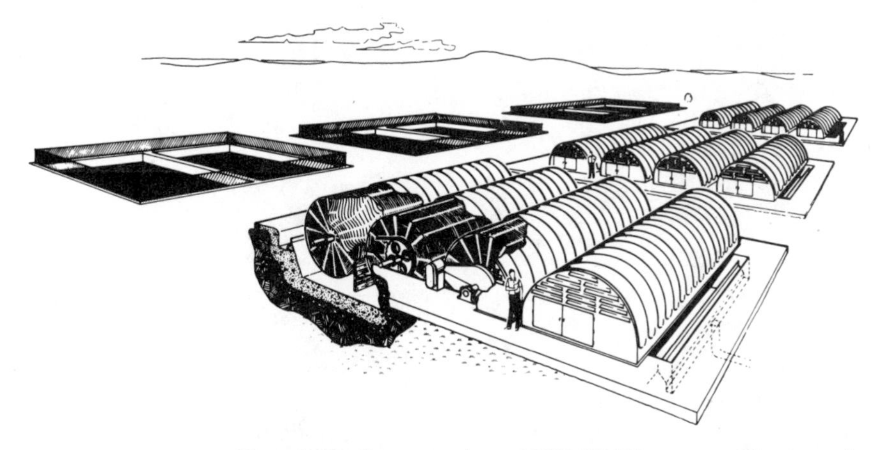

Figure 7-10. Cut-away view of BIO-SURF process. (Courtesy of Autotrol Corporation.)

process is a 25-ft-long steel shaft supporting an assembly of 12-ft-diameter polyethylene disks. The BIO-SURF media consists of alternating flat and corrugated sheets of polyethylene (see Figure 7-11). Such an arrangement provides a much larger surface area than does a simple flat disk. Selective gaps between groups of disks along a single shaft provide a series of treatment stages for small flows. However, in large installations, a 25-ft module is used

Figure 7-11. BIO-SURF media. (Courtesy of Autotrol Corporation.)

Figure 7-12. Spacing of BIO-SURF media. (Courtesy of Autotrol Corporation.)

as a single stage itself. Generally, a 25-ft by 12-ft-diameter module contains about 104,000 ft^2 of total surface area (see Figure 7-12). Each of these modules is driven by a 5-hp motor. (Autotrol also has a system called the AERO-SURF process, which is air-driven and generally more economical to operate than the mechanically driven systems).

The BIO-SURF process can be designed to produce an effluent BOD_5 of 10 mg/ℓ. The composition of effluents between 10 and 20 mg/ℓ BOD_5 generally consists of approximately 1/3 soluble and 2/3 insoluble BOD_5.

Advantages offered by a rotating biological contactor system are:

1/ Short contact periods are required because of the large active surface (generally less than 1 h).

2/ They are capable of handling a wide range of flows (less than 1 MGD to over 100 MGD).

3/ No recycle is required.

4/ Sloughed biomass generally has good settling characteristics and can easily be separated from waste stream.

5/ Operating costs are low because little skill is required in plant operation.

A disadvantage is the requirement for covering in northern climates to protect against freezing.

Design Relationships

As the rotating biological contactor system is a relatively new concept in biological wastewater treatment, few models have been proposed to describe the process. Kornegay (1975) has given the most detailed practical development to date, and it is his development that will be presented here.

The biomass that sloughs off the disks is kept in suspension by the turbulence generated by the rotating assembly. Thus, for this situation, both suspended and attached microbial growth are responsible for substrate assimilation. For the system shown in Figure 7-13, a material-balance expression for substrate entering and leaving the reactor is

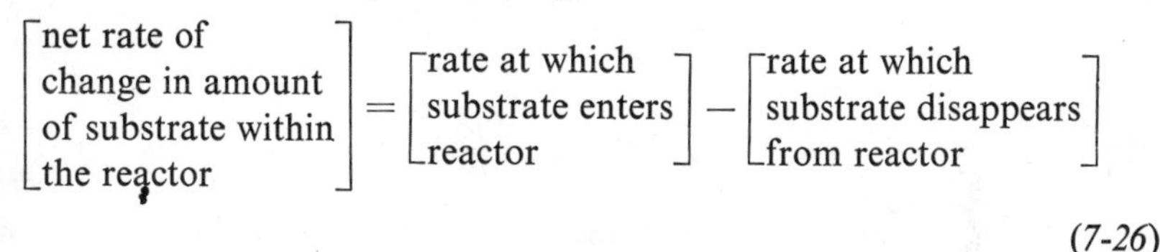

$$\begin{bmatrix}\text{net rate of}\\ \text{change in amount}\\ \text{of substrate within}\\ \text{the reactor}\end{bmatrix} = \begin{bmatrix}\text{rate at which}\\ \text{substrate enters}\\ \text{reactor}\end{bmatrix} - \begin{bmatrix}\text{rate at which}\\ \text{substrate disappears}\\ \text{from reactor}\end{bmatrix} \tag{7-26}$$

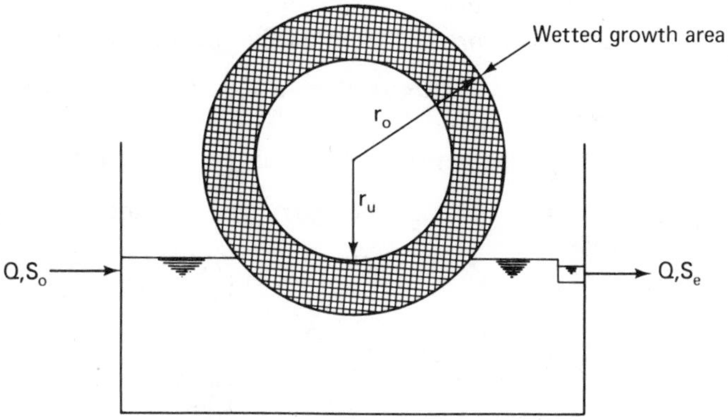

Figure 7-13. Rotating Biological Contactor. (After Kornegay, 1975.)

Substrate disappears from the reactor as a result of assimilation by both attached and suspended growth, and the amount of substrate that escapes biological utilization is removed in the effluent stream. Therefore, assuming complete mixing in the reactor, equation 7-26 can be written as

$$\left(\frac{dS}{dt}\right)V = QS_0 - \left[\left(\frac{dS}{dt}\right)_{uA} V_A + \left(\frac{dS}{dt}\right)_{uS} V_S + QS_e\right] \tag{7-27}$$

where $\left(\frac{dS}{dt}\right)_{uA}$ = rate of substrate utilization per unit volume of attached growth, mass volume^{-1} time^{-1}

$\left(\frac{dS}{dt}\right)_{uS}$ = rate of substrate utilization per unit volume of suspended growth, mass volume^{-1} time^{-1}

V_A = volume of active attached biological growth, volume
V_S = liquid volume of reactor, volume
S_0 = influent substrate concentration, mass volume^{-1}
S_e = effluent substrate concentration, mass volume^{-1}

If energy of maintenance is neglected,

$$\left(\frac{dX}{dt}\right)_g = Y_T\left(\frac{dS}{dt}\right)_u \tag{7-28}$$

or

$$\left(\frac{dX}{dt}\right)_{Ag} = Y_A\left(\frac{dS}{dt}\right)_{uA} \tag{7-29}$$

and

$$\left(\frac{dX}{dt}\right)_{Sg} = Y_S\left(\frac{dS}{dt}\right)_{uS} \tag{7-30}$$

where $\left(\frac{dX}{dt}\right)_{Ag}$ = absolute growth rate of attached biomass, mass volume^{-1} time^{-1}
Y_A = theoretical yield coefficient for attached growth
$\left(\frac{dX}{dt}\right)_{Sg}$ = absolute growth rate of suspended biomass, mass volume^{-1} time^{-1}
Y_S = theoretical yield coefficient for suspended growth

Equations 7-29 and 7-30 can be rearranged into the forms

$$\frac{(dX/dt)_{Ag}}{Y_A} = \left(\frac{dS}{dt}\right)_{uA} \tag{7-31}$$

and

$$\frac{(dX/dt)_{Sg}}{Y_S} = \left(\frac{dS}{dt}\right)_{uS} \tag{7-32}$$

Multiplying the left-hand side of equations 7-31 and 7-32 by X_f/X_f and X_s/X_s, respectively, gives

$$\frac{\frac{(dX/dt)_{Ag}}{X_f}X_f}{Y_A} = \frac{\mu_A X_f}{Y_A} = \left(\frac{dS}{dt}\right)_{uA} \tag{7-33}$$

and

$$\frac{\frac{(dX/dt)_{Sg}}{X_S}X_S}{Y_S} = \frac{\mu_S X_S}{Y_S} = \left(\frac{dS}{dt}\right)_{uS} \tag{7-34}$$

where X_f = active biomass per unit volume of attached growth
X_S = active biomass per unit volume of suspended growth
μ_A = specific growth rate of attached biomass, time^{-1}
μ_S = specific growth rate of suspended biomass, time^{-1}

Substituting from equations 7-33 and 7-34 for $(dS/dt)_{uA}$ and $(dS/dt)_{uS}$ in equation 7-27 gives

$$\left(\frac{dS}{dt}\right)V = QS_0 - QS_e - \frac{\mu_A X_f}{Y_A}V_A - \frac{\mu_S X_S}{Y_S}V_S \tag{7-35}$$

Assuming that d represents the active depth of microbial film on any rotating disk and A represents the total wetted area,

$$A = 2N\pi(r_0^2 - r_u^2) \tag{7-36}$$

where N = number of disks
r_0 = total disk radius (see Figure 7-13)
r_u = unsubmerged disk radius

Then, at steady-state, equation 7-35 can be written as

$$0 = QS_0 - QS_e - \frac{\mu_A}{Y_A}X_f d2\pi N(r_0^2 - r_u^2) - \frac{\mu_S}{Y_S}X_S V \tag{7-37}$$

where V = liquid volume of reactor

If microbial growth is assumed to follow the Monod relationship given by equation 2-12,

$$\mu = \mu_{\max}\frac{S}{K_s + S} \tag{2-12}$$

Then substituting for μ in equation 7-37 yields

$$0 = QS_0 - QS_e - \frac{(\mu_{\max})_A}{Y_A}X_f d2\pi N(r_0^2 - r_u^2)\frac{S_e}{K_s + S_e} - \frac{(\mu_{\max})_S}{Y_S}X_s V\frac{S_e}{K_s + S_e} \tag{7-38}$$

However, most rotating biological contactor systems operate at short retention times. For this situation the amount of biomass contributed by suspended growth is very small compared to the amount contributed by the attached growth so that any contribution to substrate utilization from suspended growth can be neglected. Equation 7-38 reduces to

$$Q(S_0 - S_e) = 2\frac{(\mu_{\max})_A}{Y_A}N\pi d X_f(r_0^2 - r_u^2)\frac{S_e}{K_s + S_e} \tag{7-39}$$

If the system is composed of a series of reactors, each with a series of disks on a common shaft, total substrate utilization is the sum of the utilization in each unit (reactor), or

$$U_T = U_1 + U_2 + \ldots + U_{n-1} + U_n \tag{7-40}$$

On a finite-time basis, total substrate utilization is given by

$$U_T = Q(S_0 - S_{en}) \tag{7-41}$$

where S_{en} = effluent substrate concentration from the nth reactor, mass volume^{-1}

For series operation where each unit is identical, equation 7-39 becomes

$$Q(S_0 - S_{en}) = \frac{2(\mu_{max})_A}{Y_A} N\pi X_f d(r_0^2 - r_u^2) \sum_{i=1}^{n} \frac{S_i}{K_s + S_i} \tag{7-42}$$

where S_i = substrate concentration in the effluent of a particular unit, mass volume^{-1}

If a term P is introduced and defined as

$$P = \frac{(\mu_{max})_A X_f d}{Y_A} \tag{7-43}$$

then equation 7-39 (which is applicable to a single-stage system) can be written as

$$Q(S_0 - S_e) = 2PN\pi(r_0^2 - r_u^2)\frac{S_e}{K_s + S_e} \tag{7-44}$$

while equation 7-42 (which is applicable to a multiple stage system) becomes

$$Q(S_0 - S_{en}) = 2PN\pi(r_0^2 - r_u^2) \sum_{i=1}^{n} \frac{S_i}{K_s + S_i} \tag{7-45}$$

In this equation N represents the number of disks in one stage.

Before equation 7-44 or equation 7-45 can be used for system design, the kinetic parameters K_s and P must be determined. These constants can be evaluated on the basis of laboratory or pilot-plant data. Autotrol has pilot plants of the BIO-SURF and AERO-SURF processes available in a size range of 0.5 m to full-scale.

When the system is to operate at low retention times, only substrate assimilation by the attached growth is significant. For this case equation 7-44 can be rearranged to the form

$$\frac{2N\pi(r_0^2 - r_u^2)}{Q(S_0 - S_e)} = \frac{1}{P} + \frac{K_s}{PS_e} \tag{7-46}$$

The slope of the line obtained by plotting $2N\pi(r_0^2 - r_u^2)/Q(S_0 - S_e)$ versus $1/S_e$ gives the value of K_s/P while the intercept gives the value of $1/P$. Such a plot is presented in Figure 7-14.

Example 7-3

Assume a single-stage rotating biological contactor system is to be constructed to treat a wastewater flow of 0.1 MGD with a strength of 250 mg/ℓ BOD_5. A treatability study has revealed that P = 2500 mg/ft^2-day and K_s = 100 mg/ℓ. How much surface area will be required if a soluble effluent BOD_5 of 15 mg/ℓ is desired?

solution

The surface area required is computed from equation 7-44:

$$Q(S_0 - S_e) = 2PN\pi(r_0^2 - r_u^2)\frac{S_e}{K_s + S_e}$$

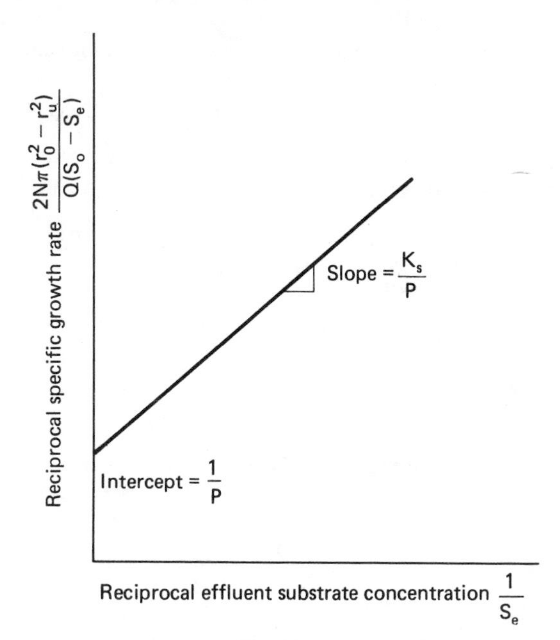

Figure 7-14. Evaluation of K_s and P for a Rotating Biological Contactor. (After Kornegay, 1975.)

Since

$$A = 2N\pi(r_0^2 - r_u^2)$$

equation 7-44 can be expressed as

$$Q(S_0 - S_e) = PA\frac{S_e}{K_s + S_e}$$

or

$$A = \frac{Q(S_0 - S_e)}{P\dfrac{S_e}{K_s + S_e}}$$

$$= \frac{(100{,}000 \text{ gal/day})(3.78\ \ell/\text{gal})(250 - 15)}{(2500)\left(\dfrac{15}{100 + 15}\right)}$$

$$= 272{,}401 \text{ ft}^2$$

Example 7-4

Calculate the surface area required in Example 7-3 if a two-stage system is to be used.

solution

A trial-and-error solution is required, using both equations 7-44 and 7-45.

1/ Develop an expression for the effluent substrate concentration from the first stage.

Equation 7-45 may be expanded into the form

$$Q(S_0 - S_e) = PA\frac{S_1}{K_s + S_1} + PA\frac{S_2}{K_s + S_2}$$

where S_1 and S_2 are the effluent substrate concentrations from the first and second stages, respectively. From equation 7-44, for the first stage,

$$Q(S_0 - S_1) = PA\frac{S_1}{K_s + S_1}$$

Therefore,

$$Q(S_0 - S_2) = Q(S_0 - S_1) + PA\frac{S_2}{K_s + S_2}$$

or

$$S_1 = \frac{PAS_2}{(K_s + S_2)Q} + S_2$$

2/ Assume a value for the surface area of each stage and solve for S_1. For the first trial, assume 68,100 ft² per stage.

$$S_1 = \frac{(2500)(68{,}100)(15)}{(100 + 15)(3.78)(100{,}000)} + 15$$
$$= 73.7\text{mg}/\ell$$

3/ Substitute the calculated value of S_1 into equation 7-44 and compute the surface area of the first stage.

$$A = \frac{(100{,}000)(3.78)(250 - 73.7)}{(2500)\dfrac{73.7}{100 + 73.7}}$$
$$= 62{,}828 \text{ ft}^2$$

Since this surface area is less than the assumed surface area, a second trial must be performed.

4/ Assuming that $A = 65{,}000$ ft² gives a calculated A of 65,184 ft². Thus, the total surface area required for equivalent treatment using a two-stage system is approximately $65{,}000 \times 2 = 130{,}000$ ft².

It is not always possible or practical to conduct a treatability study before designing a treatment system. In such a situation, surface loading may be used as the principal design criterion. As an aid in the design of a BIO-SURF process for the treatment of municipal wastewater, the Autotrol Corporation has developed the loading curves presented in Figure 7-15. The surface area required for a specific degree of treatment can be computed by selecting the appropriate surface loading from Figure 7-15 and dividing the design flow by this loading rate. The BOD removals presented in this figure are for the BIO-SURF process after secondary clarification and do not include any BOD reduction from primary treatment.

Because of its effect on microbial activity, temperature is always a factor to be considered in the design of any biological treatment process. It has been found that wastewater temperatures above 55°F do not affect the treatment

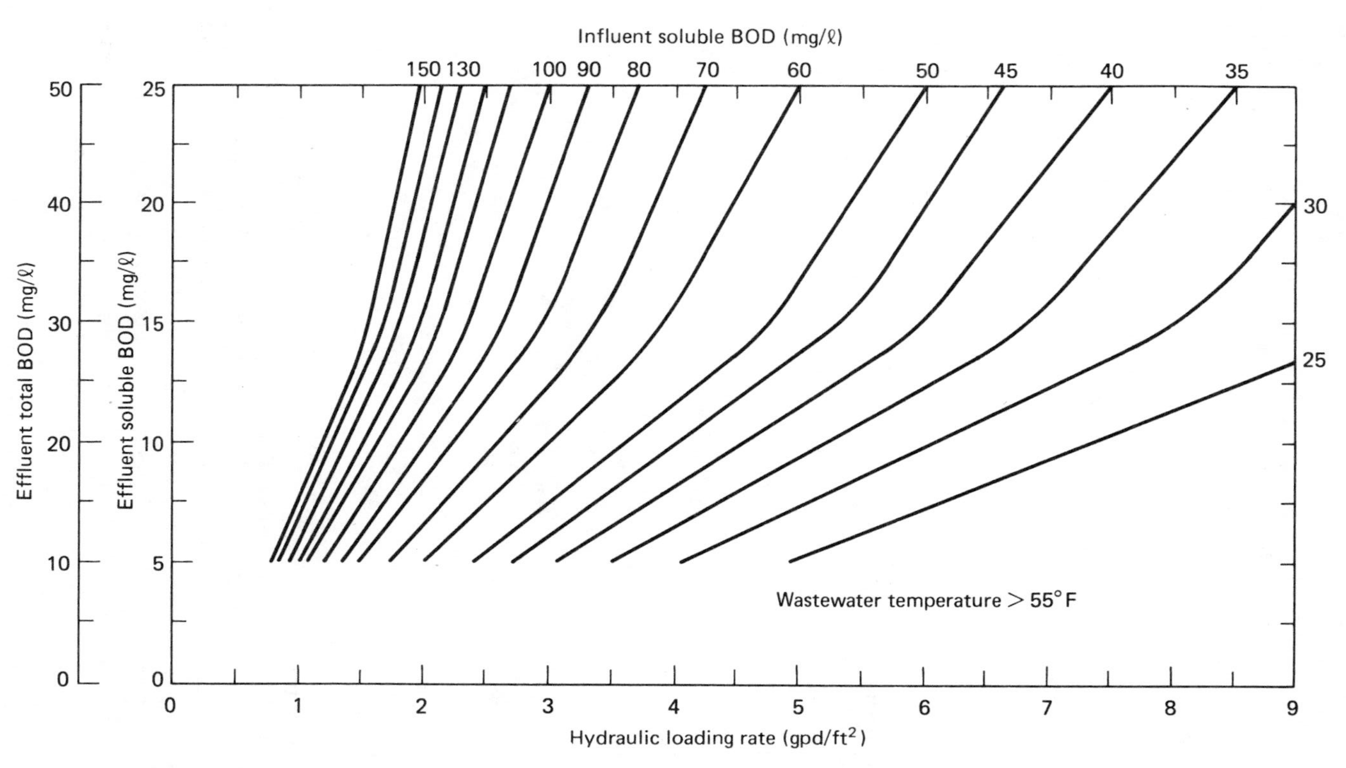

Figure 7-15. BIO-SURF Process Domestic Wastewater BOD Removal. (Courtesy of Autotrol Corporation.)

efficiency of the BIO-SURF process. However, when temperatures fall below 55°F, treatment efficiency will decrease. Figure 7-16 gives temperature-correction factors, which may be used to determine the increase in surface area required for prolonged low temperatures.

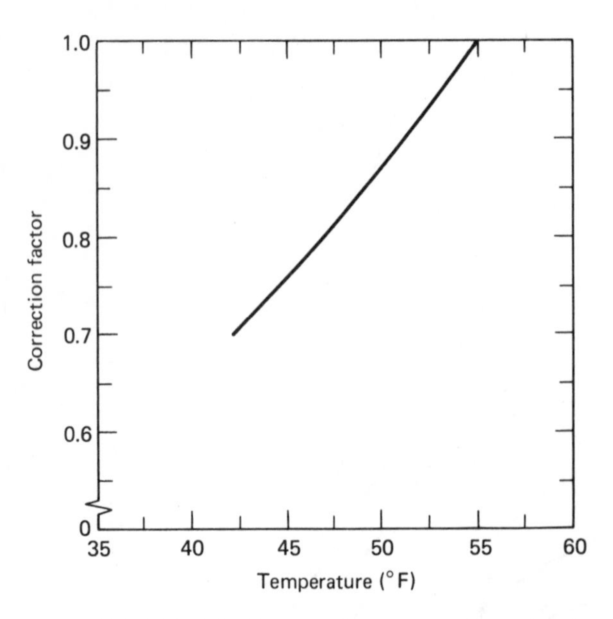

Figure 7-16. BIO-SURF Process Temperature Correction for BOD Removal. (Courtesy of Autotrol Corporation.)

7-3
Activated Biofilters

The activated biofilter (ABF) process developed and marketed by Neptune Microfloc combines an attached-growth system with a completely mixed activated sludge unit (see Figure 7-17). In this process primary effluent enters a wet well, where it is mixed with biological solids returned from the secondary clarifier and biocell recycle. This mixture is pumped to the top of the biocell, which is filled with racks of redwood slats. Oxygen is supplied by the splashing of the wastewater between layers of the redwood slats and by the movement of the wastewater in a film across the microbial layer attached to the slats. The effluent from the biocell is split, with the major portion flowing to the aeration tank and the remainder passing back to the wet well. Biological solids from the completely mixed activated sludge

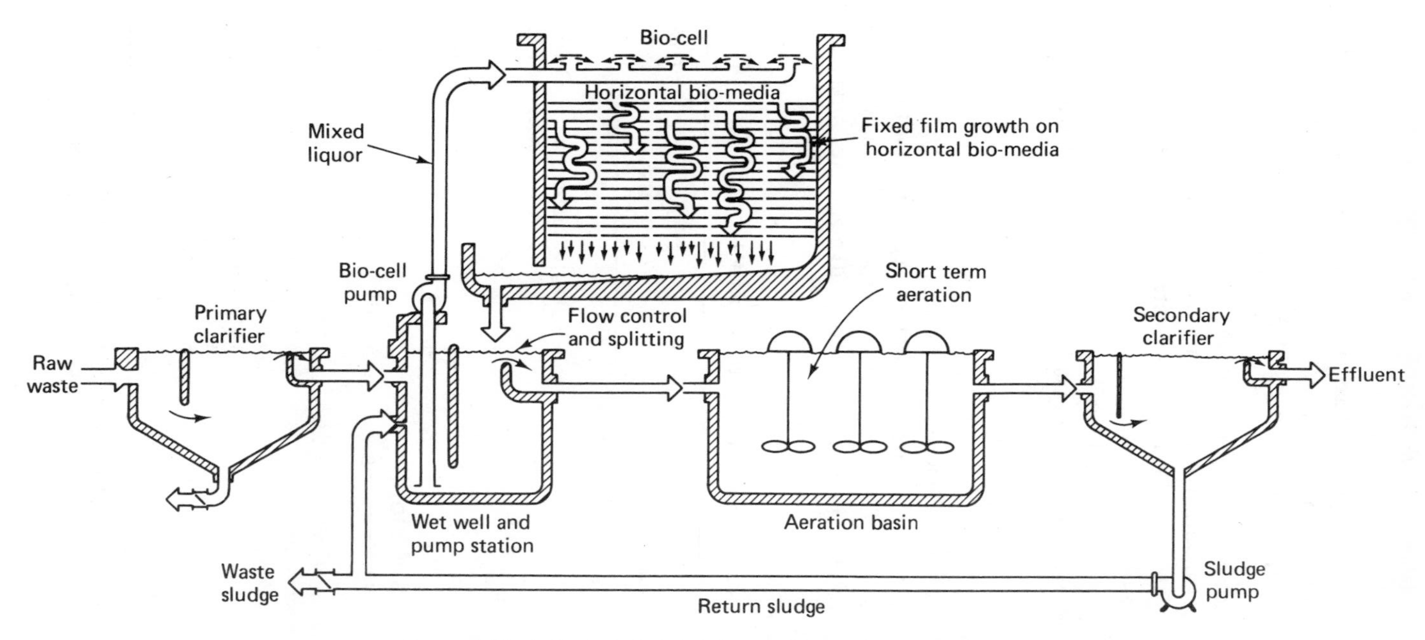

Figure 7-17. Activated Biofilter Process Flow Diagram. (Courtesy of Neptune Microfloc.)

unit are separated in the secondary clarifier and recycled to the wet well, with a portion going to waste.

The two major advantages of ABF process are:

1/ Process stability: locating the biocell ahead of the aeration tank tends to "smooth out" any variation in BOD loading. This provides a more stable operation and adds greater reliability and removal efficiency to the overall process.

2/ Flexibility: the biocell can be added ahead of existing activated sludge tanks to increase plant capacity or efficiency.

Design Considerations

The design parameters presented in Table 7-6 were reported by Hemphill (1977) for ABF systems concerned primarily with carbonaceous BOD removal. The effluent BOD_5 and suspended solids concentration from the ABF process will generally average 20 mg/ℓ. According to Dunnahoe and

TABLE 7-6

DESIGN PARAMETERS FOR ACTIVATED BIOFILTER PROCESS

Parameter	*Units*	*Typical value*	*Range*
Biocell			
Organic loading	lb BOD_5/1000 ft^3-day	200	100–300
Media depth	ft	14	5–22
BOD removal	%	65	55–85
Hydraulic			
Biocell recycle		0.4Q	0–2.0Q
Sludge recycle		0.5Q	0.3–1.0Q
Biocell flow		1.9Q	1.5–4.0Q
Biocell surface loading	gpm/ft^2	3.5	1.5–5.5
Aeration tank[a]			
Retention time[b]	h	0.8	0.5–2.0
Organic loading	lb BOD_5/1000 ft^3-day	95	50–225
F: M ratio	lb BOD_5/lb MLVSS-day	0.5	0.2–0.9
MLVSS	mg/ℓ	3000	1500–4000
MLSS	mg/ℓ	4000	2000–5000
Oxygen requirement	lb O_2/lb BOD_5	1.0	0.8–1.2
Total sludge production	lb VSS/lb BOD_5 removed	0.65	0.55–0.75

[a]Based on aeration BOD_5 loading after biocell removal.
[b]Based on design average flow and secondary influent BOD_5 = 150 mg/ℓ.
Source: After Hemphill (1977).

Hemphill (1976), the addition of a mixed-media filter will reduce this value to 10 mg/ℓ.

Heat loss through the ABF process is small because the retention time in the biocell is short. If diffused air aeration is employed in the aeration basin, total heat loss will be low [according to Boyle (1976), between 1 and 2°F]. As a result, the process is suitable for use in regions that have fairly harsh winters. The process can also be completely housed for use in extremely cold climates.

7-4 Anaerobic Filters

Anaerobic treatment of organic wastes has certain advantages that make it a more desirable waste treatment process than aerobic treatment. These advantages are (1) a usable by-product, methane gas, is produced; and (2) a high degree of waste stabilization may be accomplished while producing only a small amount of excess biomass. The latter advantage, however, is also one of the major disadvantages of anaerobic waste treatment. In Chapter 4 an equation was presented for θ_c^m which has the form

$$\theta_c^m = \frac{1}{Y_T k} \frac{K_s + S_0}{S_0} \qquad (4\text{-}174)$$

This equation shows that θ_c^m is inversely proportional to the value of the yield coefficient, Y_T, and the maximum specific substrate utilization rate, k. Values for θ_c^m determined for anaerobic processes are, therefore, much longer than those for aerobic processes, because the yield per unit of substrate utilized is much lower in anaerobic treatment. This means that the design θ_c is going to be longer in an anaerobic process because, to allow for a factor of safety, θ_c design is usually 2 to 10 times greater than θ_c^m.

Temperature is also an important consideration in anaerobic treatment because of the effect it has on the value of k. At operating temperatures below 20°C, k becomes quite small, which means that extremely long θ_c values will be required for effective waste stabilization.

In Chapter 6 it was shown that the value of θ_c is equal to the hydraulic retention time for a completely mixed treatment process operating without solids recycle. For this type of system, effluent quality is determined only by the hydraulic retention time. When anaerobic treatment is employed, the use of such a flow scheme is limited to small flows because of the extremely large volume requirements. When large flows are to be treated, solids recycle is normally employed because this makes it possible to vary θ_c independently of the hydraulic retention time so that the desired effluent quality can be achieved with a shorter retention time and hence a smaller reactor volume. The anaerobic contact process discussed in Chapter 4 is such a treatment scheme.

The anaerobic contact process is quite effective in the treatment of wastes containing a significant suspended solids concentration (e.g., meat-packing wastes). The microorganisms attach themselves to the solids and as a result are readily separated in the secondary clarifier and returned to the reactor. However, when soluble wastes are treated in the anaerobic contact process, a large number of microorganisms remain dispersed and are lost from the system in the secondary clarifier effluent. This makes it difficult to maintain a long θ_c. To circumvent this problem, McCarty (1968) recommends the use of the anaerobic filter process.

The anaerobic filter is similar to the trickling filter except that a bottom feed is employed, and as a result the filter is completely submerged in the waste (see Figure 7-18). The most common type of support media used is $1\frac{1}{2}$ to 2 in. rock. BSRTs of 100 days or more are possible because the microorganisms tend to grow on the walls of the filter as well as on the solid media and are not washed out in the effluent. Therefore, the biomass in the system becomes very large and, since it is a function of the diameter of the support media, the use of smaller particles can increase it considerably.

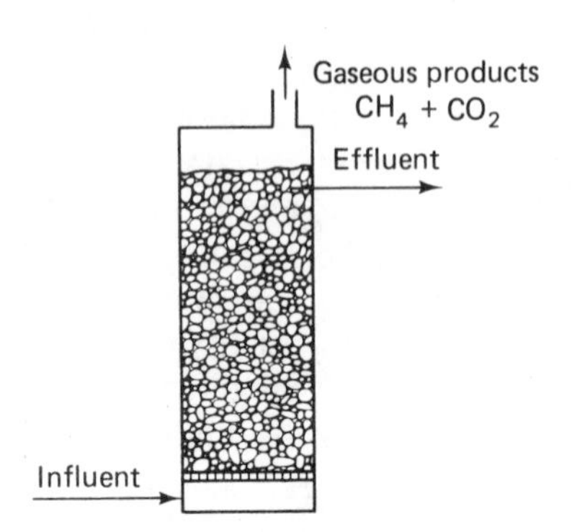

Figure 7-18. Flow Diagram for Anaerobic Filter.

Young and McCarty (1967) propose the following advantages for the anaerobic filter process:

1/ Soluble wastes can be effectively treated with an anaerobic filter.
2/ No form of recycle is required with anaerobic filters because the microorganisms remain in the filter and are not lost in the effluent.
3/ Because of the very high solids concentration maintained in the

filter, it is possible to operate at a lower temperature than would normally be possible for other types of anaerobic processes.

4/ Very low volumes of sludge are produced by the anaerobic filter.

5/ Startups and shutdowns are easier with an anaerobic filter than with other types of anaerobic processes.

There are also problems associated with the use of this type of system. Schroeder (1977) reports the following disadvantages.

1/ Anaerobic filters should only be used to treat soluble wastes because of the clogging problem that may develop when the wastes contain suspended solids.

2/ Flow distribution becomes a problem when the biological solids concentration increases to the point where it causes channelization to develop. This situation will significantly reduce the length of time between filter cleanings.

3/ Filter cleaning techniques have not been developed. Because of the size of the unit (approximately the size of a trickling filter), backwashing is not feasible.

Fluidization of the media could potentially solve, or at least reduce, these problems.

Design Considerations

Young and McCarty (1967) conducted the first intensive investigation of the anaerobic filter. In this study eight 1-ft^3 laboratory filters were utilized. The composition of the substrate used in this study is presented in Table 7-7, while the results of the anaerobic filter operation are presented in Table 7-8. Additional results were reported by McCarty (1968). The latter experiments achieved an average gas production of 12.8 mℓ/day per lb COD/1,000 ft^3/day applied to the filter. The feed mixture consisted of mixtures of methanol, acetate, and propionate, while the gas produced had an average methane content of 88.5 percent.

Jennett and Dennis (1975) employed four 0.5-ft^3 laboratory filters in the treatment of a pharmaceutical waste. Before treatment, the pH of the waste was adjusted to 6.8 with NaOH, and nitrogen and phosphorus were added to correct the nutrient deficiency that existed in the raw waste. The results of this study are given in Table 7-9. It was also noted during the investigation that the filters were able to operate over a period of 6 months without the need for solids disposal and that shock increases in organic loading did not cause process failure.

TABLE 7-7

ANAEROBIC FILTER FEED COMPOSITION

		Concentrations evaluated				
Waste composition	*pH*	*COD* (*mg/ℓ*)	NH_3 (*mg/ℓ*)	PO_4 (*mg/ℓ*)	SO_4 (*mg/ℓ*)	*Alkalinity as* $CaCO_3$ (*mg/ℓ*)
Protein–Carbohydrate	6.6–7.0	1500	120	3.5	75	700
		3000	240	6.5	75	1700, 3200
Volatile Acids	6.2	1500	15	3.5	75	950
(Acetic and propionic)		3000	30	6.5	75	1700
		6000	60	12.5	75	3200

Source: After Young and McCarty (1967).

TABLE 7-8

RESULTS OF ANAEROBIC FILTER OPERATION

Type of waste	Influent COD (mg/ℓ)	Steady-state operation period (days)	Theoretical retention time (h)	Loading rate (lb/COD/1000 ft^3/day)	Effluent quality: Suspended solids (mg/ℓ)	Soluble BOD_u (mg/ℓ)	Soluble COD (mg/ℓ)	Total COD (mg/ℓ)	Removal BOD_u (%)	Eff, COD (%)
Protein—Carbohydrate	1500	21	36	26.5	9	25	100	112	98.4	92.1
	1500	42	18	53	6	35	110	122	97.5	91.5
	1500	50	9	106	5	225	300	312	84.3	79.3
	1500	30	4.5	212	250	525	600	950	63.2	36.7
	3000	70	72	26.5	24	20	170	204	99.2	93.4
	3000	32	36	53	48	130	280	347	95.5	88.4
	3000	20	9	212	178	705	845	1105	75.4	63.0
Volatile Acids	1500	50	36	26.5	3	20	20	24	98.7	99.4
	1500	36	18	53	3	135	135	139	90.8	90.5
	1500	56	9	106	3	310	310	314	79.4	79.0
	1500	40	4.5	212	4	470	470	476	68.5	68.4
	3000	140	72	26.5	4	36	36	42	98.5	98.6
	3000	22	36	53	7	230	230	240	95.0	92.0
	6000	23	36	106	11	124	124	139	97.8	97.7
	6000	35	18	212	16	772	772	794	84.0	86.9

Source: After Young and McCarty (1967).

TABLE 7-9

SUMMARY OF STEADY-STATE FILTER PERFORMANCE UNDER VARIOUS ORGANIC LOADINGS AT 35°C

Organic load (lb/day/1000 ft³)	*Influent COD concentration (mg/ℓ)*	*Retention time (h)*	*Soluble effluent COD concentration (mg/ℓ)*	*Percentage COD removal*	*Effluent pH*	*Effluent SS (mg/ℓ)*	*Effluent volatile acids (mg/ℓ)*	*Effluent alkalinity (mg/ℓ)*
13.8	1,000	48	45	95.5	6.5	45	36	270
22.91	1,250	36	74	93.7	6.8	16	60	538
34.75	1,250	24	56.3	95.3	7.2	28	32	672
73.21	4,000	36	88	97.8	7.4	13	72	896
110	4,000	24	99	97.5	6.4	32	68	463
146.3	4,000	18	197	95.1	6.7	44	48	372
220	4,000	12	254	93.7	6.7	32	132	332
220	8,000	24	381	95.3	6.7	48	102	416
220	16,000	48	390	97.6	6.7	52	156	448

Source: After Jennett and Dennis (1975).

7-5
Nitrification in Attached-Growth Systems

Unlike the design of a suspended-growth nitrification system, designing for nitrification in an attached-growth process is mainly based on experience or operating data established during treatability studies. Mathematical models for this type of process have not been developed and refined to the point where they are readily applicable to design problems (Williamson and McCarthy, 1976a). As a result, the discussion presented in this section will be limited to a summary of the observations obtained from previous studies on nitrification in trickling filters, submerged filters, rotating biological contactors, and activated biofilters.

Nitrification in Trickling Filters

As was the case for nitrification with the activated sludge process, nitrification with trickling filters can be accomplished with a combined carbon oxidation/nitrification process or in a separate-stage nitrification process. It is important to distinguish between the two processes because the factors that limit nitrification are different for each. In combined carbon oxidation/nitrification system, Stenquist et al. (1974) suggest that organic loading limits nitrification because nitrifying bacteria are lost as a result of greater sloughing at high loadings. In their work, which considered carbon oxidation/nitrification in a plastic media trickling filter 21.5 ft deep with a recirculation ratio of 1:1, it was found that 89% ammonia/nitrogen removal (giving an effluent concentration of 2 mg/ℓ) could be achieved at an organic loading of 22 lb/BOD_5/1000 ft^3-day. Interestingly, organic nitrogen removal efficiency was low, averaging only 26% during the period of operation. The effect of temperature on nitrification was also observed and found to alter the efficiency very little. (This statement is based on observations made during a period when wastewater temperature was near 24°C and ambient air temperatures ranged between 9 and 14°C.) Conclusions from this study were:

1/ Efficient nitrification can be achieved at organic loadings up to 25 lb BOD_5/1000-ft^3-day with plastic media filters if surface loadings (including recycle) are maintained between 0.14 and 0.28 gpm/ft^2. Earlier work with rock media filters indicated that organic loadings must be less than 12 lb BOD_5/100 ft^2-day (Mohlman et al., 1946).

2/ Temperature appears to affect nitrification efficiency less in attached-growth systems than in suspended-growth systems.

3/ The organic portion of the TKN will be affected very little by treatment in a trickling filter.

Duddles et al. (1974) studied nitrification in a separate-stage process. A plastic-media trickling filter 21.5 ft deep was utilized to nitrify the effluent from a two-stage rock trickling filter plant. Influent BOD_5 to the plastic-media filter ranged from 15 to 20 mg/ℓ and the surface loading was varied from 0.5 to 2.0 gpm/ft². The second-stage nitrification pilot-plant study was conducted over a period of 18 months. Typical performance data for several of the operating periods are presented in Table 7-10. During the study period 80 to 90% nitrification was consistently achieved (a minimum residual of 1 to 1.5 mg/ℓ of ammonia nitrogen persisted in the second-stage effluent). Surface loading was found to be a factor in determining nitrification efficiency. Figure 7-19 illustrates the trend of decreasing nitrification efficiency

TABLE 7-10

PERFORMANCE DATA FOR NITRIFICATION IN A PLASTIC-MEDIA FILTER

		Flow (gpm/ft²)		*NH_3-N (mg/ℓ)*	
Operating period	*Month*	*Influent*	*Recycle*	*Tower influent*	*Tower effluent*
2	May 1971	0.5	1.0	11.3	1.3
3	June 1971	1.0	0.0	12.0	1.7
8	October 1971	0.5	1.0	16.8	1.4
11	January 1972	0.5	1.0	13.2	1.9
15	April 1972	0.71	0.0	7.5	1.2

Source: After Duddles et al. (1974).

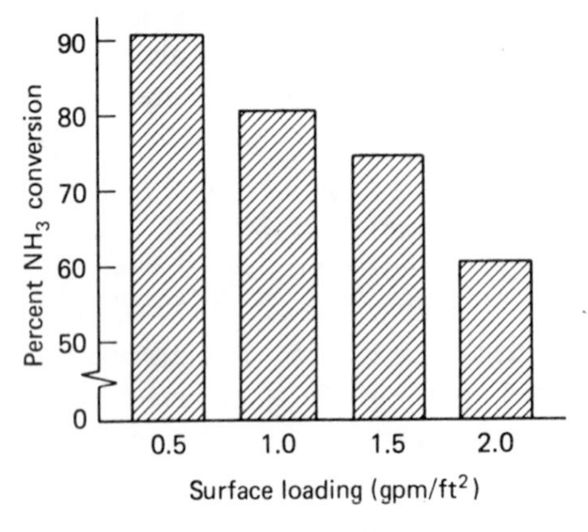

Figure 7-19. Effect of Surface Loading on Nitrification Efficiency. (After Duddles et al., 1974.)

with increasing surface loading, whereas Figure 7-20 illustrates the relationship among nitrification efficiency, temperature, and surface loading. This latter figure suggests that a high degree of nitrification can be achieved year round if the surface loading is maintained below 0.5 gpm/ft^2. A further observation from this study was that organic removal in first-stage treatment removed the requirement for a clarifier following second-stage treatment. Under such conditions, the economics of separate-stage nitrification with a plastic-media trickling filter compare favorably with other systems for nitrogen control.

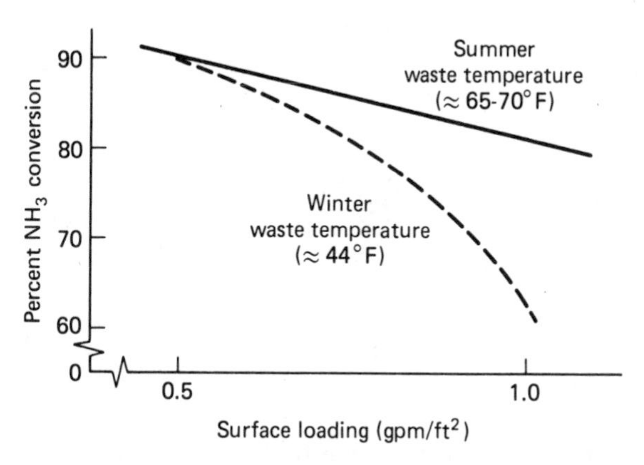

Figure 7-20. Relationship Among Temperature, Surface Loading, and Nitrification Efficiency. (After Duddles et al., 1974.)

Nitrification in Submerged Filters

McHarness and McCarty (1975) studied nitrification of an activated sludge plant effluent with the submerged filter. Figure 7-21 shows this filter to be an upflow device containing stone media 1 to 2 in. in diameter. In their study two methods of supplying oxygen to the filter were evaluated. One method involved aerating the waste flow with pure oxygen at 1 atm pressure before passing it through the filter. The second method required the bubbling of oxygen directly into the filter.

To maintain a sufficient DO level in the filter using the first method of oxygenation, it was found to be necessary to recycle a portion of the nitrified effluent. Under such conditions the following observations were made:

1/ Within the temperature range 21 to 27°C and for a 60-min retention time, influent ammonium concentrations of 14.3 ± 2.6 mg/ℓ were reduced $93 \pm 3\%$.

2/ Under the same conditions as above, BOD_5 values of 35 ± 6 mg/ℓ

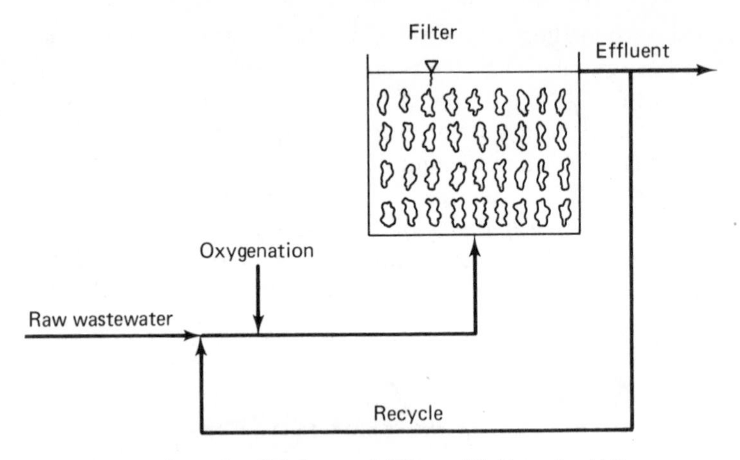

Figure 7-21. Schematic of Submerged Filter with Recycle. (After McHarness and McCarty, 1975.)

were reduced by 86% and suspended solids concentrations of 27 ± 3 mg/ℓ by 87%.

However, it was found that this method of oxygenation enhanced the clogging potential of the filter, and as a result it was necessary to clean the bed approximately twice per week.

The second method of providing oxygen to the filter which was investigated was that of bubbling oxygen through the filter itself. Using this method, the following observations were made:

1/ Nitrification efficiency closely approximated that for method 1.
2/ BOD_5 and suspended solids removal efficiencies were lower because of the turbulence generated in the filter by the gas bubbles.
3/ Clogging potential was reduced so that the filter required cleaning only once per week.

From their study McHarness and McCarty (1975) concluded that the major advantage of using a submerged filter for nitrification of secondary effluent was the simplicity in both design and operation of such a system, while the major disadvantage was filter clogging. This creates a problem in field operation: because of the size of field-scale units, backwashing is not practical; therefore, the filters have to be drained to be cleaned.

Haug and McCarty (1972) observed that maximum rates of nitrification occurred when the DO concentration was equal to or greater than the nitrification oxygen demand (on a concentration basis). Based on these findings, McHarness and McCarty (1975) suggest that for the system where the wastewater is oxygenated prior to filtration, it is safe to assume that the maximum rate of nitrification will occur when the total dissolved oxygen in the flow to the filter is equal to or greater than the total oxygen demand of the mixture secondary effluent and recycle effluent. This relationship can be expressed mathematically as

$$\text{TOA} = \frac{(Q)(\text{TOD})_0 + (RQ)(\text{TOD})_e}{Q + RQ} \tag{7-47}$$

where TOA = total DO concentration in the influent after oxygenation (see Figure 7-21), mg/ℓ

$(\text{TOD})_0$ = total oxygen demand of the secondary effluent, mg/ℓ

$(\text{TOD})_e$ = total oxygen demand of the effluent from the filter, mg/ℓ

Q = flow from secondary treatment unit, MGD

R = recycle ratio

Equation 7-47 can be solved for R to give

$$R = \frac{(\text{TOD})_0 - \text{TOA}}{\text{TOA} - (\text{TOD})_e} \tag{7-48}$$

If it is assumed that the total oxygen demand is the sum of the oxygen required for ammonium and carbon oxidation, then equation 7-48 can be expressed as

$$R = \frac{(4.57[\text{NH}_4^+\text{-N}]_0 + (\text{BOD}_u)_0) - \text{TOA}}{\text{TOA} - (4.57[\text{NH}_4^+\text{-N}]_e + (\text{BOD}_u)_e)} \tag{7-49}$$

where $[\text{NH}_4^+\text{-N}]_0$ = ammonium concentration in secondary effluent, mg/ℓ as N

$[\text{NH}_4^+\text{-N}]_e$ = ammonium concentration in effluent from filter, mg/ℓ as N

$(\text{BOD}_u)_0$ = ultimate BOD concentration in secondary effluent, mg/ℓ

$(\text{BOD}_u)_e$ = ultimate BOD concentration in effluent from filter, mg/ℓ

The DO concentration in the effluent from the oxygenation is a function of oxygen saturation and the percentage of oxygen saturation that can be achieved in the wastewater. According to McHarness and McCarty (1975), TOA may be computed from the equation

$$\text{TOA} = (F)(G)(H)\left(\frac{100}{I}\right)(J) \tag{7-50}$$

where F = atmospheric pressure, atm

G = percent oxygen saturation obtained by aerating the filter influent with pure oxygen, fraction

H = purity of oxygen used as aeration gas, fraction

I = percent of air that is composed of oxygen; this value is normally considered to be 21%

J = DO solubility in water with air at operating temperature and 1 atm pressure, mg/ℓ

For the temperature range 5 to 25°C McHarness and McCarty (1975) developed the following empirical equation to describe the relationship between retention time and nitrification efficiency:

$$t = \frac{10^b}{a(b-1)}\left[\frac{1}{[NH_4^+\text{-}N]_e^{(b-1)}} - \frac{1}{[NH_4^+\text{-}N]_a^{(b-1)}}\right](1+R) \qquad (7\text{-}51)$$

where t = retention time required to achieve the desired degree of nitrification, min

b = constant, which has a value of 1.2

$$a = 0.11(T) - 0.20 \qquad (7\text{-}52)$$

T = operating temperature, °C

$[NH_4^+\text{-}N]_e$ = ammonium concentration in effluent from filter, mg/ℓ as N

$[NH_4^+\text{-}N]_a$ = ammonium concentration in flow applied to filter, mg/ℓ as N

The ammonium concentration in the flow applied to the filter may be determined from the expression

$$[NH_4^+\text{-}N]_a = \frac{[NH_4^+\text{-}N]_0 + R[NH_4^+\text{-}N]_e}{1+R} \qquad (7\text{-}53)$$

When applying these equations to the design of a submerged filter unit, it must be remembered that they were developed for the system which oxygenates the wastewater flow and not for the system that bubbles oxygen directly into the filter.

Example 7-5

A submerged filter system, employing oxygenation of the applied wastewater, is to be used to nitrify the effluent from an activated sludge plant. The following design criteria are applicable:

1/ Q = 2MGD.
2/ $[NH_4^+\text{-}N]_0$ = 20 mg/ℓ.
3/ $[NH_4^+\text{-}N]_e$ = 2 mg/ℓ.
4/ $(BOD_u)_0$ = 30 mg/ℓ.
5/ $(BOD_u)_e$ = 3 mg/ℓ.
6/ $(TSS)_0$ = 30 mg/ℓ.
7/ $(TSS)_e$ = 3 mg/ℓ.
8/ VSS = 0.8 (TSS).
9/ Temperature = 15°C.
10/ Pressure = 1 atm.
11/ Y_T = 0.5.
12/ Y_N = 0.05.
13/ The oxygen gas that is used for aeration is 99% pure.
14/ Dissolved oxygen saturation obtained by oxygenation equals 75%.
15/ 1 to 3 in. granite will be used as the filter media.

Determine the required filter volume and daily solids accumulation in the filter.

solution

1/ From equation 7-50 compute the total DO concentration in the effluent after oxygenation. At 15°C and 1 atm pressure, the DO solubility in water with air is 10.15 mg/ℓ; therefore,

$$\text{TOA} = (1)(0.75)(0.99)(100/21)(10.15)$$
$$= 35.9 \text{ mg}/\ell$$

2/ Determine the minimum recycle ratio necessary to achieve the maximum rate of nitrification by applying equation 7-49.

$$R = \frac{[(4.57)(20) + 30] - 35.9}{35.9 - [(4.57)(2) + 3]}$$
$$= 3.6$$

3/ Estimate the ammonium concentration in flow applied to the filter using equation 7-53.

$$[NH_4^+\text{-N}]_a = \frac{20 + (3.6)(2)}{1 + 3.6}$$
$$= 5.9 \text{ mg}/\ell$$

4/ Compute the retention time required to achieve the desired degree of nitrification.

a/ Determine the coefficient, a, from equation 7-52.

$$a = (0.11)(15) - 0.20$$
$$= 1.45$$

b/ Calculate the required retention time from equation 7-51.

$$t = \frac{10^{1.2}}{(1.45)(1.2 - 1)}\left[\frac{1}{(2)^{1.2-1}} - \frac{1}{(5.9)^{1.2-1}}\right](1 + 3.6)$$
$$= 43 \text{ min}$$

5/ Determine the required filter volume.

$$t = \frac{V_v}{Q}$$

where

$$V_v = \text{filter void volume}$$

From Table 7-1 it is seen that a 1 to 3 in. granite media has 46% void space. Thus, the required filter volume is

$$V = \frac{tQ}{0.46}$$
$$= \frac{(43)(2{,}000{,}000)}{(0.46)(1440)(7.5)}$$
$$= 17{,}311 \text{ ft}^3$$

6/ Approximate the biological solids produced in the filter each day.

a/ Heterotrophic biomass production.

$$(\Delta X)_H = \frac{(0.5)(30 - 3)}{0.8}$$

$$= 17 \text{ mg}/\ell$$

b/ Autotrophic biomass production.

$$(\Delta X)_A = \frac{(0.05)(20 - 2)}{0.8}$$

$$= 1.1 \text{ mg}/\ell$$

7/ Estimate the solids accumulation in the filter each day.

$$\text{solids accumulation} = [(30 - 3) + 17 + 1.1](8.34)(2)$$

$$= 752 \text{ lb/day}$$

Nitrification with Rotating Biological Contactors

In a combined carbon oxidation/nitrification rotating biological contactor (RBC), Antonie (1970, 1972) found that nitrification begins when the BOD_5 concentration approaches 30 mg/ℓ. This implies that in a multiple-stage system, nitrification will predominate in the latter stages. Stover and Kincannon (1975) have presented data that support this observation. These workers employed a six-stage system with five 23.25-in.-diameter polystyrene disks in each stage, which were rotating at 11 rpm. A surface loading of 0.5 gpd/ft^2 was applied using a synthetic wastewater with a COD of 250 mg/ℓ and NH_3^+-N concentration of 27.6 mg/ℓ (sucrose was the sole carbon source). The hydraulic retention time was 160 min. Results of this study are presented in Figure 7-22, which shows that COD removal is virtually completed in the first stage, whereas ammonium oxidation is completed only after the fifth stage. The figure also shows that the rate of ammonium oxidation decreases after the first stage. This is attributed to the fact that nitrogen for heterotrophic cell synthesis is required mainly in the first stage. In subsequent stages ammonium removal is due almost entirely to nitrification.

Murphy et al. (1977) studied combined carbon oxidation/nitrification in an RBC that was fed screened municipal wastewater. The BOD_5 loading on the RBC varied between 1.2×10^{-3} and 2.4×10^{-3} lb BOD_5/ft^2-day. Experimental data from the study showed that the rate of nitrification was zero-order with respect to TKN concentration and could be computed from the relationship

$$K = (8.75 \times 10^{10})e^{-13,900/RT} \tag{7-54}$$

where K = unit nitrification rate, mg filterable TKN/ft^2-h

R = universal gas constant, 1.98 cal/mole-°K

T = operating temperature, °K

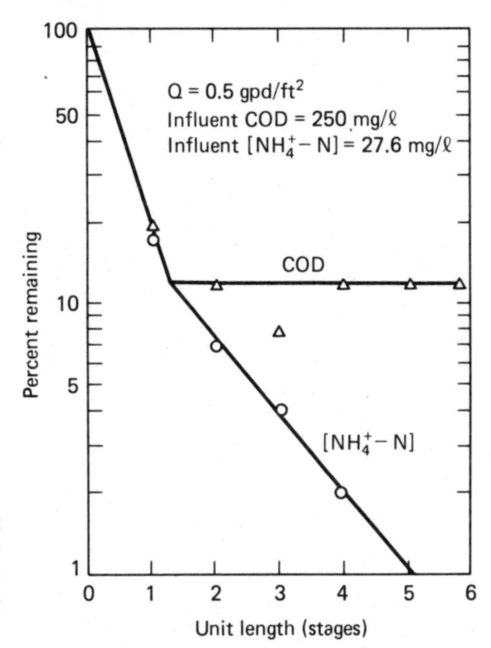

Figure 7-22. Relationship of Percent $[NH_4^+\text{-N}]$ and COD Remaining for Various Stages of the Rotating Biological Contactor Process. (After Stover and Kincannon, 1975.)

If complete mixing is assumed, a steady-state material-balance equation for TKN entering and leaving the system may be expressed as

$$0 = (Q_m)_0 - \frac{d(\text{TKN})}{dt} - (Q_m)_e \tag{7-55}$$

where $(Q_m)_0$ = mass of filterable TKN entering system per unit time, mg/h
$(Q_m)_e$ = mass of filterable TKN leaving system per unit time, mg/h
$\frac{d(\text{TKN})}{dt}$ = rate of nitrification, mg/h

For zero-order kinetics

$$\frac{d(\text{TKN})}{dt} = KA \tag{7-56}$$

where K = unit nitrification rate, mg filterable TKN/ft²-h
A = total media surface area, ft²

Substituting from equation 7-56 for $[d(\text{TKN})/dt]$ in equation 7-55 and solving for A,

$$A = \frac{(Q_m)_0 - (Q_m)_e}{K} \tag{7-57}$$

If the system is to be staged, the surface area requirements computed from equation 7-57 will be quite conservative.

Design relationships for combined carbon oxidation-nitrification in a 4-stage BIO-SURF process are presented in Figure 7-23. To apply these

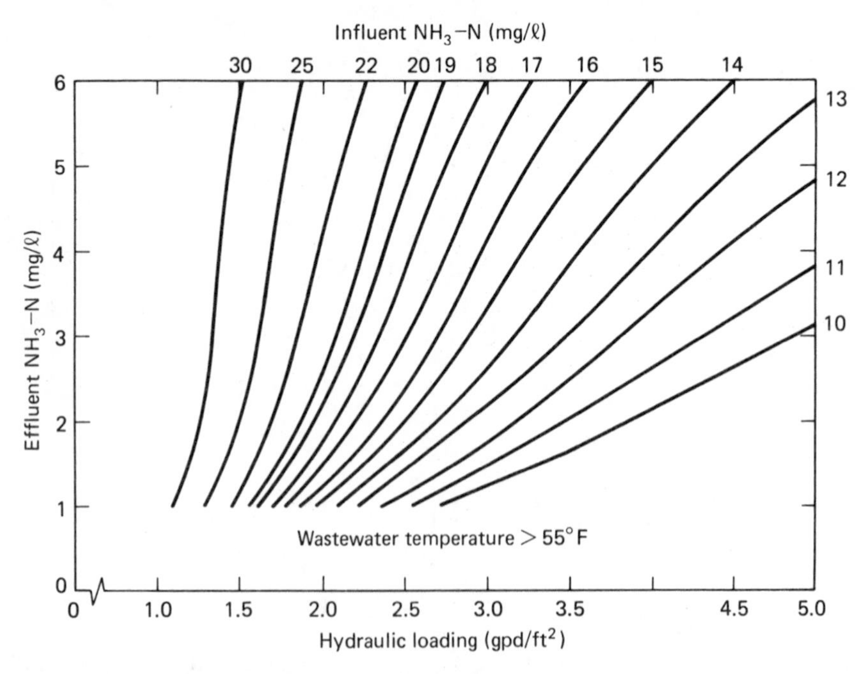

Figure 7-23. BIO-SURF Process Nitrification of Domestic Wastewater. (Courtesy of Autotrol Corporation.)

relationships in process design, the soluble BOD_5 of the wastewater must be equal to or less than 15 mg/ℓ. Temperature correction factors for nitrification in the BIO-SURF process are presented in Figure 7-24.

The growth rate of heterotrophic bacteria is much faster than that of the autotrophic bacteria. As a result, the heterotrophic organisms may displace the autotrophic organisms in the latter stages if the BOD loading to the RBC increases significantly. Thus, for process reliability large fluctuations in BOD loading should be avoided. Figure 7-25 indicates the allowable fluctuations in peak-to-average flow conditions. If this value is exceeded, either flow equalization or an increase in media surface area should be considered.

Antonie (1974) has investigated nitrification with a separate-stage RBC process. In this case pilot-plant studies employing a four-stage BIO-SURF process treating effluent from the final clarifier of different activated sludge plants across the country were studied. Results from these studies were used to construct Figure 7-26, which gives the relationship among ammonium removal, influent ammonium concentration, and surface loading for a four-stage BIO-SURF process.

In many cases where separate-stage nitrification is proposed, the final

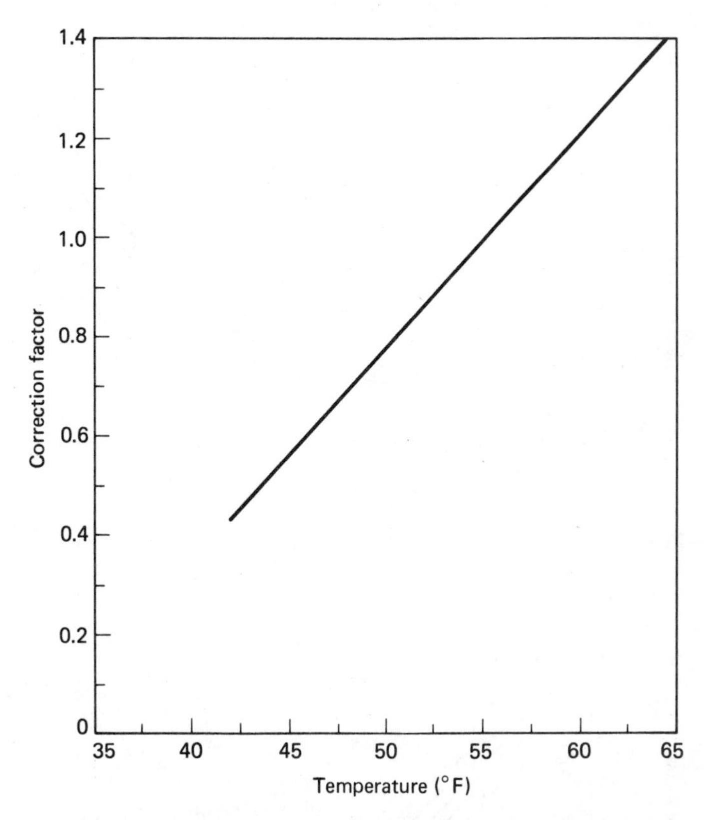

Figure 7-24. BIO-SURF Process Temperature Correction for Nitrification. (Courtesy of Autotrol Corporation.)

clarification step may be eliminated and a high-quality effluent produced by simply following the RBC with dual-media filtration.

Example 7-6

Determine the surface area required to achieve 90% nitrification and carbon removal with an RBC if the wastewater flow is 1 MGD and the influent BOD_5 and filterable TKN are 120 mg/ℓ and 20 mg/ℓ, respectively. The critical operating temperature is assumed to be 50°F.

solution

1/ Compute the unit nitrification rate from equation 7-54.

$$K = (8.75 \times 10^{10})e^{-13,900/(1.98)(283)}$$

$$= 1.47 \text{ mg/ft}^2\text{-h}$$

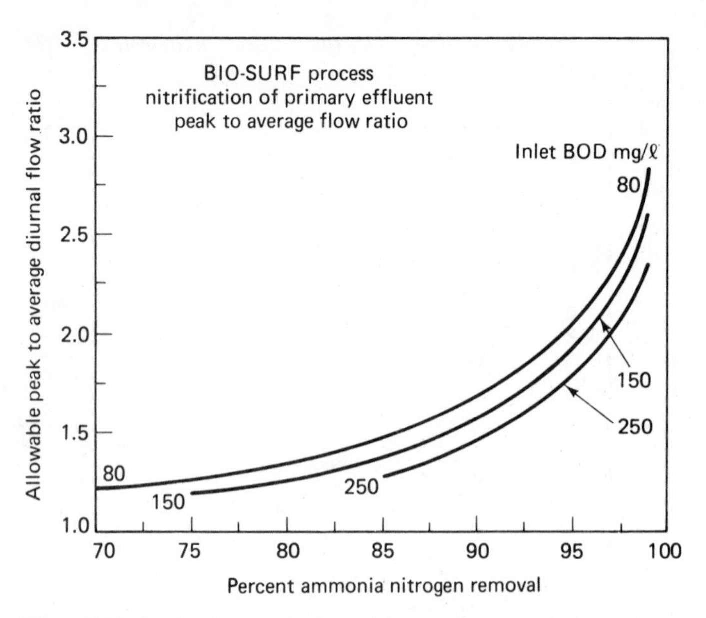

Figure 7-25. Peak Flow to Average Flow Ratios Acceptable for Nitrification in the BIO-SURF Process. (Courtesy of Autotrol Corporation.)

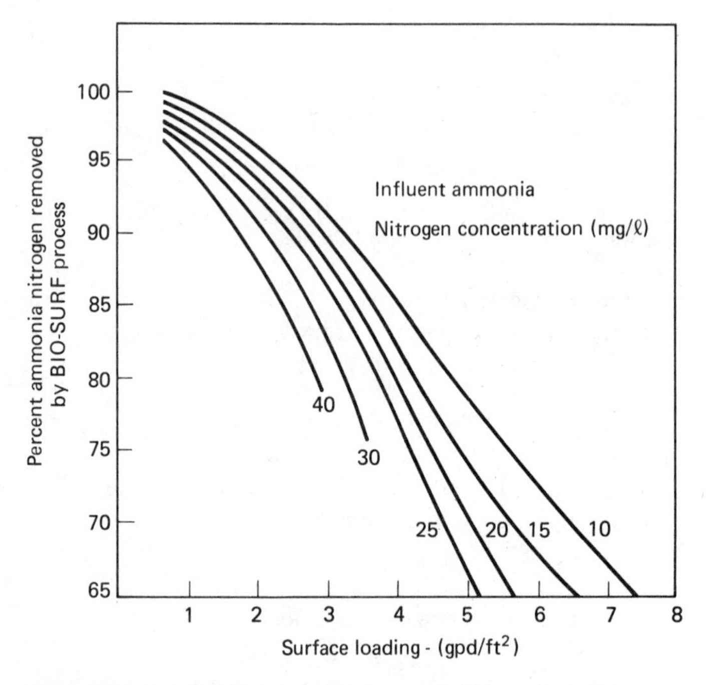

Figure 7-26. Relationship Among Ammonia Removal, Influent Ammonia Concentration, and Surface Loading for Four-Stage BIO-SURF System. (After Antonie, 1974.)

2/ Determine the mass flow rate of filterable TKN into the RBC.

$$(Q_m)_0 = \frac{(20)(1{,}000{,}000)(3.78)}{24}$$

$$= 3{,}150{,}000 \text{ mg/h}$$

3/ Determine the mass flow rate of filterable TKN out of the RBC.

$$(Q_m)_e = \frac{(0.1)(20)(1{,}000{,}000)(3.78)}{24}$$

$$= 315{,}000 \text{ mg/h}$$

4/ Calculate the surface-area requirement for 90% nitrification.

$$A = \frac{3{,}150{,}000 - 315{,}000}{1.47}$$

$$= 1{,}928{,}571 \text{ ft}^2$$

5/ Check the surface-area requirement by applying the Autotrol design curves.

a/ From Figure 7-23 the surface loading for 90% nitrification (an effluent ammonia nitrogen concentration of 2 mg/ℓ for an influent ammonia nitrogen concentration of 20 mg/ℓ) is 1.8 gpd/ft².

$$A = \frac{1{,}000{,}000}{1.8}$$

$$= 555{,}555 \text{ ft}^2$$

b/ From Figure 7-24 the media-correction factor for 50°F is 0.78.

$$A = \frac{555{,}555}{0.78}$$

$$= 712{,}250 \text{ ft}^2$$

These calculations show almost a threefold difference in the surface-area requirements determined by the zero-order nitrification-rate equation and the design curves developed by the Autotrol Corporation.

Nitrification with the Activated Biofilter

The flow scheme for the ABF nitrification system is basically the same as that given for the ABF process concerned with BOD removal (see Figure 7-17). The primary difference between the two systems is the volume of the aeration basin. When carbon removal from municipal wastewater is of primary concern, retention time in the aeration basin varies between 0.66 and 1.5 h. When nitrification is required, the volume is increased to give a retention time of 2 to 4.5 h (Dunnahoe and Hemphill, 1976). Typical design criteria for an ABF nitrification system are presented in Table 7-11. This system is capable of producing an effluent containing 20 mg/ℓ BOD_5, 20 mg/ℓ suspended solids, and 1 mg/ℓ [NH_4^+-N] (Hemphill, 1977).

TABLE 7-11

DESIGN PARAMETERS FOR ABF NITRIFICATION SYSTEM

Parameter	*Units*	*Typical value*	*Range*
Biocell			
Organic loading	lb BOD_5/1000 ft^3-day	200	100–350
Media depth	ft	14	5–22
BOD removal	%	55	50–75
Hydraulic			
Biocell recycle		0.4Q	0.0–2.0Q
Sludge recycle		0.5Q	0.3–1.0Q
Biocell flow		1.9Q	1.5–4.0Q
Biocell surface loading	gpm/ft^2	3.5	1.5–5.5
Aeration tank[a]			
Retention time[b]	h	3.5	2.5–5.0
Organic loading	lb BOD_5/1000 ft^3-day	25	20–40
Ammonia loading	lb [NH_4^+-N]/1000 ft^3-day	10	5.0–15.0
F:M ratio	lb BOD_5/lb MLVSS-day	0.13	0.1–0.2
MLVSS	mg/ℓ	3000	1500–4000
MLSS	mg/ℓ	4000	2000–5000
Carbonaceous oxygen requirement[c]	lb O_2/lb BOD_5	1.4	1.2–1.5
Total sludge production	lb VSS/lb BOD_5 removed	0.45	0.3–0.55

[a]Based on aeration BOD_5 loading after biocell removal.
[b]Based on design average flow and secondary influent BOD_5 = 150 mg/ℓ.
[c]Total oxygen requirement = carbonaceous oxygen + 4.57 (lb [NH_4^+-N] oxidized).
Source: After Hemphill (1977).

7-6

Denitrification in Attached-Growth Systems

In recent years considerable effort has been expended in developing reliable and economically feasible processes for the biological removal of nitrogen from wastewater. In biological denitrification processes, the heterotrophic organisms responsible for nitrate reduction are either suspended in the liquid phase or attached to a support medium. Suspended-growth denitrification processes were discussed in detail in Chapter 4. Process control involving solids–liquid separation, sludge recycle, and sludge wasting is a major problem associated with these systems. Attached-growth systems, however, retain the microbial solids within the boundaries of the reactor, thereby eliminating the need for solids recycle. As a result, such systems hold considerable promise as a suitable method of nitrogen removal. Attached-growth systems employed for denitrification are submerged rotating biological contactors, submerged packed columns using either high-porosity media or low-porosity fine media, and fluidized beds.

Submerged Rotating Biological Contactors

A commercial submerged RBC unit is available for denitrification, although few operational data have been published concerning its capability. Murphy et al. (1977) conducted one of the few published investigations concerning denitrification in this process. In this study nitrified effluent from an activated sludge process served as feed, and methanol was added to provide a carbon/nitrogen ratio in excess of 1.1:1. The submerged RBC was divided into four stages, with a total hydraulic volume of 400 ℓ. The theoretical retention time of the unit was 100 min. Experimental data indicated that denitrification in the presence of excess carbon was independent of the $[NO_3^- + NO_2^-\text{-N}]$ concentration; that is, the rate of denitrification followed zero-order kinetics with respect to nitrate and nitrite concentration.

Observations from a single-stage operation were employed to develop an empirical relationship that describes the rate of denitrification per unit area of disk surface between the temperature range 5 to 25°C. The equation has the form

$$K = 7.79 \times 10^{13} e^{-16,550/RT} \qquad (7\text{-}58)$$

where K = denitrification rate, mg $[NO_3^- + NO_2^-\text{-N}]/\text{ft}^2\text{-h}$
R = universal gas constant, 1.98 cal/mole-°K
T = operating temperature, °K

It was observed that the suspended solids concentration in the effluent from the submerged RBC unit was low, averaging only 21 mg/ℓ. This suggests that the final clarification step may be eliminated and a high-quality effluent produced by simply following the RBC with dual-media filtration.

Submerged High-Porosity Media Columns

The submerged packed column consists of a reactor filled with an inert packing material that is operated under conditions of saturated flow. An upflow reactor is generally used to enhance separation of suspended solids from the effluent. Flow velocities through the column are on the order of 0.1 ft/s (Regua and Schroeder, 1973).

A typical flow schematic for a denitrification system composed of submerged high-porosity media columns is presented in Figure 7-27. For effective nitrogen removal two or three columns are normally operated in series. The trade names, along with certain characteristics of the various media that have

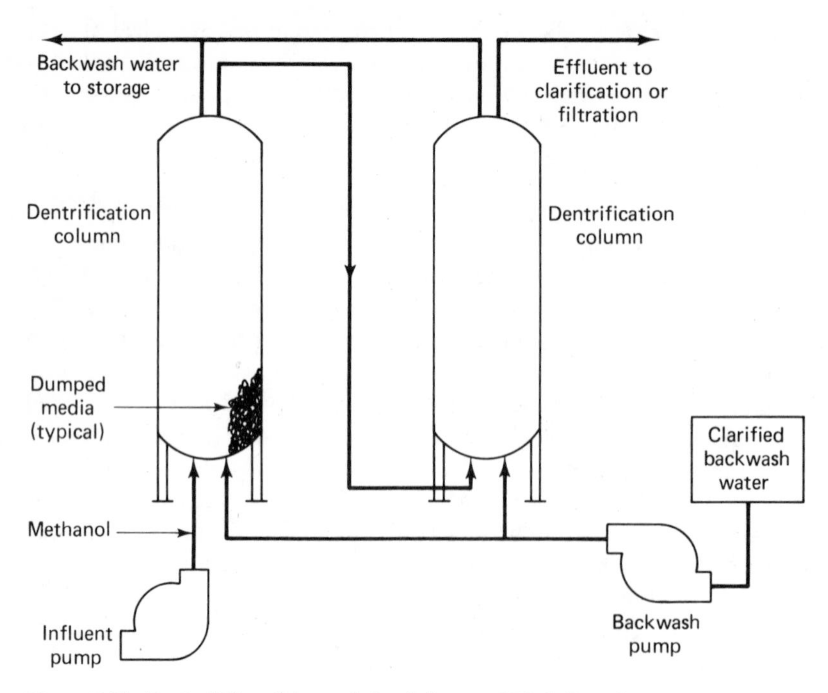

Figure 7-27. Typical Flow Schematic for Submerged High-Porosity Media Denitrification Columns. (After Brown and Caldwell, 1975.)

been used in this type system, are presented in Table 7-12. Media selection is an important consideration in column design. The larger the specific surface, the greater the surface denitrification rate, but the clogging potential of the column is also greater.

To minimize the suspended solids concentration in the effluent from the unit, it is necessary to backwash the column periodically. This removes the excess solids, which otherwise would slough into the effluent. At El Lago, Texas, where the denitrification columns were packed with Koch Flexirings, backwashing was conducted only once every month. A combined air/water backwash procedure was used where the water backwash rate was 10 gpm/ft^2 and the air backwash rate was 10 cfm/ft^2 (Brown and Caldwell, 1975). It is also good design practice to provide chlorination capabilities to aid sloughing in case a severe clogging problem develops.

Collecting experimental data from several different studies, Brown and Caldwell (1975) constructed Figure 7-28 and suggest that this figure may be used to size denitrification columns when the peak diurnal nitrate loading and minimum wastewater temperature are known.

Employing nitrified effluent from an activated sludge process and adding methanol to provide a carbon/nitrogen ratio in excess of 1.1:1, Murphy et al.

TABLE 7-12

DESIGN DATA FOR HIGH-POROSITY DENITRIFICATION COLUMNS

Parameter	*Surface Denitrification Rate (lb N Removed/ft²-day)*				
	Koch Flexirings	*Envirotech Surfpac*	*Koch Flexirings*	*Intalox saddles*	*Raschig rings*
Specific Surface (ft^2/ft^3)	*65*	*27*	*105*	*142–274*	*79*
Porosity (%)	*96*	*94*	*92*	*70–78*	*80*
Temperature, °C					
5				3.2×10^{-5}	
10				3.7×10^{-5}	
11					
12					
13	4.3×10^{-5}				
14					
15	5×10^{-5}			2.7×10^{-5}	
16					
17	5.3×10^{-5}				
18					
19					
20	13×10^{-5}			9.5×10^{-5}	
21	11×10^{-5}				
22					
23	5.9×10^{-5}				
24					
25				11×10^{-5}	

Source: After Brown and Caldwell (1975).

(1977) studied denitrification in high-porosity media columns using two different media. Operational conditions and packing for the different dentrification columns utilized in this work are presented in Table 7-13. During the course of the study it was found that backwashing the columns did not correct a persistent short-circuiting problem which caused a large variation in nitrate removal efficiencies. As a result, Murphy et al. (1977) recommend that this type of system not be used where a high-quality effluent is important.

Example 7-7

Compute the column volume required to achieve 90% nitrogen removal from a wastewater flow of 1 MGD, with a nitrate nitrogen concentration of 30 mg/ℓ. Assume that the critical wastewater temperature is 20°C and Koch Flexirings with a specific surface of 65 ft^2/ft^3 are to be used as the support medium.

TABLE 7-13

OPERATIONAL CONDITIONS AND PACKING FOR DENITRIFICATION COLUMNS

Column	*Media type*	*Packing size* (*in*)	*Porosity* (%)	*Specific surface* (ft^2/ft^3)	*Column height* (*ft*)	*Surface loading* (gpm/ft^2)	*Packed*[a] *retention time* (*min*)
1	Intalox saddles	0.37	76	241	12	3.6	16.5
						1.8	33.0
2	Intalox saddles	0.5	78	190	12	3.6	17.5
						1.8	35.0
3	Pall rings	1	90	63	8	0.7	70.0
4	Pall rings	2	92	31	8	0.7	70.0

[a]Calculated retention time based on the volume of voids.
Source: After Murphy et al. (1977).

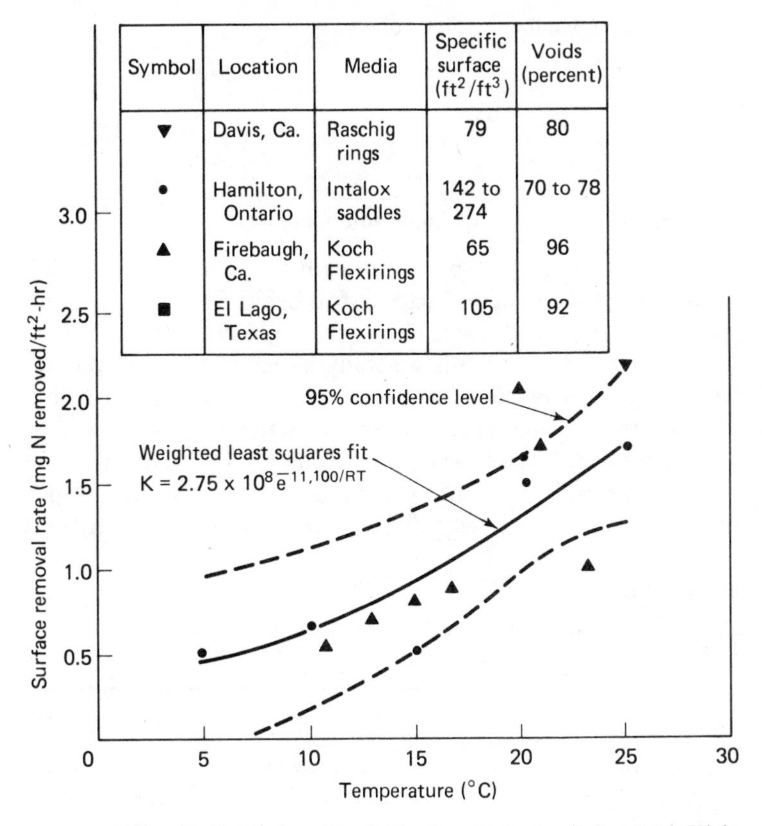

Figure 7-28. Surface Denitrification Rate for Submerged High-Porosity Media Columns. (After Brown and Caldwell, 1975.)

solution

1/ Compute the required nitrogen removal.

$$\text{nitrogen removal} = \frac{(0.9)(30)(1{,}000{,}000)(3.78)}{24}$$

$$= 4{,}252{,}500 \text{ mg/h}$$

2/ Using the empirical equation developed from the experimental data presented in Figure 7-28, compute the surface nitrogen removal rate.

$$K = 2.75 \times 10^8 e^{-11{,}100/(1.98)(293)}$$

$$= 1.35 \text{ mg N/ft}^2\text{-h}$$

3/ Calculate the media surface required.

$$A = \frac{4{,}252{,}500}{1.35}$$

$$= 3{,}150{,}000 \text{ ft}^2$$

4/ Determine the required column volume.

$$V = \frac{3{,}150{,}000}{65}$$

$$= 48{,}462 \text{ ft}^3$$

Submerged Low-Porosity Fine-Media Columns

Even though the low-porosity fine-media column is probably the most commonly used submerged packed-bed denitrification system, no procedure (rational or empirical) is available to the engineer for the design of such a unit. As a consequence, design is generally based on experimental data obtained from pilot-plant or laboratory-scale studies. For nitrified municipal wastewaters surface loadings between 0.5 to 1.5 gpm/ft² are common.

Low-porosity fine-media columns commonly serve the purpose of effluent filtration as well as denitrification of the wastewater. Suspended solids removal data for several different locations are presented in Table 7-14.

TABLE 7-14

COMPARISON OF SUSPENDED SOLIDS REMOVAL FOR SUBMERGED FINE-MEDIA DENITRIFICATION COLUMN

Location	*Sand media size*[a] *(mm)*	*Surface loading (gpm/ft²)*	*Depth (ft)*	*Influent SS (mg/ℓ)*	*Effluent SS (mg/ℓ)*
El Lago, Tex.	$d_{50} = 3$[b]	6.27	13	37	17
North Huntington Township, Pa.	$d_{10} = 2.9$	0.72	6	16	7
Tampa, Fla.	$d_{10} = 2.9$	2.5	—	20	5
Lebanon, Ohio	$d_{50} = 3.4$[b]	7.0	10	13	4
	$d_{50} = 5.9$[b]	7.0	20	13	2
	$d_{50} = 14.5$[b]	7.0	10	13	1

[a] d_{10}, effective size; d_{50}, medium particle size.
[b] Uniformly graded.
Source: After Brown and Caldwell (1975).

A combined air/water backwash is recommended for these systems. For its particular unit, Dravo suggests an air backwash rate of 6 cfm/ft² and a water backwash rate of 8 gpm/ft² (Brown and Caldwell, 1975).

The media commonly used is either sand or gravel, although activated carbon has also been utilized. Since the columns are commonly operated as upflow packed beds to reduce clogging problems, sand is the media of choice. It is preferred because it provides more surface area per cubic foot than gravel, owing to the smaller particle size, and fewer operating problems than activated carbon, because of higher specific gravity.

Tucker et al. (1974) have shown that for a given packed bed reactor and wastewater, there is an optimum hydraulic loading rate for denitrification. At lower flows, the rate of removal in the column will be limited by the nitrate loading rate, while higher flows it will be limited by the hydraulic loading rate (see Figure 7-29). However, to attain nearly complete removal of nitrogen, a single column must be operated in the nitrate-limited range. Therefore, the columns should be operated in series for the most efficient utilization.

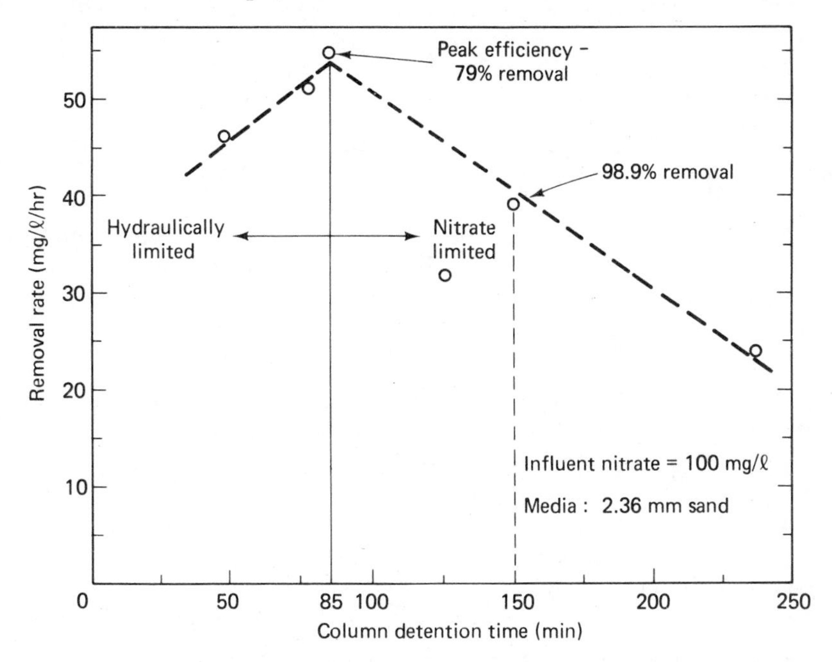

Figure 7-29. Effect of Retention Time on Packed Bed Sand Column Efficiency. (After Tucker et al., 1974.)

A comparison of maximum nitrate removal rates reported by various researchers using packed bed columns are given in Table 7-15 and compared to two fluidized bed studies. The table illustrates the effect of both particle size and fluidization on the efficiency of the reactor. The effect of both smaller particle size and fluidization is to increase the amount of biomass in the reactor per unit volume.

Fluidized Bed Denitrification Columns

In the fluidized bed denitrification column, wastewater is passed upward through a bed of fine media (e.g., activated carbon or sand) at a velocity great

TABLE 7-15

COMPARISON OF NITROGEN REMOVAL RATES IN PACKED-BED REACTORS

Media	*Type of reactor*	ft^2/ft^3	*Nitrogen removal rate* $lb/day/ft^2$	$lb/day/ft^3$
1.5-in. gravel	Packed-bed	24	1.5×10^{-4}	3.7×10^{-3}
1.0-in. gravel	Packed-bed	151	7.0×10^{-5}	10.6×10^{-3}
2.36-mm activated carbon	Packed-bed	387+	4.3×10^{-5}	16.8×10^{-3}
2.36-mm sand	Packed-bed	387	6.2×10^{-5}	24.1×10^{-3}
1.7-mm activated carbon	Fluidized	538+	1.4×10^{-4}	73.2×10^{-3}
1.18-mm activated carbon	Packed-bed	774+	1.4×10^{-5}	11.0×10^{-3}
0.6-mm sand[a]	Fluidized	1088	3.4×10^{-4}	367×10^{-3}

[a]Effective size.

enough to cause fluidization of the media. During operation the media becomes completely covered with biological growth, causing an increase in media size. In a submerged packed column, high headloss, channeling, and a reduction in nitrogen removal result from the presence of large amounts of microbial growth in the column. On the other hand, these operational problems are minimized in a fluidized bed system because sufficient voids are maintained between the particles to provide good liquid contact.

Because of the presence of large amounts of biological growth, the fluidized bed system has the highest volumetric denitrification rate of any other column configuration (this rate is shown as a function of temperature in Figure 7-30). As a consequence, greater surface loading rates may be applied to fluidized bed systems. For example, in a pilot-plant study at Nassau County, New York, a 15 gpm/ft^2 loading rate was applied to a column containing silica sand that had a rested bed depth of 6 ft and a fluidized bed depth of 12 ft (Jeris and Owens, 1975).

A schematic of a typical fluidized bed denitrification system is presented in Figure 7-31. During operation, biological growth continually builds up in the column. This results in an increase in the expanded depth of the bed, causing a continuous loss of media from the system. The sand separation unit is required if the column media inventory is to be maintained. If a high-quality effluent (low in suspended solids) is required, the system will also have to be followed by multimedia filtration.

Although the fluidized bed system has the advantage of the highest volumetric denitrification rates, the requirement for sand separation, clarification, and probably multimedia filters are disadvantages not applicable to the submerged low-porosity fine-media denitrification column.

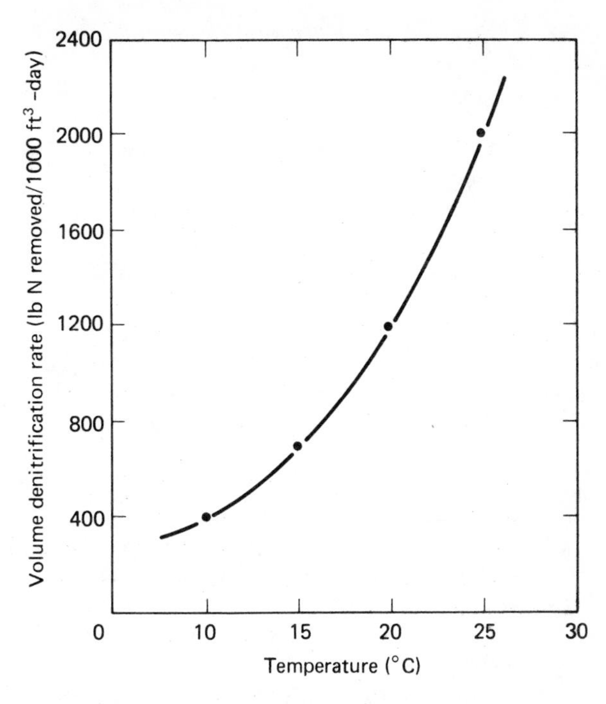

Figure 7-30. Volume Denitrification Rate for Fluidized Beds. (After Brown and Caldwell, 1975.)

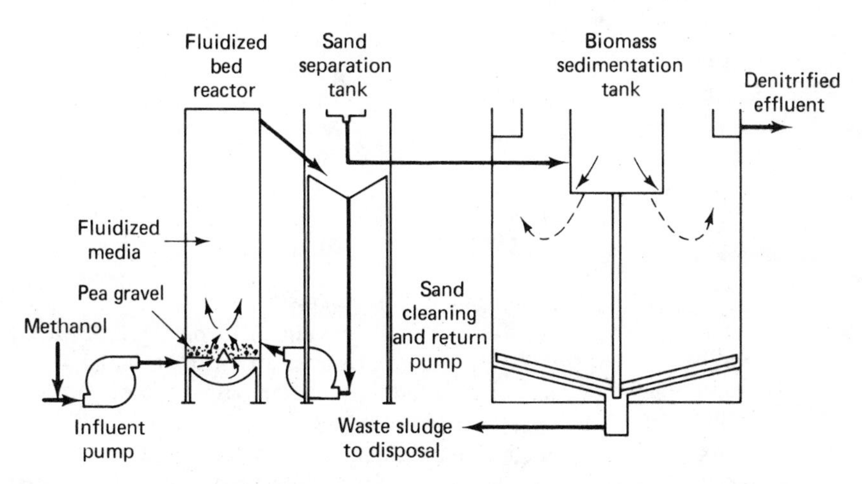

Figure 7-31. Fluidized Bed Denitrification System. (After Brown and Caldwell, 1975.)

PROBLEMS

7-1 The following laboratory data were obtained at 20°C.
Flow rate = 1.0 gpm/ft²:

Depth	% BOD_5 remaining
2.5	86
6.5	68
10.5	52
13.6	43
17.6	35
21.5	26

Flow rate = 1.5 gpm/ft²:

Depth	% BOD_5 remaining
2.5	90
6.5	75
10.5	63
13.6	55
17.6	45
21.5	38

Flow rate = 3.0 gpm/ft²:

Depth	% BOD_5 remaining
2.5	94
6.5	84
10.5	76
13.6	70
17.6	63
21.5	57

a/ Correlate the data and determine K'_0 and n.
b/ Determine the diameter of the filter required to obtain 90% BOD_5 removal from 1 MGD of municipal wastewater with an initial BOD_5 of 200 mg/ℓ. A depth of 25 ft and recirculation ratio of 2.0 will be used. The critical operating temperature is 10°C.

7-2 Compare the volume requirements given by the NRC equation and the Eckenfelder equation for a two-stage trickling filter system that is required to treat a wastewater flow of 2 MGD with a strength of 200 mg/ℓ BOD_5. Each filter is to be 8 ft deep and the critical operating temperature is expected to be 20°C. The volume requirements are to be compared at recirculation ratios of 1:1, 2:1, and 3:1, and BOD_5 removal efficiencies of 60, 70, and 80%.

On the basis of the results obtained from this problem, what can be said regarding the acceptability of these equations for design purposes?

7-3 Assume that a three-stage rotating biological contactor system is to be constructed to treat a wastewater flow of 0.1 MGD with a strength of 250 mg/ℓ BOD_5. A treatability study has revealed that $P = 2500$ mg/ft^2-day and $K_s =$ 100 mg/ℓ. How much surface area will be required if a soluble effluent BOD_5 of 15 mg/ℓ is desired?

7-4 Rework Problem 7-3 using a four-stage system and compare the surface-area requirements of a single-stage system (Example 7-3), a two-stage system (Example 7-4), a three-stage system (Problem 7-3), and a four-stage system.

7-5 Determine the media surface area necessary to treat a municipal wastewater flow of 1 MGD with a BOD_5 of 150 mg/ℓ and ammonium nitrogen concentration equal to 20 mg/ℓ. Use Figures 7-15 and 7-23. The critical operating temperature is 45°F, and 90% removal of both BOD_5 and ammonium nitrogen is required.

7-6 A plastic media filter 21 ft deep is to be used to treat 2 MGD of activated sludge effluent containing 25 mg/ℓ ammonium nitrogen. What filter volume is required for 85% ammonium conversion at 45°F?

REFERENCES

1. ANTONIE, R. L., "Application of the BIO-DISC Process to Treatment of Domestic Wastewater," paper presented at the 43rd Annual Conference of the Water Pollution Control Federation, Boston, 1970.
2. ANTONIE, R. L., "Three-Step Biological Treatment with the BIO-DISC Process," paper presented at the Spring Meeting of the New York Water Pollution Control Association, Montauk, N.Y., June 1972.
3. ANTONIE, R. L., "Nitrification of Activated Sludge Effluent: BIO-SURF Process," *Water and Sewage Works*, **44** (Nov. 1974).
4. ARCHER, E. C., AND L. R. ROBINSON, JR., "Design Considerations for Biological Filters," in *Handbook of Trickling Filter Design*, Public Works Journal Corporation, Ridgewood, N.J., 1970.
5. BAKER, J. M., AND Q. B. GRAVES, "Recent Approaches for Trickling Filter Design," *Journal of the Sanitary Engineering Division, ASCE*, **94** SA1, 65 (1968).
6. BENZIE, W. J., "The Design of High Rate Filters," in *Handbook of Trickling Filter Design*, Public Works Journal Corporation, Ridgewood, N.J., 1970.
7. BENZIE, W. J., H. O. LARKIN, AND A. F. MOORE, "Effects of Climatic and Loading Factors on Trickling Filter Performance," *Journal of the Water Pollution Control Federation*, **35**, 445 (1963).
8. BOYLE, J. D., "Biological Treatment Process in Cold Climates," *Water and Sewage Works*, **23**, R-28 (Apr. 1976).
9. BROWN AND CALDWELL, Consulting Engineers, *Process Design Manual for Nitrogen Control*, EPA Technology Transfer Series, 1975.

10. CARDENAS, R. L., JR., "Trickling Filters and the Unit Operations Laboratory," paper presented at the AAPSE Workshop on Biological Waste Treatment Processes, University of Texas, Austin, Tex., 1966.
11. CULP, G. L., "Direct Recirculation of High-Rate Trickling Filter Effluent," *Journal of the Water Pollution Control Federation*, **35**, 742 (1963).
12. DUDDLES, G. A., S. E. RICHARDSON, AND E. F. BARTH, "Plastic-Medium Trickling Filters for Biological Nitrogen Control," *Journal of the Water Pollution Control Federation*, **46**, 937 (1974).
13. DUNNAHOE, R. G. AND B. W. HEMPHILL, "The ABF Process, A Combined Fixed/Suspended Growth Biological Treatment System," paper presented at the AWWAFACE Conference, Halifax, Nova Scotia, Sept. 12–15, 1976.
14. ECKENFELDER, W. W., JR., "Trickling Filter Design and Performance," *Journal of the Sanitary Engineering Division, ASCE*, **87**, SA6, 87 (1961).
15. ECKENFELDER, W. W., JR., *Water Quality Engineering*, Cahners Books, Boston, 1970.
16. ECKENFELDER, W. W., JR., "Trickling Filters," in *Process Design in Water Quality Engineering*, ed. by E. L. Thackston and W. W. Eckenfelder, Jenkins Publishing Co., Austin, Tex. 1972.
17. ECKENFELDER, W. W., JR., AND W. BARNHART, "Performance of a High-Rate Trickling Filter Using Selected Media," *Journal of the Water Pollution Control Federation*, **35**, 1535 (1963).
18. FAIRALL, J. M., "Correlation of Trickling Filter Data," *Sewage and Industrial Wastes*, **28**, 1069 (1956).
19. GALLER, W. S., AND H. B. GOTAAS, "Analysis of Biological Filter Variables," *Journal of the Sanitary Engineering Division, ASCE*, **90**, SA6, 59 (1964).
20. Great Lakes–Upper Mississippi River Board of State Sanitary Engineers, *Recommended Standards for Sewage Works* (*Ten-State Standards*), 1971.
21. HAUG, R. T., AND P. L. MCCARTY, "Nitrification with the Submerged Filter," *Journal of the Water Pollution Control Federation*, **44**, 2086 (1972).
22. HEMPHILL, B. W., "Reliable, Cost-Effective Treatment with the ABF Process," paper presented at the Western Canada Water and Sewage Conference, Edmonton, Alberta, Sept. 28–30, 1977.
23. JANK, B. E., AND W. R., DRYNAN, "Substrate Removal Mechanism of Trickling Filters," *Journal of the Environmental Engineering Division, ASCE*, **EE3**, 187 (1973).
24. JENNETT, J. C., AND N. D. DENNIS, Jr., "Anaerobic Filter Treatment of Pharmaceutical Waste," *Journal of the Water Pollution Control Federation*, **47**, 104 (1975).
25. JERIS, J. S., "High Rate Denitrification," paper presented at the 44th Annual Conference of the Water Pollution Control Federation, Oct. 3–8, 1971.
26. JERIS, J. S., AND R. W. OWENS, "Pilot-Scale, High-Rate Biological Denitrification," *Journal of the Water Pollution Control Federation*, **47**, 2043 (1975).
27. KORNEGAY, B. H., "Modeling and Simulation of Fixed Film Biological Reactors," in *Mathematical Modeling of Water Pollution Control Processes*, ed.

by T. M. Keinath and M. Wanielista, Ann Arbor Science Publications, Inc., Ann Arbor, Mich., 1975.

28. Lipták, B. G., *Environmental Engineer's Handbook*, Vol. I, Chilton Book Company, Radnor, Pa., 1974.
29. McCarty, P. L., "Anaerobic Treatment of Soluble Wastes," in *Advances in Water Quality Improvement*, ed. by E. F. Gloyna and W. W. Eckenfelder, Jr., University of Texas Press, Austin, Tex., 1968.
30. McHarness, D. D., and P. L. McCarty, "Field Study of Nitrification with the Submerged Filter," *Journal of the Water Pollution Control Federation*, **47**, 291 (1975).
31. Metcalf and Eddy, Inc., *Wastewater Engineering*, McGraw-Hill Book Company, New York, 1972.
32. Mohlman, F. W., et al., "Sewage Treatment at Military Installations," *Sewage Works Journal*, **18**, 794 (1946).
33. Murphy, K. L., P. M. Sutton, R. W. Wilson, and B. E. Jank, "Nitrogen Control: Design Considerations for Supported Growth Systems," *Journal of the Water Pollution Control Federation*, **49**, 549 (1977).
34. Parkhurst, J. D., "Pomona Activated Carbon Pilot Plant," *Journal of the Water Pollution Control Federation*, **39**, R70 (1967).
35. Rankin, R. S., "Evaluation of the Performance of Biofiltration Plants," *Transactions, ASCE*, **120**, 823 (1955).
36. Regua, D. A., and E. D. Schroeder, "Kinetics of Packed-Bed Denitrification," *Journal of the Water Pollution Control Federation*, **45**, 1969 (1973).
37. Schroeder, E. D., *Water and Wastewater Treatment*, McGraw-Hill Book Company, New York, 1977.
38. Schulze, K. L., "Load and Efficiency of Trickling Filters," *Journal of the Water Pollution Control Federation*, **32**, 245 (1960).
39. Seidel, D. F., and R. W. Crites, "Evaluation of Anaerobic Denitrification Processes," *Journal of the Sanitary Engineering Division, ASCE*, **96**, 267 (1970).
40. Smith, R., "Cost-Effectiveness Analysis for Water Pollution Control," in *Upgrading Wastewater Stabilization Ponds to Meet New Discharge Standards*, PB-240-402, National Technical Information Service, Springfield, Va, 1974.
41. St. Amant, P. P., and P. L. McCarty, "Treatment of High Nitrate Waters," *Journal of the American Water Works Association*, **61**, 42 (1969).
42. Stack, V. T., Jr., "Theoretical Performance of the Trickling Filtration Process," *Sewage and Industrial Wastes*, **29**, 987 (1957).
43. Stenquist, R. J., D. S. Parker, and T. J. Dosh, "Carbon Oxidation-Nitrification in Synthetic Media Trickling Filters," *Journal of the Water Pollution Control Federation*, **46**, 2327 (1974).
44. Stover, E. L., and D. F. Kincannon, "One-Step Nitrification and Carbon Removal," *Water and Sewage Works*, **66** (June 1975).
45. Subcommittee on Sewage Treatment, Committee on Sanitary Engineering, National Research Council, "Sewage Treatment at Military Institutions," *Sewage Works Journal*, **18**, 787 (1946).

46. Tucker, D. O., C. W. Randall, and P. H. King, "Columnar Denitrification of a Munitions Wastewater," *Proceedings, 29th Industrial Waste Conference*, Purdue University, West Lafayette, Ind., 1974, p. 167.

47. Velz, C. J., "A Basic Law for the Performance of Biological Filters," *Sewage Works Journal*, **20**, 607 (1948).

48. Weston, Roy F., Inc., *Upgrading Existing Wastewater Treatment Plants*," EPA Technology Transfer Series, 1974.

49. Williamson, K., and P. L. McCarty, "A Model of Substrate Utilization by Bacterial Films," *Journal of the Water Pollution Control Federation*, **48**, 9 (1976a).

50. Williamson, K., and P. L. McCarty, "Verification Studies of the Biofilm Model for Bacterial Substrate Utilization," *Journal of the Water Pollution Control Federation*, **48**, 281 (1976b).

51. Young, J. C., and P. L. McCarty, "The Anaerobic Filter for Waste Treatment," *Proceedings, 22nd Industrial Waste Conference*, Purdue University, West Lafayette, Ind., 1967.

52. Zajic, J. E., *Water Pollution*, Vol. I, Marcel Dekker, Inc., New York, 1971.

8

Sludge Digestion

All conventional wastewater treatment processes produce large quantities of waste material in the form of dilute solids mixtures known as sludge. The composition and solids content of these sludges are a function of the characteristics of the raw wastewater flow and the treatment process that generated the sludge. In this regard, primary sludge produced during the treatment of municipal wastewater consists primarily of solid particles of a predominately organic nature, whereas secondary sludge consists mainly of excess biomass generated as a result of organic removal in the biological process. Raw sludges of both types are composed mainly of water with a solids content of only 0.5 to 5.0%, depending upon the origin of the solids and the method of removal.

Sludge contains a major portion of the pollutants responsible for the offensive and noxious nature of untreated wastewater and therefore must be treated or processed so that final release to the environment can be made without harmful effects. For example, during municipal wastewater treatment an average of 35% of the influent BOD_5 is removed as sludge from the primary clarifier. Then, for conventional activated sludge treatment, 30 to 40% of the biologically removed BOD_5 becomes waste activated sludge.

Thus, for an influent BOD_5 of 200 mg/ℓ, assuming 90% removel in the total system, 52 to 57% of the influent BOD_5 is removed in the form of sludge. It is not surprising, therefore, that sludge treatment accounts for 50% or more of the total capital and operating costs for most municipal plants. A variety of sludge treatment processes have been developed and applied to wastewater treatment operations. Sludge treatment processes usually fall into one of the five major categories indicated in Figure 8-1. Stabilization is one of the major categories given in this figure. The principal objectives of stabilization are to prevent nuisance-odor conditions, to reduce the pathogenic organism content of the sludge, and to reduce the liquid volume and solids quantity that must be handled subsequently.

Traditionally, sludge stabilization has been accomplished by anaerobic digestion. This process produces a stable sludge, but much of the organic material is solubilized and the resulting supernatant is very high in nutrients and organic materials. (Table 8-1 reflects anaerobic digestion supernatant

TABLE 8-1

SUPERNATANT CHARACTERISTICS FROM ANAEROBIC DIGESTERS

	Primary plants (mg/ℓ)	*Trickling[a] filters (mg/ℓ)*	*Activated sludge plants[a] (mg/ℓ)*
Suspended solids	200–1000	500–5000	5000–15,000
BOD_5	500–3000	500–5000	1000–10,000
COD	1000–5000	2000–10,000	3000–30,000
Ammonia as NH_3	300–400	400–600	500–1000
Total phosphorus as P	50–200	100–300	300–1000

[a]Includes primary sludge.
Source: After Black, Crow and Eidness (1974).

characteristics.) It has also been found that anaerobically digested secondary sludges are for the most part very difficult to dewater by mechanical processes. Furthermore, anaerobic digestion is very sensitive and upsets are frequent. Because of these undesirable characteristics, other methods of sludge stabilization have been introduced. One of these is aerobic digestion. As both the anaerobic and aerobic digestion processes are biological in nature, each will be discussed in detail in subsequent sections of this chapter.

The location of the digestion process in the overall treatment scheme is shown in Figure 4-1.

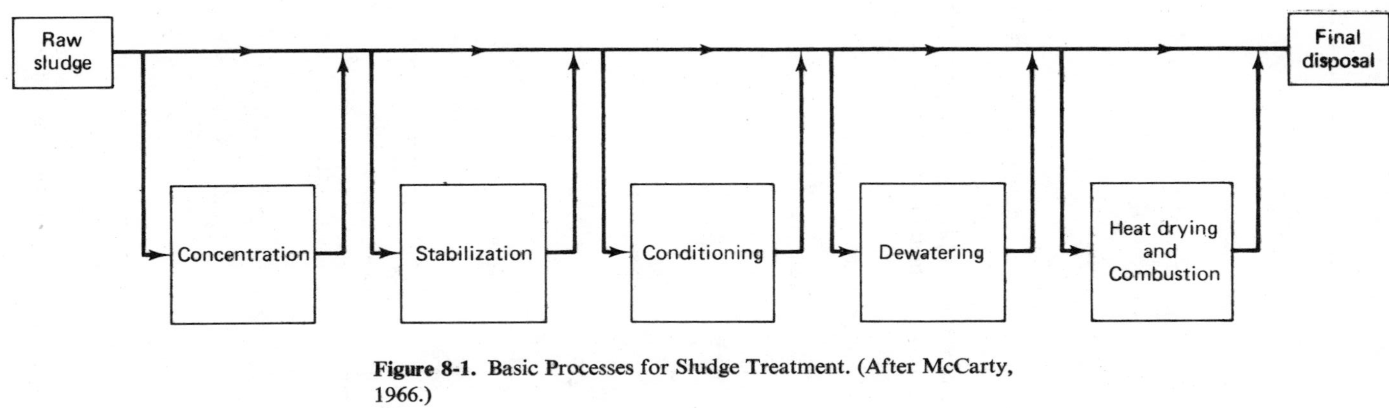

Figure 8-1. Basic Processes for Sludge Treatment. (After McCarty, 1966.)

8-1
Anaerobic Digestion

The fundamentals of anaerobic treatment, presented in Section 4-7 for the anaerobic contact process, are equally applicable to anaerobic digestion. Section 4-7 should be thoroughly understood before continuing with the discussion on anaerobic digestion in this section.

In the digestion process organic solids are converted to inoffensive end products in a manner similar to that shown in Figure 8-2. In the first step the complex organic solids are hydrolyzed by extracellular enzymes produced by the microorganisms indigenous to the process. The soluble organic material formed during hydrolysis is then metabolized by the facultative and anaerobic bacteria responsible for acid fermentation. The end products of acid fermentation (mainly short-chain acids and alcohols) are then converted to gases

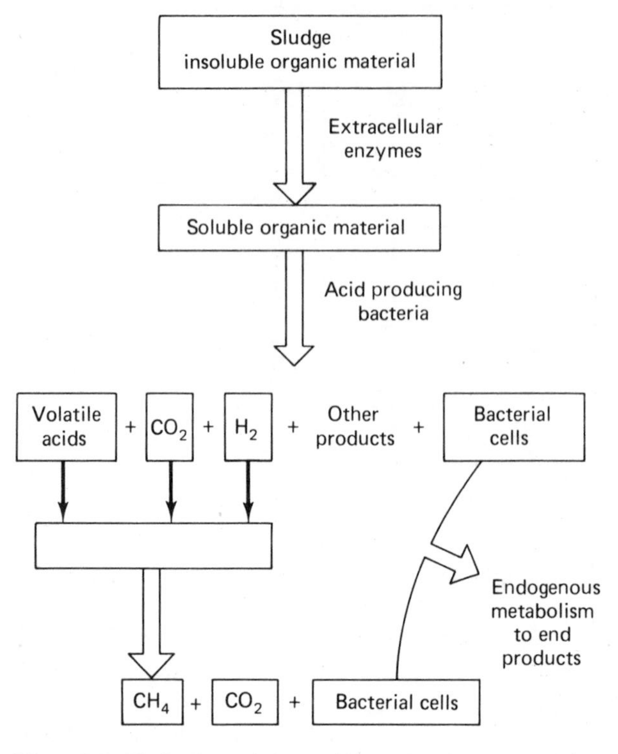

Figure 8-2. Mechanism of Anaerobic Sludge Digestion. (After Eckenfelder, 1967.)

and new bacterial cells by several different species of strictly anaerobic bacteria.

Process Description

There are essentially two types of anaerobic digestion processes used today: the standard-rate process and the high-rate process. The standard-rate process does not employ sludge mixing, but rather the digester contents are allowed to stratify into zones, as illustrated in Figures 8-3 and 8-4. Sludge feeding and withdrawal are intermittent rather than continuous. The digester

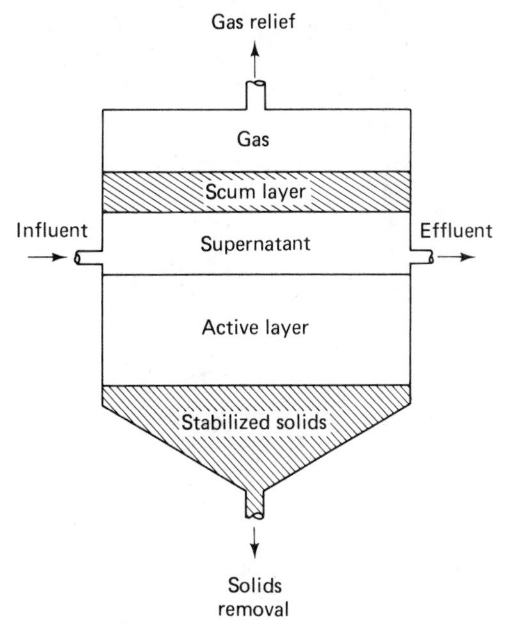

Figure 8-3. Standard-Rate Anaerobic Digestion. (After Kormanik, 1968.)

is generally heated to increase the rate of fermentation and therefore decrease the required retention time, which ranges between 30 and 60 days for heated digesters. The organic loading rate for a standard rate digester is between 0.03 and 0.1 lb total volatile solids per cubic foot of digester volume per day.

The major disadvantage of the standard-rate process is the large tank volume required because of the long retention times, the low loading rates, and the thick scum layer formation (Kormanik, 1968). Only about one-third of the tank volume is utilized in the digestion process. The remaining two-thirds of the tank volume contains the scum layer, stabilized solids, and the

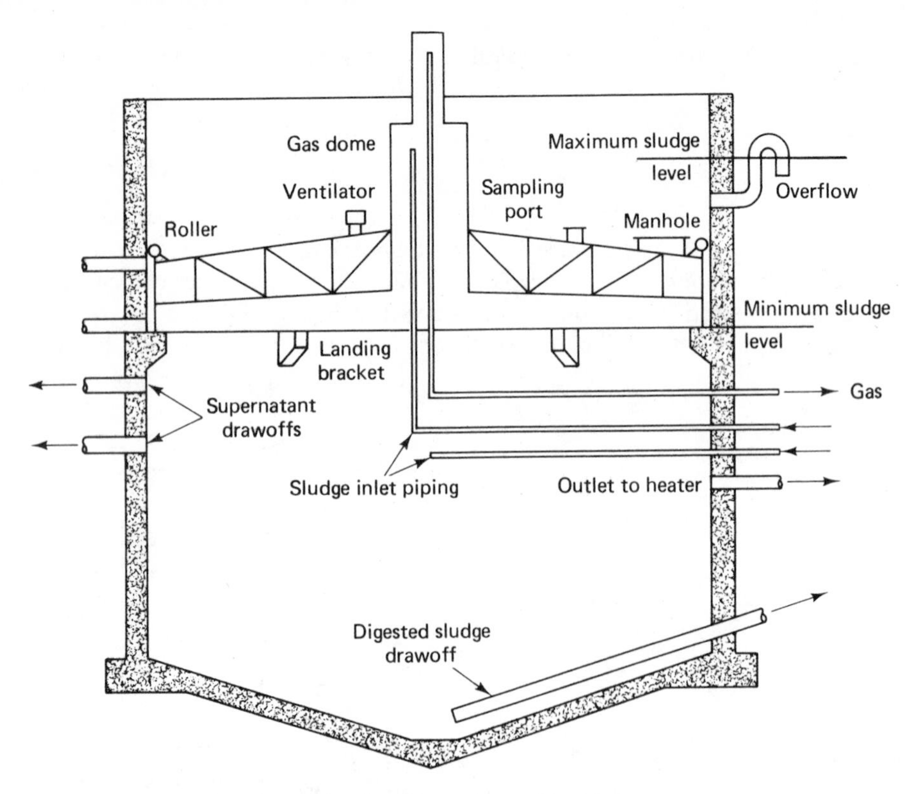

Figure 8-4. Schematic of a Standard-Rate Anaerobic Sludge Digester with a Floating Cover. (After Hammer, 1975.)

supernatant. Because of this limitation, systems of this type are generally used only at treatment plants having a capacity of 1 MGD or less.

The high-rate system evolved as a result of continuing efforts to improve the standard-rate unit. In this process, two digesters operating in series separate the functions of fermentation and solids–liquid separation (see Figure 8-5). The contents of the first-stage high-rate unit are thoroughly mixed by gas recirculation, draft-tube mixers, or pumping, and the sludge is heated to increase the rate of fermentation. Because the contents are thoroughly mixed, the temperature distribution is more uniform throughout the tank volume. Sludge feeding and withdrawal are continuous or nearly so. The retention time required for the first-stage unit is normally between 10 and 15 days. Organic loading rates vary between 0.1 and 0.14 lb total volatile solids per cubic foot of digester volume per day.

The primary functions of the second-stage digester are solids–liquid separation and residual gas extraction. Although first-stage digesters are frequently equipped with fixed covers, second-stage digester covers are often of the floating type. Second-stage units are generally not heated.

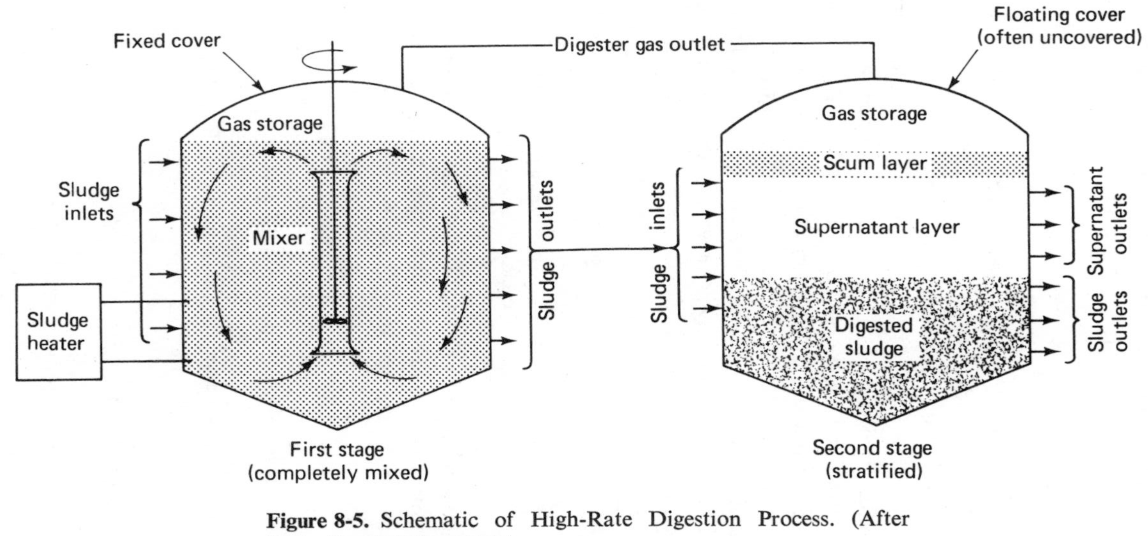

Figure 8-5. Schematic of High-Rate Digestion Process. (After Metcalf and Eddy, 1972.)

Digester tanks are usually circular, with a diameter ranging from 20 to 115 ft. The tank bottom should slope to drain toward the sludge withdrawal pipe (a rise of 1 ft/4 ft of length is common). The liquid depth at the center of the tank will ordinarily vary from 20 to 45 ft.

Kinetic Relationships

The first-stage digester of a high-rate system approximates a completely mixed reactor without solids recycle. Hence, the biological solids retention time and the hydraulic retention time are equal for this sytem. Sawyer and Roy (1955) showed that θ_c is the factor that determines the degree of volatile solids reduction during digestion. They found that essentially the same solids destruction was obtained for a given θ_c irrespective of the concentration of the feed sludge. This indicates that by concentrating the feed sludge, the same degree of solids destruction may be obtained with a reduced digester volume.

A schematic of a completely mixed reactor without solids recycle is presented in Figure 8-6. In this figure Q is the volumetric flow rate, S_0 the influent substrate concentration, X the steady-state biomass concentration, S_e the

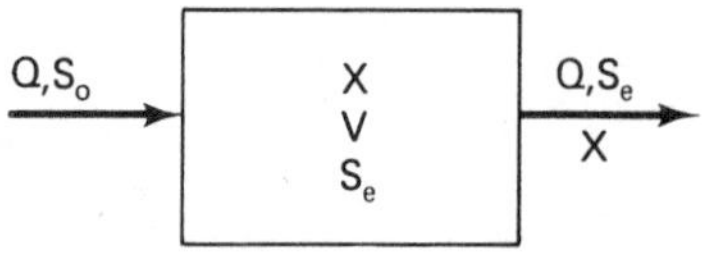

Figure 8-6. Schematic of Completely Mixed Reactor Without Solids Recycle.

effluent substrate concentration, and V the reactor volume. Assuming no biomass in the influent, a steady-state material-balance equation for biomass entering and leaving the reactor may be written as

$$0 = \left[Y_T\left(\frac{dS}{dt}\right)_u - K_dX\right]V - QX \tag{8-1}$$

Substituting for $(dS/dt)_u$ from equation 4-17 gives

$$0 = \left[\frac{Y_TQ(S_0 - S_e)}{V} - K_dX\right]V - QX \tag{8-2}$$

or

$$\frac{K_dXV}{Q} = Y_T(S_0 - S_e) - X \tag{8-3}$$

Since $\theta_c = V/Q$ for this system, equation 8-3 reduces to

$$X = \frac{Y_T(S_0 - S_e)}{1 + K_d\theta_c} \tag{8-4}$$

Assuming that the rate of substrate utilization follows the Monod relationship, the critical BSRT below which washout occurs, θ_c^m, is given by equation 4-34 for completely mixed systems without recycle.

$$\frac{1}{\theta_c^m} = Y_T \frac{kS_0}{K_s + S_0} - K_d \tag{8-5}$$

The first stage of a high-rate system is normally designed for a biological solids retention time between 2 and 10 times greater than θ_c^m (Lawrence and McCarty, 1970). For municipal wastewater sludge Lawrence (1971) reports values of 0.04 and 0.015 day^{-1} for Y_T and K_d, respectively.

For the rate-limiting-step approach as described in Section 4-7, the following kinetic equations are applicable to the design of a completely mixed anaerobic digester:

$$S_e = \frac{K_s(1 + K_d\theta_c)}{\theta_c(Y_T k - K_d) - 1} \tag{8-6}$$

$$(S_e)_{\text{overall}} = \frac{(1 + K_d\theta_c)K_c}{\theta_c(Y_T k - K_d) - 1} \tag{4-176}$$

$$(k)_T = (6.67)10^{-0.015(35-T)} \tag{4-177}$$

$$(K_c)_T = (2224)10^{0.046(35-T)} \tag{4-178}$$

Equations 4-177 and 4-178 are applicable within the temperature range 20 to 35°C.

Biological solids retention time and temperature are the key considerations in the design of anaerobic digesters. The rate of fermentation increases and decreases with temperature within certain limits (see Figure 8-7). Ther-

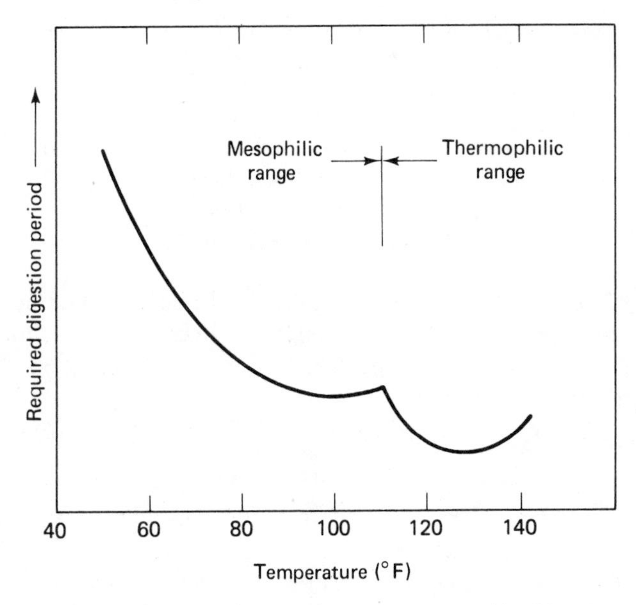

Figure 8-7. Influence of Temperature on Anaerobic Digestion Time. (After Heukelekian, 1930.)

mophilic digestion is possible and may find considerable future application (Buhr and Andrews, 1977), but it has not historically been thought of as economically feasible to heat sludge to such high temperatures. Hence, sludge digestion is generally carried out in the mesophilic range where the optimum temperature is 35°C (95°F). When sufficient data are not available to allow the engineer to compute θ_c^m from equation 8-5, values presented in Table 8-2 may be used for first-stage digester design.

TABLE 8-2

APPROXIMATE VALUES FOR θ_c^m FOR USE IN THE DESIGN OF FIRST-STAGE DIGESTERS

Operating temperature (°*F*)	θ_c^m (*days*)
65	11
75	8
85	6
95	4
105	4

Source: After McCarty (1964).

Gas Production and Heating Requirements

Digester heat loss is determined from equation 8-7:

$$H_L = UA(T_2 - T_1) \tag{8-7}$$

where H_L = heat loss, Btu/h
U = overall coefficient of heat transfer, Btu/h-ft²-°F
A = area normal to direction of heat flow, ft²
T_2 = fermentation temperature, °F
T_1 = critical winter temperature, °F

Typical values for U are given in Table 8-3. The heat loss may also be estimated as 2600 Btu/h/1000 ft³ of digester volume in the northern United States, or 1300 Btu/h/1000 ft³ of digester volume in the southern United States (Lipták, 1974).

The sludge heat requirements may be computed from the equation

$$H_R = WC(T_2 - T_1) \tag{8-8}$$

where H_R = heat necessary to bring raw sludge to fermentation temperature, Btu/day
W = average mass flow of sludge to digester, lb/day
C = mean specific heat of sludge, which is usually taken as 1.0 Btu/lb-°F

T_2 = fermentation temperature, °F
T_1 = temperature of feed sludge for critical winter conditions, °F

TABLE 8-3

OVERALL HEAT TRANSFER COEFFICIENTS FOR ANAEROBIC DIGESTERS

Digester component	*Overall heat transfer coefficient (Btu/h-ft²-°F)*
Concrete roof	0.5
Floating cover	0.24
Concrete wall air space	0.35
Concrete wall in wet earth	0.25
Concrete wall in dry earth	0.18
Floor	0.12

Source: After Lipták (1974).

If no other information is available, the feed sludge temperature may be taken as 50°F in the southern United States, 45°F in the central United States, and 40°F in the northern United States (Lipták, 1974).

Methane gas production may be estimated from equation 4-181.

$$G = G_0[\Delta S - 1.42(\Delta X)] \qquad (4\text{-}181)$$

where G = total methane produced, ft³/day
G_0 = cubic feet of methane produced per pound of degradable COD or BOD_u oxidized, ft³/lb
ΔS = degradable COD or BOD_u removed, lb/day
ΔX = biomass produced, lb/day

Since the gas produced during anaerobic digestion is only about 2/3 methane, the total volume of gas produced is given by $G/0.67$. The total available heat content can be estimated by considering that 1 ft³ of methane has a net heating value of 960 Btu at standard conditions.

Lipták (1974) reports that total gas production may be taken as 15 ft³/lb VSS destroyed and that the Btu value of this gas is between 640 and 703 Btu/ft³.

Sludge Characteristics

The volume of sludge depends on its specific gravity and water content and may be computed from the equation

$$V_s = \frac{\text{lb of dry solids}}{(S_s)(\gamma_w)(f_s)} \qquad (8\text{-}9)$$

where
V_s = daily volume of sludge produced, ft^3/day
γ_w = specific weight of water, lb/ft^3; see Table 8-4 for values
f_s = weight fraction of sludge that is solids
lb of dry solids = dry solids produced in system each day, lb/day
S_s = specific gravity of sludge

TABLE 8-4

VARIATION IN SPECIFIC WEIGHT OF WATER WITH TEMPERATURE

Temperature (°F)	*Specific weight of water (lb/ft^3)*
32	62.42
40	62.43
50	62.41
60	62.37
70	62.30
80	62.22
90	62.11
100	62.00

Source: After Metcalf and Eddy (1972).

To apply this expression, the solids content and the specific gravity of the sludge must be known. Typical concentrations of thickened and unthickened sludges are given in Table 8-5. The specific gravity of a sludge is determined

TABLE 8-5

CONCENTRATIONS OF UNTHICKENED AND THICKENED SLUDGES AND SOLIDS LOADINGS FOR MECHANICAL THICKENERS

	Sludge (% solids)		*Solids loading for mechanical thickener*
Type of sludge	*Unthickened*	*Thickened*	*(lb/ft^2-day)*
Separate sludges			
Primary	2.5–5.5	8–10	20–30
Trickling filter	4–7	7–9	8–10
Activated	0.5–1.2	2.5–3.3	4–8
Combined sludges			
Primary and trickling filter	3–6	7–9	12–20
Primary and activated	2.6–4.8	4.6–9.0	8–16

Source: After Metcalf and Eddy (1972).

from the relationship

$$\frac{1}{S_s} = \frac{f_w(1)}{S_w} + \frac{f_f(1 - f_w)}{S_f} + \frac{f_v(1 - f_w)}{S_v} \tag{8-10}$$

where S_s = specific gravity of sludge
S_w = specific gravity of water (taken as 1.0)
S_f = specific gravity of fixed solids (usually taken as 2.5)
S_v = specific gravity of volatile solids (usually taken as 1.0)
f_w = weight fraction of sludge that is water
f_f = weight fraction of solids that are fixed
f_v = weight fraction of solids that are volatile

To determine the reduction in sludge volume obtained by digestion, it is necessary to know that percent solids to which the digested sludge will concentrate as well as the specific gravity of the digested sludge. Typical values for percent solids of anaerobically digested sludges are given in Table 8-6,

TABLE 8-6

PROPORTION OF SOLIDS IN ANAEROBICALLY DIGESTED MUNICIPAL WASTEWATER SLUDGES

Type of sludge	*Percent solids*
Separate sludges	
Primary	10–15
Activated	2–3
Combined sludges	
Primary and trickling filter	10
Primary and activated	6–8

Source: After Fair and Geyer (1957).

and the specific gravity is computed from equation 8-10. To calculate the specific gravity, it is necessary to know the weight fraction of the solids that are volatile after digestion is completed. This value can be determined if the degree of volatile solids destruction is known. For feed sludges that have a volatile solids content of 65 to 70%, the volatile solids destruction during the digestion process may be estimated from the following equations (Lipták, 1974):

Standard rate digesters:

$$V_d = 30 + \frac{t}{2} \tag{8-11}$$

where V_d = volatile solids destroyed during digestion, %
t = time of digestion, days

First-stage digesters:

$$V_d = 13.7 \ln(\theta_c) + 18.94 \tag{8-12}$$

where θ_c represents the biological solids retention time expressed as days.

Design Considerations

In certain situations the second-stage digester may be eliminated by separate sludge thickening prior to digestion. The advantage of such an arrangement is the reduction in total required tank volume. A mechanical thickener may provide a thickening operation comparable to that of a second stage digester while requiring only one-tenth of the volume. The elimination of the second-stage digester, however, depends on subsequent sludge handling techniques. For example, in northern regions uncovered drying beds may not be used during winter months, which means that sludge must be stored during this period. Second-stage digesters provide the necessary volume for storage and in this case could not be eliminated.

Mixing intensity in the digester is an important design criterion. A low degree of mixing will result in a decrease in the rate of methane fermentation. When a low degree of mixing is provided, volatile acids will build up in the system and digester failure will probably result. Studies by Speece (1972) have shown that any increase in the mixing intensity, up to the point where complete mixing is achieved, will increase the rate of methane fermentation. Extrapolating laboratory data, he determined that a power level of 1.5 hp/1000 gal was required for complete mixing. As power levels of 0.03 to 0.05 hp/1000 gal are normally employed under process conditions, the mixing intensity within these units is far below that required to obtain the maximum rate of fermentation. This implies that an increase in the mixing intensity may correct an impending digester failure.

A number of different methods have been used to size anaerobic digesters. Two of the most common methods are biological solids retention time and organic loading rate. Typical values for each of these parameters, along with general operating conditions for each type of anaerobic digestion process, are presented in Table 8-7. Again, the importance of the material in Section 4-7 to the understanding of the digestion process must be emphasized.

Example 8-1

Design a high-rate anaerobic digestion system to process the sludge from a 10-MGD completely mixed activated sludge plant with a biomass production of 4879 lb/day. Also, assume that the following conditions are applicable:

1/ After grit removal the wastewater has a total suspended solids concentration of 235 mg/ℓ (of which 65% are volatile) and an ultimate BOD of 476 mg/ℓ.

TABLE 8-7

GENERAL OPERATING AND LOADING CONDITIONS FOR ANAEROBIC SLUDGE DIGESTION

Temperature	
Optimum	98°F (35°C)
General operating range	85–95°F
pH	
Optimum	7.0–7.1
General limits	6.7–7.4
Gas production	
Per pound of voltile solids added	6–8 ft^3
Per pound of volatile solids destroyed	16–18 ft^3
Gas composition	
Methane	65–69%
Carbon dioxide	31–35%
Hydrogen sulfide	Trace
Volatile acids concentration	
General operating range	200–800 mg/ℓ
Alkalinity concentration	
Normal operation	2000–3500 mg/ℓ
Volatile solids loading	
Conventional single-stage	0.02–0.05 lb VS/ft^3/day
First-stage high-rate	0.05–0.15 lb VS/ft^3/day
Volatile solids reduction	
Conventional single-stage	50–70%
First-stage high-rate	50%
Solids retention time	
Conventional single-stage	30–90 days
First-stage high-rate	10–15 days
Digester capacity based on design equivalent population	
Conventional single-stage	4–6 ft^3/PE
First-stage high-rate	0.7–1.5 ft^3/PE

Source: After Hammer (1975).

2/ The surface loading rate for the circular primary clarifiers is 600 gpd/ft^2.

3/ Relationships given in Figures 3-12, 3-13, and 3-14 are valid for this design. The sludge concentrates to 5% solids in the primary clarifier.

4/ Excess sludge from the activated sludge process has a 70% volatile fraction and is thickened to 3% solids by dissolved air flotation prior to mixing with the primary sludge.

5/ The first-stage digester temperature is 35°C.

6/ A safety factor of 3 is to be used in determining the design θ_c.

7/ Because of winter conditions, a volume sufficient for 100 days of sludge storage is provided in the second-stage digester.

8/ A power level of 150 hp/MG of digester volume is provided in the first-stage digester.

9/ The critical design temperature is 10°F. All digester sidewalls are exposed to air. The digester floor is in dry earth which has a winter temperature of 20°F.

10/ The digester depth is 25 ft.

11/ Digester sludge concentrates to 7% in the second-stage digester.

solution

1/ Compute the daily volume of sludge removed during primary clarification.

a/ With a surface loading rate of 600 gpd/ft², an 82% reduction in settleable solids is predicted from Figure 3-12.

b/ For an 82% reduction in settleable solids, a 60% reduction in total suspended solids is predicted from Figure 3-13.

c/ Total suspended solids removal is given by

$$(235)(0.6)(8.34)(10) = 11{,}759 \text{ lb/day}$$

d/ Since it is assumed that the primary sludge concentrates to 5% solids and has a volatile solids content of 65%, the specific gravity of the sludge is computed from equation 8-10.

$$\frac{1}{S_s} = \frac{(0.95)(1)}{1} + \frac{(0.35)(1-0.95)}{2.5} + \frac{(0.65)(1-0.95)}{1}$$

$$= 0.99$$

or

$$S_s = 1.01$$

e/ The daily volume of primary sludge is computed from equation 8-9.

$$V_s = \frac{11{,}759}{(1.01)(62.4)(0.05)}$$

$$= 3732 \text{ ft}^3\text{/day}$$

2/ Estimate the ultimate BOD concentration of the primary sludge.

a/ From Figure 3-14 the percent BOD removed in the primary clarifier is 40%.

b/ The daily ultimate BOD removal is

$$(0.4)(476)(8.34)(10) = 15{,}879 \text{ lb/day}$$

c/ The ultimate BOD concentration in the primary sludge is

$$\frac{15{,}879 \text{ lb/day}}{3732 \text{ ft}^3\text{/day}} \times 454{,}000 \frac{\text{mg}}{\text{lb}} \times \frac{1 \text{ ft}^3}{28.3\ \ell} = 68{,}258 \text{ mg}/\ell$$

3/ Compute the daily volume of waste activated sludge.

a/ Assuming that volatile solids is a measure of the biomass and neglecting carryover of NDVSS and FSS from the primary clarifier, the total solids produced by the activated sludge plant each day is 4879/0.7 = 6970 lb/day.

b/ Since it is assumed that the activated sludge is concentrated to 3% solids and has a volatile solids content of 70%, the specific gravity of the sludge is computed from equation 8-10.

$$\frac{1}{S_s} = \frac{(0.97)(1)}{1} + \frac{(0.3)(1-0.97)}{2.5} + \frac{(0.7)(1-0.97)}{1}$$

$$= 0.9945$$

or

$$S_s = 1.005$$

c/ The daily volume of activated sludge is computed from equation 8-9.

$$V_s = \frac{6970}{(1.005)(62.4)(0.03)}$$

$$= 3705 \text{ ft}^3/\text{day}$$

4/ Estimate the ultimate BOD concentration of the excess activated sludge.

$$\frac{4879 \text{ lb VS/day}}{3705 \text{ ft}^3/\text{day}} \times 1.42 \frac{\text{lb } O_2}{\text{lb VS}} \times 454{,}000 \frac{\text{mg}}{\text{lb}} \times \frac{1 \text{ ft}^3}{28.3\ \ell} = 30{,}000 \frac{\text{mg}}{\ell}$$

5/ Calculate the sludge flow in MGD for both primary and activated sludge.

$$\text{primary sludge flow} = \frac{(3732)(7.48)}{1{,}000{,}000} = 0.028 \text{ MGD}$$

$$\text{activated sludge flow} = \frac{(3705)(7.48)}{1{,}000{,}000} = 0.028 \text{ MGD}$$

6/ Estimate the percent volatile solids of the mixed primary and activated sludge stream.

$$\text{percent volatile solids} = \frac{(0.65)(0.028)(11{,}759) + (0.70)(0.028)(6970)}{(0.028)(11{,}759) + (0.028)(6970)}(100)$$

$$= 66.8\%$$

7/ Compute the ultimate BOD of the mixed primary and activated sludge stream.

$$\text{BOD}_u = \frac{(68{,}258)(0.028) + (30{,}000)(0.028)}{0.028 + 0.028}$$

$$= 49{,}129 \text{ mg}/\ell$$

8/ Correct k for temperature from equation 4-177.

$$k = (6.67)10^{-0.015(35-35)}$$

$$= 6.67 \text{ day}^{-1}$$

9/ Correct K_c for temperature from equation 4-178.

$$K_c = (2224)10^{0.046(35-35)}$$

$$= 2224 \text{ mg}/\ell$$

10/ Compute θ_c^m from equation 8-5.

$$\frac{1}{\theta_c^m} = (0.04)\frac{(6.67)(49{,}129)}{2224 + 49{,}129} - 0.015$$

$$= 0.24$$

or

$$\theta_c^m = 4.2 \text{ days}$$

11/ Using a safety factor of 3, calculate the design θ_c.

$$\theta_c = (4.2)(3) = 12.6 \text{ days}$$

12/ Determine the required volume of the first-stage digester.

$$V = \left(3705 \frac{\text{ft}^3}{\text{day}} + 3732 \frac{\text{ft}^3}{\text{day}}\right)(12.6 \text{ days})$$

$$= 93{,}706 \text{ ft}^3$$

13/ Determine the cross-sectional area of the first-stage digester.

$$A = \frac{93{,}706}{25}$$

$$= 3748 \text{ ft}^2$$

14/ Compute the diameter of the first-stage digester.

$$d = \left[\frac{(4)(3748)}{\pi}\right]^{1/2}$$

$$= 69 \text{ ft}$$

15/ Calculate the effluent substrate concentration from the first-stage digester using equation 4-176.

$$(S_e)_{\text{overall}} = \frac{[1 + (0.015)(12.6)](2224)}{(12.6)[(0.04)(6.67) - 0.015] - 1}$$

$$= 1{,}217 \text{ mg}/\ell$$

16/ Compute the steady-state biomass concentration in the first stage digester by applying equation 8-4.

$$X = \frac{(0.04)(49{,}129 - 1{,}217)}{1 + (0.015)(12.6)}$$

$$= 1612 \text{ mg}/\ell$$

17/ Calculate the cubic feet of methane produced per pound of ultimate BOD removed.

$$V_2 = \frac{308}{273}(22.4)$$

$$= 25.3 \ \ell$$

Therefore,

$$\text{methane production} = \frac{(25.3/64)(454)}{28.32}$$

$$= 6.34 \text{ ft}^3/\text{lb BOD}_u \text{ removed}$$

18/ Determine the total methane production from equation 4-181.

$$G = 6.34[(49{,}129 - 1{,}217)(0.028 + 0.028)(8.34) - (1.42) \times (0.028 + 0.028)(8.34)(1612)]$$
$$= 135{,}091 \text{ ft}^3/\text{day}$$

For a completely mixed system without recycle the solids lost from the system are equal to the solids contained in the process effluent.

19/ Compute the total available heat content of the digester gas.

$$\text{available heat content} = (960 \text{ Btu/ft}^3)(135{,}091 \text{ ft}^3/\text{day})(5.63/6.34)$$
$$= 115{,}164{,}112 \text{ Btu/day}$$

20/ Determine the heat losses for the first-stage digester using equation 8-7.

a/ Heat loss from the fixed concrete cover:

$$H_L = (3748)(0.5)(24)(95 - 10)$$
$$= 3{,}822{,}960 \text{ Btu/day}$$

b/ Heat loss from the sidewall area:

$$H_L = (\pi)(69)(25)(0.35)(24)(95 - 10)$$
$$= 3{,}869{,}342 \text{ Btu/day}$$

c/ Heat loss from floor area:

$$H_L = (3748)(0.12)(24)(95 - 20)$$
$$= 809{,}568 \text{ Btu/day}$$

21/ Estimate the heat required to raise the feed sludge temperature to 95°F by applying equation 8-8 and assuming that the feed sludge temperature is 50°F.

a/ Compute the total mass rate of sludge flow to the first-stage digester.

$$W = \left(3705 \frac{\text{ft}^3}{\text{day}} + 3732 \frac{\text{ft}^3}{\text{day}}\right)\left(62.4 \frac{\text{lb}}{\text{ft}^3}\right)$$
$$= 464{,}069 \text{ lb/day}$$

b/ Calculate the heat necessary to bring the raw sludge to fermentation temperature from equation 8-8.

$$H_R = (464{,}069)\frac{1 \text{ Btu}}{\text{lb-°F}}(95 - 50)$$
$$= 20{,}883{,}105 \text{ Btu/day}$$

22/ Determine the total heat requirement for the first-stage digester.

$$\text{heat requirement} = 20{,}883{,}105 + 809{,}568 + 3{,}869{,}342 + 3{,}822{,}960$$
$$= 29{,}384{,}975 \text{ Btu/day}$$

Therefore, 115,164,112 − 29,384,975 = 85,779,137 Btu/day are available for building heating, sludge incineration, and so forth. However, a large fraction of this will be lost because the heat exchanger is not 100% efficient.

23/ Estimate the volume occupied by the sludge in the second-stage digester.

a/ Compute the volatile solids destroyed during digestion from equation 8-12.

$$V_d = 13.7 \ln (12.6) + 18.94$$
$$= 53.6\%$$

b/ Calculate the percentage of volatile matter after digestion.

fraction volatile solids

$$= \frac{\text{dry wt of volatile solids}}{(\text{dry wt of fixed solids}) + (\text{dry wt of volatile solids})}$$

$$= \frac{(1 - 0.536)(0.668)(6970 + 11{,}759)}{(1 - 0.668)(6970 + 11{,}759) + (1 - 0.536)(0.668)(6970 + 11{,}759)}$$

$$= 0.48$$

c/ Calculate the specific gravity of the sludge, assuming that the digested sludge concentrates to 7% solids.

$$\frac{1}{S_s} = \frac{(0.93)(1)}{1} + \frac{(0.52)(1 - 0.93)}{2.5} + \frac{(0.48)(1 - 0.93)}{1}$$
$$= 0.98$$

or

$$S_s = 1.02$$

d/ The volume occupied by digested sludge is computed from equation 8-9.

$$V_s = \frac{[(1 - 0.668)(6970 + 11{,}759)] + [(1 - 0.536)(0.668)(6970 + 11{,}759)]}{(1.02)(62.4)(0.07)}$$

$$= 2699 \text{ ft}^3/\text{day}$$

24/ Compute the required volume of the second-stage digester.

$$V = (2699 \text{ ft}^3/\text{day})(100 \text{ days})$$
$$= 269{,}900 \text{ ft}^3$$

25/ Determine the cross-sectional area of the second-stage digester.

$$A = \frac{269{,}900}{25}$$
$$= 10{,}796 \text{ ft}^2$$

26/ Calculate the diameter of the second-stage digester.

$$d = \left[\frac{(4)(10{,}796)}{\pi}\right]^{1/2}$$
$$= 117 \text{ ft}$$

27/ Determine the horsepower required to provide a power level of 150 hp/million gal.

$$\frac{93{,}706 \text{ ft}^3 \times 7.48 \text{ gal/ft}^3}{1{,}000{,}000}(150 \text{ hp/MG}) = 105 \text{ hp}$$

Process Modeling and Control

Even though the anaerobic digestion process has several significant advantages over other methods of organic solids processing, including the formation of useful by-products such as methane gas and a humuslike slurry well suited for land reclamation, it has not enjoyed a good reputation because of its poor record with respect to useful control strategies for the process. Andrews (1977) has developed a dynamic model which can be used to predict the dynamic response of the five variables most commonly used for monitoring process stability. These variables are (1) pH, (2) volatile acids concentration, (3) alkalinity, (4) gas composition, and (5) gas flow rate. The model is summarized in Figure 8-8.

The model was developed from material balances on components in the liquid, biological, and gas phases of a CFSTR. The model indicates that there are strong interactions between the phases and uses equilibrium relationships, kinetic expressions, stoichiometric coefficients, charge balances, and mass transfer equations to reflect these interactions. The model uses an inhibition function instead of the Monod expression to relate volatile acids concentration and methane bacteria specific growth rate, and considers the un-ionized fraction of the volatile acids as both a limiting substrate and an inhibiting agent (Andrews, 1968). These modifications permit the model to predict process failure by organic overloading. The model can also predict failure brought about by toxic materials and has been modified to evaluate the effect of changes in temperature on process stability (Buhr and Andrews, 1977). Simulation studies have provided qualitative evidence for the validity of the model by predicting results similar to those commonly observed in the field. The model was developed to predict gross process failure, however, not organic solids destruction. Therefore, its usefulness is in selecting control strategies.

Graef and Andrews (1974) have shown that the type of control strategy to be initiated is dependent on the type of overloading to which the digester has been subjected. They propose a new strategy of scrubbing the carbon dioxide from the digester gas with subsequent recycle. This would provide process control through the removal of carbonic acid instead of addition of a base, as is commonly practiced. They propose that this technique should be effective in preventing failure by organic overload. They also propose the recycle of concentrated sludge from the second-stage digester, using the rate of methane production as the control signal, to prevent failure from an overload of toxic materials.

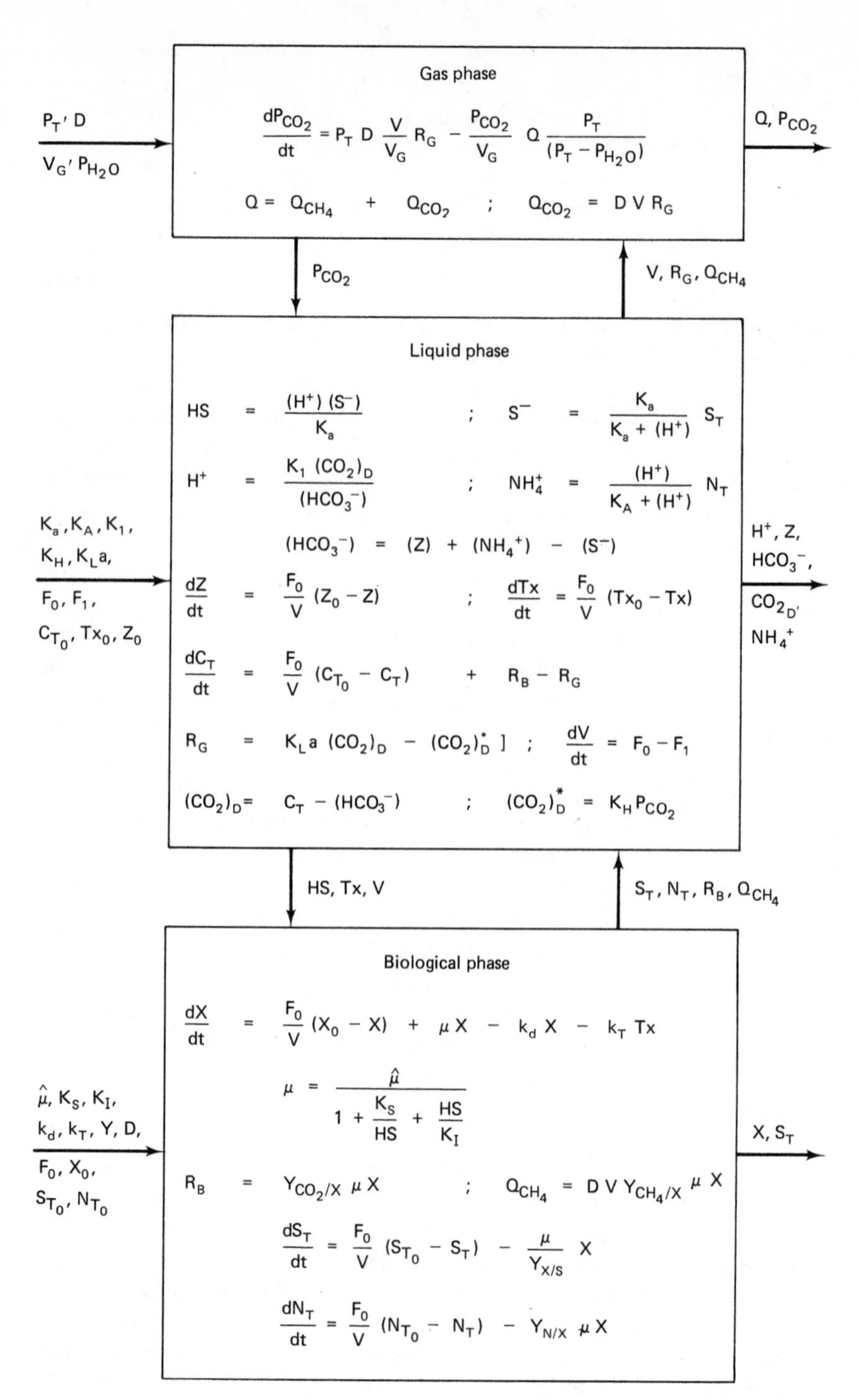

Figure 8-8. Summary of Mathematical Model for the Anaerobic Digester. (After Andrews, 1977.)

8-2
Aerobic Digestion

The aerobic digestion of biological sludges is nothing more than a continuation of the activated sludge process. As previously noted, when a culture of aerobic heterotrophs is placed in an environment containing a source of organic material, the microorganisms will remove and utilize most of this material. A fraction of the organic material removed will be utilized in the synthesis function, resulting in an increase in biomass. The remaining material will be channeled into energy metabolism and oxidized to carbon dioxide, water, and soluble inert material to provide energy for both synthesis and maintenance (life-support) functions. Once the external source of organic material has been exhausted, however, the microorganisms will enter into endogenous respiration where cellular material is oxidized to satisfy the energy of maintenance (i.e., energy for life-support requirement). If this condition is continued over an extended period of time, the total quantity of biomass will be considerably reduced, and furthermore, that portion remaining will exist at such a low energy state that it can be considered biologically stable and suitable for disposal in the environment. This forms the basic principle of aerobic digestion.

When mixtures of primary and activated sludges are digested aerobically, an additional factor has to be considered. Primary sludges, although organic and particulate in nature, contain little biomass. Most of the material represents an external food source for the active biomass contained in the biological sludge, which therefore will reduce the amount of cellular material required for energy of maintenance. Hence, longer retention times are required to achieve equivalent stabilization when primary and activated sludge mixtures are digested rather than when activated sludge is digested alone. Aerobic digestion of trickling filter humus constitutes a condition intermediate to these two extremes but can be reasonably approximated by the activated sludge reactions (Randall et al., 1974).

Aerobic digestion is a viable alternative to anaerobic digestion for sludge stabilization. The advantages and disadvantages most often claimed for aerobic digestion have been listed by Burd (1968):

Advantages:

1/ A biologically stable end product is produced.

2/ The stable end product has no odor; therefore, land disposal is feasible.

3/ Because construction is simple, capital costs for aerobic digesters are lower than for anaerobic digesters.

4/ Aerobically digested sludge generally has good dewatering properties.

5/ For biological sludges, about the same percent volatile solids reduction can be achieved in aerobic digestion as can be achieved in anaerobic digestion.

6/ Supernatant liquors from aerobic digestion have a lower BOD content than those from anaerobic digestion. The aerobic supernatant most commonly has a soluble BOD less than 100 mg/ℓ. This is important because many treatment facilities have been overloaded due to recycling of the high-BOD supernatant liquors from anaerobic digesters. Typical supernatant characteristics from aerobic digesters are listed in Table 8-8.

TABLE 8-8

CHARACTERISTICS OF AEROBIC DIGESTION SUPERNATANT

Parameter	*Typical value*
pH	7.0
BOD_5	500 mg/ℓ
Soluble BOD_5	51 mg/ℓ
COD	2600 mg/ℓ
Suspended solids	100–300 mg/ℓ
Kjeldahl nitrogen	170 mg/ℓ
Total P	98 mg/ℓ
Soluble P	26 mg/ℓ

Source: After Black, Crow and Eidness (1974).

7/ There are fewer operational problems with aerobic digestion than with the more complex anaerobic form because the system is more stable. Therefore, lower maintenance costs will be required and less skilled labor used in plant operation.

8/ Aerobically digested sludges have a higher fertilizer value than do those from anaerobic digestion.

Disadvantages:

1/ High power costs result in high operating costs, which become significant in large facilities.

2/ Solids reduction efficiency varies with temperature fluctuation.

3/ Gravity thickening following aerobic digestion generally results in a supernatant high in solids concentration.

4/ Some sludges apparently do not dewater easily by vacuum filtration after aerobic digestion.

Not only should the aerobic digestion process be considered from the standpoint of sludge stabilization, but it should also be considered as a means of conditioning sludge prior to dewatering. Randall et al. (1973) studied extensively the use of aerobic digestion as a means of sludge conditioning. The following are some of the observations from this investigation:

1/ Aerobic digestion has a considerable effect on the filtration characteristics of waste activated sludge, and it produces changes in both specific resistance and the compressibility factor.

2/ During aerobic digestion, improvement in dewaterability occurs initially, with maximum improvements after 1 to 5 days' aeration. However, further aeration will result in a worsening of dewatering properties and can produce conditions far worse than the initial state.

3/ The degree of improvement in dewaterability that occurs during aerobic digestion is a function of the origin and nature of the fresh sludge, operating biological solids retention time, rate of aeration during digestion, temperature of digestion, and time of digestion. Filterability can usually be improved by 23 to 46% by biological conditioning. However, the sludge must be filtered as soon as maximum improvement is attained in order to realize the benefit, since further digestion will worsen filterability.

4/ The rate at which sludge is mixed during aerobic digestion influences the dewatering properties of the sludge. It appears that the more rapidly a sludge is mixed, the greater the forces placed on the sludge floc particles. This results in floc disruption, creating smaller particles which in turn reduce the filterability.

5/ Dissolved oxygen concentrations above 2 mg/ℓ during digestion do not change the subsequent sludge filterability.

6/ Aerobic digestion affects the dewatering of sludge on sand drying beds. The time required to drain 750 ml from a 1-ℓ sample decreased by about 73% after 6 days of aeration, and there was an increase in the total drainable water of 2%.

7/ Changes in specific resistance and the compressibility factor are most reliably reflected by changes in median particle size. Specific resistance is inversely related to floc size, whereas the compressibility factor is directly related to floc size.

8/ The addition of artificial polymers greatly increases the median particle size and drastically reduces the specific resistance.

9/ The dewatering properties of waste activated sludge vary with the origin and degree of stabilization of the particular sludge. The conditioning effect of artificial polymers is affected by the same variables.

10/ Polymers are very specific in their conditioning action on waste activated sludge. Anionic polymers are detrimental to filterability during all phases of aerobic digestion, whereas cationic polymers improve the filterability of aerobically digested sludges.

11/ Aerobic digestion can produce a considerable decrease in polymer dose for sludge conditioning. An 80% decrease was obtained after 7 days of digestion.

Kinetic Relationships

Generally, most aerobic digesters are operated as continuous flow, completely mixed aeration units and are designed on the basis of volatile suspended solids (VSS) reduction. The model presented by Adams et al. (1974a) is probably the model most often used in design. In this model it is assumed that the loss of degradable volatile solids (proposed to be some fraction of the VSS) through endogenous respiration follows the first-order relationship

$$\left(\frac{dX_d}{dt}\right)_R = K_b X_d \tag{8-13}$$

where $\left(\frac{dX_d}{dt}\right)_R$ = rate at which degradable solids are lost as a result of endogenous respiration, mass volume^{-1} time^{-1}

K_b = reaction-rate constant for degradable VSS destruction as determined in a batch reactor, time^{-1}

X_d = degradable VSS remaining at time t, mass volume^{-1}

Considering the continuous-flow completely mixed digester shown in Figure 8-9, a material-balance expression for degradable solids entering and leaving the system is

$$\begin{bmatrix}\text{net rate of change}\\ \text{in degradable VSS}\\ \text{in digester}\end{bmatrix} = \begin{bmatrix}\text{rate at which}\\ \text{degradable VSS}\\ \text{enter the}\\ \text{digester}\end{bmatrix} - \begin{bmatrix}\text{rate at which}\\ \text{degradable VSS}\\ \text{are lost from}\\ \text{the digester}\end{bmatrix} \tag{8-14}$$

Equation 8-14 may be written in mathematical form as

$$\left[\frac{dX_d}{dt}\right] V = Q(X_d)_0 - \left[\left(\frac{dX_d}{dt}\right)_R V + Q(X_d)_e\right] \tag{8-15}$$

where Q = volumetric flow rate, volume time^{-1}

$(X_d)_0$ = degradable VSS concentration in influent, mass volume^{-1}

$(X_d)_e$ = degradable VSS concentration in effluent, mass volume^{-1}

V = digester volume

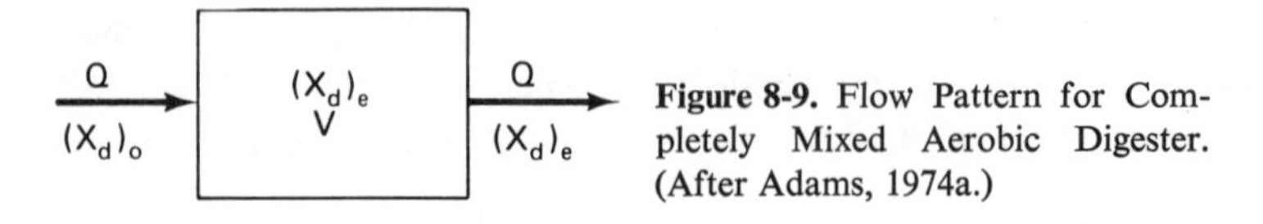

Figure 8-9. Flow Pattern for Completely Mixed Aerobic Digester. (After Adams, 1974a.)

Assuming steady-state conditions and substituting for $(dX_d/dt)_R$ from equation 8-13, equation 8-15 can be written as

$$t_d = \frac{(X_d)_0 - (X_d)_e}{K_b(X_d)_e} \tag{8-16}$$

where

$$t_d = V/Q = \text{digester retention time, time}$$

Letting

$$(X_d)_e = X_e - X_n \tag{8-17}$$

and

$$(X_d)_0 = X_0 - X_n \tag{8-18}$$

where X_d = total VSS concentration in effluent, mass volume^{-1}
X_0 = total VSS concentration in influent, mass volume^{-1}
X_n = nondegradable portion of VSS which is assumed to remain constant throughout digestion period, mass volume^{-1}

and substituting for $(X_d)_0$ and $(X_d)_e$ in equation 8-16 from equations 8-17 and 8-18, equation 8-16 becomes

$$t_d = \frac{X_0 - X_e}{K_b(X_e - X_n)} \tag{8-19}$$

Adams et al. (1974b) recommends that K_b and X_n be determined for a particular sludge through batch studies conducted in the laboratory. Figure 8-10 shows a semilog plot of degradable VSS remaining versus digestion time. The slope of the line gives a value for the constant K_b. In Figure 8-11 the same theoretical batch data are plotted arithmetically and X_n obtained.

In the model proposed by Adams et al. (1974), it is assumed that only the volatile suspended solids content of the sludge will decrease during digestion and that no destruction of fixed or nonvolatile suspended solids will occur.

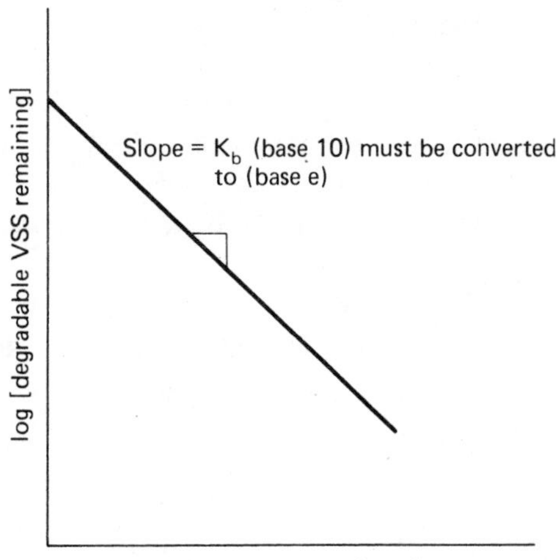

Figure 8-10. Plot of log [Degradable VSS Remaining] Versus Digestion Time. (After Adams, 1974b.)

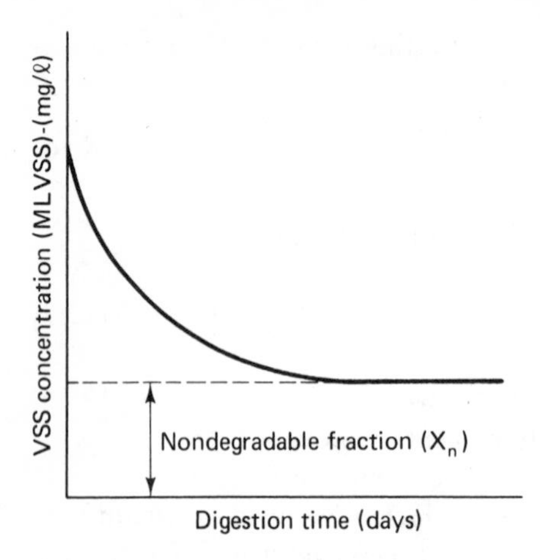

Figure 8-11. Plot of VSS Concentration Versus Digestion Time. (After Adams, 1974a.)

Randall et al. (1969, 1974, 1975a), however, observed a decrease in fixed suspended solids during digestion of waste activated sludge, implying a need for a model that more realistically describes suspended solids reduction during the aerobic digestion process. The reduction is explained as a change in the form of the fixed solids from suspended to soluble upon lysis of the microbial cells that contain the solids. Thus, active biomass and total suspended solids can be used to describe aerobic digestion. In fact, it seems logical to use total suspended solids rather than volatile suspended solids when one considers that an active cell is composed of inorganic as well as organic material, and during endogenous respiration there is solubilization of inorganic material as well as oxidation of organic material. Therefore, assuming that

1/ Total suspended solids are composed of an active fraction and an inactive fraction,
2/ The inactive fraction of the influent total suspended solids is nondegradable (i.e., material that cannot be oxidized or solubilized through microbial activity),
3/ The active fraction of the influent total suspended solids is composed of nondegradable and degradable fractions, where degradable describes material that can be oxidized or solubilized through microbial activity,
4/ Only the degradable active fraction of the total suspended solids decreases during digestion,

then a flowchart, following total suspended solids through the digestion process, can be constructed. Such a chart is shown in Figure 8-12. The following nomenclature is used in this figure:

X_0 = total suspended solids concentration in influent to digester

X_e = total suspended solids concentration in effluent from digester

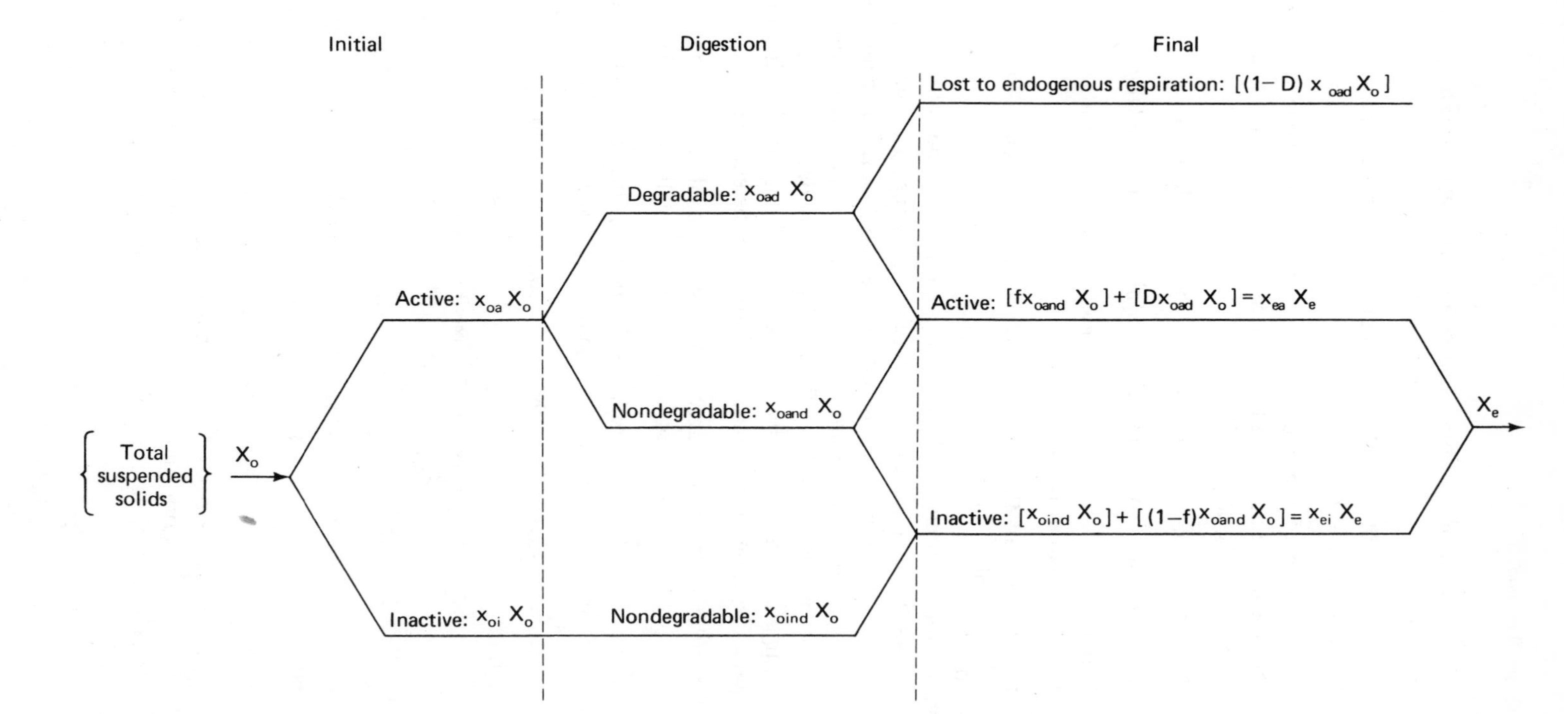

Figure 8-12. Change in Total Suspended Solids as a Result of Aerobic Digestion.

χ_{0a} = that fraction of the total suspended solids concentration in the influent which is active

χ_{0i} = that fraction of the total suspended solids concentration in the influent which is inactive

χ_{0ad} = that fraction of the active mass which is degradable (i.e., that portion of the active mass which can be oxidized or solubilized through biological activity)

χ_{0and} = that fraction of the active mass which is nondegradable

χ_{0ind} = that fraction of the inactive mass which is nondegradable

χ_{ea} = that fraction of the total suspended solids concentration in the effluent which is active

χ_{ei} = that fraction of the total suspended solids concentration in the effluent which is inactive

f = fraction of nondegradable active biomass in the influent that passes through the digestion process and appears as active nondegradable biomass in the effluent (active biomass is composed of both a degradable and a nondegradable fraction; during the digestion process both of these components decrease relative to their initial values as cellular lysis, metabolism, and solubilization take place, but a fraction is discharged in the effluent)

D = fraction of degradable active biomass in the influent that appears as degradable active biomass in effluent (usually 0.1 to 0.3)

K_d = decay rate of the degradable fraction of the active biomass approximated from the decrease in total suspended solids

An examination of Figure 8-12 shows the following relationships to be valid:

$$\chi_{0ind}X_0 + \chi_{0and}X_0 - f\chi_{0and}X_0 = \chi_{ei}X_e \tag{8-20}$$

$$f\chi_{0and}X_0 + D\chi_{0ad}X_0 = \chi_{ea}X_e \tag{8-21}$$

$$\chi_{0ind}X_0 = \chi_{0i}X_0 \tag{8-22}$$

$$\chi_{0ad}X_0 + \chi_{0and}X_0 = \chi_{0a}X_0 \tag{8-23}$$

$$\chi_{0a}X_0 + \chi_{0i}X_0 = X_0 \tag{8-24}$$

$$\chi_{ei}X_e + \chi_{ea}X_e = X_e \tag{8-25}$$

Recalling that the active degradable biomass concentration in the reactor is equal to the concentration in the effluent for a completely mixed continuous-flow system operating at steady-state, a material-balance expression for active degradable biomass gives

$$\begin{bmatrix}\text{biomass lost}\\ \text{through biological}\\ \text{activity}\end{bmatrix} = \begin{bmatrix}\text{active degradable}\\ \text{biomass in}\\ \text{influent}\end{bmatrix} - \begin{bmatrix}\text{active degradable}\\ \text{biomass in}\\ \text{effluent}\end{bmatrix} \qquad (8\text{-}26)$$

If K_d is defined as the active degradable biomass lost per unit time per unit of active degradable biomass in the system, a mathematical expression for the active degradable biomass lost through microbial activity is

$$\begin{bmatrix}\text{active degradable}\\ \text{biomass lost through}\\ \text{microbial activity}\end{bmatrix} = K_d t_d[D(\chi_{0ad})X_0] \qquad (8\text{-}27)$$

where $[D(\chi_{0ad})X_0]$ = steady-state active degradable biomass concentration in the system

It is therefore possible to represent the steady-state material-balance expression for active degradable biomass in the form

$$K_d t_d[D(\chi_{0ad})X_0] = (\chi_{0ad})X_0 - D(\chi_{0ad})X_0 \qquad (8\text{-}28)$$

which reduces to

$$K_d t_d = \frac{1 - D}{D} \qquad (8\text{-}29)$$

Equation 8-20 can be rearranged into the form

$$\chi_{0and}X_0 - f\chi_{0and}X_0 = \chi_{ei}X_e - \chi_{0ind}X_0 \qquad (8\text{-}30)$$

Substituting for $f\chi_{0and}X_0$ in equation 8-30 from equation 8-21 gives

$$\chi_{0and}X_0 - (\chi_{ea}X_e - D\chi_{0ad}X_0) = \chi_{ei}X_e - \chi_{0and}X_0 \qquad (8\text{-}31)$$

A further substitution can be made for $\chi_{0and}X_0$ from equation 8-23 and for $\chi_{0ind}X_0$ from equation 8-22.

$$\chi_{0a}X_0 - \chi_{0ad}X_0 - \chi_{ea}X_e + D\chi_{0ad}X_0 = \chi_{ei}X_e - \chi_{0i}X_0 \qquad (8\text{-}32)$$

Rearranging, this expression can be written as

$$(\chi_{0a} + \chi_{0i})X_0 + \chi_{0ad}X_0(D - 1) = (\chi_{ea} + \chi_{ei})X_e \qquad (8\text{-}33)$$

From equations 8-24 and 8-25 it can be seen that

$$\chi_{0a} + \chi_{0i} = \chi_{ea} + \chi_{ei} = 1 \qquad (8\text{-}34)$$

Thus, equation 8-33 can be expressed as

$$X_0 + \chi_{0ad}X_0(D - 1) = X_e \qquad (8\text{-}35)$$

or

$$\frac{X_0 - X_e}{X_0(\chi_{0ad})} = 1 - D \qquad (8\text{-}36)$$

Substituting for $1 - D$ from equation 8-29 and solving for t_d yields

$$t_d = \frac{X_0 - X_e}{K_d D(\chi_{0ad}) X_0} \tag{8-37}$$

Equation 8-37 reflects the importance of the physiological state of the biomass when calculating digester requirements. Upadhyaya and Eckenfelder (1975) have observed that the active fraction of sludge solids decreases with a decrease in the food/microorganism (F: M) ratio or an increase in biological solids retention time. Furthermore, Kountz and Forney (1959) have found that approximately 77% of a biological cell is degradable. This suggests that equation 8-37 can be modified to the form

$$t_d = \frac{X_0 - X_e}{K_d 0.77 D(\chi_{0a}) X_0} \tag{8-38}$$

which can be used to compute required digester retention time. It should be recognized that the value of 0.77 for the degradable fraction of a cell applies only to the active biomass, not to total or volatile suspended solids, and is the value most frequently quoted in the literature. It is reasonable to assume that this value may vary under certain conditions. When this occurs, equation 8-38 and succeeding equations should be appropriately modified.

Equation 8-38 suggests that for equivalent solids reduction when solids loading is constant, the solids retention time must be increased as the active fraction of the biomass in the feed sludge decreases. In situations where the activated sludge plant is designed to operate at long biological solids retention times (resulting in a low active fraction per unit mass of suspended solids), the solids retention time computed from equation 8-38 may be unreasonably long. In this case the engineer would decrease the desired degree of solids reduction in proportion to the decrease in active fraction.

If it is assumed that the presence of primary sludge in the aerobic digester does not result in the synthesis of new biomass but rather retards the destruction rate of cellular material by furnishing an external food source, the form of equation 8-38 remains valid. However, certain terms must be modified if the equation is to accurately describe the process. For this case equation 8-38 is modified to the form

$$t_d = \frac{(X_0)_m - (X_e)_m}{(K_d)_m 0.77 D(\chi_{0a})_m (X_0)_m} \tag{8-39}$$

where $(X_0)_m$ = total suspended solids concentration in feed to digester, mass volume^{-1}

or

$$(X_0)_m = \frac{Q_p(X_0)_p + Q_A X_0}{Q_p + Q_A} \tag{8-40}$$

where Q_p = volumetric flow rate of primary sludge, volume time^{-1}

Q_A = volumetric flow rate of activated sludge, volume time^{-1}

$(X_0)_p$ = total suspended solids concentration in primary sludge, mass volume^{-1}

$$(\chi_{0a})_m = \frac{X_0}{(X_0)_m}\chi_{0a} = \text{fraction of total} \tag{8-41}$$

solids concentration in feed to digester that is active biomass

$(X_e)_m$ = total suspended solids concentration in effluent from aerobic digester treating a mixture of primary and activated sludge, mass volume^{-1}

$(K_d)_m$ = overall decay rate of the degradable fraction of the active biomass; time^{-1}. This term accounts for the presence of the external food source in the form of primary sludge and assumes that all the external food is utilized, time^{-1}.

Recalling that K_d is defined as the active degradable biomass lost per unit time per unit of active degradable biomass in the system and that for a continuous-flow completely mixed digester operating at steady-state the total active degradable biomass in the system is given by $[D(\chi_{0ad})(X_0)V]$, $(K_d)_m$ can be expressed in the form

$$(K_d)_m = \left[\frac{\text{active degradable biomass lost through microbial activity per unit time}}{\text{active degradable biomass in system}}\right] - \left[\frac{\text{biomass saved from destruction per unit time because of presence of external food source}}{\text{active degradable biomass in system}}\right] \tag{8-42}$$

which can be expressed mathematically as

$$(K_d)_m = \frac{K_d[D(\chi_{0ad})_m(X_0)_m]V}{D(\chi_{0ad})_m(X_0)_m V} - \frac{Y_T Q S_a}{D(\chi_{0ad})_m(X_0)_m V} \tag{8-43}$$

or in the form

$$(K_d)_m = K_d - \frac{Y_T S_a}{0.77D(\chi_{0a})_m(X_0)_m t_d} \tag{8-44}$$

where Y_T = true yield coefficient representative of the organic content of the primary sludge

$$S_a = \frac{Q_P}{Q_A + Q_P}S_0 = \text{ultimate BOD of primary sludge in feed stream to digester, mass volume}^{-1} \tag{8-45}$$

S_0 = ultimate BOD of primary sludge, mass volume^{-1}

Substituting for $(K_d)_m$ in equation 8-39 from equation 8-44,

$$t_d\left[K_d - \frac{Y_T S_a}{0.77D(\chi_{0a})_m(X_0)_m t_d}\right] = \frac{(X_0)_m - (X_e)_m}{0.77D(\chi_{0a})_m(X_0)_m} \tag{8-46}$$

or

$$t_d = \frac{(X_0)_m + Y_T S_a - (X_e)_m}{K_d[0.77D(\chi_{0a})_m(X_0)_m]} \tag{8-47}$$

Equation 8-47 can then be used to estimate the size of aerobic digesters in situations where primary and activated sludge mixtures are to be processed.

Temperature Effects

Temperature affects the aerobic digestion process by altering the rate of endogenous respiration. Adams and Eckenfelder (1974) report that the rate coefficient, K_d, may be corrected for temperature by using the modified Arrenhius relationship

$$(K_d)_T = (K_d)_{20°C}\theta^{T-20} \tag{8-48}$$

where θ = temperature coefficient, which has been found to range from 1.02 to 1.11 (however, a value of 1.023 is commonly used)

Typical values for $(K_d)_{20°C}$ are given in Table 8-9.

TABLE 8-9

SOLIDS DESTRUCTION RATE COEFFICIENTS FOR AEROBIC DIGESTION AT 20°C

Type of sludge	K_d *(day⁻¹)*	*Basis*	*Reference*
Waste activated	0.12	VSS	Matsch and Drnevich (1977)
Waste activated	0.10	VSS	Andrews and Kambhu (1970)
Waste activated	0.10	VSS	Jaworski (1963)
Extended aeration	0.16	TSS	Randall et al. (1975a)
	0.18	VSS	
Trickling filter	0.04	TSS	Randall et al. (1974)
	0.05	VSS	
Primary and trickling filter	0.04	TSS	Randall et al. (1974)
	0.04	VSS	

Mavinic and Koers (1977), who studied digestion of activated sludge at temperatures of 5, 10, and 20°C, have shown that the product of temperature and digester biological solids retention time is an important design parameter. Their laboratory data, and the correlated results of two full-scale studies, indicate that the product value of 250, when temperature is in degrees Celsius, is a significant point of deflection on the suspended solids reduction curve (Figure 8-13). Above this point little additional suspended solids reduction occurs. They also found that the value of the temperature coefficient, θ, was less than 1.072 for temperatures above 15°C, and greater than 1.072 for temperatures below 15°C.

Randall et al. (1975a & b) found that the variation in K_d with temperature cannot always be described by the Arrenhius relationship above 20°C. In this work K_d was observed to vary with temperature in the manner illustrated

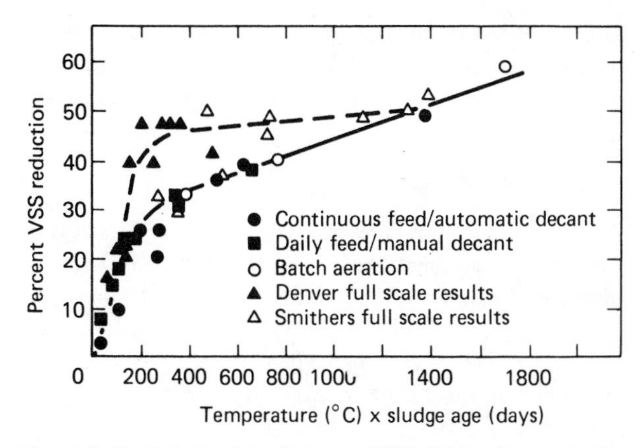

Figure 8-13. Relationship Between VSS Reduction and the Temperature-BSRT Product. (After Mavinic and Koers, 1977.)

in Figure 8-14. As a result, it is felt that laboratory studies are required to accurately obtain the variation of K_d with temperature. It should be emphasized that this variation in the rate of solids destruction with temperature

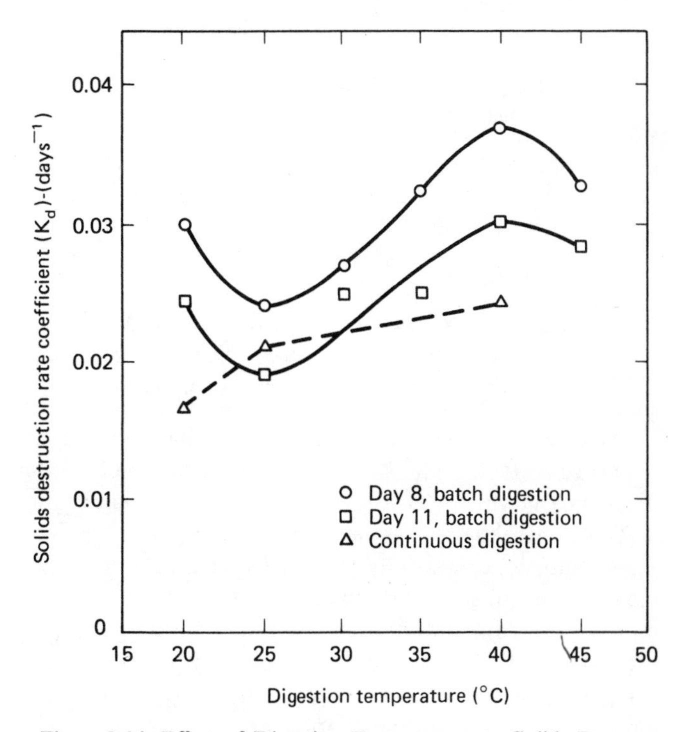

Figure 8-14. Effect of Digestion Temperature on Solids Destruction-Rate Coefficient. (After Randall et al., 1975a.)

applies only for the endogenous decay of activated sludge and not for solids destruction when the solids are a source of food for an acclimated microbial population such as occurs during thermophilic aerobic digestion.

Oxygen Requirements

If only excess activated sludge is to be aerobically digested, the recommended air requirements are 15 to 20 scfm/1000 ft³ of tank capacity. When a mixture of primary and activated sludge is to be digested, the air requirement is generally increased to 25 to 30 scfm/1000 ft³ of tank capacity.

A more rational approach to determining the oxygen requirement is to assume that the ultimate BOD from the primary sludge is satisfied during digestion, and furthermore that 1.42 lb of oxygen is required per pound of biological solids destroyed. Therefore,

$$\Delta O_2 = 1.42\begin{bmatrix}\text{pounds of biological}\\ \text{solids destroyed}\\ \text{during digestion per}\\ \text{day}\end{bmatrix} + \begin{bmatrix}\text{pounds of ultimate}\\ \text{BOD from primary}\\ \text{sludge added to}\\ \text{digester per day}\end{bmatrix}$$

which can be expressed mathematically as

$$\Delta O_2 = 1.42(8.34)[Q_0\, R(0.77)\chi_{0a}\, X_o] + (8.34)Q_p S_0 \qquad (8\text{-}49)$$

where ΔO_2 = pounds of oxygen required per day
Q_p = volumetric flow rate of primary sludge, MGD
R = reduction in active degradable biomass during digestion, fraction
χ_{0a} = fraction of TSS in digester feed which is active biomass, fraction
X_o = TSS concentration in digester feed, mg/l
Q_0 = Volumetric flow rate of digester feed, MGD
S_0 = ultimate BOD of primary sludge, mg/ℓ

In the case where only activated sludge is digested, the last term in equation 8-49 is zero. It should also be noted that equation 8-49 does not account for nitrification, which in many cases can be quite significant.

To estimate the oxygen requirement due to nitrification, not only must the ammonium concentration in the feed stream be considered, but consideration must also be given to the conversion of organic nitrogen back to the ammonium form and its ultimate release into solution during oxidation of cellular protein material. Conversion of organic nitrogen contained in the primary sludge must be considered as well. Hence, the nitrogenous oxygen demand can be estimated from the relationship

NOD =

$$\begin{bmatrix}\text{oxygen required} \\ \text{in the biologi-} \\ \text{cal oxidation of} \\ \text{ammonium con-} \\ \text{tained in the} \\ \text{influent to the} \\ \text{digester}\end{bmatrix} + \begin{bmatrix}\text{oxygen required} \\ \text{in the biologi-} \\ \text{cal oxidation of} \\ \text{ammonium released} \\ \text{during oxidation} \\ \text{of cellular} \\ \text{proteins}\end{bmatrix} + \begin{bmatrix}\text{oxygen required} \\ \text{in the biologi-} \\ \text{cal oxidation of} \\ \text{ammonium released} \\ \text{during oxidation} \\ \text{of primary} \\ \text{sludge}\end{bmatrix} \qquad (8\text{-}50)$$

which can be expressed mathematically as

$$\text{NOD} = (8.34)(4.57)\{(Q_A)[\text{NH}_4^+\text{-N}]_A + [Q_0\, R(0.77)\, \chi_{0a}\, X_o](0.122) + (Q_p)(\text{TKN})_p\} \qquad (8\text{-}51)$$

where

NOD = nitrogenous oxygen demand, lb/day

$[NH_4^+\text{-}N]_A$ = ammonium concentration in waste activated sludge, mg/ℓ as nitrogen

$[Q_0\, R(0.77)\, \chi_{0a}\, X_o]$ = biological solids destroyed per day

0.122 = nitrogen released as a result of biological solids destruction based on a cell composition of $C_{60}H_{87}O_{23}N_{12}P$

$(TKN)_p$ = total Kjeldahl nitrogen in primary sludge feed stream, mg/ℓ as nitrogen

Q_A = volumetric flow rate of excess activated sludge, MGD

Combining equations 8-49 and 8-51, the resulting equation for oxygen requirement is

$$\Delta O_2 = 1.42(8.34)[Q_0 R(0.77)\chi_{0a}X_o] + 8.34Q_pS_0 + \text{NOD} \qquad (8\text{-}52)$$

As an aid to the design engineer, Table 8-10 lists common values for the specific oxygen utilization rate for several types of sludges processed in aerobic digesters.

Mixing Requirements

In aerobic digestion adequate mixing must be achieved in order to keep the solids in suspension and to maintain optimum oxygen transfer efficiency. It is common practice to express mixing requirements in terms of power level, PL, which is defined as power per unit volume under aeration and which generally has the units horsepower per million gallons or horsepower per thousand gallons.

TABLE 8-10

SOME TYPICAL SPECIFIC OXYGEN UTILIZATION RATES FOR SLUDGES

Type of sludge	*Specific oxygen utilization rate (mg O_2/h-g VSS)*
Primary	20–40
Waste from conventional activated sludge	10–15
Waste from extended aeration activated sludge	5–8
Waste from contact-stabilization activated sludge	10
Single-stage aerobic digester	2–4
Two-state aerobic digester	0.5–2.4

Source: After Water Pollution Control Federation (1976).

When the solids level in the aerobic digester is less than 20,000 mg/ℓ, power levels between 70 and 100 hp/million gallons of digester volume are considered adequate. For a solids level greater than 20,000 mg/ℓ, power levels between 100 and 200 hp/million gallons are required.

Reynolds (1973) has developed an expression for the minimum power level required to provide adequate mixing. This expression has the form

$$\frac{P}{V} = 0.00475(\mu)^{0.3}(X)^{0.298} \qquad (8\text{-}53)$$

where $\frac{P}{V}$ = horsepower/1000 gal

μ = water viscosity, cP; see Table 8-11

X = steady-state digester TSS concentration, mg/ℓ

Equation 8-53 is valid for sludges that are between 70 and 100% digested. Power-level requirements for mixing of undigested waste activated sludges are generally twice those required for 100% digested sludges.

After the power level required for complete mixing has been determined, the required compressed air flow can be calculated from the relationship (Reynolds, 1973):

$$\frac{G_s}{V} = 50.5\frac{P/V}{\log\left(\frac{h+34}{34}\right)} \qquad (8\text{-}54)$$

where $\frac{G_s}{V}$ = cfm/1000 ft³ at operating air temperature

V = horsepower/1000 gal

h = submergence depth to the diffusers, ft

TABLE 8-11

VISCOSITY OF WATER AT DIFFERENT TEMPERATURES

Temperature (°*F*)	*Absolute viscosity* (*cP*)
0	1.7921
2	1.6740
4	1.5676
6	1.4726
8	1.3872
10	1.3097
12	1.2390
14	1.1748
16	1.1156
18	1.0603
20	1.0087
22	0.9608
24	0.9161
26	0.8746
28	0.8363
30	0.8004

Design Considerations

Criteria commonly employed in aerobic digester design are presented in Table 8-12. Digestion tanks may be open or covered. Covered tanks have been used to prevent oxygen loss, minimize heat loss, and retard freezing

TABLE 8-12

AEROBIC DIGESTER DESIGN CRITERIA

Parameter	*Value*
θ_c (days at 20°C)	
Activated sludge only	12–16
Activated sludge without primary settling	16–18
Primary plus activated or trickling filter sludge	18–22
Organic loading (lb VSS/ft^3-day)	0.024–0.14
Air requirements	
Diffused system (cfm/1000 ft^3)	
Activated sludge only	20–35
Primary plus activated sludge	> 60
Surface aeration (hp/1000 ft^3)	1.0–1.25
Dissolved oxygen concentration (mg/ℓ)	1–2
Volatile suspended solids reduction (%)	35–50

Source: After Metcalf and Eddy (1972); Black, Crow and Eidness (1974).

(see Figure 8-15). Surface aeration has become the method of choice for uncovered tanks where freezing is not a problem because the mixing qualities and oxygen transfer capability are generally superior to the diffused aeration system per unit horsepower input. Where freezing is a problem, submerged turbine or coarse bubble diffusion systems are generally used.

The required degree of digestion is established by the methods used for subsequent handling and disposal of the sludge. For example, if incineration is employed, volume reduction and concentration are of primary concern. As solids reduction is achieved in a much shorter period than stabilization, the aeration time required when incineration follows digestion is generally less than the aeration time required when land disposal is used, which requires that the control of odor production be a major consideration.

Laboratory Evaluation

In the design of an aerobic digestion process, the rate of solids destruction, the oxygen requirement, the characteristics of the supernatant and residue, and the effect of temperature variations on each of these factors are of primary concern. Generally, to obtain the desired design information, a laboratory study is initiated. Furthermore, to simplify laboratory procedures, such studies are usually conducted under batch conditions even though digesters are operated as continuous flow units in the field. However, observations made by Benefield et al. (1978) indicate that aerobic digester design should be based on data generated from continuous digestion laboratory studies, especially if temperature is an important consideration and no attempt is to be made to acclimate the feed sludge to each temperature of interest prior to initiating the digestion study. Thus, it is the continuous approach that is the recommended laboratory procedure for aerobic digestion.

Data required for the use of equation 8-38 or 8-47 in system design is obtained from a laboratory study, the procedure for which is described in the following outline:

1/ With a lab-scale continuous-flow activated sludge unit, determine the active fraction of the total suspended solids concentration over the range of biological solids retention times expected during field operation. Use either the oxygen utilization rate, adenosine triphosphate, dehydrogenase activity, or plate count method (Upadhyaya and Eckenfelder, 1975).

2/ Construct a graph relating active fraction to θ_c (see Figure 8-16).

3/ Determine the theoretical yield coefficient for the primary sludge. The required procedure for this determination is outlined in Chapter 4.

4/ Determine the desired effluent TSS concentration using a highly active sludge culture (short θ_c). For the case where only activated

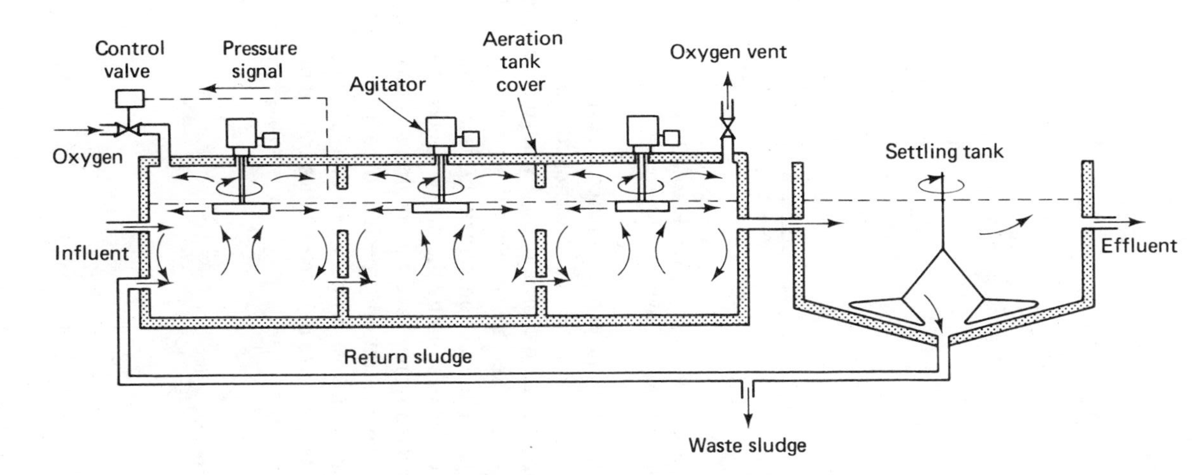

Figure 8-15. Schematic of a Pure Oxygen Aerobic Digester. (After Water Pollution Control Federation, 1976.)

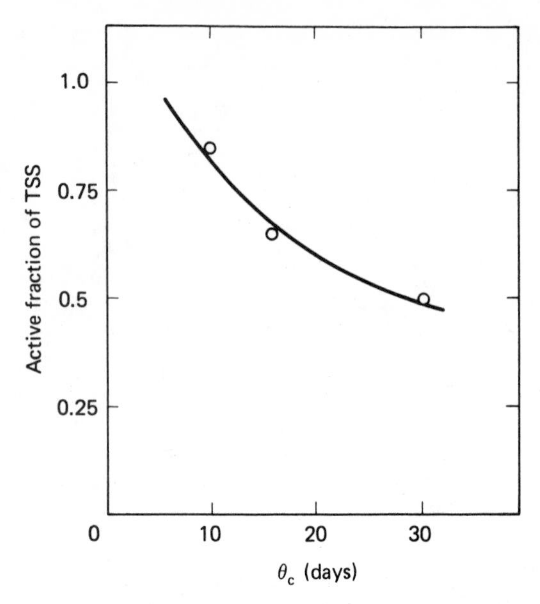

Figure 8-16. Relationship Between Active Fraction and Biological Solids Retention Time.

sludge is digested, the expected percent reduction in total suspended solids concentration can be determined from batch studies. This requires that a sample of activated sludge be aerated under batch conditions until the TSS concentration becomes fairly constant (see Figure 8-17). The fraction remaining is calculated by dividing the final TSS concentration by the initial concentration. This fraction is then subtracted from 1, and the resulting value is that fraction of the initial TSS concentration expected to be lost during digestion.

When a mixture of primary and activated sludge is to be aerobically digested, a series of batch studies should be conducted covering

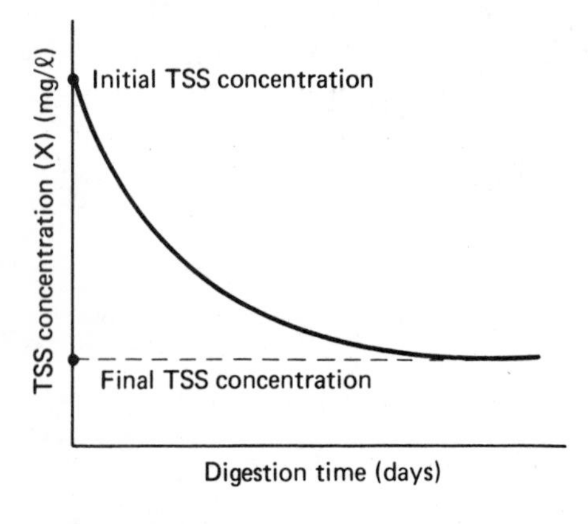

Figure 8-17. Decrease in TSS Concentration with Digestion Time.

the expected ratio range of primary sludge mass to activated sludge mass. This is required to correct for the nondegradable solids present in the primary sludge. Plot the data from each reactor as shown in Figure 8-17 and calculate the fraction of the initial TSS concentration of the mixture which is expected to be lost during digestion. Then construct a plot showing how this fraction will vary with different primary-to-activated sludge mass ratios. One possible variation is shown in Figure 8-18.

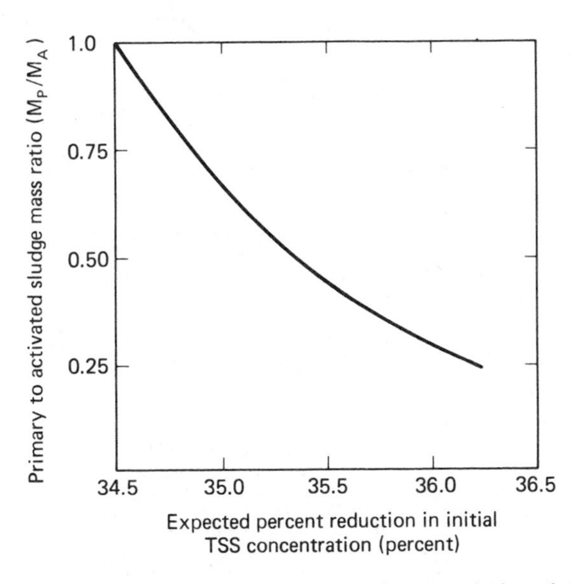

Figure 8-18. Relationship Between Primary-to-Activated Sludge Mass Ratio and Percent Reduction in Initial TSS Concentration.

5/ Determine the solids destruction rate coefficient, K_d, by aerobically digesting excess activated sludge on a continuous basis, covering the temperature range expected during actual operation. For each temperature operate digesters at several different steady-state θ_c values (e.g., 5, 10, and 15 days). K_d is given by the slope of the line obtained by plotting $[(X_0 - X_e)/X_e]$ versus digestion time, as shown in Figure 8-19.

6/ Construct a plot similar to Figure 8-14 to illustrate the variation in K_d with temperature.

7/ For each digestion time and temperature, analyze the filtrate for COD, phosphate, and nitrogen and construct plots similar to Figures 8-20, 8-21, and 8-22. Such data will generally not be used in digester design; however, they are important when characterizing the wastewater to be treated.

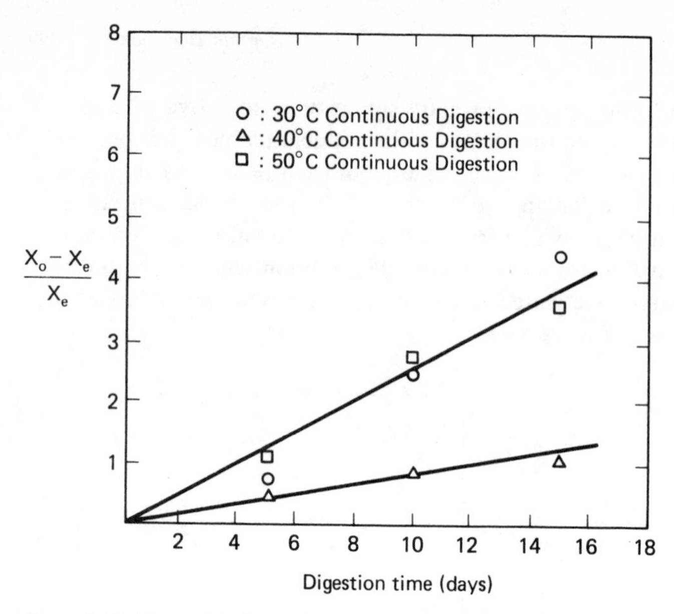

Figure 8-19. Determination of Solids Destruction-Rate Coefficient. (After Benefield et al., 1978.)

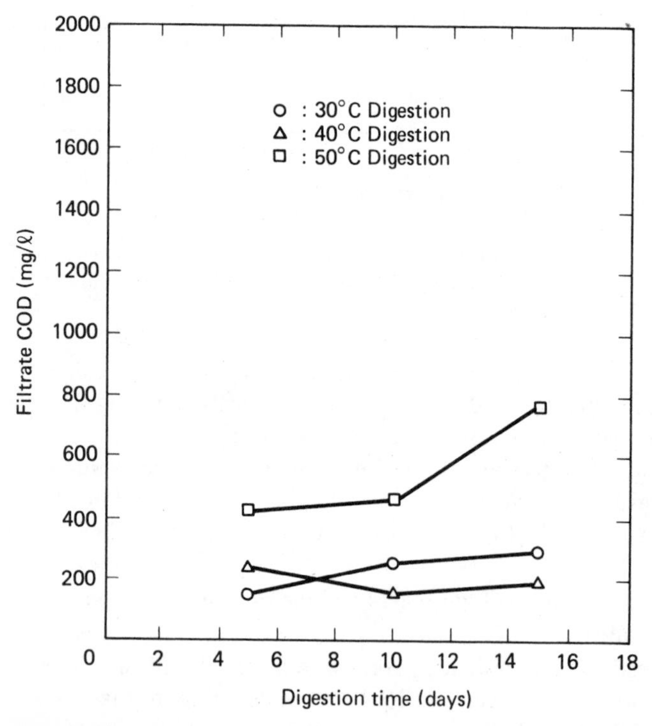

Figure 8-20. Continuous Filtrate COD Versus Digestion Time. (After Benefield et al., 1978.)

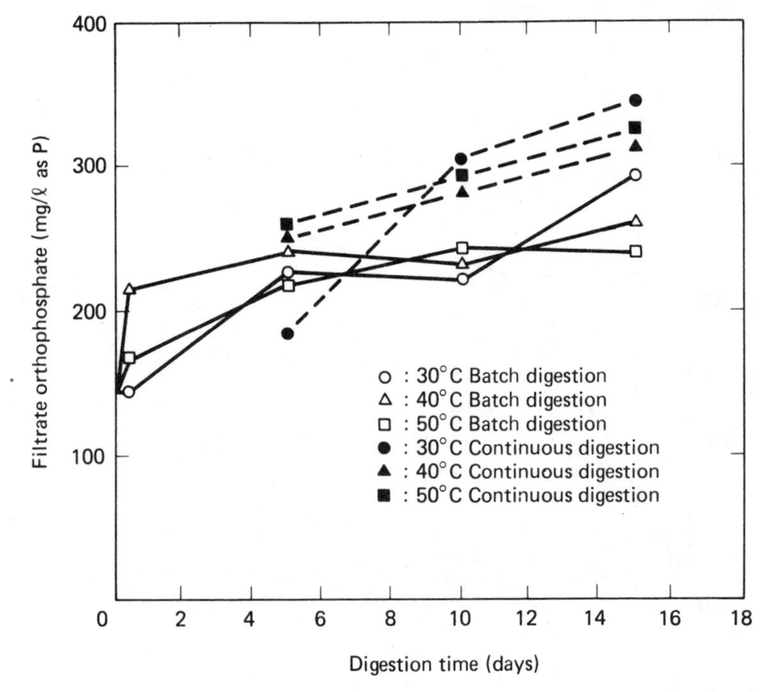

Figure 8-21. Comparison of Continuous and Batch Digestion Filtrate Orthophosphate Concentration. (After Benefield et al., 1978.)

To illustrate the design procedure, an example is presented which considers the design of an aerobic digester required to process a mixture of primary and waste activated sludge.

Example 8-2

Design an aerobic digester to process the sludge from a completely mixed activated sludge plant. The following conditions are applicable to this problem.

1/ Sludge characteristics

Winter conditions:

a/ Primary sludge:

$$\text{flow} = 0.028 \text{ MGD}$$

$$\text{temp.} = 15.5°\text{C}$$

$$\text{TKN} = 7000 \text{ mg}/\ell$$

$$\text{BOD}_u = 50{,}000 \text{ mg}/\ell$$

$$\text{TSS} = 50{,}000 \text{ mg}/\ell$$

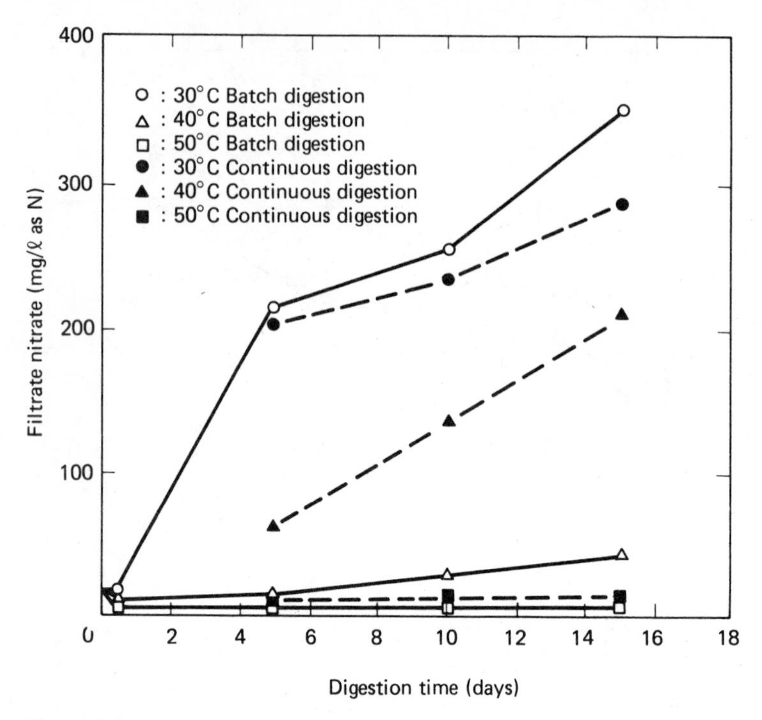

Figure 8-22. Comparison of Continuous and Batch Digestion Filtrate Nitrate Concentration. (After Benefield et al., 1978.)

b/ Activated sludge thickened to 1.5% solids:

$$\text{flow} = 0.12 \text{ MGD}$$
$$\text{temp.} = 10°\text{C}$$
$$[\text{NH}_4^+\text{-N}] = 10 \text{ mg}/\ell$$
$$\theta_c = 20 \text{ days}$$
$$\text{TSS} = 15{,}000 \text{ mg}/\ell$$

Summer conditions:

a/ Primary sludge:

$$\text{flow} = 0.028 \text{ MGD}$$
$$\text{temp.} = 24°\text{C}$$
$$\text{TKN} = 7000 \text{ mg}/\ell$$
$$\text{BOD}_u = 50{,}000 \text{ mg}/\ell$$
$$\text{TSS} = 50{,}000 \text{ mg}/\ell$$

b/ Activated sludge thickened to 1.5% solids:

$$\text{flow} = 0.12 \text{ MGD}$$
$$\text{temp.} = 27°\text{C}$$

$$[NH_4^+\text{-N}] = 10 \text{ mg}/\ell$$
$$\theta_c = 6.6 \text{ days}$$
$$TSS = 15{,}000 \text{ mg}/\ell$$

2/ Figure 8-16 is representative of the relationship between active fraction and operating θ_c in the activated sludge aeration tank.

3/ The theoretical yield coefficient for primary sludge is 0.5.

4/ Figure 8-18 is representative of the relationship between primary-to-activated sludge mass ratio and the expected percent reduction in TSS concentration for the case where the active fraction of the activated sludge approaches 100%.

5/ Figure 8-23 gives the relationship between K_d and temperature.

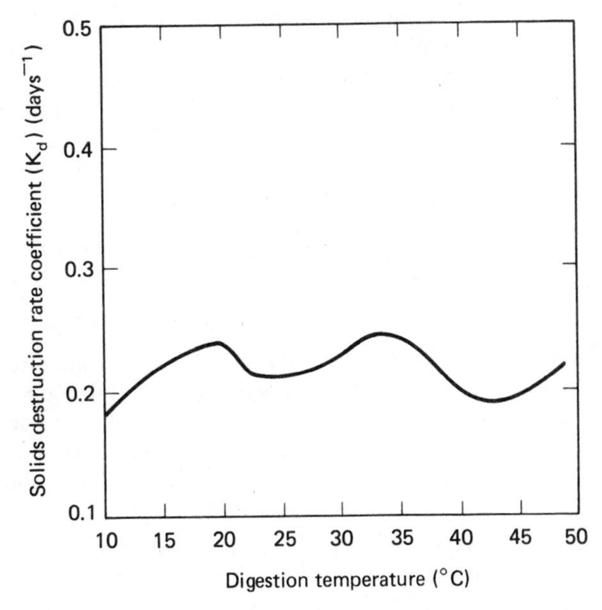

Figure 8-23. Variation in K_d with Digestion Temperature for Example 8-2.

solution

1/ Compute the digestion temperature.

Winter:

$$T = \frac{(0.028)(15.5) + (0.12)(10)}{(0.028) + (0.12)}$$
$$= 11°\text{C}$$

Summer:

$$T = \frac{(0.028)(24) + (0.12)(27)}{(0.028) + (0.12)}$$
$$= 26.4°\text{C}$$

2/ Determine the solids concentration of the sludge mixture from equation 8-40.

Winter & Summer:

$$(X_0)_m = \frac{(0.028)(50{,}000) + (0.12)(15{,}000)}{(0.028) + (0.12)}$$

$$= 21{,}622 \text{ mg}/\ell$$

3/ Compute the primary-to-activated sludge mass ratio.

$$\frac{M_p}{M_A} = \frac{(0.028)(50{,}000)}{(0.12)(15{,}000)}$$

$$= 0.78$$

4/ For $M_p/M_A = 0.78$, Figure 8-18 shows that the reduction in the TSS concentration is estimated to be 34.75% when the activated sludge portion of the mixture is from a culture approaching 100% activity per unit of biomass.

5/ Determine the active fraction for each operating θ_c from Figure 8-16.

Winter:

$$\theta_c = 20 \qquad \text{and} \qquad \chi_{0a} = 0.6$$

Summer:

$$\theta_c = 6.6 \qquad \text{and} \qquad \chi_{0a} = 0.95$$

6/ Compute the effluent TSS concentration using the relationship

$$(X_e)_{\text{design}} = MX_e \tag{8-55}$$

where $(X_e)_{\text{design}}$ = effluent TSS concentration to be used in design
X_e = effluent TSS concentration attainable only if the activated sludge solids are nearly 100% active
M = proportioning factor required to adjust for the fact that the activated sludge solids are something less than 100% active

A relationship for M can be derived by considering the basic expression for solids reduction which has the form

$$\frac{X_0 - X_e}{1.0} = \frac{X_0 - MX_e}{\chi_{0a}} \tag{8-56}$$

Solving for M gives

$$M = \frac{X_0 - \chi_{0a}(X_0 - X_e)}{X_e} \tag{8-57}$$

Hence, for this problem

Winter:

$$M = \frac{21{,}622 - (0.6)[21{,}622 - (21{,}622)(1 - 0.3475)]}{(21{,}622)(1 - 0.3475)}$$

$$= 1.2$$

Therefore,

$$(X_e)_{\text{design}} = 1.2[(21{,}622)(1 - 0.3475)]$$

$$= 16{,}930 \text{ mg}/\ell$$

Summer:

$$M = \frac{21{,}622 - (0.95)[21{,}622 - (21{,}622)(1 - 0.3475)]}{(21{,}622)(1 - 0.3475)}$$

$$= 1.03$$

Therefore,

$$(X_e)_{\text{design}} = 1.03[(21{,}622)(1 - 0.3475)]$$

$$= 14{,}532 \text{ mg}/\ell$$

7/ Calculate the active fraction of the sludge mixture from equation 8-41.

Winter:

$$(\chi_{0a})_m = \frac{15{,}000}{21{,}622}(0.6)$$

$$= 0.416$$

Summer:

$$(\chi_{0a})_m = \frac{15{,}000}{21{,}622}(0.95)$$

$$= 0.66$$

8/ Determine the ultimate BOD of the primary sludge in the feed to the digester using equation 8-45.

Winter & Summer:

$$S_a = \frac{0.028}{0.12 + 0.028}(50{,}000)$$

$$= 9460 \text{ mg}/\ell$$

9/ From Figure 8-23, the following values of K_d are determined:

Winter:

$$K_d = 0.19 \text{ day}^{-1}$$

Summer:

$$K_d = 0.22 \text{ day}^{-1}$$

10/ Compute the required digester retention time from equation 8-47, assuming that for sludge stability, the active degradable biomass must be reduced by 75% (i.e., $D = 0.25$).

Winter:

$$t_d = \frac{21{,}622 + (0.5)(9460) - 16{,}930}{(0.19)[(0.77)(0.25)(0.416)(21{,}622)]}$$

$$= 28.6 \text{ days}$$

Summer:

$$t_d = \frac{21{,}622 + (0.5)(9460) - 14{,}532}{(0.22)[(0.77)(0.25)(0.66)(21{,}622)]}$$

$$= 19.6 \text{ days}$$

11/ Calculate the required digester volume.

Winter:

$$V = (28.6)(0.028 + 0.12)$$

$$= 4.23 \text{ MG}$$

Summer:

$$V = (19.6)(0.028 + 0.12)$$
$$= 2.9 \text{ MG}$$

Therefore, winter conditions control the size of the digester, which is 4.25 MG.

12/ Determine the total oxygen requirement from equation 8-52.

Winter:

$$\begin{aligned}\Delta O_2 &= (1.42)(8.34)[(0.12 + 0.028)(0.75)(0.77)(0.416)(21{,}622)] \\ &\quad +(8.34)(0.028)(50{,}000) + (8.34)(4.57)\{(0.12)(10) \\ &\quad + [(0.12 + 0.028)(0.75)(0.77)(0.416)(21{,}622)](0.122) \\ &\quad + (0.028)(7000)\} \\ &= 31{,}871 \text{ lb/day}\end{aligned}$$

Summer:

$$\begin{aligned}\Delta O_2 &= (1.42)(8.34)[(0.12 + 0.028)(0.9)(0.77)(0.66)(21{,}622)] \\ &\quad + (8.34)(0.028)(50{,}000) + (8.34)(4.57)\{(0.12)(10) \\ &\quad + [(0.12 + 0.028)(0.9)(0.77)(0.66)(21{,}622)](0.122) \\ &\quad + (0.028)(7000)\} \\ &= 43{,}331 \text{ lb/day}\end{aligned}$$

Note: the value of 0.9 appears because it is felt that under summer conditions the large retention time will product the maximum reduction in active degradable biomass which is assumed to be 90 percent.

The total oxygen requirement is controlled by summer conditions.

13/ Assuming that the oxygen transfer rate under process conditions is 1.49 lb O_2/hp-h, compute the horsepower required for oxygen transfer.

Summer:

$$\text{hp} = \frac{43{,}331}{(1.49)(24)}$$
$$= 1212$$

14/ Calculate the power level based on the oxygen transfer requirement.

$$\frac{P}{V} = \frac{(1212)(1000)}{4{,}230{,}000}$$
$$= 0.286 \text{ hp/1000 gal}$$

15/ Compute the power level based on the mixing requirement using equation 8-53.

Winter:

$$\frac{P}{V} = (0.00475)(1.3097)^{0.3}(16{,}930)^{0.298}$$
$$= 0.094 \text{ hp/1000 gal}$$

Summer:

$$\frac{P}{V} = (0.00475)(0.8746)^{0.3}(14{,}532)^{0.298}$$

$$= 0.079 \text{ hp/1000 gal}$$

16/ Comparing the power levels computed in steps 14 and 15, it is seen that horsepower requirements for oxygen transfer control the design. The required horsepower is

$$\text{hp} = 1212$$

Autothermal Thermophilic Aerobic Digestion

Just as Buhr and Andrews (1977) have indicated that thermophilic anaerobic digestion is potentially an economical process, Gould and Drnevich (1978) have demonstrated that thermophilic aerobic digestion is feasible because, under certain operating conditions, the process can be autothermal. Using a two-step insulated reactor and pure oxygen as the aeration gas, they obtained autothermal temperatures of 50 and 52°C in the first stage and 54 and 50°C in the second stage for stage retention times of 1.4 and 3.0 days and 2.5 and 5.0 days, respectively.

The sludge solids they digested had a biodegradable fraction of 50%. They obtained 30% reduction of total volatile solids at a total retention time of 3 days, and a 40% reduction for 5 days. Thus, 80% reduction of biodegradable volatile solids was accomplished in 5 days of thermophilic digestion.

From theoretical considerations they calculated that, because of evaporative and gas-sensible heat losses brought about by the increased volume of gas required, thermophilic temperatures could not be autothermally achieved using air as the aeration gas.

Energy Considerations for Sludge Digestion

An important consideration in the processing of sludge by digestion is the amount of energy required to operate the process. A comparison quickly illustrates that aerobic digestion is energy-intensive, whereas a net energy gain can be accomplished with anaerobic digestion. Aerobic digestion requires energy input for oxygen transfer and mixing, while anaerobic digestion produces methane that can be used, not only to mix and heat the digester, but also to provide energy for other waste treatment processes within the plant. In relatively warm climates, the anaerobic digesters can produce enough energy for all uses within a conventional activated sludge plant, including generation of electricity. This has been demonstrated at Fort Worth, Texas. Such an approach requires that organic sludge generation be maximized during treatment by utilization of primary clarification and low sludge ages.

In Example 8-1 it was calculated that the anaerobic digestion of 18,729 lb/day of combined activated/primary sludge would generate 115,159,750 Btu/day, with 85,774,775 Btu/day available for uses other than heating the digester. If an aerobic digestion period of 20 days were used for the same waste sludge, the digester volume would be 1.12 MG and assuming mixing controls energy input, would require a mixing horsepower of approximately 136 hp, which is equivalent to 8,286,553 Btu/day. Thus, the energy difference between the two processes could be as much as 94,061,328 Btu/day in favor of the anaerobic digester.

PROBLEMS

8-1 Design a standard-rate anaerobic digester to process the sludge generated from a trickling filter plant treating a wastewater flow of 1 MGD. Assume that the following conditions are applicable:

a/ A high-rate, single-stage rock filter with a recirculation rate of 4:1 is used.

b/ The influent BOD_5 is 200 mg/ℓ, which is 70% of the ultimate BOD.

c/ The trickling filter is designed to achieve 90% BOD removal.

d/ The biological sludge produced from the trickling filter is given by the equation

$$W_s = 8.34 K_Y S_a Q_a$$

where W_s = biological sludge solids, lb dry wt/day

K_Y = fraction of BOD_5 applied to filter which appears as excess biological solids (use 0.2 for standard-rate filters and 0.3 for high-rate filters)

S_a = BOD_5 applied to filter, mg/ℓ

Q_a = wastewater flow applied to filter, MGD

e/ The biological sludge is 72% volatile solids, which concentrates to 2.5% in the secondary clarifier.

f/ After grit removal the wastewater has a total suspended solids concentration of 325 mg/ℓ, of which 65% is volatile.

g/ The surface loading rate for the circular primary clarifier is 800 gpd/ft².

h/ The relationships given in Figures 3-12, 3-13, and 3-14 are valid for this design. The sludge concentrates to 6% solids in the primary clarifier.

i/ The digestion temperature is 35°C.

j/ Because of winter conditions, a volume sufficient for 100 days of sludge storage must be provided.

k/ The critical design temperature is 5°F. All digester sidewalls are exposed to air. The digester floor is in dry earth which has a winter temperature of 10°F.

l/ The digester depth is 35 ft.

m/ Digested sludge concentrates to 8%.

n/ The retention time required for digestion at various temperatures is given in the following table (Metcalf and Eddy, 1972):

Temperature (°F)	Retention time (days)
50	75
60	56
70	42
80	30
90	25
100	24
110	26

o/ Provide sufficient volume for 5 days of gas storage.
p/ Increase the required active volume by 25% to allow for scum and supernatant.
q/ The feed sludge temperature is 50°F.

8-2 Size an aerobic digester to process the sludge generated from a completely mixed activated sludge plant treating a wastewater flow of 5 MGD. Assume that the following conditions are applicable:
a/ The influent BOD_5 is 250 mg/ℓ, which is 70% of the ultimate BOD. The TKN is 25 mg/ℓ.
b/ After grit removal the wastewater has a total suspended solids concentration of 300 mg/ℓ, of which 65% is volatile.
c/ The surface loading rate for the circular primary clarifier is 800 gpd/ft^2.
d/ The relationships given in Figures 3-12, 3-13, and 3-14 are valid for this design. The sludge concentrates to 5% solids in the primary clarifier.
e/ The biokinetic coefficients applicable to the aeration tank design are

$$Y_T = 0.5$$
$$K_d = 0.1 \text{ day}^{-1} \text{ at } 20°\text{C}, \theta = 1.05$$
$$K = 0.02\ \ell/\text{mg-day at } 20°\text{C}, \theta' = 1.03$$

The coefficients are based on ultimate BOD.
f/ The critical operating temperatures for both the aeration tank and digester design are 10°C and 20°C.
g/ The aeration tank design is based on a soluble effluent BOD_u of 15 mg/ℓ and an operating MLVSS of 2000 mg/ℓ.
h/ The MLSS is 72% volatile.
i/ The waste activated sludge is concentrated to 2% solids by dissolved air flotation before mixing with the primary sludge. No ammonium is present in this waste stream.
j/ The Adams approach is used in design (i.e., the required retention time is computed from equation 8-19).
k/ 40% of the volatile suspended solids in the primary sludge is nondegradable.
l/ 23% of the volatile suspended solids in the waste activated sludge is nondegradable.

m/ 75% reduction in degradable volatile suspended solids is required during digestion.

n/ The solids destruction rate coefficient is 0.05 day^{-1} at 20°C. The variation in K_b with temperature is given by the expression

$$(K_b)_T = (K_b)_{20^\circ C}(1.023)^{T-20}$$

8-3 Size an aerobic digester to process the sludge generated from a completely mixed activated sludge plant treating a wastewater flow of 10 MGD. Assume that the following conditions are applicable:

a/ After grit removal the wastewater has a total suspended solids concentration of 300 mg/ℓ, of which 65% is volatile; 40% of the VSS are non-degradable.

b/ The influent BOD_5 is 200 mg/ℓ, which is 70% of the ultimate BOD. The TKN is 25 mg/ℓ.

c/ Primary clarification is not used in this system.

d/ The biokinetic coefficients applicable to the aeration tank design are

$$Y_T = 0.5$$

$$K_d = 0.1 \text{ day}^{-1} \text{ at } 20^\circ\text{C}$$

$$K = 0.04\ \ell/\text{mg-day at } 20^\circ\text{C},\ \theta = 1.03$$

These coefficients are based on ultimate BOD.

e/ The critical operating temperatures for both the aeration tank and digester design are 10°C and 20°C.

f/ The aeration tank design is based on a soluble effluent BOD_u of 15 mg/ℓ and an operating MLVSS of 2000 mg/ℓ. In the design assume that all fixed solids enter the system in the influent and the fixed fraction of the biomass can be neglected.

g/ The waste activated sludge is concentrated to 3% solids by dissolved air flotation prior to digestion. No ammonium is present in this waste stream.

h/ The relationship between active fraction and θ_c is given by the following data (based on MLSS) and is assumed valid for both summer and winter conditions:

χ_{0a}	θ_c
0.450	3
0.400	4
0.350	5
0.325	6
0.300	7
0.275	8
0.250	9
0.240	10
0.230	11
0.220	12
0.210	13
0.200	14

i/ The solids destruction rate coefficient is 0.1 day^{-1} at 20°C. The variation in K_d with temperature is given by the expression

$$(K_d)_T = (K_d)_{20°C}(1.023)^{T-20}$$

j/ For sludge stability the active degradable biomass must be reduced by 85%.

k/ 23% of the biomass produced in the aeration tank is nondegradable.

REFERENCES

ADAMS, C., AND W. W., ECKENFELDER, *Process Design Techniques for Industrial Waste Treatment*, Enviro Press, Nashville, Tenn., 1974a.

ADAMS, C. E., JR., W. W. ECKENFELDER, JR., AND R. M. STEIN, "Modifications to Aerobic Digestor Design," *Water Research*, **8**, 213 (1974b).

ANDREWS, J. F., "A Mathematical Model for the Continuous Cultivation of Microorganisms Utilizing Inhibitory Substrates," *Biotechnology and Bioengineering*, **10**, 707 (1968).

ANDREWS, J. F., "Dynamic Models and Control Strategies for Wastewater Treatment Plants—An Overview," presented at the International Federation of Automatic Control Symposium on Environmental Systems Planning, Design and Control, Kyoto, Japan, Aug. 1–5, 1977.

ANDREWS, J. F., AND K. KAMBHU, "Thermophilic Aerobic Digestion of Organic Solid Wastes," Final Progress Report, Clemson University, Clemson, S.C., May 1970.

BENEFIELD, L. D., R. SEYFARTH, AND A. SHINDALA, "Lab Study Helps Solve Aerobic Digester Problems," *Water and Sewage Works*, ref. ed., 60 (Apr. 1978).

BLACK, CROW AND EIDNESS, CONSULTING ENGINEERS, *Process Design Manual for Sludge Treatment and Disposal*, EPA Technology Transfer Series, 1974.

BUHR, H. O., AND J. F. ANDREWS, "The Thermophilic Anaerobic Digestion Process," *Water Research*, **11**, 129 (1977).

BURD, R. S., "A Study of Sludge Handling and Disposal," *Publication WP*-20-4, U.S. Department of the Interior, Federal Water Pollution Control Administration, Office of Research and Development, Washington, D.C., May 1968.

ECKENFELDER, W. W., JR., "Mechanisms of Sludge Digestion," *Water and Sewage Works*, June 1967, p. 207.

FAIR, G. M., AND J. C. GEYER, *Elements of Water Supply and Waste-Water Disposal*, John Wiley & Sons, Inc., New York, 1957.

GOULD, M. S., AND R. F. DRNEVICH, "Autothermal Thermophilic Aerobic Digestion," *Journal of the Environmental Engineering Division, ASCE*, **104**, 259, 1978.

GRAEF, S. P., AND J. F. ANDREWS, "Stability and Control of Anaerobic Digestion," *Journal of the Water Pollution Control Federation*, **46**, 666 (1974).

HAMMER, M. J., *Water and Wastewater Technology*, John Wiley & Sons, Inc., New York, 1975.

HEUKELEKIAN, H., "Further Studies on Thermophilic Digestion of Sludge," *Sewage Works Journal*, **2**, 219 (1930).

JAWORSKI, N., "Aerobic Sludge Digestion," in *Advances in Biological Waste Treatment*, Macmillan Publishing Co., Inc., New York, 1963.

KORMANIK, R. A., "A Résumé of the Anaerobic Digestion Process," *Water and Sewage Works*, Apr. 1968, p. R-154.

KOUNTZ, R. R., AND C. FORNEY, JR., "Metabolic Energy Balances in a Total Oxidation Activated Sludge System," *Sewage and Industrial Wastes*, **31**, 819 (1959).

LAWRENCE, A. W., "Application of Process Kinetics to Design of Anaerobic Processes," in *Anaerobic Biological Treatment Processes*. F. G. Pohland, Symposium Chairman, American Chemical Society, Cleveland, Ohio, 1971.

LAWRENCE, A. W., AND P. L. MCCARTY, "A Unified Basis for Biological Treatment Design and Operation," *Journal of the Sanitary Engineering Division, ASCE*, **96**, 757, (1970).

LIPTÁK, B. G., *Environmental Engineers' Handbook*, Vol. 1, Chilton Book Company, Randor, Pa., 1974.

MATSCH, L. C., AND R. F. DRNEVICH, "Autothermal Aerobic Digestion," *Journal of the Water Pollution Control Federation*, **49**, 296 (1977).

MAVINIC, D. S., AND D. A. KOERS, "Aerobic Sludge Digestion at Cold Temperatures," *Canadian Journal of Civil Engineering*, **4**, 445, 1977.

MCCARTY, P. L., "Anaerobic Waste Treatment Fundamentals," (four parts), *Public Works*, Sept. 1964, p. 107; Oct. 1964, p. 123; Nov. 1964, p. 91; Dec. 1964, p. 95.

MCCARTY, P. L., "Sludge Concentration—Needs, Accomplishments, and Future Goals," *Journal of the Water Pollution Control Federation*, **38**, 493 (1966).

METCALF AND EDDY, INC., *Wastewater Engineering*, MCGRAW-Hill Book Company, New York, 1972.

RANDALL, C. W., F. M. SAUNDERS, AND P. H. KING, "Biological and Chemical Changes in Activated Sludge During Aerobic Digestion," in *Proceedings, 18th Southern Water Resources and Pollution Control Conference*, North Carolina State University, Raleigh, N.C., 1969.

RANDALL, C. W., D. G. PARKER, AND A. RIVERA-CORDERO, "Optimal Procedures for the Processing of Waste Activated Sludge," Virginia Water Resources Research Center, VPI-WRRC-BULL 61, Blacksburg, Va., 1973.

RANDALL, C. W., W. S. YOUNG, AND P. H. KING, "Aerobic Digestion of Trickling Filter Humus," in *Proceedings, Fourth Annual Environmental Engineering and Science Conference*, University of Louisville, Louisville, Ky., 1974.

RANDALL, C. W., R. A. GARDNER, AND P. H. KING, "The Aerobic Digestion of Activated Sludge at Elevated Temperatures," paper presented at the Fifth Annual Environmental Engineering and Science Conference, University of Louisville, Louisville, Ky., 1975a.

RANDALL, C. W., J. B. RICHARDS, AND P. H. KING, "Temperature Effects on Aerobic Digestion Kinetics," *Journal of the Environmental Engineering Division, ASCE*, **101**, EE5, 795 (1975b).

REYNOLDS, T. D., "Aerobic Digestion of Thickened Waste Activated Sludge," *Proceedings of the 28th Industrial Waste Conference*, Purdue University, West. Lafayette, Ind., 1973, p. 12.

SAWYER, C. N. AND H. K. ROY, "A Laboratory Evaluation of High-Rate Sludge Digestion," *Sewage and Industrial Wastes*, **27**, 1356 (1955).

SPEECE, R. E., "Anaerobic Treatment," in *Process Design in Water Quality Engineering*, ed. by E. L. Thackston and W. W. Eckenfelder, Jenkins Publishing Company, New York, 1972.

UPADHYAYA, A. K. AND W. W. ECKENFELDER, JR., "Biodegradable Fraction as an Activity Parameter of Activated Sludge," *Water Research*, **9**, 691 (1975).

WATER POLLUTION CONTROL FEDERATION, *Operation of Wastewater Treatment Plants A Manual of Practice*, Lancaster Press, Lancaster, Pa., 1976.

Index

A

acids
  Brönsted–Lowry theory, 78
    proton donor, 78
  equilibrium constants, 79
  titration curve, 79, 80
    Henderson–Hasselbalch equation, 80
activated biofilters, 420-423, 441, 442
  carbon removal:
    design considerations, 422
    process description, 420
  nitrification:
    design considerations, 442
activated sludge, 129-271
  anaerobic contact process, 256
  advantages, 256
  environmental factors, 264-268
    alkalinity, 265, 266
    ammonia, 267
    cation concentration, 267
    optimum conditions, 268
    pH, 265

activated sludge, *(cont.):*
  fundamental microbiology, 257-259
  gas production, 263, 264
  kinetics, 259-263
    biokinetic constants, 263
  process design considerations, 268-269
  process flow diagram, 257
biokinetic constants, 210-217
  coefficient values, 216, 217
  evaluation of, 210
    microbial decay coefficient, 213
    oxygen use coefficients, 213, 214
    settling coefficients, 214, 215
    substrate utilization rate constant, 211, 212
    yield coefficient, 213
  composition, 129
  denitrification, 249
    biomass production, 254
    description, 97, 98, 249, 250
    design criteria, 255
    kinetics, 251-253
    methanol requirement, 254

biokinetic constants *(cont.)*:
denitrification *(cont.)*:
process flow diagram, 252
reactions, 250, 251
design criteria, 187
effluent quality, 130
flow schematic, 130
kinetic model development, 131-152
assumptions, 132
dynamic versus steady-state, 152
Grau model, 145
plug-flow versus completely mixed, 143
recycle ratio, 137
specific substrate utilization rate, 136
steady-state effluent substrate concentration, 134, 149
mixing regime, 131
nitrification, 218
biokinetic constants, 229
critical BSRT, 230
description, 90, 218
design criteria, 255
distribution of organisms, 229
efficiency, 232
kinetics, 222-230
influence of DO, 223-226
influence of pH, 228
influence of temperature, 227
nitrifying bacteria, 91, 218
Nitrobacter, 40, 91, 218, 222
Nitrosomonas, 30, 91, 218, 222
organism yield, 229
predicting equilibrium pH, 218-221
process selection:
combined, 232-234
separate-stage, 232-234
reactions:
ammonia to nitrite, 91, 218
nitrite to nitrate, 91, 218
process design considerations, 184-210
effluent quality, 208-210
excess sludge production, 186-189
loading criteria, 184-186
BSRT, 185
food/microorganism ratio, 184
nutrient requirement, 190-195
oxygen requirements:
process design considerations *(cont.)*:
oxygen requirements *(cont.)*:
Herbert's theory, 192
nitrogenous, 193
Pirt's theory, 192
ten-state standards, 190
sludge viability, 189, 190
solids–liquid separation, 201-208
clarification, 201
temperature effects, 200
thickening, 201-207
temperature effects, 197-200
process modifications, 152-184
complete mix, 160, 161
contact stabilization, 163-170
conventional, 152-154
extended aeration, 161, 162
high-rate, 159, 160
oxidation ditch, 181-184
pure oxygen, 174-181
sludge reaeration, 170-173
step-aeration, 154-159
tapered aeration, 154
activation energy, 12
adenosine diphosphate (ADP), 31, 33
adenosine triphosphate (ATP), 31, 33
aeration, 281-318
diffused aerators, 294-308
coarse-bubble diffusers, 295
compressor efficiency, 306
diffuser depth, 297, 299, 301
diffuser spacing, 297, 300
effect of air flow rate, 300, 302
fine-bubble diffusers, 295
horsepower requirement, 304
one-third depth saturation DO, 303
percentage oxygen absorbed, 303
required tank geometry, 297, 301
scfm, 306
factors affecting oxygen transfer, 285-290
dissolved solids, 286
oxygen saturation, 285, 303
surface-active agents, 287
temperature, 287
turbulence, 288
fundamentals of gas transfer, 281-285
Fick's law, 282

fundamentals of gas transfer *(cont.)*:
Fick's law *(cont.)*:
overall gas transfer coefficient, 284
oxygen saturation values, 285, 290
stationary liquid film theory, 282
$K_L a$ and determinations, 290
nonsteady-state test, 290
steady-state test, 290
typical values, 292
mixing considerations, 317, 318
pumping capacity, 317
turnover time, 317
oxygen transfer rates, 289, 294, 304, 306, 309, 316
standard conditions, 289
submerged turbine aerators, 308, 309
power split, 309
surface aerators, 310-317
low speed, 310-317
motor speed, 310-312
aerated lagoons (*see* lagoons)
aerobic digestion, 479-507
design considerations, 495, 496
design criteria, 495
energy considerations, 507, 508
kinetic relationships, 482-490
solids destruction rates, 490
laboratory evaluation, 496-501
mixing requirements, 493, 494
oxygen requirements, 492, 493
utilization rates, 494
process description, 479
sludge conditioning characteristics, 481, 482
supernatant characteristics, 480
temperature effects, 490-492
thermophilic digestion, 507
aerobic lagoons, 360-382
apparent surface area, 369
depth, 361
design criteria, 375
design relationships:
biomass production, 367
effluent BOD_5, 367
oxygen requirement equations, 367
solids retention time, 365
specific substrate utilization rate, 367
temperature effects, 373
aerobic lagoons *(cont.)*:
length/width ratios, 361
mixing, 363, 364
oxygen requirements, 361, 364
power levels, 361-363
aerobic microorganisms, 29
aerobic ponds, 323-335
aerobic depth, 328
algae cells:
composition, 325
degree of reduction, 325
unit heat of combustion, 325
definition, 322
design criteria, 333
design relationships, 324-332
pond area, 325
pond depth, 331
energy conversion efficiency, 327
length/width ratios, 334
light intensity, 327, 331, 332
saturation point, 327, 331
mixing, 333
operational problems, 331
oxygenation factor, 325, 326
oxygen variation, 323
pH variation, 324
process description, 323
sludge buildup, 334
solar radiation, 325, 327-329
cloudiness correction, 327
elevation correction, 327
AERO-SURF process, 412
algae, 40, 323
cellular composition, 325
cellular unit heat of combustion, 325
degree of reduction, 325
Scenedesmus obliquus, 324
alkalinity, 82
ammonia/ammonium ion, 87, 88, 91
ammonia toxicity, 94
biological oxidation, 91
chlorine oxidation, 93
ionization constant, 95
anabolism, 31
anaerobic contact process, 256
advantages, 256
environmental factors, 264-268
alkalinity, 265, 266

environmental factors *(cont.):*
ammonia, 267
cation concentration, 267
optimum conditions, 268
pH, 265
fundamental microbiology, 257-259
gas production, 263, 264
kinetics, 259-263
biokinetic constants, 263
process design considerations, 268, 269
process flow diagram, 257
anaerobic digestion, 458-478
design considerations, 470-471
design criteria, 471
gas production, 466, 467
kinetic relationships, 464, 465
process description, 461, 464
high-rate digester, 462, 463
standard-rate digester, 461-462
process modeling, 477-478
sludge characteristics, 469
sludge heat requirements, 466
supernatant characteristics, 458
temperature effects, 465, 466
thermophilic digestion, 465
volatile solids destruction:
high-rate digester, 470
standard-rate digester, 469
anaerobic filters, 423-428
design considerations, 425
operational data, 426-428
process description, 424
anaerobic microorganisms, 29
anaerobic ponds, 353-358
definition, 323
gas production, 355
performance data, 356, 357
pond depth, 354
process description, 353, 354
sludge digestion chambers, 358
temperature effects, 354, 355
use as a pretreatment process, 354
Arrhenius equation, 11, 198
autotrophic microorganisms, 25

B

bases
Brönsted-Lowry theory, 78
bases *(cont.):*
Brönsted-Lowry theory *(cont.):*
proton acceptors, 78
best practicable treatment, 410
biochemical oxygen demand, (BOD):
bod/COD ratio, 74
BOD_5 of industrial wastewaters, 76
BOD_5 of municipal wastewaters, 77
calculation, 63
first-order representation, 66
measurement, 63
progression curve, 65
reaction rate, 66, 68
temperature correction, 71
Thomas Graphical method, 68
typical values, 69
toxicity, 72
ultimate BOD, 66, 68
Thomas Graphical method, 68
typical values, 69
biological solids retention time, (BSRT), 133
minimum BSRT, 140, 147
total system BSRT, 150
biological flocculation, 130
biokinetic constants, 210-217
coefficients, 216, 217
evaluation, 210
microbial decay coefficients, 213
oxygen use coefficients, 213, 214
settling coefficients, 214, 215
substrate utilization rate constant, 211, 212
yield coefficient, 213
BIO-SURF process, 410
breakpoint chlorination, 93, 94
Brönsted-Lowry theory, 78
buffer intensity:
carbonic acid system, 85
definition, 84
buffer solution, 80, 81
carbonic acid system, 82

C

carbonic acid system:
ionization of bicarbonate, 82
temperature correction, 83
ionization of carbonic acid, 82

carbonic acid system *(cont.)*:
ionization of carbonic acid *(cont.)*:
temperature correction, 83
catabolism, 31
chemical oxygen demand, (COD):
BOD/COD ratio, 74
chemically oxidizable material, 73
COD of industrial wastewater, 76
COD of municipal wastewater, 77
ΔCOD, 151
measurement, 73
theoretical COD, 73
chemotrophic microorganisms, 26
chemoautotrophs, 26, 40
energy reactions, 41
chemoorganotrophs, 26
Chick's law, 359
chlorine residual:
combined, 93
free, 93
chlorophyll, 41, 42
reduction by toxic chemicals, 341
citric acid cycle, 37
coarse-bubble diffuser, 295
coenzymes, 33
nicotinamide adenine dinucleotide (NAD), 33
nicotinamide adenine dinucleotide phosphate (NADP), 41
cofactor, 33
combined chlorine residual, 93
completely mixed activated sludge, 160, 161
completely mixed batch reactor, 14, 15
completely mixed reactors, 131
composite sampling, 114
contact stabilization activated sludge, 163-170
continuous-flow stirred tank reactors, 15-17
conventional activated sludge, 152-154

D

dark reaction, 41-43
dehydrogenation reaction, 33
denitrification:
activated sludge process:
biomass production, 254
denitrification *(cont.)*:
activated sludge process *(cont.)*:
description, 249, 250
design criteria, 255
kinetics, 251-253
methanol requirement, 254
process flow diagram, 252
reactions, 250, 251
fluidized bed columns:
denitrification rates, 451
process description, 449, 450
low-porosity fine-media columns:
denitrification rates, 450
performance data, 448
retention time effects, 449
microbial fundamentals, 97, 98, 249, 250
rotating biological contactors:
design relationships, 443
deoxygenation, 290
diffused aerators, 294-308
coarse-bubble diffusers, 295
compressor efficiency, 306
diffuser depth, 297, 299, 301
diffuser spacing, 297, 300
effect of air flow, 300, 302
fine-bubble diffusers, 295
horsepower requirement, 304
one-third depth saturation DO, 303
percentage oxygen absorbed, 303
required tank geometry, 297, 301
scfm, 306
diffusivity constant, 22
dispersion, 21
coefficent, 22
number, 22

E

Eckenfelder trickling filter equation, 402
electromagnetic spectrum, 41
visible region, 41
electron acceptor, 25, 34, 37, 40
electron transport system, 33, 38-40
endergonic reaction, 31
endogeneous respiration, 53
energy budget, 40
enzymes:
activity, 27

enzymes *(cont.)*:
coenzyme, 33
cofactor, 33
definition, 26
hydrolytic, 26, 27
intracellular, 27
oxidative, 26
dehydrogenases, 27
enzyme/substrate complex, 7
equilibrium constants:
ammonium ion, 95
carbonic acid system, 83
hypochlorus acid, 93
various acids, 79
water, 95
eutrophication, 97
limiting nitrogen concentration, 249
limiting phosphorus concentration, 249
exergonic reaction, 31
exponential growth, 45
extended aeration activated sludge, 161, 162

F

facultative lagoons, 382-385
design considerations, 383-385
design criteria, 385
oxygen requirement, 384
facultative microorganisms, 29
psychrophiles, 28
thermophiles, 28
facultative ponds, 338-353
definition, 322
design criteria, 346, 347
design relationships, 340-345
BOD removal rate constant:
Mara equation, 344
Marais chart, 343
Thirumurthi chart, 343
Gloyna equation, 340
Marais and Shaw equation, 342
maximum BOD_5 loading, 344
flow patterns, 348, 349
performance, 351-353
process description, 339, 340
fermentation, 33-35
pathway, 35
Fick's law, 282
fine-bubble diffuser, 295
food/microorganism ratio (F:M), 185
free chlorine residual, 93
free energy, 31

G

Galler–Gotaas equation, 400
gas transfer, 281-285
Fick's law, 282
overall gas transfer coefficient, 282, 290
stationary liquid film theory, 282
Gibbs free energy, 31
grab sampling, 114
Grau model for substrate utilization, 57, 145
growth curve, 43, 44
acceleration phase, 43
declining phase, 43
endogenous phase, 44
exponential phase, 43
stationary phase, 44
growth rate, 45
exponential, 45
limiting nutrient, 46
Monod relationship, 46, 55
specific growth rate constant, 46
growth yield, 47
observed yield coefficient, 57, 58, 188
temperature effects, 199
true yield coefficient, 51
yield constant, 47

H

heat balance, 368, 369
heat gain by solar radiation, 368
heat loss by convection, 368
heat loss by evaporation, 368
heat loss by radiation, 368
Henderson–Hasselbalch equation, 80
Henry's law, 82
Herbert's theory for energy of maintenance, 53
heterotrophic microorganisms, 25, 323
high-rate activated sludge, 159, 160

hydraulic retention time, 17
hydrolysis reaction, 31

I

industrial wastewater:
brewery, 110
canning, 106
dairy, 110
petroleum, 107
poultry, 109
pulp and paper, 108
slaughterhouse, 109
steel mill, 111
tannery, 109
textile, 111
infiltration rates, 112

K

$K_L a$ values, 290
Kjeldahl nitrogen, 88
Krebs cycle, 34, 37, 39, 89

L

lagoons, 359-385
aerated lagoons, 359-385
aerobic lagoons, 360-382
apparent surface area, 369
depth, 361
design criteria, 375
design relationships, 365-373
length/width ratios, 361
mixing, 363, 364
oxygen requirements, 361, 364
power levels, 361, 363
definition, 323
facultative lagoons, 382-385
design considerations, 383-385
design criteria, 385
oxygen requirement, 384
latent heat of vaporization, 369
light intensity, 327, 331, 332
saturation point, 327, 331
light reaction, 41-43
loading criteria for activated sludge, 184-186
low speed surface aerators, 310-317
power-level, 312, 315, 316
low speed surface aerators *(cont.)*:
power-level *(cont.)*:
pumping capacity, 317
speed and submergence, 316
turnover time, 317
luxury uptake, 99

M

maturation ponds, 323
maximum specific substrate utilization rate, 49
mean cell residence time, 133
mesophilic microorganisms, 28
metabolism:
energy, 33-40
aerobic respiration, 37
citric acid cycle, 37
Krebs cycle, 37
tricarboxylic acid cycle, 37
anaerobic respiration, 39
autotrophs, 40
chemoautotrophs, 40
budget, 40
electron transport system, 33, 38, 40
fermentation, 33-35
pathway, 35
oxidative phosphorylation, 33, 39
substrate-level phosphorylation, 33, 34
photosynthesis, 41
methemoglobinemia, 249
Michaelis-Menten equation, 9-11
microaerophilic microorganisms, 29
microbial decay coefficient, 54
evaluation, 213
temperature effects, 199
microbial growth:
energy and carbon source requirements, 49
endogenous respiration, 53
energy of maintenance, 50
Herbert's theory, 53
microbial decay coefficient, 54
Pirt's theory, 50
growth curve, 43, 44
acceleration phase, 43
declining phase, 43
endogenous phase, 44

microbial growth *(cont.):*
growth curve *(cont.):*
exponential phase, 43
lag phase, 43
stationary phase, 44
growth rate, 45, 46
exponential, 45
limiting nutrient, 46
Monod relationship, 46
specific growth rate, 46
growth yield, 47
observed yield, 57
true growth yield, 51
yield coefficient, 47
oxygen requirement, 29
pH effects, 30, 31
temperature effects, 28
optimum, 28
microbial reactions, 26
microorganisms:
algae, 40
autotrophic, 25
chemotrophic, 26
chemoautotrophs, 26
chemoorganotrophs, 26
electron acceptors, 25, 34, 38-40
heterotrophic, 25, 323
metabolism:
anabolism, 31
catabolism, 31
energy, 33-40
Nitrobacter, 40, 91, 218, 222
Nitrosomonas, 30, 91, 218, 222
nutrient requirements, 25
oxygen classification, 29
aerobes, 29
anaerobes, 29
facultative, 29
microaerophiles, 29
obligate aerobes, 29
obligate anaerobes, 29
phototrophic, 25
sulfur bacteria, 347, 348
temperature classification, 28
mesophilic, 28
psychrophilic, 28
thermophilic, 28
Thiobacillus thiooxidans, 30
monochloramine, 93
Monod equation, 46
motor speed surface aerator, 310-317
municipal wastewater, 62

N

nicotinamide adenine dinucleotide (NAD), 33, 34
nicotinamide adenine dinucleotide phosphate (NADP), 41
nitrification:
description, 90, 218
nitrifying bacteria, 91, 218
Nitrobacter, 40, 91, 218, 222
Nitrosomonas, 30, 91, 218, 222
reactions:
ammonia to nitrite, 9, 218
nitrite to nitrate, 91, 218
nitrogen:
ammonia/ammonium ion distribution, 88
ammonia-chlorine reaction, 93
deamination reactions, 90
nitrogen forms:
ammonia/ammonium ion, 87
nitrate, 87
nitrite, 87
organic, 87
nitrogen removal, 88
spatial distribution, 92
total Kjeldahl nitrogen, 88
nominal hydraulic retention time, 17
nonbiodegradable organic fraction, 102, 189
NRC equation, 396
nutrient requirements, 195-197

O

obligate microorganisms:
aerobic, 29
anaerobic, 29
psychrophiles, 29
thermophiles, 29
optimum growth temperature, 28
oxidation ditches, 181-184
oxidation pond, 322
oxidative phosphorylation, 33, 39

oxygen requirements, 190-195
Herbert's theory, 192
nitrogenous, 193
oxygen use coefficients, 192
evaluation, 213, 214
Pirt's theory, 50
ten-state standards, 190
oxygen saturation, 285-303
oxygen transfer, 285-290
valves, 290
factors affecting:
dissolved solids, 286
oxygen saturation, 285-303
rates, 289, 294, 304, 306, 309, 316
standard conditions, 289
surface-active agents, 287
temperature, 287
turbulence, 288
$K_L a$ values, 290
oxygenation factor, 325, 326

P

P-700, 42
pH:
definition, 78
effect on $HOCl/OCl^-$ distribution, 93
effect on microbial growth rate, 30
effect on NH_4^+/NH_3 distribution, 88, 94
scale, 78
phosphorus:
luxury uptake, 99
phosphorus forms:
condensed phosphate, 98
organic phosphate, 98
orthophosphate, 98
photosynthesis, 41, 323, 324
phototrophic microorganisms, 25
Pirt's theory for energy of maintenance, 50
plug flow reactors, 19-21, 131
ponds, 322-359
aerobic ponds, 323-335
aerobic depth, 328
algae cells, 325
definition, 322
design criteria, 333
design relationships, 324-332
energy conversion efficiency, 327
ponds *(cont.)*:
aerobic ponds *(cont.)*:
length/width ratios, 334
light intensity, 327, 331, 332
mixing, 333
operational problems, 331
oxygen variation, 323
pH variation, 324
process description, 323
sludge buildup, 334
solar radiation, 325, 327-329
anaerobic ponds, 353-358
definition, 323
gas production, 355
performance data, 356, 357
pond depth, 354
process description, 353, 354
sludge digestion chambers, 358
temperature effects, 354, 355
use as a pretreatment process, 354
faculatative ponds, 338-353
definition, 322
design criteria, 346, 347
design relationships, 340-345
flow patterns, 348, 349
performance, 351-353
process description, 339, 340
maturation ponds, 323
polishing ponds, 358-359
bacterial die-off, 359
definition, 323
effluent quality, 359
recommended pond arrangement, 359
tertiary ponds, 323
probability plotting, 117
proton acceptor, 79
proton donor, 79
psychrophilic micoorganisms, 28
pumping capacity, 317
pure oxygen activated sludge, 174-181
pyruvate, 34, 36

R

Rankin temperature scale, 305
reactions:
activation energy, 12
BOD, 66

reactions *(cont.)*:
  dark, 41-43
  dehydrogenation, 33
  endergonic, 31
  enzymatic, 7-11
  exergonic, 31
  light, 41-43
  order, 2
    first-order, 4, 5, 56
    mixed-order, 10
    second-order, 5, 6
    zero-order, 3, 4, 56
  temperature effects, 11-13, 198
reactors:
  completely mixed batch, 14, 15
  continuous-flow stirred tank, 15-17
    series arrangement, 18, 19
  definition, 1
  driving force, 16
  operating characteristics, 21
  plug flow, 19-21
    dispersion, 21, 22
  retention time, 17
rotating biological contactors, 410-420, 436-441, 443
  carbon removal:
    design relationships, 413-420
      design curves, 419
    process description, 410-412
  denitrification:
    design relationships, 443
  nitrification:
    design relationships, 436, 437
      design curves, 438, 439

S

sample preservation, 115
saturated water vapor pressure, 370, 371
saturation constant, 9
sewage lagoons, 332
sloughing, 392
sludge age, 133
sludge concentration, 468
sludge production, 186-189
sludge-reaeration activated sludge, 170-173
sludge stabilization, 457
sludge viability, 189, 190
sludge volume index (SVI), 137, 200, 207, 208
solar radiation, 325, 327-329
  cloudiness correction, 327
  elevation correction, 327
sodium sulfite, 291
solids content of wastewater:
  distribution, 100
  fixed, 99
  nonbiodegradable, 102
  settleable, 99
  solids removal, 101, 102
  suspended, 99
  total, 99
  volatile, 99
solids–liquid separation, 201-208
  clarification, 201
  temperature effects, 200
  thickening, 201-207
    limiting flux, 203
    solids flux, 201
specific gravity of sludge, 469
specific growth rate, 46, 133
specific substrate utilization rate, 49, 135
specific weight of water, 305, 468
standard conditions for oxygen transfer, 289
scfm, 306
standard-rate trickling filter, 393
stationary liquid film theory, 282
steady-state conditions, 8, 17, 19
step-aeration activated sludge, 154-159
submerged turbine aerators, 308, 309
  power split, 309
submerged filters, 431-434
  nitrification:
    design equations, 433, 434
    process description, 431
substrate-level phosphorylation, 33, 34
substrate utilization rate constant, 56
  evaluation, 211, 212
  temperature effects, 198
surface-active agents, 287
surface aerators, 310-317
  low speed, 310-317
    power level, 312, 315, 316
    pumping capacity, 317
    speed and submergence, 316

surface aerators *(cont.)*
low speed *(cont.):*
turnover time, 317
motor speed, 310-312

T

tapered aeration activated sludge, 154
temperature characteristic, 12
temperature effects:
aerobic digestion, 490-492
anaerobic digestion, 490-492
equilibrium constants:
ammonium ion, 95
carbonic acid system, 83
water, 95
methane fermentation, 355
microbial decay coefficient, 200
microbial growth, 28, 29
nitrification in trickling filters, 431
oxygen transfer rates, 289
reaction rates, 11-13
sludge settleability, 200
substrate utilization, 198
yield coefficient, 199
tertiary ponds, 323
thermophilic microorganisms, 28
Thiobacillus thiooxidans, 30
titration curve, 80
total organic carbon, (TOC):
COD/TOC ration, 76
TOC of industrial wastewaters, 76
TOC of municipal wastewaters, 77
total organic carbon analyzer, 74
tricarboxylic acid cycle, 37
trickling filters, 391-410, 429-431
carbon removal:
applicability of process, 409, 410
design equations:
Eckenfelder equation, 402
equation limitations, 408, 409
Galler-Gotaas equation, 400
NRC equation, 396
filter media, 393, 394
process description, 391, 392
recirculation patterns, 395
sloughing, 392
types of filters:
trickling filters *(cont.):*
carbon removal *(cont.):*
high-rate filters, 393
standard-rate filters, 393, 394
nitrification:
performance data, 430
surface loading effects, 430
temperature effects, 431
turbulence, 288
turnover time, 317

V

van't Hoff rule, 11
vapor pressure of water, 286
viability, 189, 190
viscosity of water, 495

W

waste stabilization, 322
wastewater characteristics, 62-103
alkalinity, 82
biochemical oxygen demand, 63
chemical oxygen demand, 73
industrial:
brewery, 110
canning, 106
diary, 110
petroleum, 107
poultry, 109
pulp and paper, 108
slaughterhouse, 109
steel mill, 111
tannery, 109
textile, 111
nitrogen forms, 88
pH, 78
phosphorus forms, 99
solids content, 99, 100
total organic carbon, 74
typical composition, 103
wastewater flows:
industrial, 106-111
infiltration rates, 112
flow variations, 116
municipal, 104, 105

wastewater sampling:
- composite sampling, 114
- data analysis, 117-124
- grab sampling, 114
- sample preservation, 115

water vapor pressure, 286
water viscosity, 495

Y

yield coefficient, 47
- evaluation, 213

yield coefficient *(cont.)*·
- observed yield coefficient, 57, 58, 188
- temperature effects, 199
- true yield coefficient, 51

Z

zone settling velocity, 207